BARRON'S

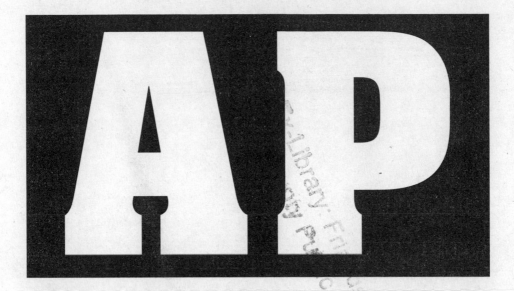

AP CALCULUS

ADVANCED PLACEMENT EXAMINATION

REVIEW OF CALCULUS AB AND CALCULUS BC

7TH EDITION

Shirley O. Hockett
Professor Emerita of Mathematics
Ithaca College
Ithaca, New York

and

David Bock
Ithaca High School
Ithaca, New York

All inquiries should be addressed to:
Barron's Educational Series, Inc.
250 Wireless Boulevard
Hauppauge, New York 11788
http://www.barronseduc.com

Library of Congress Catalog Card No. 2001043300
International Standard Book No. 0-7641-1790-4

Library of Congress Cataloging-in-Publication Data

Hockett, Shirley O.
    AP calculus: advanced placement examination: review of calculus AB
and calculus BC / Shirley O. Hockett and David Bock.—7th ed.
        p.   cm.
    At head of title: Barron's.
    Includes index.
    ISBN 0-7641-1790-4
    1. Calculus—Examinations, questions, etc.   2. College entrance
achievement tests—Study guides.   I. Title: Barron's AP calculus.
II. Bock, David.   III. Title.

QA309 .H585   2002
515'.076—dc21   2001043300

PRINTED IN THE UNITED STATES OF AMERICA
9 8 7 6 5 4 3 2 1

Contents

Topical Review and Practice: Contents _____

*Additional topic intended for BC course only.

*Additional topic intended for BC course only.

*Additional topic intended for BC course only.

Preface

The Calculus reform movement that began a dozen years ago continues to have a major impact on how calculus is taught and learned. David Bock and I hope this seventh edition of Barron's *How to Prepare for the AP Calculus (Advanced Placement Examination)* embodies the numerous important changes in content and approach that characterize the reform.

We are very fortunate in having had, once again, the assistance of Judith Broadwin. She evaluated the sixth edition carefully and thoroughly, and made many helpful suggestions on modifying and improving this new edition. As noted in our last preface: "Her many years as a respected teacher, consultant, and lecturer on AP Calculus have made her contributions invaluable."

David Bock has been a superb teacher of AP Calculus at Ithaca High School for more than twenty years, and has kept pace with calculus reform. This is the third edition on which we have collaborated; his contribution has increased with each one. I have continued to find it a pleasure to work with him.

I repeat here my appreciation to publisher Robert Kreiger for his permission to use material from *Applied Calculus* by Hockett and Sternstein.

Once again Dave and I thank Barron's Senior Editor Max Reed for her sensitive and knowledgeable management of this project, and for her persistent good humor throughout.

In the preface of the sixth edition I wrote the following:

In each of the first five editions I expressed appreciation to my "competent, skillful husband, who alone, double-handedly, typed the entire manuscript." It is time to point out that Chas has been more than a typist. It was he who encouraged me to write the book to begin with—thirty years ago. For each of the six editions he has been a discussant, evaluator, typist, copy editor, copier, proofreader, critic, and cheerleader. I could not have done it without him.

It saddens me to report that Chas died on November 3, 2000. This seventh edition is lovingly dedicated to him.

Shirley O. Hockett

Introduction

This book is intended for students who are preparing to take one of the two Advanced Placement Examinations in Mathematics offered by the College Entrance Examination Board, and for their teachers. It is based on the May 2002/May 2003 course description published by the College Board, and covers the topics listed there for both Calculus AB and Calculus BC.

Candidates who are planning to take the CLEP Examination on Calculus with Elementary Functions are referred to the section of this Introduction on that examination on page xvii.

The Courses

Calculus AB and BC are both full-year courses in the calculus of functions of a single variable. Calculus BC includes all the topics in the AB course and additional ones such as sequences and series. Both courses are intended for students who have already studied college-preparatory mathematics; algebra, geometry, trigonometry, analytic geometry, and elementary functions (linear, polynomial, and rational; exponential and logarithmic; trigonometric and inverse trigonometric; and piecewise).

Comparison of Calculus AB and Calculus BC Courses

The BC course assumes that students already have a thorough knowledge of the elementary functions noted above. The AB topical course outline that follows can be covered in a full high-school academic year even if some time is allotted to elementary functions.

In the past, Calculus BC was presented as an enrichment of Calculus AB and as a more challenging course requiring deeper understanding. Both courses are now described as "challenging and demanding," with the BC course viewed as an "extension" of the AB course.

The "New" AP Calculus Courses

The calculus reform movement that began more than a dozen years ago has produced numerous significant changes in the teaching and learning of calculus, including

(1) increased emphasis on the concepts, with decreased emphasis on manipulation and memorization;

(2) regular use of technology;

Quotes are from the College Board's 2002–2003 AP Course Description.

(3) increased attention to applications;

(4) developing the student's ability to express functions, concepts, problems, and conclusions "graphically, numerically, analytically, and verbally," and to understand how these are related.

We have tried to embody these principles in the seventh edition of this book so that it reflects these changes.

Topics That May Be Tested on the Calculus AB Exam

1. **Functions and Graphs**
   Rational, trigonometric, inverse trigonometric, exponential, and logarithmic functions.

2. **Limits and Continuity**
   Intuitive definitions; one-sided limits; functions becoming infinite; asymptotes and graphs; limit of a quotient; $\lim_{\theta \to 0} \dfrac{\sin \theta}{\theta}$.

   Definition of continuity; kinds of discontinuities; theorems about continuous functions; Rational Function, Extreme Value, and Intermediate Value Theorems.

3. **Differentiation**
   Definition of derivative as the limit of a difference quotient and as instantaneous rate of change; formulas; composite functions and the chain rule; differentiability and continuity; estimating a derivative numerically and graphically; implicit differentiation; derivative of the inverse of a function; the Mean Value Theorem; recognizing a given limit as a derivative.

4. **Applications of Derivatives**
   Slope; critical points; average velocity; tangents and normals; increasing and decreasing functions; using the first and second derivative for the following: local max or min, inflection points, curve sketching, global max or min and optimization problems; relating a function and its derivatives graphically; motion along a line; local linearization and its use in approximating a function; related rates.

5. **The Definite Integral**
   Definite integral as the limit of a Riemann sum; area; definition of definite integral; properties of the definite integral; Riemann sums using rectangles or trapezoids on equal subdivisions; comparing approximating sums; average value of a function; Fundamental Theorem of Calculus; graphing a function from its derivative (revisited); interpreting ln x as an area.

6. **Integration**
   Antiderivatives and basic formulas; applications of antiderivatives; differential equations; motion problems.

7. **Applications of Integration to Geometry**
   Area of a region, including between two curves; volume of a solid of known cross-section, including a solid of revolution.

8. **Further Applications of Integration and Riemann Sums**
   Velocity and distance problems involving motion along a line; other applications involving the use of integrals of rates as net change or the use of integrals as accumulation functions.

9. **Differential Equations**
   Basic definitions; geometric interpretations using slope fields*; solving first-order separable differential equations analytically; exponential growth and decay.

Topics That May Be Tested on the Calculus BC Exam

Any of the topics listed above for the Calculus AB exam may be tested on the BC exam. The following additional topics are restricted to the BC exam.

1. **Functions and Graphs**
   Parametrically defined functions; polar functions.

2. **Limits and Continuity**
   No additional topics.

3. **Differentiation**
   Derivatives of polar and parametrically defined functions; indeterminate forms; L'Hôpital's Rule.

4. **Applications of Derivatives**
   Tangents to parametrically defined curves; slopes of polar curves.

5. **The Definite Integral**
   Integrals involving parametrically defined functions.

6. **Integration**
   By parts; by partial fractions (involving nonrepeating linear factors only).

7. **Applications of Integration to Geometry**
   Area of a region bounded by parametrically defined or polar curves; arc length; improper integrals.

8. **Further Applications of Integration and Riemann Sums**
   Velocity and distance problems involving motion along a planar curve; velocity and acceleration vectors.

---

* Slope fields will *first* be listed as a topic for Calculus AB in the 2003–2004 Course Description. The topic will not be tested on the AB Examination before 2004.

9. **Differential Equations**
    Slope fields[†]; Euler's Method; applications of differential equations, including logistic growth.

10. **Sequences and Series**
    Definition of series as a sequence of partial sums and of its convergence as the limit of that sequence; harmonic, geometric, and p-series; integral, ratio, and comparison tests for convergence; alternating series and error bound; conditional and absolute convergence; power series, including interval and radius of convergence; Taylor polynomials and graphs; finding a power series for a function; MacLaurin and Taylor series; Lagrange error bound for Taylor polynomials; computations using series.

The Examinations

The Calculus AB and BC Examinations and the course descriptions are prepared by committees of teachers from colleges or universities and from secondary schools. The examinations are intended to determine the extent to which a student has mastered the subject-matter of the course.

Each examination is to be three hours and fifteen minutes long, as follows:

Section I will have two parts. Part A will have 28 multiple-choice questions for which 55 minutes will be allowed. The use of calculators is *not* permitted in Part A.

Part B will have 17 multiple-choice questions, for which 50 minutes will be allowed. Some of the questions in Part B will require the use of a graphing calculator.

Section II, the free-response section, will contain two parts in both the AB and BC examinations. *Each part will have three questions*, as follows:

Part A will *require* a graphing calculator for some questions or parts of questions. After 45 minutes, however, *you will no longer be permitted to use a calculator.*

Part B will also be allotted 45 minutes, but *you will not be allowed to use a calculator.* You may use some of this time to work further on the questions of Part A, but the calculator may not be used for any of the work done during the second 45-minute period.

The section that follows on the Graphing Calculator gives important information on its use (and misuse!).

The Graphing Calculator: Using Your Graphing Calculator on the AP Exam

The Four Calculator Procedures

Each student is expected to bring a graphing calculator to the AP Exam. Different models of calculators vary in their features and capabilities; however, there are four procedures you must be able to perform on your[‡] calculator:

---

†See footnote on page x.
‡Occasional references to a calculator in this book will be to a TI-83.

    C1. Produce the graph of a function within an arbitrary viewing window;

    C2. Solve an equation numerically;

    C3. Compute the derivative of a function numerically; and

    C4. Compute definite integrals numerically.

Guidelines for Calculator Use

    1. On multiple-choice questions in Section I, Part B, *you may use any feature or program on your calculator.* Warning: don't use it too much! Only a few of these questions require the calculator, and in some cases using it may be too time-consuming or otherwise disadvantageous.

    2. On the free-response questions of Section II Part A:

    (a) You may use the calculator to perform any of the four listed procedures. When you do, you need only write the equation, derivative, or definite integral (called the "setup") that will produce the solution, then write the calculator result to the required degree of accuracy (three places after the decimal point unless otherwise specified). Note especially that a setup must be presented in standard algebraic or calculus notation, not just in calculator syntax. For example, while you may or may not choose to write "fnInt(cos(T),T,0, π)", you *must* include in your work the setup $\int_0^\pi \cos t \, dt$.

    (b) For a solution for which you use a calculator capability other than the four listed above, you must write down the mathematical steps that yield the answer. A correct answer alone will not earn full credit.

    (c) When asked to "justify," you must provide *mathematical reasoning* to support your answer. Calculator results alone will not be sufficient.

The Procedures Explained

    Here is more detailed guidance for the four allowed procedures.

    C1. "Produce the graph of a function within an arbitrary viewing window." Be sure that you create the graph *in the window specified,* then copy it carefully onto your exam paper. If no window is prescribed in the question, clearly indicate the window dimensions you have used.

    C2. "Solve an equation numerically" is equivalent to "Find the zeros of a function." Remember, you must first show your setup—write the equation out algebraically; then it is sufficient just to write down the calculator solution.

    C3. "Compute the derivative of a function numerically." When you seek the value of the derivative of a function at a specific point, you may use your calculator. First, indicate what you are finding—for example, $f'(6)$—then write the numerical answer obtained from your calculator. Note that if you need to find the derivative itself, rather than its value at a particular point, you must show how you obtained it and what it is, even though some calculators are able to perform symbolic operations.

    C4. "Compute definite integrals numerically." If, for example, you need to find the area under a curve, you must first show your setup. Write the complete integral, including the integrand in terms of a single variable and with the limits of integration. You may then simply write the calculator answer; you need not compute an antiderivative.

Sample Solutions of Free-Response Questions

The following set of examples illustrate proper use of your calculator on the examination. In all of these examples, the function is

$$f(x) = \frac{10x}{x^2 + 4} \quad \text{for} \quad 0 \leq x \leq 4$$

1. Graph f in [0,4] × [0,3].
    Using Procedure C1, we let
        $Y_1 = 10X/(X^2+4)$.
    Set the calculator window to the
    dimensions printed in your exam paper.
    Copy your graph carefully into
    the window on the exam paper.

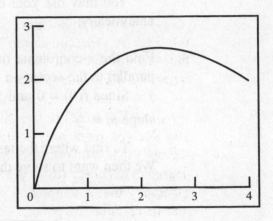

Viewing window [0,4] × [0,3]

2. Write the local linearization for $f(x)$ near $x = 1$.
    Note that $f(1) = 2$. Then, using procedure C3, calculate $f'(1)$. On a TI-83, for example, we keyed in nDeriv(Y,X,1), getting $f'(1) = 1.2$
    Then write the tangent-line (or local linear) approximation

    $$f(x) \simeq f(1) + f'(1)(x - 1)$$

    $$f(x) \simeq 2 + 1.2(x - 1) = 1.2x + 0.8$$

    You need not simplify, as we have, after the last equals sign just above.

3. Find the coordinates of any maxima of f. Justify your answer.
    Since finding a maximum is not one of the four allowed procedures, you must use calculus and show your work, writing the derivative algebraically and setting it equal to zero to find any critical numbers:

    $$f'(x) = \frac{(x^2 + 4)10 - 10x(2x)}{(x^2 + 4)^2} = \frac{40 - 10x^2}{(x^2 + 4)^2}$$

    $$= \frac{10(2 - x)(2 + x)}{(x^2 + 4)^2}$$

    $f'(x) = 0$ at $x = 2$ and at $x = -2$; but -2 is not in the specified domain.
    We analyze the signs of f' (which is easier here than it would be to use the second-derivative test) to assure that $x = 2$ does yield a maximum for f:

    $$
    \begin{array}{c}
    f \quad\quad \text{incr} \quad\quad\quad \text{decr} \\
    \hline
    0 \quad\quad\quad 2 \quad\quad\quad\quad 4 \\
    f' \quad\quad\quad + \quad\quad\quad\quad\quad -
    \end{array}
    $$

This implies that f does have a maximum at

$$\left(2, \frac{10(2)}{2^2 + 4}\right) = \left(2, \frac{5}{2}\right)$$

—but you may leave $f(2)$ in its unsimplified form, without evaluating to $\frac{5}{2}$.

Note that the signs analysis provides the justification required by the question.

You may use your calculator's maximum-finder to *verify* the result you obtain analytically.

4.   Find the x-coordinate of the point where the line tangent to the curve $y = f(x)$ is parallel to the secant on the interval $[0,4]$.

Since $f(0) = 0$ and $f(4) = 2$, the secant passes through $(0,0)$ and $(4,2)$ and has slope $m = \frac{1}{2}$.

To find where the tangent is parallel to the secant, we find $f'(x)$ as in Example 3. We then want to solve the equation

$$f'(x) = \frac{40 - 10x^2}{(x^2 + 4)^2} = \frac{1}{2}$$

The last equality above is the setup; we use procedure C2 to solve the equation: $x = 1.458$ is the desired answer.

5.   Estimate the area under the curve $y = f(x)$ using the Trapezoid Rule with four subintervals.

$$T = \frac{h}{2}[f(0) + 2f(1) + 2f(2) + 2f(3) + f(4)]$$

$$= \frac{1}{2}\left[0 + 2(2) + 2\left(\frac{5}{2}\right) + 2\left(\frac{30}{13}\right) + 2\right]$$

You may leave the answer in this form or simplify it to 7.808. If your calculator has a program for the Trapezoid Rule, you may use it to *complete* the computation after you have shown the setup as in the two centered equations above. If you omit them you will lose credit.

6.   Find the volume of the solid generated when the curve $y = f(x)$ on $[0,4]$ is rotated about the x-axis.

Using disks, we have

$$\Delta V = \pi R^2 H = \pi y^2 \Delta x$$

$$V = \pi \int_0^4 y^2 dx$$

Note that the equation above is not yet the setup: the definite integral must be in terms of x alone:

$$V = \pi \int_0^4 \left(\frac{10x}{x^2 + 4}\right)^2 dx$$

Now we have shown the setup. Using procedure C4, we can evaluate V:

$$V = 55.539$$

A Warning about Solutions in This Book

Students should be aware that *throughout the text* we frequently do *not* observe the restrictions cited above on the use of the calculator in answering free-response questions. In providing explanations for solutions to illustrative examples or to miscellaneous exercises we often exploit the capabilities of the calculator to the fullest. Indeed, students are encouraged to do just that on any question of Section I, Part B, of the AP examination for which they use a calculator. However, they must carefully observe the restrictions imposed on when and how the calculator may be used in answering questions in Section II of the examination, if they do not want to risk losing credit.

Additional Notes and Reminders

IMPROPER INTEGRALS. Most calculators can compute only definite integrals. Avoid using yours to obtain improper integrals, such as

$$\int_0^\pi \frac{1}{x^2}\,dx \qquad \text{or} \qquad \int_0^2 \frac{dx}{(x-1)^{2/3}}$$

The TI-82 and TI-83 both produce error messages "ERR: TOL NOT MET" or "Err: Divide by 0," but can often be quite slow about it.

ROUNDING-OFF ERRORS. To achieve three-place accuracy in a final answer, do *not* round off variables at intermediate steps, since this is likely to produce error-accumulations. If necessary, store longer intermediate answers internally in the calculator; do *not* copy them down on paper (storing is faster and avoids transcription errors). Round off only after your calculator produces the final answer.

ROUNDING THE FINAL ANSWER: UP OR DOWN? In rounding to three decimal places, remember that whether one rounds down or up depends on the nature of the problem. The mechanical rule followed in accounting (by which anything less than 0.0005 is rounded down, anything equal to or greater than 0.0005 is rounded up) does not apply.

Suppose, for example, that a problem seeks the largest k, to three decimal places, for which a condition is met, and the unrounded answer is 0.1239 . . . . Then 0.124 is too large: it does not meet the condition. The rounded answer must be 0.123. However, suppose that an otherwise identical problem seeks the smallest k for which a condition is met; in this case 0.1239 meets the condition but 0.1238 does not. So the rounding must be up, to 0.124.

FINAL ANSWERS TO SECTION II QUESTIONS. Although we usually express a final answer in this book in simplest form (often evaluating it on the calculator), this is hardly ever necessary on Section II questions of the AP Examination. According to the directions printed on the exam, "unless otherwise specified" (1) you need not simplify algebraic or numerical answers; (2) answers involving decimals should be correct to three places after the decimal point. However, be aware that if you try to simplify, you must do so correctly or you will lose credit.

USE YOUR CALCULATOR WISELY. Bear in mind that you will not be allowed to use your calculator at all on Part A of Section I. In Part B of Section I and part of Section II *only a few questions* will require one. As repeated often in this section, the questions that require a calculator *will not be identified.* You will have to be sensitive not only to when it is necessary to use the calculator but also to when it is efficient to do so.

The calculator is a marvelous tool, capable of illustrating complicated concepts with detailed pictures and of performing tasks that would otherwise be excessively time-consuming—or even impossible. But the completion of calculations and the displaying of graphs on the calculator is sometimes slow. Sometimes it is faster to find an answer using arithmetic, algebra, and analysis without recourse to the calculator. Before you start pushing buttons, take a few seconds to decide on the best way to attack a problem.

Grading the Examinations

Each completed AP examination paper receives a grade according to the following five-point scale:

5. Extremely well qualified
4. Well qualified
3. Qualified
2. Possibly qualified
1. No recommendation

Many colleges and universities accept a grade of 3 or better for credit or advanced placement or both; some also consider a grade of 2. More than 63 percent of the candidates who took the 2000 Calculus AB Examination earned grades of 3, 4, or 5. More than 78 percent of the 2000 BC candidates earned 3 or better. More than 171,400 students altogether took the 2000 mathematics examination.

The multiple-choice questions in Section I are scored by machine. To compensate for wild guessing on these questions, one-fourth of the number of incorrect answers is subtracted from the number of correct ones. Blind or haphazard guessing is therefore likely to lower a student's grade. However, if one or more of the choices given for a question can be eliminated as clearly incorrect, then the chance of guessing the correct answer from among the remaining ones is increased.

The problems in Section II are graded by college and high-school teachers called "readers." The answers in any one examination booklet are evaluated by different readers, and for each reader all scores given by preceding readers are concealed, as are the student's name and school. Readers are provided sample solutions for each problem, with detailed scoring scales and point distributions that allow partial credit for correct portions of a student's answer. Problems in Section II are all counted equally.

In the determination of the overall grade for each examination, the two sections are given equal weight. The total raw score is then converted into one of the five grades described above.

Students who take the BC examination will be given not only a Calculus-BC grade but also a Calculus-AB subscore grade. The latter will be based on the part of the BC examination dealing with topics in the AB syllabus.

In general, students will not be expected to answer all the questions correctly in either Section I or II.

Great care is taken by all involved in the scoring and reading of papers to make certain that they are graded consistently and fairly so that a student's overall AP grade reflects as accurately as possible his or her achievement in calculus.

The CLEP Calculus Examination

Many colleges grant credit to students who perform acceptably on tests offered by the College Level Examination Program (CLEP). Their calculus examination is one such test.

The College Board's *CLEP Official Study Guide: 2001 Edition* provides descriptions of all CLEP examinations, test-taking tips, and suggestions on reference and supplementary materials. According to the *Guide*, the calculus examination covers topics usually taught in a one-semester college calculus course. It is assumed that students taking the exam will have studied college-preparatory mathematics (algebra, plane and solid geometry, analytic geometry, and trigonometry).

There are 45 multiple-choice questions on the CLEP calculus exam for which 90 minutes are allowed. A calculator may *not* be used during the examination.

Approximately 5 percent of the questions are on the topic limits, about 55 percent on differential calculus, and about 40 percent on integral calculus. The specific topics that may be tested on the CLEP calculus exam are essentially those on pages ix and x under the heading "Topics That May Be Tested on the Calculus AB Exam." (L'Hôpital's Rule is listed as a CLEP calculus topic but only as a BC topic for the AP exam. And the *only* topics listed as applications of the definite integral for the CLEP calculus test are "average value of a function on an interval" and "area.")

Since any topic that may be tested on the CLEP calculus exam is included in this book on the AP Exam, a candidate who plans to take the CLEP exam will benefit from a review of the AB topics covered here. The multiple-choice questions in Part A of Chapter 11 and in Part A of Section I of each of the four AB Practice Examinations will provide good models for questions on the CLEP calculus test.

A complete description of the knowledge and skills required and of the specific topics that may be tested on the CLEP exam can be downloaded from the College Board's web site at *www.collegeboard.com/clep*.

This Review Book

This book consists of the following parts:

The Topical Review includes 10 chapters with notes on the main topics of the Calculus AB and BC syllabi and with numerous carefully worked-out examples. Each chapter concludes with a set of multiple-choice questions, followed immediately by answers and solutions.

This review is followed by further practice: (1) Chapter 11, which includes a set of multiple-choice questions on miscellaneous topics and an answer key; (2) Chapter 12, a set of miscellaneous free-response problems that are intended to be similar to those in Section II of the AP examinations. They are followed by solutions. These problems were designed especially to reflect changes discussed earlier.

The next part of the book titled Practice Examinations has four AB and four BC practice exams, both Section I and II to simulate the actual AP examinations. Each is followed by answers and explanations.

In this book, review material on topics covered only in Calculus BC is preceded by an asterisk (*), as are both multiple-choice questions and free-response-type problems that are likely to occur only on a BC Examination.

THE TEACHER WHO USES THIS BOOK WITH A CLASS may profitably do so in any of several ways. If the book is used throughout a year's course, the teacher can assign all or part of each set of multiple-choice questions and some miscellaneous exercises after the topic has been covered. These sets can also be used for review purposes shortly before examination time. The Practice Examinations will also be very helpful in reviewing toward the end of the year. The teacher who wishes to give take-home examinations throughout the year can assemble such examinations by choosing appropriate problems, on the material to be covered, from the sample Miscellaneous Questions or Exercises, or Practice Examinations.

STUDENTS WHO USE THIS BOOK INDEPENDENTLY will improve their performance by studying the illustrative examples carefully and trying to complete practice problems before referring to the solution keys.

Since many FIRST-YEAR MATHEMATICS COURSES IN COLLEGES follow syllabi much like that proposed by the College Board for high-school Advanced Placement courses, college students and teachers may also find the book useful.

TOPICAL REVIEW
AND PRACTICE

Chapter 1

Functions

Review of Definitions and Properties[†]

A. Definitions

A1. A *function f* is a correspondence that associates with each element *a* of a set called the *domain* one and only one element *b* of a set called the *range*. We write $f(a) = b$; *b* is the *value* of *f* at *a*. The elements in the domain are called *inputs,* and those in the range are called *outputs.*

A function is often represented by an equation, a graph, or a table.

A vertical line cuts the graph of a function in at most one point.

Example 1. The domain of $f(x) = x^2 - 2$ is the set of all real numbers; its range is the set of all reals greater than or equal to –2. Note that

$$f(0) = 0^2 - 2 = -2, \quad f(-1) = (-1)^2 - 2 = -1,$$
$$f(\sqrt{3}) = (\sqrt{3})^2 - 2 = 1, \quad f(c) = c^2 - 2,$$
$$f(x + h) - f(x) = [(x + h)^2 - 2] - [x^2 - 2]$$
$$= x^2 + 2hx + h^2 - 2 - x^2 + 2 = 2hx + h^2.$$

Example 2. The domain of $f(x) = \dfrac{4}{x - 1}$ is the set of all reals except $x = 1$ (which we shorten to "$x \neq 1$").

The domain of $g(x) = \dfrac{x}{x^2 - 9}$ is $x \neq 3, -3$.

The domain of $h(x) = \dfrac{\sqrt{4 - x}}{x}$ is $x \leq 4$, $x \neq 0$ (which is a short way of writing $\{x \mid x \text{ is real}, x < 0 \text{ or } 0 < x \leq 4\}$).

[†]Because students in AP courses are assumed to have a firm control of precalculus topics, such topics are not directly tested on the AP examinations. Nevertheless, some of this chapter's multiple-choice questions deal with those topics to reinforce important basic principles.

A2. Two functions f and g with the same domain may be combined to yield their sum and difference: $f(x) + g(x)$ and $f(x) - g(x)$,

also written $(f+g)\,(x)$ and $(f-g)\,(x)$ respectively;

their product and quotient: $f(x)g(x)$ and $f(x)/g(x)$,

also written respectively as $(fg)(x)$ and $(f/g)\,(x)$.

The quotient is defined for all x in the shared domain except those values for which $g(x)$, the denominator, equals zero.

Example 3.   If $f(x) = x^2 - 4x$ and $g(x) = x + 1$, then

$$\frac{f(x)}{g(x)} = \frac{x^2 - 4x}{x+1} \quad \text{and has domain } x \neq -1;$$

$$\frac{g(x)}{f(x)} = \frac{x+1}{x^2 - 4x} = \frac{x+1}{x(x-4)} \quad \text{and has domain } x \neq 0, 4.$$

A3. The *composition* (or *composite*) of f with g, written $f(g(x))$, read as "f of g of x," is the function obtained by replacing x wherever it occurs in $f(x)$ by $g(x)$. We also write $(f \circ g)\,(x)$ for $f(g(x))$. The domain of $(f \circ g)\,(x)$ is the set of all x in the domain of g for which $g(x)$ is in the domain of f.

Example 4a.   If $f(x) = 2x - 1$ and $g(x) = x^2$, then

$$f(g(x)) = 2(x^2) - 1 = 2x^2 - 1;$$

but

$$g(f(x)) = (2x - 1)^2 = 4x^2 - 4x + 1.$$

In general, $f(g(x)) \neq g(f(x))$.

Example 4b.   If $f(x) = 4x^2 - 1$ and $g(x) = \sqrt{x}$, then

$$f(g(x)) = 4x - 1 \quad (x \geq 0);$$

whereas

$$g(f(x)) = \sqrt{4x^2 - 1} \quad (|x| \geq \tfrac{1}{2}).$$

A4. A function f is $\genfrac{}{}{0pt}{}{odd}{even}$ if, for all x in the domain of f, $\genfrac{}{}{0pt}{}{f(-x) = -f(x)}{f(-x) = f(x)}$.

The graph of an odd function is symmetric about the origin; the graph of an even function is symmetric about the y-axis.

Example 5. The graphs of $f(x) = \frac{1}{2}x^3$ and $g(x) = 3x^2 - 1$ are shown in Figure N1–1; $f(x)$ is odd, $g(x)$ even.

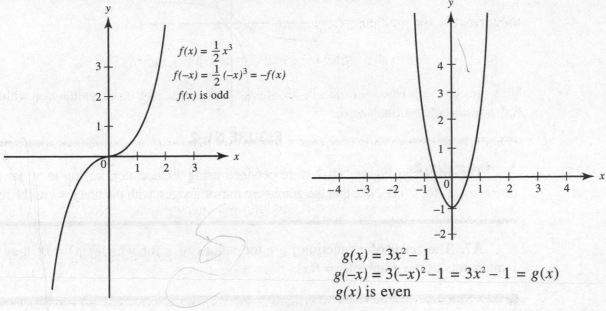

FIGURE N1–1

A5. If a function f yields a single output for each input and also yields a single input for every output, then f is said to be *one-to-one*. Geometrically, this means that any horizontal line cuts the graph of f in at most one point. The function sketched at the left in Figure N1–1 is one-to-one; the function sketched at the right is not. A function that is increasing (or decreasing) on an interval I is one-to-one on that interval (see pages 93–96 for definitions of increasing and decreasing functions).

A6. If f is one-to-one with domain X and range Y, then there is a function f^{-1}, with domain Y and range X, such that

$$f^{-1}(y_0) = x_0 \quad \text{if and only if} \quad f(x_0) = y_0.$$

The function f^{-1} is the *inverse* of f. It can be shown that f^{-1} is also one-to-one and that its inverse is f. The graphs of a function and its inverse are symmetric with respect to the line $y = x$.

> To find the inverse of $y = f(x)$,
> solve for x in terms of y,
> then interchange x and y.

Example 6. Find the inverse of the one-to-one function $f(x) = x^3 - 1$.

Solve the equation for x: $\qquad x = \sqrt[3]{y+1}$.

Interchange x and y: $\qquad y = \sqrt[3]{x+1} = f^{-1}(x)$.

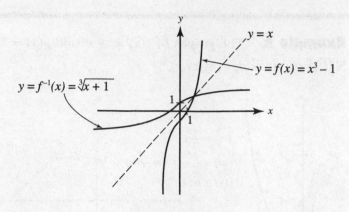

FIGURE N1–2

The graphs in Figure N1–2 were obtained using a calculator, keying in $\mathtt{Y_1}$ for $f(x)$ and $\mathtt{DrawInv\ Y_1}$ for $f^{-1}(x)$. Note that the graphs are mirror images, with the line $y = x$ as the mirror.

A7. The *zeros* of a function f are the values of x for which $f(x) = 0$; they are the x-intercepts of the graph of $y = f(x)$.

Example 7.   The zeros of $f(x) = x^4 - 2x^2$ are the x's for which $x^4 - 2x^2 = 0$. The function has three zeros, since $x^4 - 2x^2 = x^2(x^2 - 2)$ equals zero if $x = 0$, $+\sqrt{2}$, or $-\sqrt{2}$.

B.   Special Functions

The *absolute-value* function $f(x) = |x|$ and the *greatest-integer* function $g(x) = [x]$ are sketched in Figure N1–3.

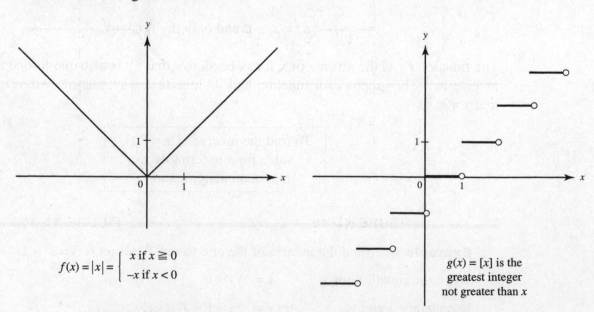

$$f(x) = |x| = \begin{cases} x & \text{if } x \geq 0 \\ -x & \text{if } x < 0 \end{cases}$$

$g(x) = [x]$ is the greatest integer not greater than x

Absolute-value function   Greatest-integer function

FIGURE N1–3

Example 8. A function f is defined on the interval $[-2, 2]$ and has the graph shown in Figure N1-4.

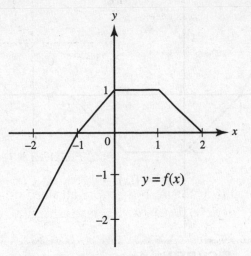

FIGURE N1–4

(a) Sketch the graph of $y = |f(x)|$.
(b) Sketch the graph of $y = f(|x|)$.
(c) Sketch the graph of $y = -f(x)$.
(d) Sketch the graph of $y = f(-x)$.

The graphs are shown in Figures N1–4a through N1–4d.

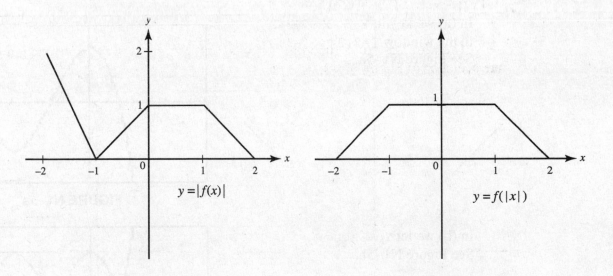

FIGURE N1–4a

FIGURE N1–4b

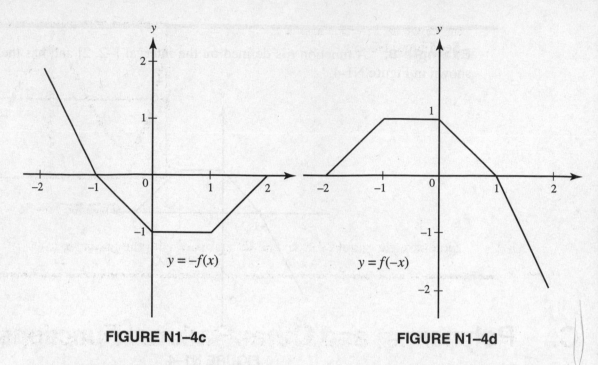

FIGURE N1–4c	**FIGURE N1–4d**

Note that graph (c) of $y = -f(x)$ is the reflection of graph (a) of $y = f(x)$ in the x-axis, whereas graph (d) of $y = f(-x)$ is the reflection of graph (b) of $y = f(x)$ in the y-axis. How do the graphs of $|f(x)|$ and $f(|x|)$ compare with the graph of $f(x)$?

Example 9.   Let $f(x) = x^3 - 3x^2 + 2$. Graph the following functions on your calculator:

(a) $y = f(x)$ (b) $y = |f(x)|$ (c) $y = f(|x|)$

In (a) we key in $\text{Y}_1 = \text{X}^3 - 3\text{X}^2 + 2$
in the window $[-3,3] \times [-3,3]$.
See Figure N1–5a.

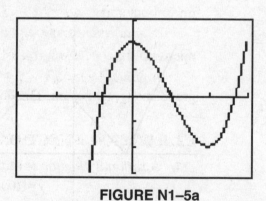

FIGURE N1–5a

In (b) we let $\text{Y}_2 = \text{abs } \text{Y}_1$.
See Figure N1–5b.

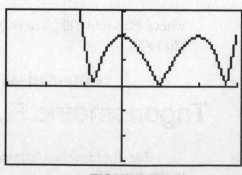

FIGURE N1–5b

In (c) we let Y$_3$ = Y$_1$(abs X).
See Figure N1–5c.

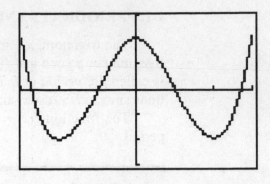

FIGURE N1–5c

Note how the graphs for (b) and (c) compare with the graph for (a).

C.   Polynomial and Other Rational Functions____

C1. POLYNOMIAL FUNCTIONS.

A *polynomial function* is of the form

$$f(x) = a_0x^n + a_1x^{n-1} + a_2x^{n-2} + \cdots + a_{n-1}x + a_n,$$

where n is a positive integer or zero, and the a's, the *coefficients,* are constants. If $a_0 \neq 0$, the degree of the polynomial is n.

A *linear* function, $f(x) = mx + b$, is of the first degree; its graph is a straight line with slope m, the constant rate of change of $f(x)$ (or y) with respect to x, and b is the line's y-intercept.

A *quadratic* function, $f(x) = ax^2 + bx + c$, has degree 2; its graph is a parabola that opens up if $a > 0$, down if $a < 0$, and whose axis is the line $x = -\dfrac{b}{2a}$.

A *cubic,* $f(x) = a_0x^3 + a_1x^2 + a_2x + a_3$, has degree 3; calculus enables us to sketch its graph easily; and so on. The domain of every polynomial is the set of all reals.

C2. RATIONAL FUNCTIONS.

A *rational function* is of the form

$$f(x) = \frac{P(x)}{Q(x)},$$

where $P(x)$ and $Q(x)$ are polynomials. The domain of f is the set of all reals for which $Q(x) \neq 0$.

D.   Trigonometric Functions_____

The fundamental trigonometric identities, graphs, and reduction formulas are given in the Appendix.

D1. PERIODICITY AND AMPLITUDE.

The trigonometric functions are periodic. A function f is *periodic* if there is a positive number p such that $f(x + p) = f(x)$ for each x in the domain of f. The smallest such p is called the *period* of f. The graph of f repeats every p units along the x-axis. The functions $\sin x$, $\cos x$, $\csc x$, and $\sec x$ have period 2π; $\tan x$ and $\cot x$ have period π.

The function $f(x) = A \sin bx$ has amplitude A and period $\frac{2\pi}{b}$; $g(x) = \tan cx$ has period $\frac{\pi}{c}$.

Example 10. (a) For what value of k does $f(x) = \frac{1}{k} \cos kx$ have period 2? (b) What is the amplitude of f for this k?

(a) Function f has period $\frac{2\pi}{k}$; since this must equal 2, we solve the equation $\frac{2\pi}{k} = 2$, getting $k = \pi$.

(b) It follows that the amplitude of f that equals $\frac{1}{k}$ has a value of $\frac{1}{\pi}$.

Example 11. Find (a) the period and (b) the maximum value of $f(x) = 3 - \sin \frac{\pi x}{3}$. (c) What is the smallest positive x for which f is a maximum? (d) Sketch the graph. (e) Verify your sketch with a graphing calculator.

(a) The period of f is $2\pi \div \frac{\pi}{3}$, or 6.

(b) Since the maximum value of $-\sin x$ is $-(-1)$ or $+1$, the maximum value of f is $3 + 1$ or 4.

(c) $-\left(\sin \frac{\pi x}{3}\right)$ equals $+1$ when $\sin \frac{\pi x}{3} = -1$, that is, when $\frac{\pi x}{3} = \frac{3\pi}{2}$. Solving yields $x = \frac{9}{2}$.

(d) See Figure N1–6a.

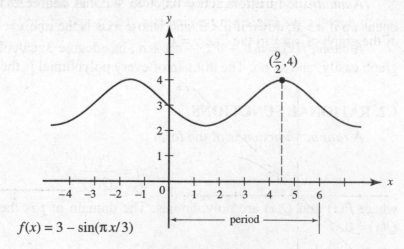

$f(x) = 3 - \sin(\pi x/3)$

FIGURE N1–6a

(e) On the calculator, we enter Y₁=3-sin (πX/3) in the window [-6,6] × [0,5]. We then graph Y₁, as shown in Figure N1–6b.

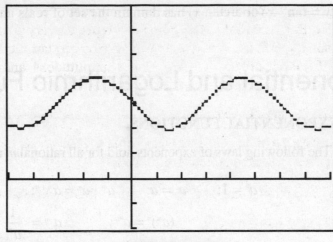

FIGURE N1–6b

D2. INVERSES.

We obtain *inverses* of the trigonometric functions by limiting the domains of the latter so each trigonometric function is one-to-one over its restricted domain. For example, we restrict

$$\sin x \text{ to } -\frac{\pi}{2} \leqq x \leqq \frac{\pi}{2},$$

$$\cos x \text{ to } 0 \leqq x \leqq \pi,$$

$$\tan x \text{ to } -\frac{\pi}{2} < x < \frac{\pi}{2}.$$

The graphs of $f(x) = \sin x$ on $\left[-\frac{\pi}{2}, \frac{\pi}{2}\right]$ and of its inverse $f^{-1}(x) = \sin^{-1}x$ are shown in Figure N1–7. The inverse trigonometric function $\sin^{-1}x$ is also commonly denoted by arcsin x, which denotes *the* angle whose sine is x. The graph of $\sin^{-1}x$ is, of course, the reflection of the graph of $\sin x$ in the line $y = x$.

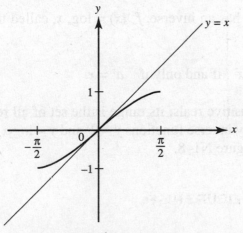

$y = \sin x$
domain: $-\frac{\pi}{2} \leqq x \leqq \frac{\pi}{2}$
range: $-1 \leqq y \leqq 1$

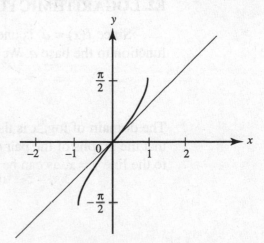

$y = \sin^{-1}x = \arcsin x$
domain: $-1 \leqq x \leqq 1$
range: $\frac{\pi}{2} - \leqq y \leqq \frac{\pi}{2}$

FIGURE N1–7

Also, for other inverse trigonometric functions,

$y = \cos^{-1} x$ (or arccos x) has domain $-1 \leqq x \leqq 1$ and range $0 \leqq y \leqq \pi$;

$y = \tan^{-1} x$ (or arctan x) has domain the set of reals and range $-\dfrac{\pi}{2} < y < \dfrac{\pi}{2}$.

E.   Exponential and Logarithmic Functions _____

E1. EXPONENTIAL FUNCTIONS.

The following laws of exponents hold for all rational m and n, provided that $a > 0$, $a \neq 1$:

$$a^0 = 1; \qquad a^1 = a; \qquad a^m \cdot a^n = a^{m+n}; \qquad a^m \div a^n = a^{m-n};$$

$$(a^m)^n = a^{mn}; \qquad a^{-m} = \frac{1}{a^m}.$$

These properties define a^x when x is rational. When x is irrational, we note that x is an infinite nonrepeating decimal; we then define a^x as the limit of an infinite sequence of rational numbers. For example,

$$3^{\sqrt{2}} = 3^{1.4142135...}$$

is the limit of the unending sequence of rational numbers

$$3^1, \; 3^{1.4}, \; 3^{1.41}, \; 3^{1.414}, \; \ldots .$$

The exponential function $f(x) = a^x$ ($a > 0$, $a \neq 1$) is thus defined for all real x; its domain is the set of positive reals. The graph of $y = a^x$, when $a = 2$, is shown in Figure N1–8.

Of special interest and importance in the calculus is the exponential function $f(x) = e^x$, where e is an irrational number whose decimal approximation to five decimal places is 2.71828. We define e on page 32.

E2. LOGARITHMIC FUNCTIONS.

Since $f(x) = a^x$ is one-to-one it has an inverse, $f^{-1}(x) = \log_a x$, called the *logarithmic function* to the base a. We note that

$$y = \log_a x \quad \text{if and only if} \quad a^y = x.$$

The domain of $\log_a x$ is the set of positive reals; its range is the set of all reals. It follows that the graphs of the pair of mutually inverse functions $y = 2^x$ and $y = \log_2 x$ are symmetric to the line $y = x$, as can be seen in Figure N1–8.

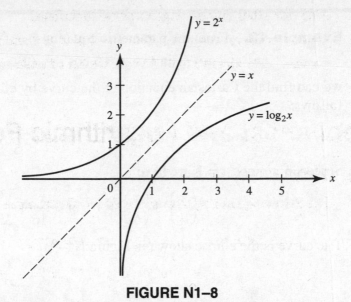

FIGURE N1–8

The logarithmic function $\log_a x$ $(a > 0, a \neq 1)$ has the following properties:

$$\log_a 1 = 0; \qquad \log_a a = 1; \qquad \log_a mn = \log_a m + \log_a n;$$

$$\log_a \frac{m}{n} = \log_a m - \log_a n; \qquad \log_a x^m = m \log_a x.$$

The logarithmic base e is so important and convenient in calculus that we use a special symbol:

$$\log_e x = \ln x.$$

Logarithms to the base e are called *natural logarithms*. The domain of $\ln x$ is the set of positive reals; its range is the set of all reals. The graphs of the mutually inverse functions $\ln x$ and e^x are given in the Appendix.

*F.   Parametrically Defined Functions_____

If the x- and y-coordinates of a point on a graph are given as functions f and g of a third variable, say t, then

$$x = f(t), \qquad y = g(t)$$

are called *parametric equations* and t is called the *parameter.* When t represents time, as it often does, then we can view the curve as that followed by a moving particle as the time varies.

*An asterisk denotes a topic covered only in Calculus BC.

Example 12. From the parametric equations

$$x = 4 \sin t, \qquad y = 5 \cos t \qquad (0 \leqq t \leqq 2\pi)$$

we can find the Cartesian equation of the curve by eliminating the parameter t as follows:

$$\sin t = \frac{x}{4}, \qquad \cos t = \frac{y}{5}.$$

Since $\sin^2 t + \cos^2 t = 1$, we have

$$\left(\frac{x}{4}\right)^2 + \left(\frac{y}{5}\right)^2 = 1 \quad \text{or} \quad \frac{x^2}{16} + \frac{y^2}{25} = 1.$$

The curve is the ellipse shown in Figure N1–9a.

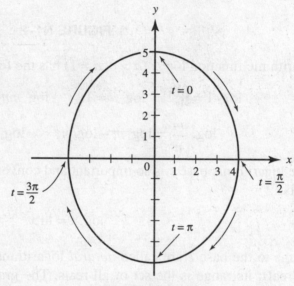

FIGURE N1–9a

The calculator graphs parametric equations in rectangular format. For the equations in Example 12, we let

$$X_{1T} = 4\sin T$$
$$Y_{1T} = 5\cos T$$

and graph in the window $[-5, 5] \times [-5, 5]$ with T in $[0, 2\pi]$ and the standard T-step, as shown in Figure N1–9b.

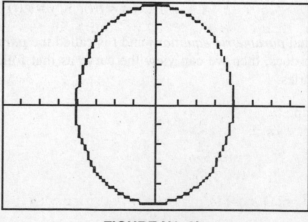

FIGURE N1–9b

Note that, as t increases from 0 to 2π, a particle moving in accordance with the given parametric equations starts at point $(0, 5)$ (when $t = 0$) and travels in a clockwise direction along the ellipse, returning to $(0, 5)$ when $t = 2\pi$.

Example 13.   For the pair of parametric equations

$$x = 1 - t, \qquad y = \sqrt{t} \qquad (t \geqq 0)$$

we can eliminate t by squaring the second equation and substituting for t in the first; then we have

$$y^2 = t \qquad \text{and} \qquad x = 1 - y^2.$$

We see the graph of the equation $x = 1 - y^2$ on the left in Figure N1–10. At the right we see only the upper part of this graph, the part defined by the parametric equations for which t and y are both restricted to nonnegative numbers.

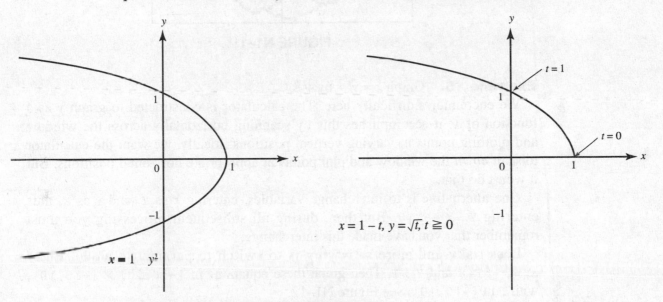

FIGURE N1–10

The function defined by the parametric equations here is $y = F(x) = \sqrt{1 - x}$, whose graph is at the right above; its domain is $x \leqq 1$ and its range is the set of nonnegative reals.

   Verify the graph of the parametric equations on your calculator in the window $[-5,2] \times [-1,3]$.

Example 14.   A satellite is in orbit around a planet that is orbiting around a star. The satellite makes 12 orbits each year. Its path is given by the parametric equations

$$x = 4 \cos t + \cos 12t,$$
$$y = 4 \sin t + \sin 12t.$$

Graph the satellite's path.

   In parametric mode, we let

$$X_{1T} = 4\cos T + \cos 12T,$$
$$Y_{1T} = 4\sin T + \sin 12T,$$

and graph these in $[-7,7] \times [-5,5]$ with T in $[0,2\pi]$.

For the graph shown in Figure N1–11, we used the T-step equal to 0.02 rather than the standard step to get a smoother curve (though one that takes longer to generate).

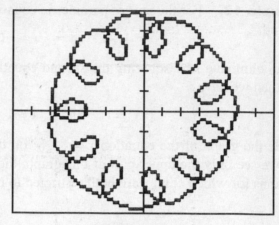

FIGURE N1–11

Example 15.   Graph $x = y^2 - 6y + 8$.

We encounter a difficulty here. The calculator is constructed to graph y as a function of x: it accomplishes this by scanning horizontally across the window and plotting points in varying vertical positions. Ideally, we want the calculator to scan *down* the window and plot points at appropriate horizontal positions. But it won't do that.

One alternative is to interchange variables, entering x as $\mathtt{Y_1}$ and y as $\mathtt{X}$, thus entering $\mathtt{Y_1=X^2-6X+8}$. But then, during all subsequent processing you must remember that you have made this interchange.

Less risky and more satisfying is to switch to parametric mode: Enter $\mathtt{X_{1T}=T^2-6T+8}$ and $\mathtt{Y_{1T}=T}$. Then graph these equations in $\mathtt{[-10,10]} \times \mathtt{[-10,10]}$, with $\mathtt{T}$ in $\mathtt{[-10,10]}$. See Figure N1–12.

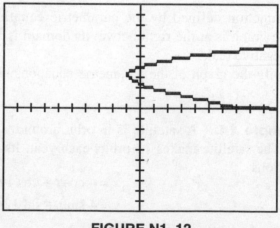

FIGURE N1–12

Example 16.   Let $f(x) = x^3 + x$; graph $f^{-1}(x)$.

Once again we use parametric mode, entering $X_{1T} = T$; $Y_{1T} = T^3 + T$. We then graph

$$X_{2T} = Y_{1T}, \qquad Y_{2T} = X_{1T}.$$

Figure N1-13 shows both $f(x)$ and $f^{-1}(x)$.

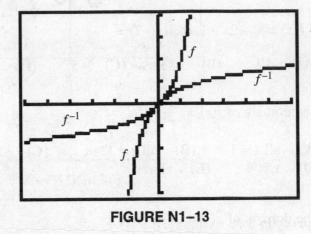

FIGURE N1–13

Parametric equations give rise to vector functions, which will be discussed in connection with motion along a curve in Chapter 4.

Set 1: Multiple-Choice Questions on Functions: Review†

1. If $f(x) = x^3 - 2x - 1$, then $f(-2) =$

    (A) -17 **(B)** -13 **(C)** -5 **(D)** -1 **(E)** 3

2. The domain of $f(x) = \dfrac{x-1}{x^2+1}$ is

    (A) all $x \neq 1$ **(B)** all $x \neq 1, -1$ **(C)** all $x \neq -1$
    (D) $x \geqq 1$ **(E)** all reals

3. The domain of $g(x) = \dfrac{\sqrt{x-2}}{x^2-x}$ is

    (A) all $x \neq 0, 1$ **(B)** $x \leqq 2, x \neq 0, 1$ **(C)** $x \leqq 2$
    (D) $x \geqq 2$ **(E)** $x > 2$

4. If $f(x) = x^3 - 3x^2 - 2x + 5$ and $g(x) = 2$, then $g(f(x)) =$

    (A) $2x^3 - 6x^2 - 2x + 10$ **(B)** $2x^2 - 6x + 1$ **(C)** -6
    (D) -3 **(E)** 2

5. With the functions and choices as in Question 4, which choice is correct for $f(g(x))$?

6. If $f(x) = x^3 + Ax^2 + Bx - 3$ and if $f(1) = 4$ and $f(-1) = -6$, what is the value of $2A + B$?

    (A) 12   **(B)** 8   **(C)** 0   **(D)** -2
    (E)   It cannot be determined from the given information.

7. Which of the following equations has a graph that is symmetric with respect to the origin?

    (A) $y = \dfrac{x-1}{x}$ **(B)** $y = 2x^4 + 1$ **(C)** $y = x^3 + 2x$

    (D) $y = x^3 + 2$ **(E)** $y = \dfrac{x}{x^3+1}$

8. Let g be a function defined for all reals. Which of the following conditions is not sufficient to guarantee that g has an inverse function?
    (A) $g(x) = ax + b, a \neq 0$.   **(B)** g is strictly decreasing.
    (C) g is symmetric to the origin.   **(D)** g is strictly increasing.
    (E) g is one-to-one.

---

†Because students in AP courses are assumed to have firm control of precalculus topics, such topics are not directly tested on the AP examinations. Nevertheless, some of this chapter's multiple-choice questions deal with those topics to reinforce important basic principles.

9. Let $y = f(x) = \sin(\arctan x)$. Then the range of f is

(A) $\{y \mid 0 < y \leqq 1\}$ (B) $\{y \mid -1 < y < 1\}$ (C) $\{y \mid -1 \leqq y \leqq 1\}$

(D) $\left\{y \mid -\dfrac{\pi}{2} < y < \dfrac{\pi}{2}\right\}$ (E) $\left\{y \mid -\dfrac{\pi}{2} \leqq y \leqq \dfrac{\pi}{2}\right\}$

10. Let $g(x) = |\cos x - 1|$. The maximum value attained by g on the closed interval $[0, 2\pi]$ is for x equal to

(A) -1 (B) 0 (C) $\dfrac{\pi}{2}$ (D) 2 (E) π

11. Which of the following functions is not odd?

(A) $f(x) = \sin x$ (B) $f(x) = \sin 2x$ (C) $f(x) = x^3 + 1$

(D) $f(x) = \dfrac{x}{x^2 + 1}$ (E) $f(x) = \sqrt[3]{2x}$

12. The roots of the equation $f(x) = 0$ are 1 and -2. The roots of $f(2x) = 0$ are

(A) 1 and -2 (B) $\dfrac{1}{2}$ and -1 (C) $-\dfrac{1}{2}$ and 1

(D) 2 and -4 (E) -2 and 4

13. The set of zeros of $f(x) = x^3 + 4x^2 + 4x$ is

(A) $\{-2\}$ (B) $\{0, -2\}$ (C) $\{0, 2\}$ (D) $\{2\}$ (E) $\{2, 2\}$

14. The values of x for which the graphs of $y = x + 2$ and $y^2 = 4x$ intersect are

(A) -2 and 2   (B) -2 (C) 2   (D) 0   (E) none of these

15. The function whose graph is a reflection in the y-axis of the graph of $f(x) = 1 - 3^x$ is

(A) $g(x) = 1 - 3^{-x}$ (B) $g(x) = 1 + 3^x$ (C) $g(x) = 3^x - 1$
(D) $g(x) = \ln_3 (x - 1)$ (E) $g(x) = \ln_3 (1 - x)$

16. Let $f(x)$ have an inverse function $g(x)$. Then $f(g(x)) =$

(A) 1   (B) x (C) $\dfrac{1}{x}$ (D) $f(x) \cdot g(x)$ (E) none of these

17. The function $f(x) = 2x^3 + x - 5$ has exactly one real zero. It is between

(A) -2 and -1 (B) -1 and 0   (C) 0 and 1
(D) 1 and 2   (E) 2 and 3

18. The period of $f(x) = \sin \frac{2\pi}{3} x$ is

(A) $\frac{1}{3}$ (B) $\frac{2}{3}$ (C) $\frac{3}{2}$ (D) 3   (E) 6

19. The range of $y = f(x) = \ln (\cos x)$ is

(A) $\{y \mid -\infty < y \leq 0\}$ (B) $\{y \mid 0 < y \leq 1\}$ (C) $\{y \mid -1 < y < 1\}$

(D) $\left\{y \mid -\frac{\pi}{2} < y < \frac{\pi}{2}\right\}$ (E) $\{y \mid 0 \leq y \leq 1\}$

20. If $\log_b (3^b) = \frac{b}{2}$, then $b =$

(A) $\frac{1}{9}$ (B) $\frac{1}{3}$ (C) $\frac{1}{2}$ (D) 3   (E) 9

21. Let f^{-1} be the inverse function of $f(x) = x^3 + 2$. Then $f^{-1}(x) =$

(A) $\frac{1}{x^3 - 2}$ (B) $(x + 2)^3$ (C) $(x - 2)^3$

(D) $\sqrt[3]{x + 2}$ (E) $\sqrt[3]{x - 2}$

22. The set of x-intercepts of the graph of $f(x) = x^3 - 2x^2 - x + 2$ is

(A) $\{1\}$ (B) $\{-1, 1\}$ (C) $\{1, 2\}$
(D) $\{-1, 1, 2\}$ (E) $\{-1, -2, 2\}$

23. If the domain of f is restricted to the open interval $\left(-\frac{\pi}{2}, \frac{\pi}{2}\right)$, then the range of $f(x) = e^{\tan x}$ is

(A) the set of all reals   (B) the set of positive reals
(C) the set of nonnegative reals   (D) $\{y \mid 0 < y \leq 1\}$
(E) none of these

24. Which of the following is a reflection of the graph of $y = f(x)$ in the x-axis?

(A) $y = -f(x)$ (B) $y = f(-x)$ (C) $y = |f(x)|$
(D) $y = f(|x|)$ (E) $y = -f(-x)$

25. The smallest positive x for which the function $f(x) = \sin\left(\frac{x}{3}\right) - 1$ is a maximum is

(A) $\frac{\pi}{2}$ (B) π (C) $\frac{3\pi}{2}$ (D) 3π (E) 6π

26. $\tan\left(\arccos\left(-\frac{\sqrt{2}}{2}\right)\right) =$

   (A)   –1   (B) $-\frac{\sqrt{3}}{3}$ (C) $-\frac{1}{2}$ (D) $\frac{\sqrt{3}}{3}$ (E)   1

27. If $f^{-1}(x)$ is the inverse of $f(x) = 2e^{-x}$, then $f^{-1}(x) =$

   (A) $\ln\left(\frac{2}{x}\right)$ (B) $\ln\left(\frac{x}{2}\right)$ (C) $\left(\frac{1}{2}\right)\ln x$

   (D) $\sqrt{\ln x}$ (E) $\ln(2 - x)$

28. Which of the following functions does not have an inverse function?

   (A) $y = \sin x \left(-\frac{\pi}{2} \leqq x \leqq \frac{\pi}{2}\right)$ (B) $y = x^3 + 2$ (C) $y = \frac{x}{x^2 + 1}$

   (D) $y = \frac{1}{2}e^x$ (E) $y = \ln(x - 2)$ (where $x > 2$)

29. Suppose that $f(x) = \ln x$ for all positive x and $g(x) = 9 - x^2$ for all real x. The domain of $f(g(x))$ is

   (A) $\{x \mid x \leqq 3\}$ (B) $\{x \mid |x| \leqq 3\}$ (C) $\{x \mid |x| > 3\}$
   (D) $\{x \mid |x| < 3\}$ (E) $\{x \mid 0 < x < 3\}$

30. Suppose (as in Question 29) that $f(x) = \ln x$ for all positive x and $g(x) = 9 - x^2$ for all real x. The range of $y = f(g(x))$ is

   (A) $\{y \mid y > 0\}$ (B) $\{y \mid 0 < y \leqq \ln 9\}$ (C) $\{y \mid y \leqq \ln 9\}$
   (D) $\{y \mid y < 0\}$ (E)   none of these

ANSWERS AND SOLUTIONS TO MULTIPLE-CHOICE QUESTIONS

For each chapter set (Set 1 through Set 10), and for the set of miscellaneous multiple-choice questions in Chapter 11, we give first a table showing the correct answers. Explanations for the individual questions follow.

Answers for Set 1: Functions

1.	C	7.	C	13.	B	19.	A	25.	C
2.	E	8.	C	14.	E	20.	E	26.	A
3.	D	9.	B	15.	A	21.	E	27.	A
4.	E	10.	E	16.	B	22.	D	28.	C
5.	D	11.	C	17.	D	23.	B	29.	D
6.	B	12.	B	18.	D	24.	A	30.	C

1.   C. $f(-2) = (-2)^3 - 2(-2) - 1 = -5.$

2.   E.   The denominator, $x^2 + 1$, is never 0.

3.   D.   Since $x - 2$ may not be negative, $x \geqq 2$. The denominator equals 0 at $x = 0$ and $x = 1$, but these values are not in the interval $x \geq 2$.

4.   E.   Since $g(x) = 2$, g is a constant function. Thus, for all $f(x)$, $g(f(x)) = 2$.

5.   D. $f(g(x)) = f(2) = -3.$

6.   B.   Solve the pair of equations
$$\left\{ \begin{array}{l} 4 = \ 1 + A + B - 3 \\ -6 = -1 + A - B - 3 \end{array} \right\}.$$
Add to get A; substitute in either equation to get B. $A = 2$ and $B = 4$.

7.   C.   The graph of $f(x)$ is symmetric to the origin if $f(-x) = -f(x)$. When we replace x by $-x$, we obtain $-y$ only in (C).

8.   C.   For g to have an inverse function it must be one-to-one. Note, on pages 268–271, that although the graph of $y = xe^{-x^2}$ is symmetric to the origin, it is not one-to-one.

9.   B.   Note that $-\dfrac{\pi}{2} < \arctan x < \dfrac{\pi}{2}$ and that the sine function varies from -1 to 1 as the argument varies from $-\dfrac{\pi}{2}$ to $\dfrac{\pi}{2}$.

10.   E.   The maximum value of g is 2, attained when $\cos x = -1$. On $[0, 2\pi]$, $\cos x = -1$ for $x = \pi$.

11.   C. f is odd if $f(-x) = -f(x)$.

12. **B.** Since $f(q) = 0$ if $q = 1$ or $q = -2$, $f(2x) = 0$ if $2x$, a replacement for q, equals 1 or -2.

13. **B.** $f(x) = x(x^2 + 4x + 4) = x(x + 2)^2$.

14. **E.** Solving simultaneously yields $(x + 2)^2 = 4x$; $x^2 + 4x + 4 = 4x$; $x^2 + 4 = 0$. There are no real solutions.

15. **A.** The reflection of $y = f(x)$ in the y-axis is $y = f(-x)$.

16. **B.** If g is the inverse of f, then f is the inverse of g. This implies that the function f assigns to each value $g(x)$ the number x.

17. **D.** Since f is continuous (see page 36), then, if f is negative at a and positive at b, f must equal 0 at some intermediate point. Since $f(1) = -2$ and $f(2) = 13$, this point is between 1 and 2.

18. **D.** The function $\sin bx$ has period $\frac{2\pi}{b}$. Then $2\pi \div \frac{2\pi}{3} = 3$.

19. **A.** Since $\ln q$ is defined only if $q > 0$, the domain of $\ln \cos x$ is the set of x for which $\cos x > 0$, that is, when $0 < \cos x \leqq 1$. Thus $-\infty < \ln \cos x \leqq 0$.

20. **E.** $\log_b 3^b = \frac{b}{2}$ implies $b \log_b 3 = \frac{b}{2}$. Then $\log_b 3 = \frac{1}{2}$ and $3 = b^{1/2}$. So $3^2 = b$.

21. **E.** Letting $y = f(x) = x^3 + 2$, we get
$$x^3 = y - 2, \; x = \sqrt[3]{y - 2} = f^{-1}(y).$$
So $f^{-1}(x) = \sqrt[3]{x - 2}$.

22. **D.** Since $f(1) = 0$, $x - 1$ is a factor of f. Since $f(x)$ divided by $x - 1$ yields $x^2 - x - 2$, therefore
$$f(x) = (x - 1)(x + 1)(x - 2).$$

23. **B.** If $-\frac{\pi}{2} < x < \frac{\pi}{2}$, then $-\infty < \tan x < \infty$ and $0 < e^{\tan x} < \infty$.

24. **A.** The reflection of $f(x)$ in the x-axis is $-f(x)$.

25. **C.** $f(x)$ attains its maximum when $\sin\left(\frac{x}{3}\right)$ does. The maximum value of the sine function is 1; the smallest positive argument is $\frac{\pi}{2}$. Set $\frac{x}{3}$ equal to $\frac{\pi}{2}$.

26. **A.** $\arccos\left(\frac{-\sqrt{2}}{2}\right) = \frac{3\pi}{4}$; $\tan\left(\frac{3\pi}{4}\right) = -1$.

27. **A.** Let $y = f(x) = 2e^{-x}$; then $\frac{y}{2} = e^{-x}$ and $\ln \frac{y}{2} = -x$. So
$$x = -\ln \frac{y}{2} = \ln \frac{2}{y} = f^{-1}(y).$$
Thus $f^{-1}(x) = \ln \frac{2}{x}$.

28. **C.** The function in (C) is not one-to-one since, for each y between $-\frac{1}{2}$ and $\frac{1}{2}$ (except 0), there are two x's in the domain.

29. **D.** The domain of the $\ln$ function is the set of positive reals. The function $g(x) > 0$ if $x^2 < 9$.

30. **C.** Since the domain of $f(g)$ is $(-3, 3)$, $\ln (9 - x^2)$ takes on every real value less than or equal to $\ln 9$.

Chapter 2

Limits and Continuity

Review of Definitions and Methods

A1.   Definitions and Examples _____

The number L is the *limit of the function* $f(x)$ as x approaches c if, as the values of x get arbitrarily close (but not equal) to c, the values of $f(x)$ approach (or equal) L. We write

$$\lim_{x \to c} f(x) = L.$$

In order for $\lim_{x \to c} f(x)$ to exist, the values of f must tend to the same number L as x approaches c from either the left or the right. We write

$$\lim_{x \to c^-} f(x)$$

for the *left-hand limit* of f at c (as x approaches c through values *less* than c), and

$$\lim_{x \to c^+} f(x)$$

for the *right-hand limit* of f at c (as x approaches c through values *greater* than c).

Example 1.   The greatest-integer function $g(x) = [x]$, shown in Figure N2–1, has different left-hand and right-hand limits at *every* integer. For example,

$$\lim_{x \to 1^-}[x] = 0 \quad \text{but} \quad \lim_{x \to 1^+}[x] = 1.$$

This function, therefore, does not have a limit at $x = 1$ or, by the same reasoning, at any other integer.

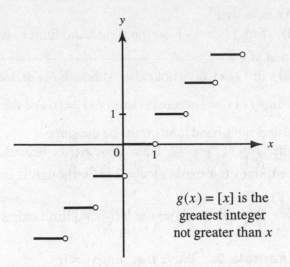

$g(x) = [x]$ is the greatest integer not greater than x

FIGURE N2–1

However, $[x]$ does have a limit at every nonintegral real number. For example,

$$\lim_{x \to 0.6}[x] = 0; \qquad \lim_{x \to 0.99}[x] = 0; \qquad \lim_{x \to 2.01}[x] = 2; \qquad \lim_{x \to 2.95}[x] = 2$$

$$\lim_{x \to -3.1}[x] = -4; \qquad \lim_{x \to -2.9}[x] = -3; \qquad \lim_{x \to -0.9}[x] = -1; \qquad \lim_{x \to -0.01}[x] = -1.$$

Example 2.   Suppose the function $y = f(x)$, graphed in Figure N2–2, is defined as follows:

$$f(x) = \begin{cases} x+1 & (-2 < x < 0) \\ 2 & (x = 0) \\ -x & (0 < x < 2) \\ 0 & (x - 2) \\ x-4 & (2 < x \le 4) \end{cases}$$

We want to decide which limits of f exist, if any, at

(i) $x = -2$, (ii) $x = 0$,

(iii) $x = 2$, (iv) $x = 4$.

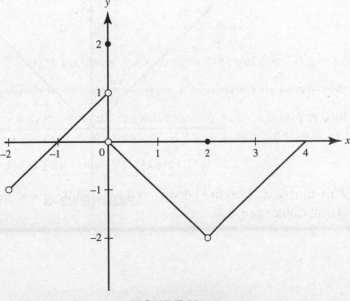

FIGURE N2–2

We note that

(i) $\lim\limits_{x \to -2^+} f(x) = -1$, so the right-hand limit exists at $x = -2$, even though f is not defined at $x = -2$.

(ii) $\lim\limits_{x \to 0} f(x)$ does not exist. Although f is defined at $x = 0$ ($f(0) = 2$), we observe that $\lim\limits_{x \to 0^-} f(x) = 1$ whereas $\lim\limits_{x \to 0^+} f(x) = 0$. For the limit to exist at a point, the left-hand and right-hand limits must be the same.

(iii) $\lim\limits_{x \to 2} f(x) = -2$. This limit exists because $\lim\limits_{x \to 2^-} f(x) = \lim\limits_{x \to 2^+} f(x) = -2$. Indeed, the limit exists at $x = 2$ even though it is different from the value of f at 2 ($f(2) = 0$).

(iv) $\lim\limits_{x \to 4^-} f(x) = 0$, so the left-hand limit exists at $x = 4$.

Example 3. Prove that $\lim\limits_{x \to 0} |x| = 0$.

The graph of $|x|$ is shown in Figure N2–3.

We examine both left- and right-hand limits of the absolute-value function as $x \to 0$. Since

$$|x| = \begin{cases} -x & \text{if } x < 0 \\ x & \text{if } x > 0 \end{cases}$$

it follows that

$$\lim_{x \to 0^-} |x| = \lim_{x \to 0^-} (-x) = 0$$

and

$$\lim_{x \to 0^+} |x| = \lim_{x \to 0^+} x = 0.$$

Since the left-hand and right-hand limits both equal 0, $\lim\limits_{x \to 0} |x| = 0$.

Note that $\lim\limits_{x \to c} |x| = c$ if $c > 0$ but equals $-c$ if $c < 0$.

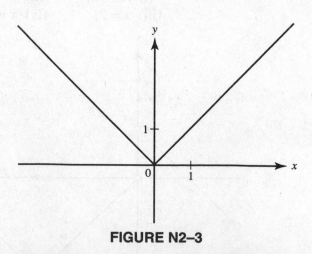

FIGURE N2–3

Definition

The function $f(x)$ is said to *become infinite* (positively or negatively) as x approaches c if $f(x)$ can be made arbitrarily large (positively or negatively) by taking x sufficiently close to c. We write

$$\lim_{x \to c} f(x) = +\infty \quad (\text{or } \lim_{x \to c} f(x) = -\infty).$$

Since for the limit to exist it must be a finite number, neither of the preceding limits exists.

This definition can be extended to include x approaching c from the left or from the right. The following examples illustrate these definitions.

Example 4. The graph of $f(x) = \frac{1}{x}$ (Figure N2–4) shows that

$$\lim_{x \to 0^-} \frac{1}{x} = -\infty,$$

$$\lim_{x \to 0^+} \frac{1}{x} = +\infty.$$

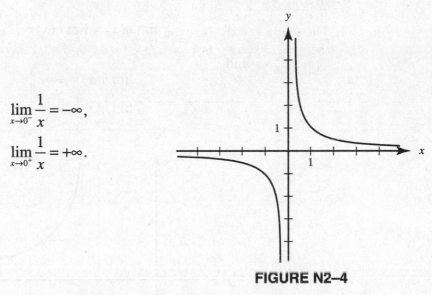

FIGURE N2–4

Example 5. From the graph of $g(x) = \dfrac{1}{(x-1)^2}$ (Figure N2–5) we see that

$$\lim_{x \to 1^-} g(x) = \lim_{x \to 1^+} g(x) = \infty.$$

We may also say that $\lim_{x \to 1} g(x) = \infty$.

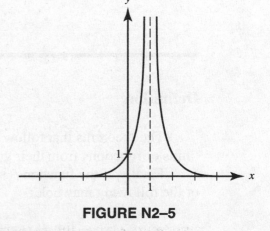

FIGURE N2–5

Remember that none of the limits in Examples 4 and 5 exists!

Definition

We write

$$\lim_{x\to\infty} f(x) = L \quad (\text{or } \lim_{x\to-\infty} f(x) = L)$$

if the difference between $f(x)$ and L can be made arbitrarily small by making x sufficiently large positively (or negatively).

In Examples 4 and 5, note that $\lim_{x\to\infty} f(x) = \lim_{x\to-\infty} f(x) = 0$ and $\lim_{x\to\infty} g(x) = \lim_{x\to-\infty} g(x) = 0$.

Example 6. From the graph of $h(x) = 1 + \dfrac{3}{x-2} = \dfrac{x+1}{x-2}$ (Figure N2–6), we can conjecture that

$$\lim_{x\to-\infty} h(x) = \lim_{x\to+\infty} h(x) = 1$$

and

$$\lim_{x\to2^-} h(x) = -\infty,$$

$$\lim_{x\to2^+} h(x) = +\infty.$$

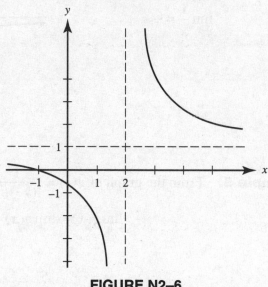

FIGURE N2–6

Definition

The theorems that follow in §B of this chapter confirm the conjectures made about limits of functions from their graphs.

Finally, if the function $f(x)$ becomes infinite as x becomes infinite, then one or more of the following may hold:

$$\lim_{x\to+\infty} f(x) = +\infty \text{ or } -\infty \quad \text{or} \quad \lim_{x\to-\infty} f(x) = +\infty \text{ or } -\infty .$$

Behavior of Polynomials

Every polynomial whose degree is greater than or equal to 1 becomes infinite as x does. It becomes positively or negatively infinite, depending only on the sign of the leading coefficient and the degree of the polynomial.

Example 7.

(a) Let $f(x) = x^3 - 3x^2 + 7x + 2$. Then

$$\lim_{x \to +\infty} f(x) = +\infty, \qquad \lim_{x \to -\infty} f(x) = -\infty.$$

(b) If $g(x) = -4x^4 + 1{,}000{,}000x^3 + 100$, then

$$\lim_{x \to +\infty} g(x) = -\infty, \qquad \lim_{x \to -\infty} g(x) = -\infty.$$

(c) Suppose $f(x) = -5x^3 + 3x^2 - 4\pi + 8$. Then

$$\lim_{x \to +\infty} h(x) = -\infty, \qquad \lim_{x \to -\infty} h(x) = +\infty.$$

(d) For $k(x) = \pi - 0.001x$, we see that

$$\lim_{x \to +\infty} k(x) = -\infty, \qquad \lim_{x \to -\infty} k(x) = +\infty.$$

It's easy to write down rules for the behavior of a polynomial as x becomes infinite!

A2.   Asymptotes

The line $y = b$ is a *horizontal asymptote* of the graph of $y = f(x)$ if

$$\lim_{x \to \infty} f(x) = b \quad \text{or} \quad \lim_{x \to -\infty} f(x) = b.$$

The graph of $f(x) = \frac{1}{x}$ (Figure N2–4) on page 25 has the x-axis ($y = 0$) as horizontal asymptote.

So does the graph of $g(x) = \frac{1}{(x-1)^2}$ (Figure N2–5) on page 25.

The graph of $h(x) = \frac{x+1}{x-2}$ has the line $y = 1$ as horizontal asymptote, as shown at the right.

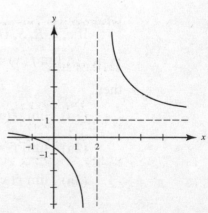

The line $x = a$ is a *vertical asymptote* of the graph of $y = f(x)$ if one or more of the following holds:

$$\lim_{x \to a^-} f(x) = +\infty \quad \text{or} \quad \lim_{x \to a^-} f(x) = -\infty$$

or

$$\lim_{x \to a^+} f(x) = +\infty \quad \text{or} \quad \lim_{x \to a^+} f(x) = -\infty.$$

The graph of $f(x) = \frac{1}{x}$ (Figure N2–4) has $x = 0$ (the y-axis) as vertical asymptote.

The graph of $g(x) = \frac{1}{(x-1)^2}$ (Figure N2–5) has $x = 1$ as vertical asymptote.

The graph of $h(x) = \frac{x+1}{x-2}$ (Figure N2–6) has the line $x = 2$ as vertical asymptote.

Example 8. From the graph of $k(x) = \frac{2x-4}{x-3}$ in Figure N2–7, we see that $y = 2$ is a horizontal asymptote, since

$$\lim_{x \to +\infty} k(x) = \lim_{x \to -\infty} k(x) = 2.$$

Also, $x = 3$ is a vertical asymptote; the graph shows that

$$\lim_{x \to 3^-} k(x) = -\infty$$

and

$$\lim_{x \to 3^+} k(x) = +\infty.$$

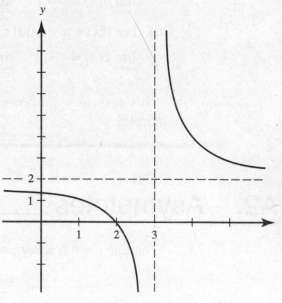

FIGURE N2–7

B. Theorems on Limits

If c, k, R, S, U, and V are finite numbers and if

$$\lim_{x \to c} f(x) = R, \qquad \lim_{x \to c} g(x) = S, \qquad \lim_{x \to \infty} f(x) = U, \qquad \lim_{x \to \infty} g(x) = V,$$

then:

(1a) $\lim_{x \to c} kf(x) = kR$ **(1b)** $\lim_{x \to \infty} kf(x) = kU$

(2a) $\lim_{x \to c} [f(x) + g(x)] = R + S$ **(2b)** $\lim_{x \to \infty} [f(x) + g(x)] = U + V$

(3a) $\lim_{x \to c} f(x)g(x) = RS$ **(3b)** $\lim_{x \to \infty} f(x)g(x) = UV$

(4a) $\lim_{x \to c} \dfrac{f(x)}{g(x)} = \dfrac{R}{S}$ (if $S \neq 0$)   **(4b)** $\lim_{x \to \infty} \dfrac{f(x)}{g(x)} = \dfrac{U}{V}$ (if $V \neq 0$)

(5a) $\lim\limits_{x\to c} k = k$ **(5b)** $\lim\limits_{x\to\infty} k = k$

(6) *The Squeeze or Sandwich Theorem.* If $f(x) \leqslant g(x) \leqslant h(x)$ and if $\lim\limits_{x\to c} f(x) = \lim\limits_{x\to c} h(x) = L$, then $\lim\limits_{x\to c} g(x) = L$. Figure N2–8 illustrates this theorem.

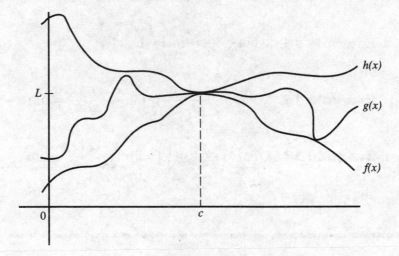

FIGURE N2–8

Squeezing function g between functions f and h forces g to have the same limit L at $x = c$ as do f and g.

Example 9. $\lim\limits_{x\to 2}(5x^2 - 3x + 1) = 5\lim\limits_{x\to 2} x^2 - 3\lim\limits_{x\to 2} x + \lim\limits_{x\to 2} 1$

$$= 5\cdot 4 \quad -3\cdot 2 \quad +1$$
$$= 15.$$

Example 10. $\lim\limits_{x\to 0}(x\,\cos 2x) = \lim\limits_{x\to 0} x \cdot \lim\limits_{x\to 0}(\cos 2x)$

$$= 0 \quad \cdot 1$$
$$= 0.$$

Example 11. $\lim\limits_{x\to -1}\dfrac{3x^2 - 2x - 1}{x^2 + 1} = \lim\limits_{x\to -1}(3x^2 - 2x - 1) \div \lim\limits_{x\to -1}(x^2 + 1)$

$$= (3 + 2 - 1) \quad \div (1 + 1)$$
$$= 2.$$

Example 12. $\lim\limits_{x\to 3}\dfrac{x^2 - 9}{x - 3} = \lim\limits_{x\to 3}\dfrac{(x-3)(x+3)}{x-3}$. Since

$$\frac{(x-3)(x+3)}{x-3} = x + 3 \quad (x \neq 3)$$

and since, by the definition of $\lim\limits_{x\to c} f(x)$ in §A1, x must be different from 3 as $x \to 3$, it follows that

$$\lim\limits_{x\to 3}\frac{(x-3)(x+3)}{x-3} = \lim\limits_{x\to 3}(x+3) = 6,$$

where the factor $x - 3$ is removed *before* taking the limit.

Example 13. $\lim\limits_{x \to -2} \dfrac{x^3+8}{x^2-4} = \lim\limits_{x \to -2} \dfrac{(x+2)(x^2-2x+4)}{(x+2)(x-2)} = \lim\limits_{x \to -2} \dfrac{x^2-2x+4}{x-2} = \dfrac{4+4+4}{-4} = -3.$

Example 14. $\lim\limits_{x \to 0} \dfrac{x}{x^3} = \lim\limits_{x \to 0} \dfrac{1}{x^2} = \infty.$ As $x \to 0$, the numerator approaches 1 while the denominator approaches 0; the limit does *not* exist.

Example 15. $\lim\limits_{x \to 1} \dfrac{x^2-1}{x^2-1} = \lim\limits_{x \to 1} 1 = 1.$

Example 16. $\lim\limits_{\Delta x \to 0} \dfrac{(3+\Delta x)^2 - 3^2}{\Delta x} = \lim\limits_{\Delta x \to 0} \dfrac{6\Delta x + \Delta x^2}{\Delta x} = \lim\limits_{\Delta x \to 0} 6 + \Delta x = 6.$

Example 17. $\lim\limits_{h \to 0} \dfrac{1}{h}\left(\dfrac{1}{2+h} - \dfrac{1}{2} \right) = \lim\limits_{h \to 0} \dfrac{2-(2+h)}{2h(2+h)} = \lim\limits_{h \to 0} \dfrac{-h}{2h(2+h)} =$

$\lim\limits_{h \to 0} -\dfrac{1}{2(2+h)} = -\dfrac{1}{4}.$

C. Limit of a Quotient of Polynomials

To find $\lim\limits_{x \to \infty} \dfrac{P(x)}{Q(x)}$, where $P(x)$ and $Q(x)$ are polynomials in x, we can divide both numerator and denominator by the highest power of x that occurs and use the fact that $\lim\limits_{x \to \infty} \dfrac{1}{x} = 0.$

Example 18. $\lim\limits_{x \to \infty} \dfrac{3-x}{4+x+x^2} = \lim\limits_{x \to \infty} \dfrac{\dfrac{3}{x^2} - \dfrac{1}{x}}{\dfrac{4}{x^2} + \dfrac{1}{x} + 1} = \dfrac{0-0}{0+0+1} = 0.$

Example 19. $\lim\limits_{x \to \infty} \dfrac{4x^4 + 5x + 1}{37x^3 - 9} = \lim\limits_{x \to \infty} \dfrac{4 + \dfrac{5}{x^3} + \dfrac{1}{x^4}}{\dfrac{37}{x} - \dfrac{9}{x^4}} = \infty$ (no limit).

Example 20. $\lim\limits_{x \to \infty} \dfrac{x^3 - 4x^2 + 7}{3 - 6x - 2x^3} = \lim\limits_{x \to \infty} \dfrac{1 - \dfrac{4}{x} + \dfrac{7}{x^3}}{\dfrac{3}{x^3} - \dfrac{6}{x^2} - 2} = \dfrac{1-0+0}{0-0-2} = -\dfrac{1}{2}.$

The Rational Function Theorem

We see from Examples 18, 19, and 20 that: if the degree of $P(x)$ is less than that of $Q(x)$, then $\lim\limits_{x \to \infty} \dfrac{P(x)}{Q(x)} = 0$; if the degree of $P(x)$ is higher than that of $Q(x)$, then $\lim\limits_{x \to \infty} \dfrac{P(x)}{Q(x)} = \infty$ or $-\infty$

(i.e., does not exist); and if the degrees of $P(x)$ and $Q(x)$ are the same, then $\lim\limits_{x \to \infty} \dfrac{P(x)}{Q(x)} = \dfrac{a_n}{b_n}$, where a_n and b_n are the coefficients of the highest powers of x in $P(x)$ and $Q(x)$ respectively. These conclusions constitute the *Rational Function Theorem.*

This theorem holds also when we replace "$x \to \infty$" by "$x \to -\infty$." Verify this statement in Examples, 18, 19, and 20. Note also that:

(i)   when $\lim\limits_{x \to \pm\infty} \dfrac{P(x)}{Q(x)} = 0$, then $y = 0$ is a horizontal asymptote of the graph of $y = \dfrac{P(x)}{Q(x)}$.

(ii)   when $\lim\limits_{x \to \pm\infty} \dfrac{P(x)}{Q(x)} = +\infty$ or $-\infty$, then the graph of $y = \dfrac{P(x)}{Q(x)}$ has no horizontal asymptotes.

(iii)   when $\lim\limits_{x \to \pm\infty} \dfrac{P(x)}{Q(x)} = \dfrac{a_n}{b_n}$, then $y = \dfrac{a_n}{b_n}$ is a horizontal asymptote of the graph of $y = \dfrac{P(x)}{Q(x)}$.

Example 21. $\lim\limits_{x \to \infty} \dfrac{100x^2 - 19}{x^3 + 5x^2 + 2} = 0$; $\lim\limits_{x \to -\infty} \dfrac{x^3 - 5}{1 + x^2} = -\infty$ (no limit); $\lim\limits_{x \to \infty} \dfrac{x - 4}{13 + 5x} = \dfrac{1}{5}$;

$\lim\limits_{x \to -\infty} \dfrac{4 + x^2 - 3x^3}{x + 7x^3} = -\dfrac{3}{7}$; $\lim\limits_{x \to \infty} \dfrac{x^3 + 1}{2 - x^2} = -\infty$ (no limit).

D.   Other Basic Limits

D1. The basic trigonometric limit is the following:

$$\lim_{\theta \to 0} = \frac{\sin \theta}{\theta} = 1 \text{ if } \theta \text{ is measured in radians.}$$

This statement can be proved using geometry and the Squeeze Theorem.

We can also use a graphing calculator to find values of $Y_1 = \dfrac{\sin X}{X}$ for X's close to 0, both positive and negative. How close can we get Y_1 to 1 on the calculator?

Example 22.   Prove that $\lim\limits_{x \to \infty} \frac{\sin x}{x} = 0$.

Since, for all x, $-1 \leqslant \sin x \leqslant 1$, it follows that, if $x > 0$, then $-\dfrac{1}{x} \leqslant \dfrac{\sin x}{x} \leqslant \dfrac{1}{x}$. But as $x \to \infty$, $-\dfrac{1}{x}$ and $\dfrac{1}{x}$ both approach 0; therefore $\dfrac{\sin x}{x}$ must also approach 0. To obtain graphical confirmation of this fact, and of the additional fact that $\lim\limits_{x \to -\infty} \dfrac{\sin x}{x}$ also equals 0, graph

$$Y_1 = \frac{\sin X}{X}, \ Y_2 = \frac{1}{X}, \text{ and } Y_3 = -\frac{1}{X}$$

in $[-4\pi, 4\pi] \times [-1, 1]$. Observe, as $x \to \pm\infty$, that Y_2 and Y_3 approach 0 and that Y_1 is squeezed between them.

D2. The number e can be defined as follows:

$$e = \lim_{n \to \infty} \left(1 + \frac{1}{n}\right)^n.$$

The value of e can be obtained on a graphing calculator to a large number of decimal places by keying in

$$Y_1 = \left(1 + \frac{1}{X}\right)^X$$

then evaluating Y_1 for larger and larger X.

In Chapter 3 we will show that

$$\lim_{h \to 0} \frac{e^h - 1}{h} = 1.$$

Use a graphing calculator to evaluate $\frac{e^x - 1}{x}$ for inputs of x closer and closer to 0, both positive and negative.

E. Continuity

If a function is continuous over an interval, we can draw its graph without lifting pencil from paper. The graph has no holes, breaks, or jumps on the interval.

If $f(x)$ is continuous at a point $x = c$, then the closer x is to c, the closer $f(x)$ gets to $f(c)$. This is made precise by the following definition:

The function $y = f(x)$ is continuous at $x = c$ if

$$\lim_{x \to c} f(x) = f(c).$$

All three of the following statements must be true for $f(x)$ to be continuous at $x = c$:

(1) $f(c)$ exists, that is, is defined (c is in the domain of f);

(2) $\lim_{x \to c} f(x)$ exists;

(3) $\lim_{x \to c} f(x)$ equals $f(c)$.

A function is continuous over the closed interval $[a,b]$ if it is continuous at each x such that $a \leqslant x \leqslant b$.

A function that is not continuous at $x = c$ is said to be discontinuous at that point. We call the latter a *point of discontinuity*.

Continuous Functions

Polynomials are continuous everywhere; namely, at every real number.

Rational functions, $\frac{P(x)}{Q(x)}$, are continuous at each point in their domain; that is,

except where $Q(x) = 0$. The function $f(x) = \frac{1}{x}$, for example, is continuous except at $x = 0$, where f is not defined.

The absolute value function $f(x) = |x|$ (sketched in Figure N2–3, page 24) is continuous everywhere.

The trigonometric, inverse trigonometric, exponential, and logarithmic functions are continuous at each point in their domains.

Functions of the type $\sqrt[n]{x}$ (where n is a positive integer $\geqslant 2$) are continuous at each x for which $\sqrt[n]{x}$ is defined.

The greatest-integer function $f(x) = [x]$ (Figure N2–1, page 23) is discontinuous at each integer, since it does not have a limit at any integer.

Kinds of Discontinuities

In Example 2, page 23, $y = f(x)$ is defined as follows:

$$f(x) = \begin{cases} x+1 & (-2 < x < 0) \\ 2 & (x = 0) \\ -x & (0 < x < 2) \\ 0 & (x = 2) \\ x-4 & (2 < x \leqslant 4) \end{cases}$$

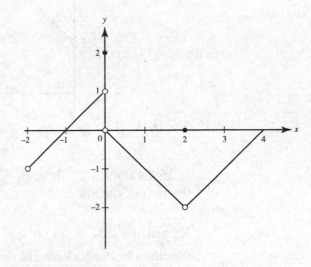

The graph shown at the right is the same as the graph (Figure N2–2) for Example 2 on page 23.

We observe that f is not continuous at $x = -2$, $x = 0$, or $x = 2$. At $x = -2$, f is not defined; at $x = 0$, although f is defined, $\lim\limits_{x \to 0} f(x)$ does not exist. At $x = 2$, f is defined $(f(2) = 0)$, and $\lim\limits_{x \to 2} f(x)$ exists $(\lim\limits_{x \to 2^+} f(x) = \lim\limits_{x \to 2^-} f(x) = -2)$, but $\lim\limits_{x \to 2} f(x) \neq f(2)$.

A discontinuity of the type found at $x = 0$ above is called a *jump discontinuity*. Although both left- and right-hand limits exist, they are different. The greatest-integer function, $g(x) = [x]$, has a jump discontinuity at every integer.

In the example, the discontinuity at $x = 2$ is called *removable*. (We usually represent a removable discontinuity on a graph by a small hollow circle.) If we define f at $x = 2$ to be -2 (which is $\lim\limits_{x \to 2} f(x)$), then the redefined (new) function no longer has a discontinuity at $x = 2$. We cannot, however, "remove" a jump discontinuity by any redefinition whatsoever.

Whenever the graph of a function $f(x)$ has the line $x = a$ as a vertical asymptote, then $f(x)$ becomes positively or negatively infinite as $x \to a^+$ or as $x \to a^-$. The function is then said to have an *infinite discontinuity*. See, for example, Figure N2–4 (page 25) for $f(x) = \frac{1}{x}$, Figure N2–5 (page 25) for $g(x) = \frac{1}{(x-1)^2}$, or Figure N2–7 (page 28) for $k(x) = \frac{2x-4}{x-3}$. Each of these functions exhibits an infinite discontinuity.

Example 23. $f(x) = \dfrac{x-1}{x^2+x} = \dfrac{x-1}{x(x+1)}$ is not continuous at $x = 0$ or $= -1$, since the function is not defined for either of these numbers. Note also that neither $\lim\limits_{x \to 0} f(x)$ nor $\lim\limits_{x \to -1} f(x)$ exists.

Example 24.   In Figure N2–9, $f(x)$ is continuous on $[(0,1), (1,3),$ and $(3,5)]$. The discontinuity at $x = 1$ is removable; the one at $x = 3$ is not. (Note that f is continuous from the right at $x = 0$ and from the left at $x = 5$.)

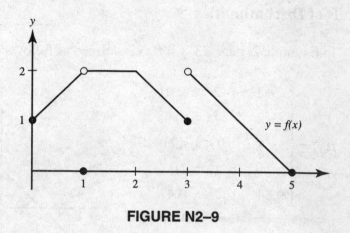

FIGURE N2–9

In Examples 25 through 30, we determine if the functions are continuous at the points specified:

Example 25. $f(x) = \frac{1}{2}x^4 - \sqrt{3}x^2 + 7$ at $x = -1$. Since f is a polynomial, it is continuous everywhere, including, of course, at $x = -1$.

Example 26. $g(x) = \dfrac{1}{x-3}$ (a) at $x = 3$; (b) at $x = 0$. This function is continuous except where the denominator equals 0 (where g has an infinite discontinuity). It is not continuous at $x = 3$, but is continuous at $x = 0$.

Example 27. $h(x) = \begin{cases} \dfrac{4}{x-2} & \text{if } x \neq 2 \\ 1 & \text{if } x = 2 \end{cases}$

(a) at $x = 2$; (b) at $x = 3$.

$h(x)$ has an infinite discontinuity at $x = 2$; it is continuous at $x = 3$ and at every other point different from 2. The discontinuity at $x = 2$ is not removable. See Figure N2–10.

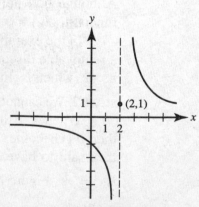

FIGURE N2–10

Example 28. $k(x) = \frac{x^2-4}{x-2}$ $(x \neq 2)$

at $x = 2$.

Note that $k(x) = x + 2$ for all $x \neq 2$. The function is continuous everywhere except at $x = 2$, where k is not defined. The discontinuity at 2 is removable. If we redefine $f(2)$ to equal 4, the new function will be continuous everywhere. See Figure N2–11.

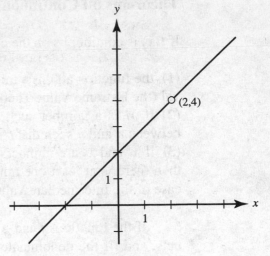

FIGURE N2–11

Example 29. $f(x) = \begin{cases} x^2 + 2 & x \leq 1 \\ 4 & x > 1 \end{cases}$

at $x = 1$.

$f(x)$ is not continuous at $x = 1$ since $\lim_{x \to 1^-} f(x) = 3 \neq \lim_{x \to 1^+} f(x) = 4$. This function has a jump discontinuity at $x = 1$ (which cannot be removed). See Figure N2–12.

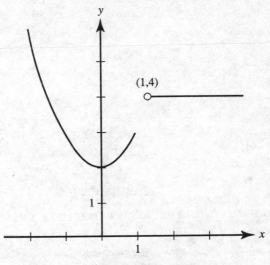

FIGURE N2–12

Example 30. $g(x) = \begin{cases} x^2 & x \neq 2 \\ 1 & x = 2 \end{cases}$

at $x = 2$.

$g(x)$ is not continuous at $x = 2$ since $\lim_{x \to 2} g(x) = 4 \neq g(2) = 1$. This discontinuity can be removed by redefining $g(2)$ to equal 4. See Figure N2–13.

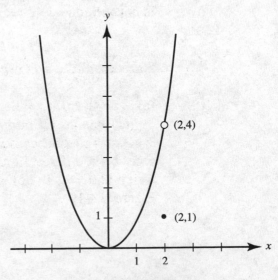

FIGURE N2–13

Theorems on Continuous Functions

If $f(x)$ is continuous on the closed interval $[a,b]$, then it follows that

(1) the function attains a minimum value and a maximum value somewhere on the interval (the Extreme Value Theorem);

(2) if M is a number such that $f(a) \leq M \leq f(b)$, then there is at least one number c between a and b such that $f(c) = M$ (the Intermediate Value Theorem);

(3) if x_1 and x_2 are numbers between a and b and if $f(x_1)$ and $f(x_2)$ have opposite signs, then there is at least one number c between x_1 and x_2 such that $f(c) = 0$ (this is a special case of the Intermediate Value Theorem).

If the functions f and g are continuous at $x = c$, so are their sums, differences, products, and (if the denominator is not zero at c) their quotients; also kf, where k is a constant, is continuous at c.

Set 2: Multiple-Choice Questions on Limits and Continuity

1. $\lim\limits_{x \to 2} \dfrac{x^2 - 4}{x^2 + 4}$ is

   (A) 1   **(B)** 0   **(C)** $-\dfrac{1}{2}$ **(D)** –1   **(E)** ∞

2. $\lim\limits_{x \to \infty} \dfrac{4 - x^2}{x^2 - 1}$ is

   (A) 1   **(B)** 0   **(C)** –4   **(D)** –1   **(E)** ∞

3. $\lim\limits_{x \to 3} \dfrac{x - 3}{x^2 - 2x - 3}$ is

   (A) 0   **(B)** 1   **(C)** $\dfrac{1}{4}$ **(D)** ∞ **(E)** none of these

4. $\lim\limits_{x \to 0} \dfrac{x}{x}$ is

   (A) 1   **(B)** 0   **(C)** ∞ **(D)** –1   **(E)** nonexistent

5. $\lim\limits_{x \to 2} \dfrac{x^3 - 8}{x^2 - 4}$ is

   (A) 4   **(B)** 0   **(C)** 1   **(D)** 3   **(E)** ∞

6. $\lim\limits_{x \to \infty} \dfrac{4 - x^2}{4x^2 - x - 2}$ is

   (A) –2   **(B)** $-\dfrac{1}{4}$ **(C)** 1   **(D)** 2   **(E)** nonexistent

7. $\lim\limits_{x \to -\infty} \dfrac{5x^3 + 27}{20x^2 + 10x + 9}$ is

   (A) $-\infty$ **(B)** –1   **(C)** 0   **(D)** 3   **(E)** ∞

8. $\lim\limits_{x \to \infty} \dfrac{3x^2 + 27}{x^3 - 27}$ is

   (A) 3   **(B)** ∞ **(C)** 1   **(D)** –1   **(E)** 0

9. $\lim\limits_{x \to \infty} \dfrac{2^{-x}}{2^x}$ is

   (A) –1   **(B)** 1   **(C)** 0   **(D)** ∞ **(E)** none of these

10. $\lim\limits_{x \to -\infty} \dfrac{2^{-x}}{2^x}$ is

(A) −1   (B) 1   (C) 0   (D) ∞   (E) none of these

11. If $[x]$ is the greatest integer not greater than x, then $\lim\limits_{x \to 1/2}[x]$ is

(A) $\dfrac{1}{2}$ (B) 1   (C) nonexistent   (D) 0   (E) none of these

12. (With the same notation) $\lim\limits_{x \to -2}[x]$ is

(A) −3   (B) −2   (C) −1   (D) 0   (E) none of these

13. The graph of $y = \arctan x$ has

(A) vertical asymptotes at $x = 0$ and $x = \pi$
(B) horizontal asymptotes at $y = \pm\dfrac{\pi}{2}$
(C) horizontal asymptotes at $y = 0$ and $y = \pi$
(D) vertical asymptotes at $x = \pm\dfrac{\pi}{2}$
(E) none of these

14. $\lim\limits_{x \to \infty} \sin x$

(A) is −1   (B) is infinity   (C) oscillates between −1 and 1
(D) is zero   (E) is 1

15. The graph of $y = \dfrac{x^2 - 9}{3x - 9}$ has

(A) a vertical asymptote at $x = 3$ (B) a horizontal asymptote at $y = \dfrac{1}{3}$
(C) a removable discontinuity at $x = 3$ (D) an infinite discontinuity at $x = 3$
(E) none of these

16. The function $f(x) = \begin{cases} x^2/x & (x \neq 0) \\ 0 & (x = 0) \end{cases}$

(A) is continuous everywhere
(B) is continuous except at $x = 0$
(C) has a removable discontinuity at $x = 0$
(D) has an infinite discontinuity at $x = 0$
(E) has $x = 0$ as a vertical asymptote

17. $\lim\limits_{x \to 0} \dfrac{\sin x}{x^2 + 3x}$ is

(A) 1   (B) $\dfrac{1}{3}$ (C) 3   (D) ∞   (E) $\dfrac{1}{4}$

18. $\lim\limits_{x\to 0}\sin\frac{1}{x}$ is

(A) ∞ (B) 1   (C) nonexistent   (D) –1   (E) none of these

19. Which statement is true about the curve $y=\dfrac{2x^2+4}{2+7x-4x^2}$?

(A) The line $x=-\frac{1}{4}$ is a vertical asymptote.
(B) The line $x=1$ is a vertical asymptote.
(C) The line $y=-\frac{1}{4}$ is a horizontal asymptote.
(D) The graph has no vertical or horizontal asymptotes.
(E) The line $y=2$ is a horizontal asymptote.

Questions 20 through 24 are based
on the function f shown in the graph
and defined below:

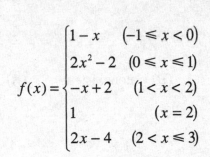

$$f(x)=\begin{cases}1-x & (-1\leqslant x<0)\\ 2x^2-2 & (0\leqslant x\leqslant 1)\\ -x+2 & (1<x<2)\\ 1 & (x=2)\\ 2x-4 & (2<x\leqslant 3)\end{cases}$$

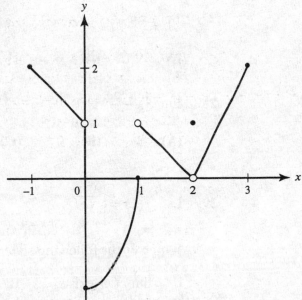

20. $\lim\limits_{x\to ?}f(x)$

(A) equals 0   (B) equals 1   (C) equals 2
(D) does not exist   (E) none of these

21. The function f is defined on $[-1,3]$

(A) if $x\neq 0$ (B) if $x\neq 1$ (C) if $x\neq 2$
(D) if $x\neq 3$ (E) at each x in $[-1,3]$

22. The function f has a removable discontinuity at

(A) $x=0$ (B) $x=1$ (C) $x=2$ (D) $x=3$ (E) none of these

23. On which of the following intervals is f continuous?

(A) $-1\leqslant x\leqslant 0$ (B) $0<x<1$ (C) $1\leqslant x\leqslant 2$
(D) $2\leqslant x\leqslant 3$ (E) none of these

24. The function f has a jump discontinuity at

    (A) $x = -1$ **(B)** $x = 1$ **(C)** $x = 2$
    (D) $x = 3$ **(E)** none of these

25. $\displaystyle\lim_{x\to\infty}\frac{2x^2+1}{(2-x)(2+x)}$ is

    (A) -4 **(B)** -2 **(C)** 1 **(D)** 2 **(E)** nonexistent

26. $\displaystyle\lim_{x\to 0}\frac{|x|}{x}$ is

    (A) 0 **(B)** nonexistent   **(C)** 1 **(D)** -1 **(E)** none of these

27. $\displaystyle\lim_{x\to\infty} x\sin\frac{1}{x}$ is

    (A) 0 **(B)** ∞ **(C)** nonexistent   **(D)** -1 **(E)** 1

28. $\displaystyle\lim_{x\to\pi}\frac{\sin(\pi-x)}{\pi-x}$ is

    (A) 1 **(B)** 0 **(C)** ∞ **(D)** nonexistent   **(E)** none of these

29. Let $f(x)=\begin{cases}\dfrac{x^2-1}{x-1} & \text{if } x \neq 1 \\ 4 & \text{if } x = 1\end{cases}$

Which of the following statements is (are) true?

    I. $\displaystyle\lim_{x\to 1} f(x)$ exists.   II. $f(1)$ exists.   III. f is continuous at $x = 1$.

    (A) I only   **(B)** II only   **(C)** I and II
    (D) none of them   **(E)** all of them

30. If $\begin{cases} f(x) = \dfrac{x^2-x}{2x} & \text{for } x \neq 0, \\ f(0) = k, \end{cases}$

and if f is continuous at $x = 0$, then $k =$

    (A) -1 **(B)** $-\dfrac{1}{2}$ **(C)** 0 **(D)** $\dfrac{1}{2}$ **(E)** 1

31. Suppose $\begin{cases} f(x) = \dfrac{3x(x-1)}{x^2-3x+2} & \text{for } x \neq 1, 2, \\ f(1) = -3, \\ f(2) = 4. \end{cases}$

Then $f(x)$ is continuous

    (A) except at $x = 1$ **(B)** except at $x = 2$ **(C)** except at $x = 1$ or 2
    (D) except at $x = 0, 1$, or 2 **(E)** at each real number

32. The graph of $f(x) = \dfrac{4}{x^2 - 1}$ has

   (A) one vertical asymptote, at $x = 1$
   (B) the y-axis as vertical asymptote
   (C) the x-axis as horizontal asymptote and $x = \pm 1$ as vertical asymptotes
   (D) two vertical asymptotes, at $x = \pm 1$, but no horizontal asymptote
   (E) no asymptote

33. Suppose $\lim\limits_{x \to -3^-} f(x) = -1$, $\lim\limits_{x \to -3^+} f(x) = -1$, and $f(-3)$ is not defined. Which of the following statements is (are) true?

   I. $\lim\limits_{x \to -3} f(x) = -1$.
   II. f is continuous everywhere except at $x = -3$.
   III. f has a removable discontinuity at $x = -3$.

   (A) None of them   (B) I only   (C) III only
   (D) I and III only   (E) All of them

34. The graph of $y = \dfrac{2x^2 + 2x + 3}{4x^2 - 4x}$ has

   (A) a horizontal asymptote at $y = +\dfrac{1}{2}$ but no vertical asymptotes
   (B) no horizontal asymptotes but two vertical asymptotes, at $x = 0$ and $x = 1$
   (C) a horizontal asymptote at $y = \dfrac{1}{2}$ and two vertical asymptotes, at $x = 0$ and $x = 1$
   (D) a horizontal asymptote at $x = 2$ but no vertical asymptotes
   (E) a horizontal asymptote at $y = \dfrac{1}{2}$ and two vertical asymptotes, at $x = \pm 1$

35. Let $f(x) = \begin{cases} \dfrac{x^2 + x}{x} & \text{if } x \neq 0 \\ 1 & \text{if } x = 0 \end{cases}$.

   Which of the following statements is (are) true?

   I. $f(0)$ exists.   II. $\lim\limits_{x \to 0} f(x)$ exists.   III. f is continuous at $x = 0$.

   (A) I only   (B) II only   (C) I and II only
   (D) all of them   (E) none of them

36. If $y = \dfrac{1}{2 + 10^{\frac{1}{x}}}$, then $\lim\limits_{x \to 0} y$ is

   (A) 0   (B) $\dfrac{1}{12}$ (C) $\dfrac{1}{2}$ (D) $\dfrac{1}{3}$ (E) nonexistent

37. $\lim\limits_{x \to 0} \sqrt{3 + \arctan \dfrac{1}{x}}$ is

   (A) $-\infty$ (B) $\sqrt{3 - \dfrac{\pi}{2}}$ (C) $\sqrt{3 + \dfrac{\pi}{2}}$

   (D) ∞ (E) none of these

Answers for Set 2: Limits and Continuity

1. B	9. C	17. B	24. B	31. B
2. D	10. D	18. C	25. B	32. C
3. C	11. D	19. A	26. B	33. D
4. A	12. E	20. A	27. E	34. C
5. D	13. B	21. E	28. A	35. D
6. B	14. C	22. C	29. C	36. E
7. A	15. C	23. B	30. B	37. E
8. E	16. A			

1. B.   The limit as $x \to 2$ is $0 \div 8$.

2. D.   Use the Rational Function Theorem (pages 30 and 31). The degrees of $P(x)$ and $Q(x)$ are the same.

3. C.   Remove the common factor $x - 3$ from numerator and denominator.

4. A.   The fraction equals 1 for all nonzero x.

5. D.   Note that $\dfrac{x^3-8}{x^2-4} = \dfrac{(x-2)(x^2+2x+4)}{(x-2)(x+2)}$.

6. B.   Use the Rational Function Theorem.

7. A.   Use the Rational Function Theorem.

8. E.   Use the Rational Function Theorem.

9. C.   The fraction is equivalent to $\dfrac{1}{2^{2x}}$; the denominator approaches ∞.

10. D.   Since $\dfrac{2^{-x}}{2^{x}} = 2^{-2x}$, therefore, as $x \to -\infty$, the fraction $\to +\infty$.

11. D.   See Figure N2–1 on page 23.

12. E.   Note, from Figure N2–1, that $\lim\limits_{x\to-2^-}[x] = -3$ but $\lim\limits_{x\to-2^+}[x] = -2$.

13. B.   Graph $\tan^{-1} x$ on your calculator in $[-5, 5] \times [-\frac{\pi}{2}, \frac{\pi}{2}]$.

14. C.   As $x \to \infty$, the function $\sin x$ does, indeed, oscillate between -1 and 1.

15. C.   Since $\dfrac{x^2-9}{3x-9} = \dfrac{(x-3)(x+3)}{3(x-3)} = \dfrac{x+3}{3}$ (provided $x \neq 3$), y can be defined to be equal to 2 at $x = 3$, removing the discontinuity at that point.

16. A.   Note that $\dfrac{x^2}{x} = x$ if $x \neq 0$ and that $\lim\limits_{x\to 0} f = 0$.

17. B.   Note that $\dfrac{\sin x}{x^2+3x} = \dfrac{\sin x}{x(x+3)} = \dfrac{\sin x}{x} \cdot \dfrac{1}{x+3} \to 1 \cdot \dfrac{1}{3}$.

18. C.   As $x \to 0$, $\frac{1}{x}$ takes on varying finite values as it increases. Since the sine function repeats, $\sin \frac{1}{x}$ oscillates, taking on, infinitely many times, each value between -1 and 1. The calculator graph of $Y_1 = \sin(1/X)$ exhibits this oscillating discontinuity at $x = 0$.

19. A.   Note that, since $y = \dfrac{2x^2 + 4}{(2-x)(1+4x)}$, both $x = 2$ and $x = -\dfrac{1}{4}$ are vertical asymptotes. Also, $y = -\dfrac{1}{2}$ is a horizontal asymptote.

20. A. $\lim\limits_{x \to 2^-} f(x) = \lim\limits_{x \to 2^+} f(x) = 0$.

21. E.   Verify that f is defined at $x = 0, 1, 2,$ and 3 (as well as at all other points in $[-1,3]$).

22. C.   Note that $\lim\limits_{x \to 2^-} f(x) = \lim\limits_{x \to 2^+} f(x) = 0$. However, $f(2) = 1$. Redefining $f(2)$ as 0 removes the discontinuity.

23. B.   The function is not continuous at $x = 0, 1,$ or 2.

24. B. $\lim\limits_{x \to 1^-} f(x) = 0 \neq \lim\limits_{x \to 1^+} f(x) = 1$.

25. B. $\dfrac{2x^2 + 1}{(2-x)(2+x)} = \dfrac{2x^2 + 1}{4 - x^2}$. Use the Rational Function Theorem (pages 30 and 31).

26. B.   Since $|x| = x$ if $x > 0$ but equals $-x$ if $x < 0$, $\lim\limits_{x \to 0^+} \dfrac{|x|}{x} = \lim\limits_{x \to 0^+} \dfrac{x}{x} = 1$ while $\lim\limits_{x \to 0^-} \dfrac{|x|}{x} = \lim\limits_{x \to 0^-} \dfrac{-x}{x} = -1$.

27. E.   Note that $x \sin \frac{1}{x}$ can be rewritten as $\dfrac{\sin \frac{1}{x}}{\frac{1}{x}}$ and that, as $x \to \infty$, $\frac{1}{x} \to 0$.

28. A.   As $x \to \pi$, $(\pi - x) \to 0$.

29. C.   Since $f(x) = x + 1$ if $x \neq 1$, $\lim\limits_{x \to 1} f(x)$ exists (and is equal to 2).

30. B. $f(x) = \dfrac{x(x-1)}{2x} = \dfrac{x-1}{2}$, for all $x \neq 0$. For f to be continuous at $x = 0$, $\lim\limits_{x \to 0} f(x)$ must equal $f(0)$. $\lim\limits_{x \to 0} f(x) = -\dfrac{1}{2}$.

31. B.   Only $x = 1$ and $x = 2$ need be checked. Since $f(x) = \dfrac{3x}{x-2}$ for $x \neq 1, 2$, and $\lim\limits_{x \to 1} f(x) = -3 = f(1)$, f is continuous at $x = 1$. Since $\lim\limits_{x \to 2} f(x)$ does not exist, f is not continuous at $x = 2$.

32. C.   As $x \to \pm\infty$, $y = f(x) \to 0$, so the x-axis is a horizontal asymptote. Also, as $x \to \pm 1$, $y \to \infty$, so $x = \pm 1$ are vertical asymptotes.

33. D.   No information is given about the domain of f except in the neighborhood of $x = -3$.

34. C.   As $x \to \infty$, $y \to \dfrac{1}{2}$; the denominator (but not the numerator) of y equals 0 at $x = 0$ and at $x = 1$.

35. D.   The function is defined at 0 to be 1, which is also $\lim\limits_{x \to 0} \dfrac{x^2 + x}{x} = \lim\limits_{x \to 0} (x+1)$.

36. E.   As $x \to 0^+$, $10^{\frac{1}{x}} \to \infty$ and therefore $y \to 0$. As $x \to 0^-$, $\frac{1}{x} \to -\infty$, so

$10^{\frac{1}{x}} \to 0$ and therefore $\to \frac{1}{2}$. Because the two one-sided limits are not equal, the limit does not exist.

37. E.   As $x \to 0^-$, arctan $\frac{1}{x} \to -\frac{\pi}{2}$, so $y \to \sqrt{3 - \frac{\pi}{2}}$. As $x \to 0^+$, $y \to \sqrt{3 + \frac{\pi}{2}}$.

The graph has a jump discontinuity at $x = 0$. (Verify with a calculator.)

Chapter 3

Differentiation

Review of Definitions and Methods

A. Definition of Derivative _____

At any x in the domain of the function $y = f(x)$, the *derivative* is defined to be

$$\lim_{\Delta x \to 0} \frac{f(x + \Delta x) - f(x)}{\Delta x} \text{ or } \lim_{\Delta x \to 0} \frac{\Delta y}{\Delta x}. \tag{1}$$

The function is said to be *differentiable* at every x for which this limit exists, and its derivative may be denoted by $f'(x)$, y', $\frac{dy}{dx}$, or $D_x y$. Frequently Δx is replaced by h or some other symbol.

The derivative of $y = f(x)$ at $x = a$, denoted by $f'(a)$ or $y'(a)$, may be defined as follows:

$$f'(a) = \lim_{h \to 0} \frac{f(a + h) - f(a)}{h}. \tag{2}$$

The fraction $\frac{f(a+h) - f(a)}{h}$ is called the *difference quotient for f at a* and represents the *average rate of change of f from a to a + h*. Geometrically, it is the slope of the secant PQ to the curve $y = f(x)$ through the points $P(a, f(a))$ and $Q(a + h, f(a + h))$. The limit, $f'(a)$, of the difference quotient is the (*instantaneous*) *rate of change of f* at point a. Geometrically, the derivative $f'(a)$ is the limit of the slope of secant PQ as Q approaches P; that is, as h approaches zero. This limit is the *slope of the curve at P*. The *tangent to the curve at P* is the line through P with this slope.

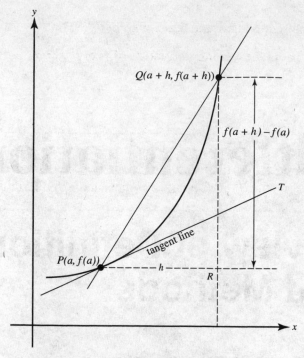

FIGURE N3–1a

In Figure N3–1a, PQ is the secant line through $(a, f(a))$ and $(a + h, f(a + h))$. The average rate of change from a to $a + h$ equals $\dfrac{RQ}{PR}$, which is the slope of secant PQ.

PT is the tangent to the curve at P. As h approaches zero, point Q approaches point P along the curve, PQ approaches PT, and the slope of PQ approaches the slope of PT, which equals $f'(a)$.

If we replace $(a + h)$ by x, in (2) above, so that $h = x - a$, we get the equivalent expression

$$f'(a) = \lim_{x \to a} \frac{f(x) - f(a)}{x - a}. \tag{3}$$

See Figure N3–1b.

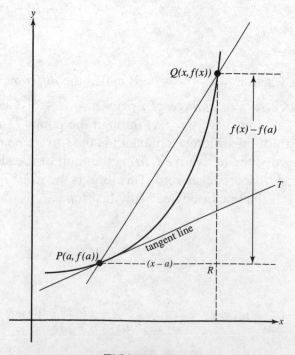

FIGURE N3–1b

The second derivative, denoted by $f''(x)$ or $\dfrac{d^2y}{dx^2}$ or y'', is the (first) derivative of $f'(x)$. Also, $f''(a)$ is the second derivative of $f(x)$ at $x = a$.

B.   Formulas

The formulas in this section for finding derivatives are so important that familiarity with them is essential. If a and n are constants and u and v are differentiable functions of x, then:

$$\frac{da}{dx} = 0 \tag{1}$$

$$\frac{d}{dx}\,au = a\,\frac{du}{dx} \tag{2}$$

$$\frac{d}{dx}\,u^n = nu^{n-1}\,\frac{du}{dx} \quad \text{(the power rule);} \quad \frac{d}{dx}\,x^n = nx^{n-1} \tag{3}$$

$$\frac{d}{dx}\,(u+v) = \frac{d}{dx}\,u + \frac{d}{dx}\,v;\quad \frac{d}{dx}\,(u-v) = \frac{d}{dx}\,u - \frac{d}{dx}\,v \tag{4}$$

$$\frac{d}{dx}\,(uv) = u\,\frac{dv}{dx} + v\,\frac{du}{dx} \quad \text{(the product rule)} \tag{5}$$

$$\frac{d}{dx}\left(\frac{u}{v}\right) = \frac{v\,\frac{du}{dx} - u\,\frac{dv}{dx}}{v^2} \quad (v \neq 0) \quad \text{(the quotient rule)} \tag{6}$$

$$\frac{d}{dx}\,\sin u = \cos u\,\frac{du}{dx} \tag{7}$$

$$\frac{d}{dx}\,\cos u = -\sin u\,\frac{du}{dx} \tag{8}$$

$$\frac{d}{dx}\,\tan u = \sec^2 u\,\frac{du}{dx} \tag{9}$$

$$\frac{d}{dx}\,\cot u = -\csc^2 u\,\frac{du}{dx} \tag{10}$$

$$\frac{d}{dx}\,\sec u = \sec u\,\tan u\,\frac{du}{dx} \tag{11}$$

$$\frac{d}{dx}\,\csc u = -\csc u\,\cot u\,\frac{du}{dx} \tag{12}$$

$$\frac{d}{dx}\,\ln u = \frac{1}{u}\,\frac{du}{dx} \tag{13}$$

$$\frac{d}{dx}\,e^u = e^u\,\frac{du}{dx} \tag{14}$$

$$\frac{d}{dx}\,a^u = a^u\,\ln a\,\frac{du}{dx} \tag{15}$$

$$\frac{d}{dx}\,\sin^{-1} u = \frac{d}{dx}\,\arcsin u = \frac{1}{\sqrt{1-u^2}}\,\frac{du}{dx} \quad (-1 < u < 1) \tag{16}$$

$$\frac{d}{dx}\cos^{-1} u = \frac{d}{dx} \arccos u = -\frac{1}{\sqrt{1-u^2}}\frac{du}{dx} \quad (-1 < u < 1) \tag{17}$$

$$\frac{d}{dx}\tan^{-1} u = \frac{d}{dx} \arctan u = \frac{1}{1+u^2}\frac{du}{dx} \tag{18}$$

$$\frac{d}{dx}\cot^{-1} u = \frac{d}{dx} \text{arccot } u = -\frac{1}{1+u^2}\frac{du}{dx} \tag{19}$$

$$\frac{d}{dx}\sec^{-1} u = \frac{d}{dx} \text{arcsec } u = \frac{1}{|u|\sqrt{u^2-1}}\frac{du}{dx} \quad (|u| > 1) \tag{20}$$

$$\frac{d}{dx}\csc^{-1} u = \frac{d}{dx} \text{arccsc } u = -\frac{1}{|u|\sqrt{u^2-1}}\frac{du}{dx} \quad (|u| > 1) \tag{21}$$

If c is in the domain of functions f and f', then the value of the derivative at c equals $f'(c)$.

C. The Chain Rule; the Derivative of a Composite Function

Formula (3) on pages 46 and 47 says

$$\frac{d}{dx} u^n = nu^{n-1}\frac{du}{dx}.$$

This formula is an application of the *chain rule*. For example, if we use formula (3) to find the derivative of $(x^2 - x + 2)^4$, we get

$$\frac{d}{dx}(x^2 - x + 2)^4 = 4(x^2 - x + 2)^3 \cdot (2x - 1).$$

In this last equation, if we let $y = (x^2 - x + 2)^4$ and let $u = x^2 - x + 2$, then $y = u^4$. The preceding derivative now suggests one form of the chain rule:

$$\frac{dy}{dx} = \frac{dy}{du} \cdot \frac{du}{dx} = 4u^3 \cdot \frac{du}{dx} = 4(x^2 - x + 2)^3 \cdot (2x - 1)$$

as before. Formula (3) on pages 46 and 47 gives the general case where $y = u^n$ and u is a differentiable function of x:

Suppose we think now of y as the composite function $f(g(x))$, where $y = f(u)$ and $u = g(x)$ are differentiable functions. Then

$$\begin{aligned}(f(g(x)))' &= f'(g(x)) \cdot g'(x) \\ &= f'(u) \cdot g'(x) \qquad \bigstar \\ &= \frac{dy}{du} \cdot \frac{du}{dx}.\end{aligned}$$

as we obtained above. The starred form ($\bigstar$) of the chain rule tells us how to differentiate the composite function: "Find the derivative of the 'outside' function first, then multiply by the derivative of the 'inside' one."

For example:

$$\frac{d}{dx}(x^3+1)^{10} = 10(x^3+1)^9 \cdot 3x^2 = 30x^2(x^3+1)^9,$$

$$\frac{d}{dx}\sqrt{7x-2} = \frac{d}{dx}(7x-2)^{\frac{1}{2}} = \frac{1}{2}(7x-2)^{-\frac{1}{2}} \cdot 7,$$

$$\frac{d}{dx}\frac{3}{(2-4x^2)^4} = \frac{d}{dx}3(2-4x^2)^{-4} = 3\cdot(-4)(2-4x^2)^{-5}\cdot(-8x),$$

$$\frac{d}{dx}\sin\left(\frac{\pi}{2}-x\right) = \cos\left(\frac{\pi}{2}-x\right)\cdot(-1),$$

$$\frac{d}{dx}\cos^3 2x = \frac{d}{dx}(\cos 2x)^3 = 3(\cos 2x)^2\cdot(-\sin 2x\cdot 2).$$

Many of the formulas listed above in §B and most of the illustrative examples that follow use the chain rule. Often the chain rule is used more than once in finding a derivative. Note that the algebraic simplifications that follow are included only for completeness.

Example 1.   If $y = 4x^3 - 5x + 7$, then

$$y' = \frac{dy}{dx} = 12x^2 - 5 \quad \text{and} \quad y'' = \frac{d^2y}{dx^2} = 24x.$$

Note that $y'(1) = 7$ and $y''(1) = 24$.

Example 2.   If $f(x) = (3x+2)^5$, then $f'(x) = 5(3x+2)^4 \cdot 3 = 15(3x+2)^4$.
Note that $f'(-1) = 15(-3+2)^4 = 15$.

Example 3.   If $y = \sqrt{3-x-x^2}$, then $y = (3-x-x^2)^{\frac{1}{2}}$ and

$$\frac{dy}{dx} = \frac{1}{2}(3-x-x^2)^{-\frac{1}{2}}(-1-2x) = -\frac{1+2x}{2\sqrt{3-x-x^2}}.$$

Example 4.   If $y = \dfrac{5}{\sqrt{(1-x^2)^3}}$, then $y = 5(1-x^2)^{-\frac{3}{2}}$ and

$$\frac{dy}{dx} = \frac{-15}{2}(1-x^2)^{-\frac{5}{2}}(-2x) = \frac{15x}{(1-x^2)^{\frac{5}{2}}}.$$

Example 5.   If $s(t) = (t^2+1)(1-t)^2$, then
$$s'(t) = (t^2+1)\cdot 2(1-t)(-1) + (1-t)^2\cdot 2t$$
$$= 2(1-t)(-1+t-2t^2).$$
Also, $s'(1) = 2(0)(-2) = 0$.

Example 6.   If $f(t) = e^{2t}\sin 3t$, then
$$f'(t) = e^{2t}(\cos 3t\cdot 3) + \sin 3t(e^{2t}\cdot 2)$$
$$= e^{2t}(3\cos 3t + 2\sin 3t).$$
Also, $f'(0) = 1(3\cdot 1 + 2\cdot 0) = 3$.

Example 7. If $f(v) = \dfrac{2v}{1-2v^2}$, then

$$f'(v) = \frac{(1-2v^2)\cdot 2 - 2v(-4v)}{(1-2v^2)^2} = \frac{2+4v^2}{(1-2v^2)^2} .$$

It follows that $f'(-1) = \dfrac{6}{(1-2)^2} = 6$. Note that neither $f(v)$ nor $f'(v)$ exists where the denominator equals zero, namely, where $1 - 2v^2 = 0$ or where v equals $\pm\dfrac{\sqrt{2}}{2}$.

Example 8. If $f(x) = \dfrac{\sin x}{x^2}$, $x \neq 0$, then

$$f'(x) = \frac{x^2 \cos x - \sin x \cdot 2x}{x^4} = \frac{x \cos x - 2\sin x}{x^3} .$$

Example 9. If $y = \tan(2x^2 + 1)$, then

$$y' = 4x \sec^2(2x^2 + 1).$$

Example 10. If $x = \cos^3(1 - 3\theta)$, then

$$\frac{dx}{d\theta} = -3\cos^2(1-3\theta)\sin(1-3\theta)(-3)$$

$$= 9\cos^2(1-3\theta)\sin(1-3\theta).$$

Example 11. If $y = e^{(\sin x)+1}$, then

$$\frac{dy}{dx} = \cos x \cdot e^{(\sin x)+1} .$$

Example 12. If $y = (x + 1)\ln^2(x + 1)$, then

$$\frac{dy}{dx} = (x+1)\frac{2\ln(x+1)}{x+1} + \ln^2(x+1) = 2\ln(x+1) + \ln^2(x+1).$$

Example 13. If $g(x) = (1 + \sin^2 3x)^4$, then

$$g'(x) = 4(1 + \sin^2 3x)^3(2\sin 3x \cos 3x)\cdot(3)$$

$$= 24(1 + \sin^2 3x)^3(\sin 3x \cos 3x).$$

Then $g'\left(\dfrac{\pi}{2}\right) = 24(1+(-1)^2)^3(-1\cdot 0) = 24\cdot 8\cdot 0 = 0.$

Example 14. If $y = \sin^{-1} x + x\sqrt{1-x^2}$, then

$$y' = \frac{1}{\sqrt{1-x^2}} + \frac{x(-2x)}{2\sqrt{1-x^2}} + \sqrt{1-x^2}$$

$$= \frac{1-x^2 + 1 - x^2}{\sqrt{1-x^2}} = 2\sqrt{1-x^2} .$$

Example 15.   If $u = \ln \sqrt{v^2 + 2v - 1}$, then $u = \dfrac{1}{2} \ln (v^2 + 2v - 1)$ and

$$\frac{du}{dv} = \frac{1}{2}\frac{2v+2}{v^2+2v-1} = \frac{v+1}{v^2+2v-1}.$$

Example 16.   If $s = e^{-t}(\sin t - \cos t)$, then

$$s' = e^{-t}(\cos t + \sin t) + (\sin t - \cos t)(-e^{-t})$$
$$= e^{-t}(2\cos t) = 2e^{-t}\cos t.$$

Example 17.   Let $y = 2u^3 - 4u^2 + 5u - 3$ and $u = x^2 - x$. Then

$$\frac{dy}{dx} = (6u^2 - 8u + 5)(2x - 1) = [6(x^2 - x)^2 - 8(x^2 - x) + 5](2x - 1).$$

Example 18.   If $y = \sin (ax + b)$, with a and b constants, then

$$\frac{dy}{dx} = [\cos(ax + b)] \cdot a = a \cos(ax + b) .$$

Example 19.   If $f(x) = ae^{kx}$ (with a and k constants), then

$$f'(x) = kae^{kx} \quad \text{and} \quad f'' = k^2 ae^{kx}.$$

Example 20.   Also, if $y = \ln (kx)$, where k is a constant, we can use both formula (13), pages 46 and 47, and the chain rule to get

$$\frac{dy}{dx} = \frac{1}{kx} \cdot k = \frac{1}{x}.$$

Alternatively, we can rewrite the given function using a property of logarithms: $\ln (kx) = \ln k + \ln x$. So

$$\frac{dy}{dx} = 0 + \frac{1}{x} = \frac{1}{x},$$

as before.

Example 21.   If $f(u) = u^2 - u$ and $u = g(x) = x^3 - 5$, we can let $F(x) = f(g(x))$, then evaluate $F'(2)$ as follows:

$$F'(2) = f'(g(2))g'(2) = f'(3) \cdot (12) = 5 \cdot 12 = 60,$$

where we note, since $g'(x) = 3x^2$, that $g'(2) = 12$, and, since $f'(u) = 2u - 1$, that $f'(3) = 5$.

Of course, we get exactly the same answer as follows.
Since $F(x) = (x^3 - 5)^2 - (x^3 - 5)$, we have

$$F'(x) = 2(x^3 - 5) \cdot 3x^2 - 3x^2,$$
$$F'(2) = 2 \cdot (3) \cdot 12 - 12 = 60.$$

D.   Differentiability and Continuity

If a function f has a derivative at $x = c$, then f is continuous at $x = c$.

This statement is an immediate consequence of the definition of the derivative of $f'(c)$ in the form

$$f'(c) = \lim_{x \to c} \frac{f(x) - f(c)}{x - c}.$$

If $f'(c)$ exists, then it follows that $\lim_{x \to c} f(x) = f(c)$, which guarantees that f is continuous at $x = c$.

If f is differentiable at c, its graph cannot have a hole or jump at c, nor can $x = c$ be a vertical asymptote of the graph. The tangent to the graph of f cannot be vertical at $x = c$; there cannot be a corner or cusp at $x = c$.

Each of the "prohibitions" in the preceding paragraph (each "cannot") tells how a function may fail to have a derivative at c. These cases are illustrated in Figures N3–2 (a) through (f).

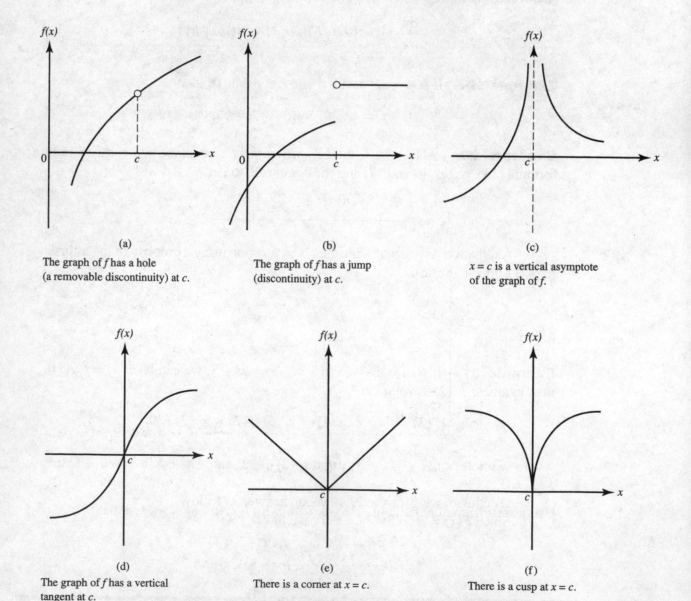

(a)

The graph of f has a hole (a removable discontinuity) at c.

(b)

The graph of f has a jump (discontinuity) at c.

(c)

$x = c$ is a vertical asymptote of the graph of f.

(d)

The graph of f has a vertical tangent at c.

(e)

There is a corner at $x = c$.

(f)

There is a cusp at $x = c$.

FIGURE N3–2

The graph in (e) is for the absolute function, $f(x) = |x|$. Since $f'(x) = -1$ for all negative x but $f'(x) = +1$ for all positive x, $f'(0)$ does not exist.

We may conclude from the preceding discussion that, although differentiability implies continuity, the converse is false. The functions in (d), (e), and (f) in Figure N3–2 are all continuous at $x = 0$ but not one of them is differentiable at the origin.

E.   Estimating a Derivative

E1. NUMERICALLY.

Example 22. The table shown gives the temperatures of a polar bear on a very cold arctic day (t = minutes; T = degrees Fahrenheit):

t	0	1	2	3	4	5	6	7	8
T	98	94.95	93.06	91.90	91.17	90.73	90.45	90.28	90.17

Our task is to estimate the derivative of T numerically at various times. The graph of $T(t)$ is sketched in Figure N3–3, but we shall use only the data from the table.

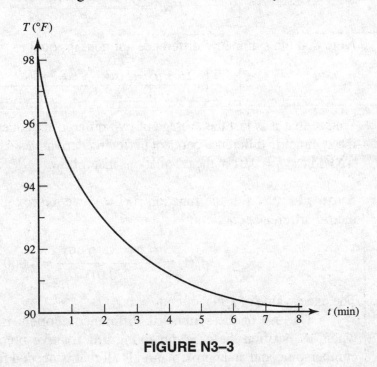

FIGURE N3–3

Using the difference quotient $\dfrac{T(t+h) - T(t)}{h}$ with h equal to 1, we see that

$$T'(0) \simeq \frac{T(1) - T(0)}{1} = -3.05°/\,\text{min}.$$

Also,

$$T'(1) \simeq \frac{T(2) - T(1)}{1} = -1.89^\circ / \text{min},$$

$$T'(2) \simeq \frac{T(3) - T(2)}{1} = -1.16^\circ / \text{min},$$

$$T'(3) \simeq \frac{T(4) - T(3)}{1} = -0.73^\circ / \text{min},$$

and so on.

The following table shows the *approximate* values of $T'(t)$ obtained from the difference quotients above:

t	0	1	2	3	4	5	6	7
$T'(t)$	−3.05	−1.89	−1.16	−0.73	−0.47	−0.28	−0.17	−0.11

The entries for $T'(t)$ represent the approximate slopes of the T curve at times 0.5, 1.5, 2.5, and so on.

From a Symmetric Difference Quotient

In Example 22 we approximated a derivative numerically from a table of values. We can also estimate $f'(a)$ numerically using the *symmetric difference quotient*, which is defined as follows:

$$f'(a) \simeq \frac{f(a+h) - f(a-h)}{2h}.$$

Note that the symmetric difference quotient is equal to

$$\frac{1}{2}\left[\frac{f(a+h) - f(a)}{h} + \frac{f(a) - f(a-h)}{h} \right].$$

We see that it is just the average of two difference quotients. Many calculators use the symmetric difference quotient in finding derivatives. Unless specified otherwise, we will use $h = 0.01$ in the calculations that follow.

Example 23. For the function $f(x) = x^4$ we approximate $f'(1)$ using the symmetric difference quotient:

$$f'(1) \simeq \frac{(1.01)^4 - (0.99)^4}{2(0.01)} = 4.0004.$$

The exact value of $f'(1)$, of course, is 4.

The use of the symmetric difference quotient is particularly convenient when, as is often the case, obtaining a derivative precisely (with formulas) is cumbersome and an approximation is all that is needed for practical purposes.

A word of caution is in order. Sometimes a wrong result is obtained using the symmetric difference quotient. On pages 52–53 we noted that $f(x) = |x|$ does not have a derivative at $x = 0$, since $f'(x) = -1$ for all $x < 0$ but $f'(x) = 1$ for all $x > 0$. Our calculator (which uses the symmetric difference quotient) tells us (incorrectly!) that $f'(0) = 0$. Note that, if $f(x) = |x|$, the symmetric difference quotient gives 0 for $f'(0)$ for every $h \neq 0$. If, for example, $h = 0.01$, then we get

$$f'(0) \simeq \frac{|0.01| - |-0.01|}{0.02} = \frac{0}{0.02} = 0,$$

which, as previously noted, is incorrect. The calculator graph of `nDeriv(abs X)`, which we see in Figure N3–4, shows that $f'(0)$ does not exist.

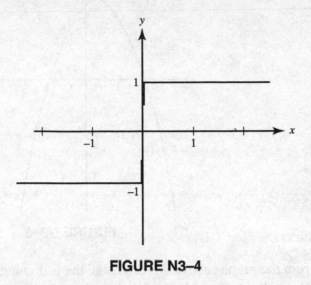

FIGURE N3–4

E2. GRAPHICALLY.

If we have the graph of a function $f(x)$, we can use it to graph $f'(x)$. We accomplish this by estimating the slope of the graph of $f(x)$ at enough points to assure a smooth curve for $f'(x)$. The grid lines should be spaced evenly, both vertically and horizontally. In Figure N3–5 we see the graph of $y = f(x)$. Below it is a table of the approximate slopes read from the graph.

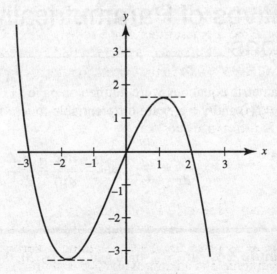

FIGURE N3–5

x	–3	–2.5	–2	–1.5	–1	0	0.5	1	1.5	2	2.5
$f'(x)$	–6	–3	–0.8	0.9	2	2.4	1.7	0.4	1.5	–4	–7

Figure N3–6 was obtained by plotting the points from the table of slopes above and drawing a smooth curve through these points. The result is the graph of $y = f'(x)$.

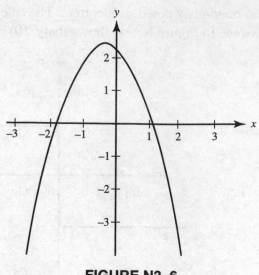

FIGURE N3–6

From the graphs above we can make the following observations:

(1) At the points where the slope of f (in Figure N3–5) equals 0, the graph of f' (Figure N3–6) has x-intercepts: approximately $x = -1.8$ and $x = 1.1$. We've drawn horizontal broken lines on the curve at these points.

(2) On intervals where f $\frac{\text{decreases}}{\text{increases}}$, the derivative is $\frac{\text{negative}}{\text{positive}}$. We see here that f decreases for $x < -1.8$ (approximately) and for $x > 1.1$ (approximately), and that x increases for $-1.8 < x < 1.1$ (approximately). In Chapter 4 we discuss other behaviors of f that are reflected in the graph of f'.

*F.   Derivatives of Parametrically Defined Functions

Parametric equations were defined on page 11.

If $x = f(t)$ and $y = g(t)$ are differentiable functions of t, then

$$\frac{dy}{dx} = \frac{\dfrac{dy}{dt}}{\dfrac{dx}{dt}} \quad \text{and} \quad \frac{d^2y}{dx^2} = \frac{d}{dx}\left(\frac{dy}{dx}\right) = \frac{\dfrac{d}{dt}\left(\dfrac{dy}{dx}\right)}{\dfrac{dx}{dt}}.$$

Example 24.   If $x = 2\sin\theta$ and $y = \cos 2\theta$, then

$$\frac{dx}{d\theta} = 2\cos\theta, \quad \frac{dy}{d\theta} = -2\sin 2\theta,$$

$$\frac{dy}{dx} = \frac{\dfrac{dy}{d\theta}}{\dfrac{dx}{d\theta}} = \frac{-2\sin 2\theta}{2\cos\theta} = -\frac{2\sin\theta\cos\theta}{\cos\theta} = -2\sin\theta.$$

*An asterisk denotes a topic covered only in Calculus BC.

Also,

$$\frac{d^2y}{dx^2} = \frac{\dfrac{d}{d\theta}\left(\dfrac{dy}{dx}\right)}{\dfrac{dx}{d\theta}} = \frac{-2\cos\theta}{2\cos\theta} = -1.$$

Example 25. Find the equation of the tangent to the curve in Example 24 for $\theta = \dfrac{\pi}{6}$.

When $\theta = \dfrac{\pi}{6}$, the slope of the tangent, $\dfrac{dy}{dx}$, equals $-2\sin\left(\dfrac{\pi}{6}\right) = -1$. Since $x = 2\sin\left(\dfrac{\pi}{6}\right) = 1$ and $y = \cos\left(2 \bullet \dfrac{\pi}{6}\right) = \cos\dfrac{\pi}{3} = \dfrac{1}{2}$, the equation is

$$y - \frac{1}{2} = -1(x-1) \quad \text{or} \quad y = -x + \frac{3}{2}.$$

Example 26. Suppose two objects are moving in the plane during the time interval $0 \leqslant t \leqslant 4$. Their positions at time t are described by the parametric equations

$$x_1 = 2t, \qquad y_1 = 4t - t^2 \qquad \text{and}$$
$$x_2 = t + 1, \qquad y_2 = 4 - t.$$

(a) Find all collision points. Justify your answer.
(b) Use a calculator to help you sketch the paths of the objects, indicating the direction in which each object travels.

  (a) Equating x_1 and x_2 yields $t = 1$. When $t = 1$, both y_1 and y_2 equal 3. So $t = 1$ yields a *true* collision point (not just an intersection point) at (2,3). (An *intersection point* is any point that is on both curves, but not necessarily at the same time.)

  (b) Using parametric mode, we enter X_{1T} and Y_{1T} for the first curve, X_{2T} and Y_{2T} for the second, with T in $[0,4]$, in the window $[0,3] \times [0,5]$. The calculator graph we obtain is shown in Figure N3–7.

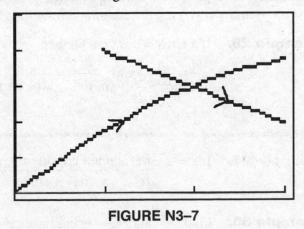

FIGURE N3–7

We've inserted arrows to indicate the direction of motion.

  If we change the calculator from sequential to simultaneous graphing we can watch the objects as they move, seeing that they do indeed pass through the intersection point at the same time. (It may help to choose a small T-step, say 0.02, in order to slow the motion enough to follow easily.)

G. Implicit Differentiation

When a functional relationship between x and y is defined by an equation of the form $F(x,y) = 0$ we say that the equation defines y *implicitly* as a function of x. Some examples are $x^2 + y^2 - 9 = 0$, $y^2 - 4x = 0$, and $\cos(xy) = y^2 - 5$ (which can be written $\cos(xy) - y^2 + 5 = 0$). Sometimes two (or more) explicit functions are defined by $F(x,y) = 0$. For example, $x^2 + y^2 - 9 = 0$ defines the two functions $y_1 = +\sqrt{9 - x^2}$ and $y_2 = -\sqrt{9 - x^2}$, the upper and lower halves, respectively, of the circle centered at the origin with radius 3. Each function is differentiable except at the points where $x = 3$ and $x = -3$.

Implicit differentiation is the technique we use to find a derivative when y is not defined explicitly in terms of x but is differentiable.

In the following examples, we differentiate both sides with respect to x, using appropriate formulas, and then solve for $\dfrac{dy}{dx}$.

Example 27. If $x^2 + y^2 - 9 = 0$, then

$$2x + 2y\frac{dy}{dx} = 0 \quad \text{and} \quad \frac{dy}{dx} = -\frac{x}{y}.$$

Note that the derivative above holds for every point on the circle, and exists for all y different from 0 (where the tangents to the circle are vertical).

Example 28. If $x^2 - 2xy + 3y^2 = 2$, then

$$2x - 2\left(x\frac{dy}{dx} + y\cdot 1\right) + 6y\frac{dy}{dx} = 0 \quad \text{and} \quad \frac{dy}{dx}(6y - 2x) = 2y - 2x,$$

so

$$\frac{dy}{dx} = \frac{y - x}{3y - x}.$$

Example 29. If $x \sin y = \cos(x + y)$, then

$$x \cos y\frac{dy}{dx} + \sin y = -\sin(x + y)\left(1 + \frac{dy}{dx}\right),$$

so

$$\frac{dy}{dx} = -\frac{\sin y + \sin(x + y)}{x \cos y + \sin(x + y)}.$$

Example 30. Find $\dfrac{dy}{dx}$ and $\dfrac{d^2y}{dx^2}$ using implicit differentiation on the equation $x^2 + y^2 = 1$.

$$2x + 2y\frac{dy}{dx} = 0 \quad \rightarrow \quad \frac{dy}{dx} = -\frac{x}{y}. \tag{1}$$

Then

$$\frac{d^2y}{dx^2} = -\frac{y \cdot 1 - x\left(\dfrac{dy}{dx}\right)}{y^2} = -\frac{y - x\left(-\dfrac{x}{y}\right)}{y^2} \tag{2}$$

$$= -\frac{y^2 + x^2}{y^3} = -\frac{1}{y^3}, \tag{3}$$

where we substituted for $\dfrac{dy}{dx}$ from (1) in (2), then used the given equation to simplify in (3).

Example 31. To verify the formula for the derivative of the inverse sine function, $y = \sin^{-1} x = \arcsin x$, with domain $[-1,1]$ and range $\left[-\dfrac{\pi}{2}, \dfrac{\pi}{2}\right]$, we proceed as follows:

$$y = \sin^{-1} x \quad \leftrightarrow \quad x = \sin y.$$

Now we differentiate with respect to x:

$$1 = \cos y \frac{dy}{dx},$$

$$\frac{dy}{dx} = \frac{1}{\cos y} = \frac{1}{+\sqrt{1 - \sin^2 y}} = \frac{1}{\sqrt{1 - x^2}},$$

where we chose the positive sign for $\cos y$ since $\cos y$ is nonnegative if $-\dfrac{\pi}{2} < y < \dfrac{\pi}{2}$. Note that this derivative exists only if $-1 < x < 1$.

Generalization of Formula (3), pages 46–47.
If $y = u^{p/q}$ with u a differentiable function of x and p/q a rational number, then implicit differentiation can be used to show that

$$\frac{dy}{dx} = \frac{p}{q}(u^{(p/q)-1})\frac{du}{dx}.$$

This generalizes formula (3) so that, whereas as originally given n had to be an integer, now it can be any rational number.

H. Derivative of the Inverse of a Function

Suppose f and g are inverse functions. What is the relationship between their derivatives? Recall that the graphs of inverse functions are the reflections of each other in the line $y = x$, and that at corresponding points their x- and y-coordinates are interchanged.

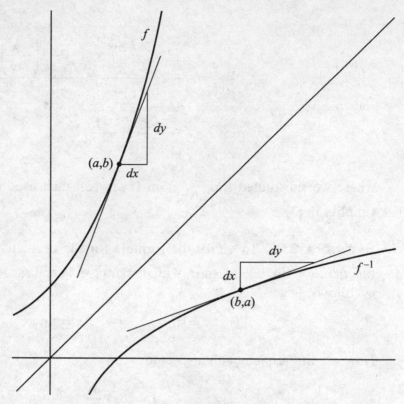

FIGURE N3–8

Figure N3–8 shows a function f passing through point (a,b) and the line tangent to f at that point. The slope of the curve there, $f'(a)$, is represented by the ratio of the legs of the triangle, $\dfrac{dy}{dx}$. When this figure is reflected across the line $y = x$, we obtain the graph of f^{-1}, passing through point (b,a), with the horizontal and vertical sides of the slope triangle interchanged. Note that the slope of the line tangent to the graph of f^{-1} at $x = b$ is represented by $\dfrac{dx}{dy}$, the reciprocal of the slope of f at $x = a$. We have, therefore,

$$\left(f^{-1}\right)'(b) = \frac{1}{f'(a)} \qquad \text{or} \qquad \left(f^{-1}\right)'(x) = \frac{1}{f'\left(f^{-1}(x)\right)}.$$

Simply put, the derivative of the inverse of a function at a point is the *reciprocal* of the derivative of the function *at the corresponding point*.

Implicit differentiation can also be used to find the derivative of the inverse of a function. If $y = f(x)$ is differentiable and its inverse is written as $f^{-1}(y)$, then

$$x = f^{-1}(y).$$

We now differentiate implicitly, with respect to x:

$$1 = (f^{-1})'(y)\frac{dy}{dx},$$

which yields

$$(f^{-1})'(y) = \frac{1}{\frac{dy}{dx}} .$$

We can replace the left side in the last equation by $\frac{dx}{dy}$, getting

$$\frac{dx}{dy} = \frac{1}{\frac{dy}{dx}} .$$

Thus, the derivatives of mutually inverse differentiable functions are reciprocals.

If we denote the inverse of $f(x)$ by $f^{-1}(x)$, then

$$\frac{d}{dx}(f^{-1}(x)) \text{ at } x = f(a) \text{ equals } \frac{1}{\frac{d}{dx}(f(x))} \text{ at } x = a.$$

Example 32.   If $f(3) = 8$ and $f'(3) = 5$, what do we know about f^{-1}?

Since f passes through the point $(3,8)$, f^{-1} must pass through the point $(8,3)$. Furthermore, since the graph of f has slope 5 at $(3,8)$, the graph of f^{-1} must have slope $\frac{1}{5}$ at $(8,3)$.

Example 33.   A function f and its derivative take on the values shown in the table. If g is the inverse of f, find $g'(6)$.

To find the slope of g at the point where $x = 6$, we must look at the point on f where $y = 6$, namely, $(2,6)$. Since $f'(2) = \frac{1}{3}$, $g'(6) = 3$.

x	$f(x)$	$f'(x)$
2	6	$\frac{1}{3}$
6	8	$\frac{3}{2}$

Example 34.   Let $y = f(x) = x^3 + x - 2$, and let g be the inverse function. Evaluate $g'(0)$.

Since $f'(x) = 3x^2 + 1$, $g'(y) = \frac{1}{3x^2 + 1}$. To find x when $y = 0$, we must solve the equation $x^3 + x - 2 = 0$. It is clear that $x = 1$. So

$$g'(0) = \frac{1}{3(1)^2 + 1} = \frac{1}{4} .$$

Example 35.   To find where the tangent to the curve $4x^2 + 9y^2 = 36$ is vertical, we differentiate the equation implicitly to get $\frac{dy}{dx}$: $8x + 18y\frac{dy}{dx} = 0$; so $\frac{dy}{dx} = -\frac{4x}{9y}$.

Since the tangent line to a curve is vertical when $\frac{dx}{dy} = 0$, we conclude that $-\frac{9y}{4x}$ must equal zero; that is, y must equal zero. When we substitute $y = 0$ in the original equation, we get $x = \pm3$. The points $(\pm3,0)$ are the ends of the major axis of the ellipse, where the tangents are indeed vertical.

I. The Mean Value Theorem

If the function $f(x)$ is continuous at each point on the closed interval $a \leqslant x \leqslant b$ and has a derivative at each point on the open interval $a < x < b$, then there is at least one number c, $a < c < b$, such that $\dfrac{f(b) - f(a)}{b - a} = f'(c)$. This important theorem, which relates average rate of change and instantaneous rate of change, is illustrated in Figure N3–9. For the function sketched in the figure there are two numbers, c_1 and c_2, between a and b where the slope of the curve equals the slope of the chord PQ (i.e., where the tangent to the curve is parallel to the secant line).

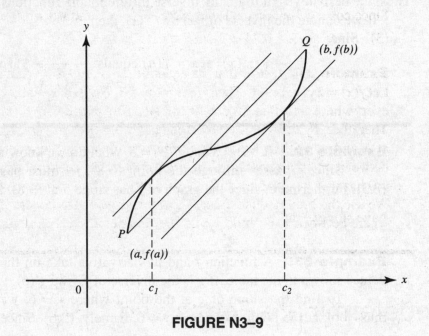

FIGURE N3–9

We will often refer to the Mean Value Theorem by its initials, MVT.

If, in addition to the hypotheses of the MVT, it is given that $f(a) = f(b) = 0$, then there is a number c between a and b such that $f'(c) = 0$. This special case of the MVT is called Rolle's theorem (see Figure N3–10).

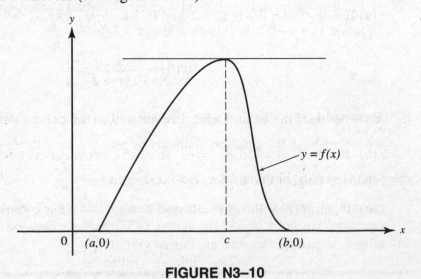

FIGURE N3–10

The Mean Value Theorem is one of the most useful laws when properly applied.

Example 36.   If $x > 0$, prove that $\sin x < x$.
Note that $\sin x \leqq 1$ for all x and consider the following cases if $x > 0$:

(1) $x > 1$. Here $\sin x < x$, since $\sin x \leqq 1$ and $1 < x$.
(2) $0 < x < 1$. Since $\sin x$ is continuous on $[0, 1]$ and since its derivative $\cos x$ exists at all x and therefore on the open interval $(0, 1)$, the Mean Value Theorem can be applied to the function, with $a = 0$ and $b = x$. Thus there is a number c, $0 < c < x < 1$, such that

$$\frac{\sin x - \sin 0}{x - 0} = \cos c.$$

Since $\cos c < 1$ if $0 < x < 1$, therefore $\dfrac{\sin x}{x} < 1$; and since $x > 0$, therefore $\sin x < x$.
(3) Since $\sin 1 \approx 0.84 < 1$, $\sin x < x$ if $x = 1$.

Example 37.   Show that the equation $2x^3 + 3x + 1 = 0$ has exactly one root.
Let $f(x) = 2x^3 + 3x + 1$. Then $f'(x) = 6x^2 + 3$. Note that f and f' are both continuous everywhere and that $f'(x) > 0$ for all x; in particular, $f'(x)$ never equals zero. Therefore, by the MVT, $f(x) = 0$ can have at most one root; if it had two roots, say at a and at b, then Rolle's theorem would imply the existence of a number c between a and b such that $f'(c)$ is zero, which is a contradiction. To show that $f(x) = 0$ has one root, we need only note that $f(-1)$ is negative whereas $f(0)$ is positive. We use the Intermediate Value Theorem (page 36) to conclude that f is zero somewhere between $x = -1$ and $x = 0$. The equation therefore has exactly one root.

*J.   Indeterminate Forms and L'Hôpital's Rule

To find the limit of an indeterminate form of the type $\dfrac{0}{0}$ or $\dfrac{\infty}{\infty}$, we apply L'Hôpital's rule, which involves taking derivatives. In the following, a is a finite number. The rule has several parts:

(a) If $\displaystyle\lim_{x\to a} f(x) = \lim_{x\to a} g(x) = 0$ and if $\displaystyle\lim_{x\to a}\frac{f'(x)}{g'(x)}$ exists, then[†]

$$\lim_{x\to a}\frac{f(x)}{g(x)} = \lim_{x\to a}\frac{f'(x)}{g'(x)}\,;$$

if $\displaystyle\lim_{x\to a}\frac{f'(x)}{g'(x)}$ does not exist, then L'Hôpital's rule cannot be applied.

(b) If $\displaystyle\lim_{x\to a} f(x) = \lim_{x\to a} g(x) = \infty$, the same consequences follow as in case (a). The rules in (a) and (b) both hold for one-sided limits.

(c) If $\displaystyle\lim_{x\to\infty} f(x) = \lim_{x\to\infty} g(x) = 0$ and if $\displaystyle\lim_{x\to\infty}\frac{f'(x)}{g'(x)}$ exists, then

$$\lim_{x\to\infty}\frac{f(x)}{g(x)} = \lim_{x\to\infty}\frac{f'(x)}{g'(x)}\,;$$

*An asterisk denotes a topic covered only in Calculus BC.
†The limit can be finite or infinite ($+\infty$ or $-\infty$).

if $\lim_{x \to \infty} \dfrac{f'(x)}{g'(x)}$ does not exist, then L'Hôpital's rule cannot be applied. (Here the notation "$x \to \infty$" represents either "$x \to +\infty$" or "$x \to -\infty$.")

(d) If $\lim_{x \to \infty} f(x) = \lim_{x \to \infty} g(x) = \infty$, the same consequences follow as in case (c).

In applying any of the above rules, if we obtain $\dfrac{0}{0}$ or $\dfrac{\infty}{\infty}$ again, we can apply the rule once more, repeating the process until the form we obtain is no longer indeterminate.

Example 38. $\lim\limits_{x \to 3} \dfrac{x^2 - 9}{x - 3}$ is of type $\dfrac{0}{0}$ and thus equals $\lim\limits_{x \to 3} \dfrac{2x}{1} = 6$. (Compare with Example 12, page 29.)

Example 39. $\lim\limits_{x \to 0} \dfrac{\tan x}{x}$ is of type $\dfrac{0}{0}$ and therefore equals $\lim\limits_{x \to 0} \dfrac{\sec^2 x}{1} = 1$.

Example 40. $\lim\limits_{x \to -2} \dfrac{x^3 + 8}{x^2 - 4}$ (Example 13, page 30) is of type $\dfrac{0}{0}$ and thus equals

$\lim\limits_{x \to -2} \dfrac{3x^2}{2x} = -3$, as before. Note that $\lim\limits_{x \to -2} \dfrac{3x^2}{2x}$ is *not* the limit of an indeterminate form!

Example 41. $\lim\limits_{h \to 0} \dfrac{e^h - 1}{h}$ is of type $\dfrac{0}{0}$ and therefore equals $\lim\limits_{h \to 0} \dfrac{e^h}{1} = 1$.

Example 42. $\lim\limits_{x \to \infty} \dfrac{x^3 - 4x^2 + 7}{3 - 6x - 2x^3}$ (Example 20, page 30) is of type $\dfrac{\infty}{\infty}$, so that it

equals $\lim\limits_{x \to \infty} \dfrac{3x^2 - 8x}{-6 - 6x^2}$, which is again of type $\dfrac{\infty}{\infty}$. Apply L'Hôpital's rule twice more:

$$\lim_{x \to \infty} \frac{6x - 8}{-12x} = \lim_{x \to \infty} \frac{6}{-12} = -\frac{1}{2}.$$

For this problem, it is easier and faster to apply the Rational Function Theorem!

Example 43. $\lim\limits_{x \to \infty} \dfrac{\ln x}{x}$ is of type $\dfrac{\infty}{\infty}$ and equals $\lim\limits_{x \to \infty} \dfrac{1/x}{1} = 0$.

Example 44. Here's an example showing what happens if we apply the rule erroneously. Noting that the leftmost limit below is of type $\dfrac{0}{0}$, we have

$$\lim_{x \to 0} \frac{x^2 \sin(1/x)}{\tan x} = \lim_{x \to 0} \frac{x^2 \cos(1/x)(-1/x^2) + 2x \sin(1/x)}{\sec^2 x}$$

$$= \lim_{x \to 0} \frac{-\cos(1/x) + 2x \sin(1/x)}{\sec^2 x},$$

which *does not exist* because, as $x \to 0$, $-\cos \dfrac{1}{x}$ oscillates between -1 and $+1$,

while $2x \sin \dfrac{1}{x} \to 0$, and $\sec^2 x \to 1$. We can draw no conclusion whatsoever!

But if we rewrite the original fraction, we do not need to apply L'Hôpital's rule at all:

$$\lim_{x\to 0}\frac{x^2\sin(1/x)}{\tan x}=\lim_{x\to 0}\frac{x\sin(1/x)}{(\tan x)/x}=\frac{0}{1}=0.$$

We showed that $\lim_{x\to 0}((\tan x)/x)=1$ in Example 39.

L'Hôpital's rule can also be applied to indeterminate forms of the types $0\cdot\infty$ and $\infty-\infty$, if they can be transformed to either $\dfrac{0}{0}$ or $\dfrac{\infty}{\infty}$.

Example 45. $\lim_{x\to\infty}x\sin\dfrac{1}{x}$ is of the type $\infty\cdot 0$. Since $x\sin\dfrac{1}{x}=\dfrac{\sin 1/x}{1/x}$ and, as

$x\to\infty$, the latter is the indeterminate form $\dfrac{0}{0}$, we see that

$$\lim_{x\to\infty}x\sin\frac{1}{x}=\lim_{x\to\infty}\frac{-\dfrac{1}{x^2}\cos\dfrac{1}{x}}{-\dfrac{1}{x^2}}=\lim_{x\to\infty}\cos\frac{1}{x}=1.$$

Other indeterminate forms, such as 0^0, 1^∞, and ∞^0, may be resolved by taking the natural logarithm and then applying L'Hôpital's rule.

Example 46. $\lim_{x\to 0}(1+x)^{1/x}$ is of type 1^∞. Let $y=(1+x)^{1/x}$, so that $\ln y=$

$\dfrac{1}{x}\ln(1+x)$. Then $\lim_{x\to 0}\ln y=\lim_{x\to 0}\dfrac{\ln(1+x)}{x}$, which is of type $\dfrac{0}{0}$. Thus,

$$\lim_{x\to 0}\ln y=\lim_{x\to 0}\frac{\dfrac{1}{1+x}}{1}=\frac{1}{1}=1,$$

and since $\lim_{x\to 0}\ln y=1$, $\lim_{x\to 0}y=e^1=e$.

Example 47. $\lim_{x\to\infty}x^{1/x}$ is of type ∞^0. Let $y=x^{1/x}$, so that $\ln y=\dfrac{1}{x}\ln x=\dfrac{\ln x}{x}$

(which as $x\to\infty$, is of type $\dfrac{\infty}{\infty}$). Then $\lim_{x\to\infty}\ln y=\dfrac{1/x}{1}=0$, and $\lim_{x\to\infty}y=e^0=1$.

Redo Set 2, pages 37–41, applying L'Hôpital's rule wherever possible.

K.   Recognizing a Given Limit as a Derivative____

It is often extremely useful to evaluate a limit by recognizing that it is merely an expression for the definition of the derivative of a specific function (often at a specific point). The relevant definition is

$$f'(c) = \lim_{h \to 0} \frac{f(c+h) - f(c)}{h}.$$

Example 48. $\lim\limits_{h \to 0} \dfrac{(2+h)^4 - 2^4}{h}$ is the derivative of $f(x) = x^4$ at the point where $x = 2$. Since $f'(x) = 4x^3$, the value of the given limit is $f'(2) = 4(2^3) = 32$.

Example 49. $\lim\limits_{h \to 0} \dfrac{\sqrt{9+h} - 3}{h} = f'(9)$, where $f(x) = \sqrt{x}$. The value of the limit is $\dfrac{1}{2} x^{-1/2}$ when $x = 9$, or $\dfrac{1}{6}$.

Example 50. $\lim\limits_{h \to 0} \dfrac{1}{h}\left(\dfrac{1}{2+h} - \dfrac{1}{2}\right) = f'(2)$, where $f(x) = \dfrac{1}{x}$. Verify that $f'(2) = -\dfrac{1}{4}$ and compare with Example 17, page 30.

Example 51. $\lim\limits_{h \to 0} \dfrac{e^h - 1}{h} = f'(0)$, where $f(x) = e^x$. The limit has value e^0 or 1 (see also Example 41, on page 64).

Example 52. $\lim\limits_{x \to 0} \dfrac{\sin x}{x}$ is $f'(0)$, where $f(x) = \sin x$, because we can write

$$f'(0) = \lim_{x \to 0} \frac{\sin(0 + x) - \sin 0}{x} = \lim_{x \to 0} \frac{\sin x}{x}.$$

The answer is 1, since $f'(x) = \cos x$ and $f'(0) = \cos 0 = 1$. Of course, we already know that the given limit is the basic trigonometric limit with value 1. Also, L'Hôpital's rule yields 1 as the answer immediately.

Set 3: Multiple-Choice Questions on Differentiation

In each of Questions 1–5 a function is given. Choose the alternative that is the derivative, $\dfrac{dy}{dx}$, of the function.

1. $y = (4x + 1)(1 - x)^3$

    (A) $-12(1 - x)^2$ **(B)** $(1 - x)^2(1 + 8x)$ **(C)** $(1 - x)^2(1 - 16x)$

    (D) $3(1 - x)^2(4x + 1)$ **(E)** $(1 - x)^2(16x + 7)$

2. $y = \dfrac{2 - x}{3x + 1}$

    (A) $-\dfrac{7}{(3x + 1)^2}$ **(B)** $\dfrac{6x - 5}{(3x + 1)^2}$ **(C)** $-\dfrac{9}{(3x + 1)^2}$

    (D) $\dfrac{7}{(3x + 1)^2}$ **(E)** $\dfrac{7 - 6x}{(3x + 1)^2}$

3. $y = \sqrt{3 - 2x}$

    (A) $\dfrac{1}{2\sqrt{3 - 2x}}$ **(B)** $-\dfrac{1}{\sqrt{3 - 2x}}$ **(C)** $-\dfrac{(3 - 2x)^{3/2}}{3}$

    (D) $-\dfrac{1}{3 - 2x}$ **(E)** $\dfrac{2}{3}(3 - 2x)^{3/2}$

4. $y = \dfrac{2}{(5x + 1)^3}$

    (A) $-\dfrac{30}{(5x + 1)^2}$ **(B)** $-30(5x + 1)^{-4}$ **(C)** $\dfrac{-6}{(5x + 1)^4}$

    (D) $-\dfrac{10}{3}(5x + 1)^{-4/3}$ **(E)** $\dfrac{30}{(5x + 1)^4}$

5. $y = 3x^{2/3} - 4x^{1/2} - 2$

    (A) $2x^{1/3} - 2x^{-1/2}$ **(B)** $3x^{-1/3} - 2x^{-1/2}$ **(C)** $\dfrac{9}{5}x^{5/3} - 8x^{3/2}$

    (D) $\dfrac{2}{x^{1/3}} - \dfrac{2}{x^{1/2}} - 2$ **(E)** $2x^{-1/3} - 2x^{-1/2}$

In Questions 6–13, differentiable functions f and g have the values shown in the table.

x	f	f'	g	g'
0	2	1	5	–4
1	3	2	3	–3
2	5	3	1	–2
3	10	4	0	–1

6. If $A = f + 2g$, then $A'(3) =$

(A) –2   (B) 2   (C) 7   (D) 8   (E) 10

7. If $B = f \cdot g$, then $B'(2) =$

(A) –20   (B) –7   (C) –6   (D) –1   (E) 13

8. If $D = \dfrac{1}{g}$, then $D'(1) =$

(A) $-\dfrac{1}{2}$ (B) $-\dfrac{1}{3}$ (C) $-\dfrac{1}{9}$ (D) $\dfrac{1}{9}$ (E) $\dfrac{1}{3}$

9. If $H(x) = \sqrt{f(x)}$, then $H'(3) =$

(A) $\dfrac{1}{4}$ (B) $\dfrac{1}{2\sqrt{10}}$ (C) 2   (D) $\dfrac{2}{\sqrt{10}}$ (E) $4\sqrt{10}$

10. If $K(x) = \left(\dfrac{f}{g}\right)(x)$, then $K'(0) =$

(A) $\dfrac{-13}{25}$ (B) $-\dfrac{1}{4}$ (C) $\dfrac{13}{25}$ (D) $\dfrac{13}{16}$ (E) $\dfrac{22}{25}$

11. If $M(x) = f(g(x))$, then $M'(1) =$

(A) –12   (B) –6   (C) 4   (D) 6   (E) 12

12. If $P(x) = f(x^3)$, then $P'(1) =$

(A) 2   (B) 6   (C) 8   (D) 12   (E) 54

13. If $S(x) = f^{-1}(x)$, then $S'(3) =$

(A) –2   (B) $-\dfrac{1}{25}$ (C) $\dfrac{1}{4}$ (D) $\dfrac{1}{2}$ (E) 2

In Questions 14–21 find y'.

14. $y = 2\sqrt{x} - \dfrac{1}{2\sqrt{x}}$

(A) $x + \dfrac{1}{x\sqrt{x}}$ (B) $x^{-1/2} + x^{-3/2}$ (C) $\dfrac{4x-1}{4x\sqrt{x}}$

(D) $\dfrac{1}{\sqrt{x}} + \dfrac{1}{4x\sqrt{x}}$ (E) $\dfrac{4}{\sqrt{x}} + \dfrac{1}{x\sqrt{x}}$

15. $y = \sqrt{x^2 + 2x - 1}$

(A) $\dfrac{x+1}{y}$ (B) $4y(x+1)$ (C) $\dfrac{1}{2\sqrt{x^2+2x-1}}$

(D) $-\dfrac{x+1}{(x^2+2x-1)^{3/2}}$ (E) none of these

16. $y = \dfrac{x}{\sqrt{1-x^2}}$

(A) $\dfrac{1-2x^2}{(1-x^2)^{3/2}}$ (B) $\dfrac{1}{1-x^2}$ (C) $\dfrac{1}{\sqrt{1-x^2}}$

(D) $\dfrac{1-2x^2}{(1-x^2)^{1/2}}$ (E) none of these

17. $y = \ln \dfrac{e^x}{e^x - 1}$

(A) $x - \dfrac{e^x}{e^x - 1}$ (B) $\dfrac{1}{e^x - 1}$ (C) $-\dfrac{1}{e^x - 1}$

(D) 0 (E) $\dfrac{e^x - 2}{e^x - 1}$

18. $y = \tan^{-1} \dfrac{x}{2}$

(A) $\dfrac{4}{4+x^2}$ (B) $\dfrac{1}{2\sqrt{4-x^2}}$ (C) $\dfrac{2}{\sqrt{4-x^2}}$

(D) $\dfrac{1}{2+x^2}$ (E) $\dfrac{2}{x^2+4}$

19. $y = \ln (\sec x + \tan x)$

    (A) $\sec x$ **(B)** $\dfrac{1}{\sec x}$ **(C)** $\tan x + \dfrac{\sec^2 x}{\tan x}$

    (D) $\dfrac{1}{\sec x + \tan x}$ **(E)** $-\dfrac{1}{\sec x + \tan x}$

20. $y = \dfrac{e^x - e^{-x}}{e^x + e^{-x}}$

    (A) 0 **(B)** 1 **(C)** $\dfrac{2}{(e^x + e^{-x})^2}$

    (D) $\dfrac{4}{(e^x + e^{-x})^2}$ **(E)** $\dfrac{1}{e^{2x} + e^{-2x}}$

21. $y = \ln (x\sqrt{x^2 + 1})$

    (A) $1 + \dfrac{x}{x^2 + 1}$ **(B)** $\dfrac{1}{x\sqrt{x^2 + 1}}$ **(C)** $\dfrac{2x^2 + 1}{x\sqrt{x^2 + 1}}$

    (D) $\dfrac{2x^2 + 1}{x(x^2 + 1)}$ **(E)** none of these

22. The graph of g' is shown here. Which of the following statements is (are) true of g at $x = a$?

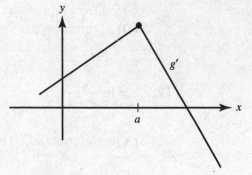

    I. g is continuous.
    II. g is differentiable.
    III. g is increasing.

    (A) I only   **(B)** III only   **(C)** I and III only
    (D) II and III only   **(E)** I, II, and III

23. A function f has the derivative shown. Which of the following statements must be false?

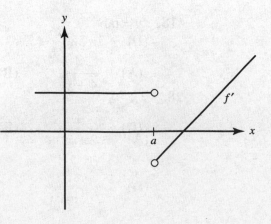

    (A) f is continuous at $x = a$.
    (B) $f(a) = 0$.
    (C) f has a vertical asymptote at $x = a$.
    (D) f has a jump discontinuity at $x = a$.
    (E) f has a removable discontinuity at $x = a$.

24. The function f whose graph is shown has $f' = 0$ at $x =$

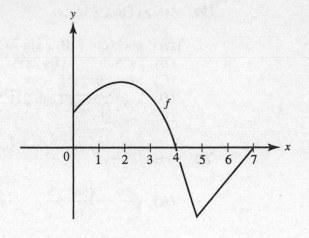

(A) 2 only
(B) 2 and 5
(C) 4 and 7
(D) 2, 4, and 7
(E) 2, 4, 5, and 7

25. A differentiable function f has the values shown. Estimate $f'(1.5)$.

x	1.0	1.2	1.4	1.6
$f(x)$	8	10	14	22

(A) 8   (B) 12   (C) 18   (D) 40   (E) 80

26. Water is poured into a conical reservoir at a constant rate. If $h(t)$ is the rate of change of the depth of the water, then h is

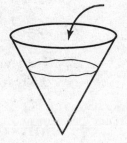

(A) constant
(B) linear and increasing
(C) linear and decreasing
(D) nonlinear and increasing
(E) nonlinear and decreasing

In Questions 27–33, find $\dfrac{dy}{dx}$.

27. $y = x^2 \sin \dfrac{1}{x}$ $(x \neq 0)$

(A) $2x \sin \dfrac{1}{x} - x^2 \cos \dfrac{1}{x}$ (B) $-\dfrac{2}{x} \cos \dfrac{1}{x}$ (C) $2x \cos \dfrac{1}{x}$

(D) $2x \sin \dfrac{1}{x} - \cos \dfrac{1}{x}$ (E) $-\cos \dfrac{1}{x}$

28. $y = \dfrac{1}{2 \sin 2x}$

(A) $-\csc 2x \cot 2x$ (B) $\dfrac{1}{4 \cos 2x}$ (C) $-4 \csc 2x \cot 2x$

(D) $\dfrac{\cos 2x}{2\sqrt{\sin 2x}}$ (E) $-\csc^2 2x$

29. $y = e^{-x} \cos 2x$

    (A) $-e^{-x}(\cos 2x + 2 \sin 2x)$
    (B) $e^{-x}(\sin 2x - \cos 2x)$
    (C) $2e^{-x} \sin 2x$
    (D) $-e^{-x}(\cos 2x + \sin 2x)$
    (E) $-e^{-x} \sin 2x$

30. $y = \sec^2 \sqrt{x}$

    (A) $\dfrac{\sec \sqrt{x} \tan \sqrt{x}}{\sqrt{x}}$
    (B) $\dfrac{\tan \sqrt{x}}{\sqrt{x}}$
    (C) $2 \sec \sqrt{x} \tan^2 \sqrt{x}$

    (D) $\dfrac{\sec^2 \sqrt{x} \tan \sqrt{x}}{\sqrt{x}}$
    (E) $2 \sec^2 \sqrt{x} \tan \sqrt{x}$

31. $y = x \ln^3 x$

    (A) $\dfrac{3 \ln^2 x}{x}$
    (B) $3 \ln^2 x$
    (C) $3x \ln^2 x + \ln^3 x$

    (D) $3(\ln x + 1)$
    (E) none of these

32. $y = \dfrac{1+x^2}{1-x^2}$

    (A) $-\dfrac{4x}{(1-x^2)^2}$
    (B) $\dfrac{4x}{(1-x^2)^2}$
    (C) $\dfrac{-4x^3}{(1-x^2)^2}$

    (D) $\dfrac{2x}{1-x^2}$
    (E) $\dfrac{4}{1-x^2}$

33. $y = \sin^{-1} x - \sqrt{1-x^2}$

    (A) $\dfrac{1}{2\sqrt{1-x^2}}$
    (B) $\dfrac{2}{\sqrt{1-x^2}}$
    (C) $\dfrac{1+x}{\sqrt{1-x^2}}$

    (D) $\dfrac{x^2}{\sqrt{1-x^2}}$
    (E) $\dfrac{1}{\sqrt{1+x}}$

Use the graph to answer Questions 34–36. It consists of two line segments and a semicircle.

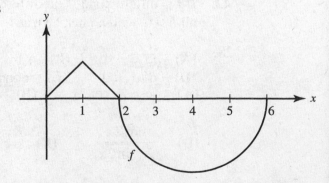

34. $f'(x) = 0$ for $x =$

    (A)   1 only
    (B)   2 only
    (C)   4 only
    (D)   1 and 4
    (E)   2 and 6

35. $f'(x)$ does not exist for $x =$

   (A)   1 only   (B)   2 only   (C)   1 and 2
   (D)   2 and 6   (E)   1, 2, and 6

36. $f'(5) =$

   (A) $\dfrac{1}{2}$ (B) $\dfrac{1}{\sqrt{3}}$ (C)   1   (D)   2   (E) $\sqrt{3}$

37. At how many points on the interval $[-5,5]$ is a tangent to $y = x + \cos x$ parallel to the secant line?

   (A)   none   (B)   1   (C)   2   (D)   3   (E)   more than 3

38. From the values of f shown, estimate $f'(2)$.

x	1.92	1.94	1.96	1.98	2.00
$f(x)$	6.00	5.00	4.40	4.10	4.00

   (A)   −0.10   (B)   −0.20   (C)   −5   (D)   −10   (E)   −25

39. Using the values shown in the table for Question 38, estimate $(f^{-1})'(4)$.

   (A)   −0.2   (B)   −0.1   (C)   −5   (D)   −10   (E)   −25

40. The "left half" of the parabola defined by $y = x^2 - 8x + 10$ for $x \leq 4$ is a one-to-one function; therefore its inverse is also a function. Call that inverse g. Find $g'(3)$.

   (A) $-\dfrac{1}{2}$ (B) $-\dfrac{1}{6}$ (C) $\dfrac{1}{6}$ (D) $\dfrac{1}{2}$ (E) $\dfrac{11}{2}$

41. For $f(x) = 5^x$, what is the estimate of $f'(2)$ obtained by using the symmetric difference quotient with $h = 0.03$?

   (A)   25.029   (B)   40.236   (C)   40.252   (D)   41.223   (E)   80.503

42. If f is differentiable and difference quotients overestimate the slope of f at $x = a$ for all $h > 0$, which must be true?

   (A) $f'(a) > 0$ (B) $f'(a) < 0$ (C) $f''(a) > 0$
   (D) $f''(a) < 0$ (E)   none of these

In each of Questions 43–46, y is a differentiable function of x. Choose the alternative that is the derivative $\dfrac{dy}{dx}$.

43. $x^3 - xy + y^3 = 1$

(A) $\dfrac{3x^2}{x - 3y^2}$ (B) $\dfrac{3x^2 - 1}{1 - 3y^2}$ (C) $\dfrac{y - 3x^2}{3y^2 - x}$

(D) $\dfrac{3x^2 + 3y^2 - y}{x}$ (E) $\dfrac{3x^2 + 3y^2}{x}$

44. $x + \cos(x + y) = 0$

(A) $\csc(x + y) - 1$ (B) $\csc(x + y)$ (C) $\dfrac{x}{\sin(x + y)}$

(D) $\dfrac{1}{\sqrt{1 - x^2}}$ (E) $\dfrac{1 - \sin x}{\sin y}$

45. $\sin x - \cos y - 2 = 0$

(A) $-\cot x$ (B) $-\cot y$ (C) $\dfrac{\cos x}{\sin y}$

(D) $-\csc y \cos x$ (E) $\dfrac{2 - \cos x}{\sin y}$

46. $3x^2 - 2xy + 5y^2 = 1$

(A) $\dfrac{3x + y}{x - 5y}$ (B) $\dfrac{y - 3x}{5y - x}$ (C) $3x + 5y$

(D) $\dfrac{3x + 4y}{x}$ (E) none of these

Individual instructions are given in full for each of the remaining questions of this set.

*47. If $x = t^2 - 1$ and $y = t^4 - 2t^3$, then, when $t = 1$, $\dfrac{d^2y}{dx^2}$ is

(A) 1 (B) -1 (C) 0 (D) 3 (E) $\dfrac{1}{2}$

48. If $f(x) = x^4 - 4x^3 + 4x^2 - 1$, then the set of values of x for which the derivative equals zero is

(A) $\{1, 2\}$ (B) $\{0, -1, -2\}$ (C) $\{-1, +2\}$
(D) $\{0\}$ (E) $\{0, 1, 2\}$

49. If $f(x) = 16\sqrt{x}$, then $f''(4)$ is equal to

(A) -32 (B) -16 (C) -4 (D) -2 (E) $-\dfrac{1}{2}$

*An asterisk denotes a topic covered only in Calculus BC.

50. If $f(x) = \ln x^3$, then $f''(3)$ is

(A) $-\dfrac{1}{3}$ (B) -1 (C) -3 (D) 1 (E) none of these

51. If a point moves on the curve $x^2 + y^2 = 25$, then, at $(0, 5)$, $\dfrac{d^2y}{dx^2}$ is

(A) 0 (B) $\dfrac{1}{5}$ (C) -5 (D) $-\dfrac{1}{5}$ (E) nonexistent

52. If $y = a \sin ct + b \cos ct$, where a, b, and c are constants, then $\dfrac{d^2y}{dt^2}$ is

(A) $ac^2(\sin t + \cos t)$ (B) $-c^2y$ (C) $-ay$
(D) $-y$ (E) $a^2c^2 \sin ct - b^2c^2 \cos ct$

53. If $f(u) = \sin u$ and $u = g(x) = x^2 - 9$, then $(f \circ g)'(3)$ equals

(A) 0 (B) 1 (C) 6 (D) 9 (E) none of these

54. If $f(x) = 5^x$ and $5^{1.002} \approx 5.016$, which is closest to $f'(1)$?

(A) 0.016 (B) 1.0 (C) 5.0 (D) 8.0 (E) 32.0

55. If $f(x) = \dfrac{x}{(x-1)^2}$, then the set of x's for which $f'(x)$ exists is

(A) all reals
(B) all reals except $x = 1$ and $x = -1$
(C) all reals except $x = 1$
(D) all reals except $x = \dfrac{1}{3}$ and $x = -1$
(E) all reals except $x = 1$

56. If $y = e^x(x - 1)$, then $y''(0)$ equals

(A) -2 (B) -1 (C) 0 (D) 1 (E) none of these

57. If $y = \sqrt{x^2 + 1}$, then the derivative of y^2 with respect to x^2 is

(A) 1 (B) $\dfrac{x^2 + 1}{2x}$ (C) $\dfrac{x}{2(x^2 + 1)}$ (D) $\dfrac{2}{x}$ (E) $\dfrac{x^2}{x^2 + 1}$

58. If $f(x) = \dfrac{1}{x^2 + 1}$ and $g(x) = \sqrt{x}$, then the derivative of $f(g(x))$ is

(A) $\dfrac{-\sqrt{x}}{(x^2 + 1)^2}$ (B) $-(x + 1)^{-2}$ (C) $\dfrac{-2x}{(x^2 + 1)^2}$

(D) $\dfrac{1}{(x + 1)^2}$ (E) $\dfrac{1}{2\sqrt{x}(x + 1)}$

*59. If $x = e^\theta \cos \theta$ and $y = e^\theta \sin \theta$, then, when $\theta = \dfrac{\pi}{2}$, $\dfrac{dy}{dx}$ is

   (A) 1   (B) 0   (C) $e^{\pi/2}$ (D) nonexistent   (E) −1

*60. If $x = \cos t$ and $y = \cos 2t$, then $\dfrac{d^2y}{dx^2}$ $(\sin t \neq 0)$ is

   (A) $4 \cos t$ (B) 4   (C) $\dfrac{4y}{x}$ (D) −4   (E) $-4 \cot t$

61. If $y = x^2 + x$, then the derivative of y with respect to $\dfrac{1}{1-x}$ is

   (A) $(2x + 1)(x - 1)^2$ (B) $\dfrac{2x+1}{(1-x)^2}$ (C) $2x + 1$

   (D) $\dfrac{3-x}{(1-x)^3}$ (E) none of these

62. $\lim\limits_{h \to 0} \dfrac{(1+h)^6 - 1}{h}$ is

   (A) 0   (B) 1   (C) 6   (D) ∞ (E) nonexistent

63. $\lim\limits_{h \to 0} \dfrac{\sqrt[3]{8+h} - 2}{h}$ is

   (A) 0   (B) $\dfrac{1}{12}$ (C) 1   (D) 192   (E) ∞

64. $\lim\limits_{h \to 0} \dfrac{\ln(e+h) - 1}{h}$ is

   (A) 0   (B) $\dfrac{1}{e}$ (C) 1   (D) e (E) nonexistent

65. $\lim\limits_{x \to 0} \dfrac{\cos x - 1}{x}$ is

   (A) −1   (B) 0   (C) 1   (D) ∞ (E) none of these

66. The function $f(x) = x^{2/3}$ on $[-8, 8]$ does not satisfy the conditions of the Mean Value Theorem because

   (A) $f(0)$ is not defined   (B) $f(x)$ is not continuous on $[-8, 8]$
   (C) $f'(-1)$ does not exist   (D) $f(x)$ is not defined for $x < 0$
   (E) $f'(0)$ does not exist

67. If $f(a) = f(b) = 0$ and $f(x)$ is continuous on $[a, b]$, then

   (A) $f(x)$ must be identically zero
   (B) $f'(x)$ may be different from zero for all x on $[a, b]$
   (C) there exists at least one number c, $a < c < b$, such that $f'(c) = 0$
   (D) $f'(x)$ must exist for every x on (a, b)
   (E) none of the preceding is true

*An asterisk denotes a topic covered only in Calculus BC.

68. If $f(x) = 2x^3 - 6x$, at what point on the interval $0 \leqslant x \leqslant \sqrt{3}$, if any, is the tangent to the curve parallel to the secant line?

(A) 1   (B) –1   (C) $\sqrt{2}$ (D) 0   (E)   nowhere

69. If h is the inverse function of f and if $f(x) = \dfrac{1}{x}$, then $h'(3) =$

(A) –9   (B) $-\dfrac{1}{9}$ (C) $\dfrac{1}{9}$ (D) 3   (E) 9

70. Suppose $y = f(x) = 2x^3 - 3x$. If $h(x)$ is the inverse function of f, then $h'(-1) =$

(A) –1   (B) $\dfrac{1}{5}$ (C) $\dfrac{1}{3}$ (D) 1   (E) 3

71. Suppose $f(1) = 2$, $f'(1) = 3$, and $f'(2) = 4$. Then $(f^{-1})'(2)$

(A) equals $-\dfrac{1}{3}$ (B) equals $-\dfrac{1}{4}$ (C) equals $\dfrac{1}{4}$

(D) equals $\dfrac{1}{3}$ (E)   cannot be determined

72. If $f(x) = x^3 - 3x^2 + 8x + 5$ and $g(x) = f^{-1}(x)$, then $g'(5) =$

(A) 8   (B) $\dfrac{1}{8}$ (C) 1   (D) $\dfrac{1}{53}$ (E) 53

*73. $\displaystyle\lim_{x \to \infty} \dfrac{e^x}{x^{50}}$ equals

(A) 0   (B) 1   (C) $\dfrac{1}{50!}$ (D) ∞ (E)   none of these

74. Suppose $\displaystyle\lim_{x \to 0} \dfrac{g(x) - g(0)}{x} = 1$. It follows necessarily that

(A) g is not defined at $x = 0$
(B) g is not continuous at $x = 0$
(C)   the limit of $g(x)$ as x approaches 0 equals 1
(D) $g'(0) = 1$
(E) $g'(1) = 0$

75. If $\sin(xy) = x$, then $\dfrac{dy}{dx} =$

(A) $\sec(xy)$ (B) $\dfrac{\sec(xy)}{x}$ (C) $\dfrac{\sec(xy) - y}{x}$

(D) $-\dfrac{1 + \sec(xy)}{x}$ (E) $\sec(xy) - 1$

---

*An asterisk denotes a topic covered only in Calculus BC.

Use this graph of $y = f(x)$ for Questions 76 and 77.

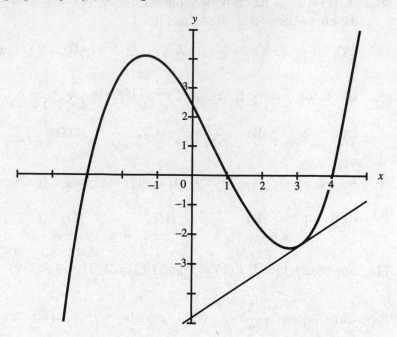

76. $f'(3)$ is most closely approximated by

(A) 0.3   (B) 0.8   (C) 1.5   (D) 1.8   (E) 2

77. The rate of change of $f(x)$ is least at $x \simeq$

(A) –3   (B) –1.3   (C) 0   (D) 0.7   (E) 2.7

Use the following definition of the *symmetric difference quotient for $f'(x_0)$* for Questions 78–81: for small values of h,

$$f'(x_0) = \frac{f(x_0 + h) - f(x_0 - h)}{2h}.$$

78. $f'(x_0)$ equals the exact value of the derivative at $x = x_0$

(A) only when f is linear
(B) whenever f is quadratic
(C) if and only if $f'(x_0)$ exists
(D) whenever $|h| < 0.001$.
(E) none of these

79. To how many places is the symmetric difference accurate when it is used to approximate $f'(0)$ for $f(x) = 4^x$ and $h = 0.08$?

(A) 1   (B) 2   (C) 3   (D) 4   (E) more than 4

80. To how many places is $f'(x_0)$ accurate when it is used to approximate $f'(0)$ for $f(x) = 4^x$ and $h = 0.001$?

(A) 1   (B) 2   (C) 3   (D) 4   (E) more than 4

81. The value of $f'(0)$ obtained using the symmetric difference quotient with $f(x) = |x|$ and $h = 0.001$ is

   (A) –1   (B) 0   (C) ±1   (D) 1   (E) indeterminate

82. If $\dfrac{d}{dx} f(x) = g(x)$ and $h(x) = \sin x$, then $\dfrac{d}{dx} f(h(x))$ equals

   (A) $g(\sin x)$ (B) $\cos x \cdot g(x)$ (C) $g'(x)$
   (D) $\cos x \cdot g(\sin x)$ (E) $\sin x \cdot g(\sin x)$

83. $\displaystyle\lim_{x \to 0} \dfrac{\sin 2x}{x}$ is

   (A) 1   (B) 2   (C) $\dfrac{1}{2}$ (D) 0   (E) ∞

84. $\displaystyle\lim_{x \to 0} \dfrac{\sin 3x}{\sin 4x}$ is

   (A) 1   (B) $\dfrac{4}{3}$ (C) $\dfrac{3}{4}$ (D) 0   (E) nonexistent

85. $\displaystyle\lim_{x \to 0} \dfrac{1 - \cos x}{x}$ is

   (A) nonexistent   (B) 1   (C) 2   (D) ∞ (E) none of these

86. $\displaystyle\lim_{x \to 0} \dfrac{\tan \pi x}{x}$ is

   (A) $\dfrac{1}{\pi}$ (B) 0   (C) 1   (D) π (E) ∞

87. $\displaystyle\lim_{x \to \infty} x^2 \sin \dfrac{1}{x}$

   (A) is 1   (B) is 0   (C) is ∞
   (D) oscillates between –1 and 1   (E) is none of these

88. Let $f(x) = 3^x - x^3$. The tangent to the curve is parallel to the secant through $(0,1)$ and $(3,0)$ for x equal

   (A) only to 0.984   (B) only to 1.244   (C) only to 2.727
   (D) to 0.984 and 2.804   (E) to 1.244 and 2.727

Questions 89–93 are based on the following graph of $f(x)$, sketched on $-6 \leqslant x \leqslant 7$. Assume the horizontal and vertical grid lines are equally spaced at unit intervals.

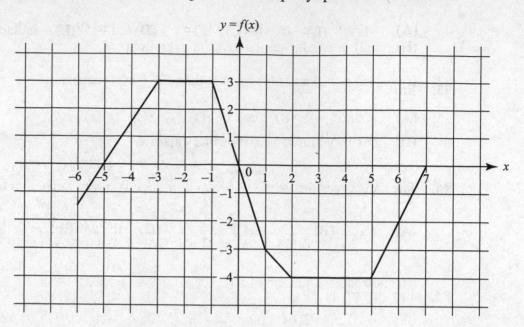

89. On the interval $1 < x < 2$, $f(x)$ equals

(A) $-x - 2$ (B) $-x - 3$ (C) $-x - 4$ (D) $-x + 2$ (E) $x - 2$

90. Over which of the following intervals does $f'(x)$ equal zero?

I. $(-6, -3)$ II. $(-3, -1)$ III. $(2, 5)$

(A) I only (B) II only (C) I and II only
(D) I and III only (E) II and III only

91. How many points of discontinuity does $f'(x)$ have on the interval $-6 < x < 7$?

(A) none (B) 2 (C) 3 (D) 4 (E) 5

92. For $-6 < x - 3$, $f'(x)$ equals

(A) $-\dfrac{3}{2}$ (B) -1 (C) 1 (D) $\dfrac{3}{2}$ (E) 2

93. Which of the following statements about the graph of $f'(x)$ is false?

(A) It consists of six horizontal segments.
(B) It has four jump discontinuities.
(C) $f'(x)$ is discontinuous at each x in the set $\{-3, -1, 1, 2, 5\}$.
(D) $f'(x)$ ranges from -3 to 2.
(E) On the interval $-1 < x < 1$, $f'(x) = -3$.

*94. The graph in the *xy*-plane represented by $x = 3 + 2 \sin t$ and $y = 2 \cos t - 1$, for $-\pi \leqslant t \leqslant \pi$, is

  (A)   a semicircle   (B)   a circle   (C)   an ellipse
  (D)   half of an ellipse   (E)   a hyperbola

95. $\lim\limits_{x \to 0} \dfrac{\sec x - \cos x}{x^2}$ equals

  (A)   0   (B) $\dfrac{1}{2}$ (C)   1   (D)   2   (E)   none of these

96. The table gives the values of a function f that is differentiable on the interval [0,1]:

x	0.10	0.20	0.30	0.40	0.50	0.60
$f(x)$	0.171	0.288	0.357	0.384	0.375	0.336

The best approximation of $f'(0.10)$ according to this table is

  (A)   0.12   (B)   1.08   (C)   1.17   (D)   1.77   (E)   2.88

97. At how many points on the interval [*a,b*] does the function graphed satisfy the Mean Value Theorem?

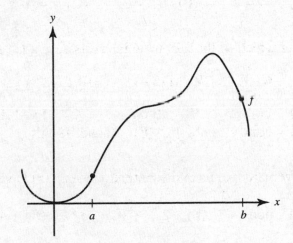

  (A)   none   (B)   1   (C)   2   (D)   3   (E)   4

In each of Questions 98–101 a pair of equations that represents a curve parametrically is given. Choose the alternative that is the derivative $\dfrac{dy}{dx}$.

*98. $x = t - \sin t$ and $y = 1 - \cos t$

  (A) $\dfrac{\sin t}{1 - \cos t}$ (B) $\dfrac{1 - \cos t}{\sin t}$ (C) $\dfrac{\sin t}{\cos t - 1}$

  (D) $\dfrac{1 - x}{y}$ (E) $\dfrac{1 - \cos t}{t - \sin t}$

*An asterisk denotes a topic covered only in Calculus BC.

*99. $x = \cos^3 \theta$ and $y = \sin^3 \theta$

     (A) $\tan^3 \theta$ **(B)** $-\cot \theta$ **(C)** $\cot \theta$ **(D)** $-\tan \theta$ **(E)** $-\tan^2 \theta$

*100. $x = 1 - e^{-t}$ and $y = t + e^{-t}$

     (A) $\dfrac{e^{-t}}{1 - e^{-t}}$ **(B)** $e^{-t} - 1$ **(C)** $e^{t} + 1$ **(D)** $e^{t} - e^{-2t}$ **(E)** $e^{t} - 1$

*101. $x = \dfrac{1}{1-t}$ and $y = 1 - \ln(1 - t)$ $(t < 1)$

     (A) $\dfrac{1}{1-t}$ **(B)** $t - 1$ **(C)** $\dfrac{1}{x}$ **(D)** $\dfrac{(1-t)^2}{t}$ **(E)** $1 + \ln x$

---

*An asterisk denotes a topic covered only in Calculus BC.

Answers for Set 3: Differentiation

1.	C	22.	E	42.	C	62.	C	82.	D
2.	A	23.	C	43.	C	63.	B	83.	B
3.	B	24.	A	44.	A	64.	B	84.	C
4.	B	25.	D	45.	D	65.	B	85.	E
5.	E	26.	E	46.	B	66.	E	86.	D
6.	B	27.	D	47.	E	67.	B	87.	C
7.	B	28.	A	48.	E	68.	A	88.	E
8.	E	29.	A	49.	E	69.	B	89.	A
9.	D	30.	D	50.	A	70.	C	90.	E
10.	C	31.	E	51.	D	71.	D	91.	E
11.	A	32.	B	52.	B	72.	B	92.	D
12.	B	33.	C	53.	C	73.	D	93.	B
13.	D	34.	C	54.	D	74.	D	94.	B
14.	D	35.	E	55.	E	75.	C	95.	C
15.	A	36.	B	56.	D	76.	B	96.	C
16.	E	37.	D	57.	A	77.	D	97.	D
17.	C	38.	C	58.	B	78.	B	98.	A
18.	E	39.	A	59.	E	79.	B	99.	D
19.	A	40.	B	60.	B	80.	E	100.	E
20.	D	41.	C	61.	A	81.	B	101.	C
21.	D								

Many of the explanations provided include intermediate steps that would normally be reached on the way to a final algebraically simplified result. You may not need to reach the final answer.

NOTE: the formulas or rules in parentheses referred to in the explanations are given on pages 47 and 48.

1. C.   By the product rule, (5),

$$y' = (4x+1)[3(1-x)^2(-1)] + (1-x)^3 \cdot 4$$
$$= (1-x)^2(-12x-3+4-4x)$$
$$= (1-x)^2(1-16x).$$

2. A.   By the quotient rule, (6),

$$y' = \frac{(3x+1)(-1)-(2-x)(3)}{(3x+1)^2} = -\frac{7}{(3x+1)^2}.$$

3. B.   Since $y = (3-2x)^{1/2}$, by the power rule, (3),

$$y' = \frac{1}{2}(3-3x)^{-1/2} \cdot (-2) = -\frac{1}{\sqrt{3-2x}}.$$

4. B.   Since $y = 2(5x+1)^{-3}, y' = -6(5x+1)^{-4}(5).$

5. E. $y' = 3\left(\dfrac{2}{3}\right)x^{-1/3} - 4\left(\dfrac{1}{2}\right)x^{-1/2}$

6. B. $(f + 2g)'(3) = f'(3) + 2g'(3) = 4 + 2(-1)$

7. B. $(f \cdot g)'(2) = f(2) \cdot g'(2) + g(2) \cdot f'(2) = 5(-2) + 1(3)$

8. E. $\left(\dfrac{1}{g}\right)'(1) = -1 \cdot \dfrac{1}{[g(1)]^2} \cdot g'(1) = -1 \cdot \dfrac{1}{3^2}(-3)$

9. D. $(\sqrt{f})'(3) = \dfrac{1}{2}[f(3)]^{-1/2} \cdot f'(3) = \dfrac{1}{2}(10^{-1/2}) \cdot 4$

10. C. $\left(\dfrac{f}{g}\right)'(0) = \dfrac{g(0) \cdot f'(0) - f(0) \cdot g'(0)}{[g(0)]^2} = \dfrac{5(1) - 2(-4)}{5^2}$

11. A. $M'(1) = f'(g(1)) \cdot g'(1) = f'(3)g'(1) = 4(-3)$

12. B. $[f(x^3)]' = f'(x^3) \cdot 3x^2$, so $P'(1) = f'(1^3) \cdot 3 \cdot 1^2 = 2 \cdot 3$.

13. D. $f(S(x)) = x$ implies that $f'(S(x) \cdot S'(x) = 1$, so

$$S'(3) = \dfrac{1}{f'(S(3))} = \dfrac{1}{f'(f^{-1}(3))} = \dfrac{1}{f'(1)}.$$

14. D.   Rewrite: $y = 2x^{1/2} - \dfrac{1}{2}x^{1/2}$, so $y' = x^{-1/2} + \dfrac{1}{4}x^{-3/2}$.

15. A.   Rewrite: $y = (x^2 + 2x - 1)^{1/2}$. (Use rule (3).)

16. E.   Use the quotient rule:

$$y = \dfrac{\sqrt{1-x^2} \cdot 1 - x \cdot \dfrac{1 \cdot -2x}{2\sqrt{1-x^2}}}{1-x^2} = \dfrac{\dfrac{1-x^2+x^2}{\sqrt{1-x^2}}}{1-x^2}$$

$$= \dfrac{1}{(1-x^2)^{3/2}}.$$

17. C.   Since

$$y = \ln e^x - \ln(e^x - 1)$$
$$= x - \ln(e^x - 1),$$

then

$$y' = 1 - \dfrac{e^x}{e^x - 1} = \dfrac{e^x - 1 - e^x}{e^x - 1} = -\dfrac{1}{e^x - 1}.$$

18. E.   Use formula (18): $y' = \dfrac{\dfrac{1}{2}}{1 + \dfrac{x^2}{4}}$.

19. **A.** Use formulas (13), (11), and (9):

$$y' = \frac{\sec x \tan x + \sec^2 x}{\sec x + \tan x} = \frac{\sec x(\tan x + \sec x)}{\sec x + \tan x}.$$

20. **D.** By the quotient rule,

$$y' = \frac{(e^x + e^{-x})(e^x + e^{-x}) - (e^x - e^{-x})(e^x - e^{-x})}{(e^x + e^{-x})^2}$$

$$= \frac{(e^{2x} + 2 + e^{-2x}) - (e^{2x} - 2 + e^{-2x})}{(e^x + e^{-x})^2} = \frac{4}{(e^x + e^{-x})^2}.$$

21. **D.** Since $y = \ln x + \dfrac{1}{2} \ln (x^2 + 1)$,

$$y' = \frac{1}{x} + \frac{1}{2} \cdot \frac{2x}{x^2 + 1} = \frac{2x^2 + 1}{x(x^2 + 1)}.$$

22. **E.** Since $g'(a)$ exists, g is differentiable and thus continuous; $g'(a) > 0$.

23. **C.** Near a vertical asymptote the slopes must approach $\pm\infty$.

24. **A.** There is only one horizontal tangent.

25. **D.** Use the symmetric difference quotient; then

$$f'(1.5) = \frac{f(1.6) - f(1.4)}{1.6 - 1.4} = \frac{8}{0.2}.$$

26. **E.** Since the water level rises more slowly as the cone fills, the rate of depth change is decreasing, as in (C) and (E). However, at every instant the portion of the cone containing water is similar to the entire cone; the volume is proportional to the cube of the depth of the water. The rate of change of depth (the derivative) is therefore not linear.

27. **D.** $y' = x^2 \cos \dfrac{1}{x} \left(-\dfrac{1}{x^2} \right) + \sin \dfrac{1}{x} (2x).$

28. **A.** Since $y = \dfrac{1}{2} \csc 2x$, $y' = \dfrac{1}{2} (-\csc 2x \cot 2x \cdot 2).$

29. **A.** $y' = e^{-x}(-2\sin 2x) + \cos 2x(-e^{-x}).$

30. **D.** Use formulas (3) and (11):

$$y' = 2\sec\sqrt{x} \cdot \sec\sqrt{x} \tan\sqrt{x} \cdot \left(\frac{1}{2\sqrt{x}} \right).$$

31. **E.** $y' = \dfrac{x(3 \ln^2 x)}{x} + \ln^3 x.$ The correct answer is $3 \ln^2 x + \ln^3 x.$

32. **B.** $y' = \dfrac{(1 - x^2)(2x) - (1 + x^2)(-2x)}{(1 - x^2)^2}.$

33. **C.** $y' = \dfrac{1}{\sqrt{1 - x^2}} - \dfrac{1 \cdot (-2x)}{2\sqrt{1 - x^2}}.$

34. C.   The only horizontal tangent is at $x = 4$. Note that $f'(1)$ does not exist.

35. E.   The graph has corners at $x = 1$ and $x = 2$; the tangent line is vertical at $x = 6$.

36. B.   Consider triangle ABC: $AB = 1$; radius $AC = 2$; thus, $BC = \sqrt{3}$ and AC has $m = -\sqrt{3}$. The tangent line is perpendicular to the radius.

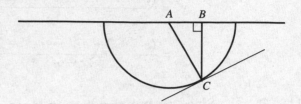

37. D.   The graph of $y = x + \cos x$ is shown in window $[-5,5] \times [-6,6]$. The average rate of change is represented by the slope of secant segment $\overline{AB}$. There appear to be 3 points at which tangent lines are parallel to $\overline{AB}$.

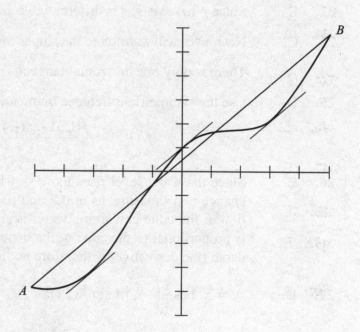

38. C. $f'(2) \simeq \dfrac{f(2) - f(1.98)}{2 - 1.98} = \dfrac{4.00 - 4.10}{0.02}$

39. A.   Since an estimate of the answer for Question 38 is $f'(2) \approx -5$, then

$$\left(f^{-1}\right)'(4) = \frac{1}{f'(2)} \approx \frac{1}{-5} = -0.2.$$

40. B.   When $x = 3$ on g^{-1}, $y = 3$ on the original half-parabola. $3 = x^2 - 8x + 10$ at $x = 1$ (and at $x = 7$, but that value is not in the given domain).

$$g'(3) = \frac{1}{y'(1)} = \frac{1}{2x - 8}\bigg|_{x=1} = -\frac{1}{6}.$$

41. C. $\dfrac{5^{2.03} - 5^{1.97}}{2.03 - 1.97} \approx 40.25158$

42. C. The diagrams show secant lines (whose slope is the difference quotient) with greater slopes than the tangent line. In both cases, f is concave upward.

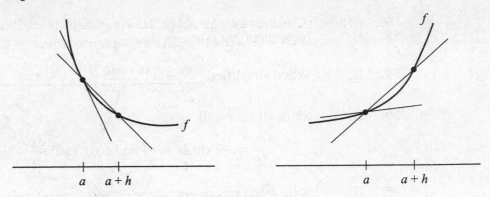

43. C. Let y' be $\dfrac{dy}{dx}$; then $3x^2 - (xy' + y) + 3y^2 y' = 0$; $y'(3y^2 - x) = y - 3x^2$.

44. A. $1 - \sin(x + y)(1 + y') = 0$; $\dfrac{1 - \sin(x + y)}{\sin(x + y)} = y'$.

45. D. $\cos x + \sin y \cdot y' = 0$; $y' = -\dfrac{\cos x}{\sin y}$.

46. B. $6x - 2(xy' + y) + 10yy' = 0$; $y'(10y - 2x) = 2y - 6x$.

47. E. $\dfrac{dy}{dx} = \dfrac{4t^3 - 6t^2}{2t} = 2t^2 - 3t(t \neq 0)$; $\dfrac{d^2 y}{dx^2} = \dfrac{4t - 3}{2t}$. Replace t by 1.

48. E. $f'(x) = 4x^3 - 12x^2 + 8x = 4x(x - 1)(x - 2)$.

49. E. $f'(x) = 8x^{-1/2}; f''(x) = -4x^{-3/2} = -\dfrac{4}{x^{3/2}}$; $f''(4) = -\dfrac{4}{8}$.

50. A. $f(x) = 3 \ln x; f'(x) = \dfrac{3}{x}$; $f''(x) = \dfrac{-3}{x^2}$. Replace x by 3.

51. D. $2x + 2yy' = 0$; $y' = -\dfrac{x}{y}$; $y'' = -\dfrac{y - xy'}{y^2}$. At $(0,5)$, $y'' = -\dfrac{5 - 0}{25}$.

52. B. $y' = ac \cos ct - bc \sin ct$;
$y'' = -ac^2 \sin ct - bc^2 \cos ct$.

53. C. $|(f \circ g)'$ at $x = 3$ equals $f'(g(3)) \cdot g'(3)$ equals $\cos u$ (at $u = 0$) times $2x$
(at $x = 3$) = $1 \cdot 6 = 6$.

54. D. $f'(1) \approx \dfrac{5^{1.002} - 5^1}{0.002} = \dfrac{5.016 - 5}{0.002}$.

55. E. Here $f'(x)$ equals $\dfrac{-x^2 - x - 1}{(x - 1)^3}$.

56. D. $y' = e^x \cdot 1 + e^x(x - 1) = xe^x$
$y'' = xe^x + e^x$ and $y''(0) = 0 \cdot 1 + 1 = 1$.

57.   A. $\dfrac{dy^2}{dx^2} = \dfrac{\frac{dy^2}{dx}}{\frac{dx^2}{dx}}$. Since $y^2 = x^2 + 1,\ \dfrac{dy^2}{dx^2} = \dfrac{2x}{2x}$.

58.   B.   Note that $f(g(x)) = \dfrac{1}{x+1}$.

59.   E.   When simplified, $\dfrac{dy}{dx} = \dfrac{\cos\theta + \sin\theta}{\cos\theta - \sin\theta}$.

60.   B.   Since (if $\sin t \neq 0$)

$$\dfrac{dy}{dt} = -2\sin 2t = -4\sin t\cos t \text{ and } \dfrac{dx}{dt} = -\sin t,$$

then $\dfrac{dy}{dx} = 4\cos t$. Thus:

$$\dfrac{d^2y}{dx^2} = -\dfrac{4\sin t}{-\sin t}.$$

61.   A. $\dfrac{dy}{d\left(\frac{1}{1-x}\right)} = \dfrac{\frac{dy}{dx}}{\frac{d\left(\frac{1}{1-x}\right)}{dx}} = \dfrac{2x+1}{\frac{1}{(1-x)^2}}$.

NOTE: Since each of the limits in Questions 62 through 65 yields an indeterminate form of the type $\dfrac{0}{0}$, we can apply L'Hôpital's rule in each case, getting identical answers.

62.   C.   The given limit is the derivative of $f(x) = x^6$ at $x = 1$.

63.   B.   The given limit is the definition for $f'(8)$, where $f(x) = \sqrt[3]{x}$;

$$f'(x) = \dfrac{1}{3x^{2/3}}.$$

64.   B.   The given limit is $f'(e)$, where $f(x) = \ln x$.

65.   B.   The given limit is the derivative of $f(x) = \cos x$ at $x = 0$; $f'(x) = -\sin x$.

66.   E.   Since $f'(x) = \dfrac{2}{3x^{1/3}}$, $f'(0)$ is not defined; $f'(x)$ must be defined on $(-8,8)$.

67.   B.   Sketch the graph of $f(x) = 1 - |x|$; note that $f(-1) = f(1) = 0$ and that f is continuous on $[-1,1]$. Only (B) holds.

68.   A.   Note that $f(0) = f(\sqrt{3}) = 0$ and that $f'(x)$ exists on the given interval. By the MVT, there is a number c in the interval such that $f'(c) = 0$. If $c = 1$, then $6c^2 - 6 = 0$. (-1 is not in the interval.)

69.   B.   Since the inverse, h, of $f(x) = \dfrac{1}{x}$ is $h(x) = \dfrac{1}{x}$, then $h'(x) = -\dfrac{1}{x^2}$. Replace x by 3.

70. C.   Since $f'(x) = 6x^2 - 3$, therefore $h'(x) = \dfrac{1}{6x^2-3}$; also, $f(x)$, or $2x^3 - 3x$,

equals -1, by observation, for $x = 1$. So $h'(-1)$ or $\dfrac{1}{6x^2-3}$ (when $x = 1$)

equals $\dfrac{1}{6-3} = \dfrac{1}{3}$.

71. D. $\left(f^{-1}\right)'(2) = \dfrac{1}{f'(1)} = \dfrac{1}{3}.$

72. B.   Since $f(0) = 5$, $g'(5) = \dfrac{1}{f'(0)} = \dfrac{1}{3x^2-6x+8}\Big|_{x=0} = \dfrac{1}{8}.$

73. D.   After 50(!) applications of L'Hôpital's rule we get $\lim\limits_{x\to\infty} \dfrac{e^x}{50!}$, which "equals" ∞. A perfunctory examination of the limit, however, shows immediately that the answer is ∞. In fact, $\lim\limits_{x\to\infty} \dfrac{e^x}{x^n}$ for any positive integer n, no matter how large, is ∞.

74. D.   The given limit is the derivative of $g(x)$ at $x = 0$.

75. C. $\cos(xy)(xy' + y) = 1;\ x\cos(xy)y' = 1 - y\cos(xy);$

$$y' = \frac{1 - y\cos(xy)}{x\cos(xy)}.$$

76. B.   The tangent line appears to contain $(3,-2.6)$ and $(4,-1.8)$.

77. D. $f'(x)$ is least at the point of inflection of the curve, at about 0.7.

78. B.   Note that $\dfrac{(x+h)^2-(x-h)^2}{2h} = 2x$, the derivative of $f(x) = x^2$. You should be able to confirm that the symmetric difference quotient works for the general quadratic $g(x) = ax^2 + bx + c$.

79. B.   By calculator, $f'(0) = 1.386294805$ and $\dfrac{4^{0.08}-4^{-0.08}}{0.16} = 1.3891\ldots.$

80. E.   Now $\dfrac{4^{0.001}-4^{-0.001}}{0.002} = 1.386294805.$

81. B.   Note that any line determined by two points equidistant from the origin will necessarily be horizontal.

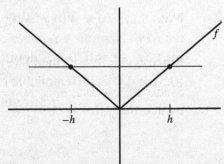

82. D.   Note that $\dfrac{d}{dx}\, f(h(x)) = f'(h(x))\cdot h'(x) = g(h(x))\cdot h'(x) = g(\sin x)\cdot \cos x.$

NOTE: In Questions 83 through 87 the limits are all indeterminate forms of the type $\dfrac{0}{0}$. We have therefore applied L'Hôpital's rule in each one. The indeterminacy can also be resolved by introducing $\dfrac{\sin a}{a}$, which approaches 1 as a approaches 0. The latter technique is presented in square brackets.

83. B. $\displaystyle\lim_{x\to 0} \frac{\sin 2x}{x} = \lim_{x\to 0} \frac{2\cos 2x}{1} = \frac{2\cdot 1}{1} = 2.$

[Using $\sin 2x = 2\sin x \cos x$ yields $\displaystyle\lim_{x\to 0} 2\left(\frac{\sin x}{x}\right)\cos x = 2\cdot 1\cdot 1 = 2.$]

84. C. $\displaystyle\lim_{x\to 0} \frac{\sin 3x}{\sin 4x} = \lim_{x\to 0} \frac{3\cos 3x}{4\cos 4x} = \frac{3\cdot 1}{4\cdot 1} = \frac{3}{4}.$

[We rewrite $\dfrac{\sin 3x}{\sin 4x}$ as $\dfrac{\sin 3x}{3x}\cdot\dfrac{4x}{\sin 4x}\cdot\dfrac{3}{4}$. As $x \to 0$, so do $3x$ and $4x$; the fraction approaches $1\cdot 1\cdot \dfrac{3}{4}$.]

85. E. $\displaystyle\lim_{x\to 0} \frac{1-\cos x}{x} = \lim_{x\to 0} \frac{\sin x}{1} = 0.$

[We can replace $1 - \cos x$ by $2\sin^2 \dfrac{x}{2}$, getting

$$\lim_{x\to 0} \frac{2\sin^2 \dfrac{x}{2}}{x} = \lim_{x\to 0} \frac{\sin^2 \dfrac{x}{2}}{\dfrac{x}{2}} = \lim_{x\to 0} \sin\frac{x}{2}\left(\frac{\sin\dfrac{x}{2}}{\dfrac{x}{2}}\right) = 0\cdot 1.]$$

86. D. $\displaystyle\lim_{x\to 0} \frac{\tan \pi x}{x} = \lim_{x\to 0} \frac{(\sec^2 \pi x)\cdot \pi}{1} = 1\cdot \pi = \pi.$

[$\dfrac{\tan \pi x}{x} = \dfrac{\sin \pi x}{x\cos \pi x} = \pi\cdot\dfrac{\sin \pi x}{\pi x}\cdot\dfrac{1}{\cos x}$; as x (or πx) approaches 0, the original fraction approaches $\pi\cdot 1\cdot \dfrac{1}{1} = \pi.]$

87. C.   The limit is easiest to obtain here if we rewrite:

$$\lim_{x\to\infty} x^2 \sin\frac{1}{x} = \lim_{x\to\infty} x\frac{\sin(1/x)}{(1/x)} = \infty\cdot 1 = \infty.$$

88. E.   Since $f(x) = 3^x - x^3$, then $f'(x) = 3^x \ln 3 - 3x^2$. Furthermore, f is continuous on $[0,3]$ and f' is differentiable on $(0,3)$, so the MVT applies. We therefore seek c such that $f'(c) = \dfrac{f(3)-f(0)}{3} = -\dfrac{1}{3}$. Using a graphing calculator, we key in

$\texttt{Y}_1 = 3^x \ln 3 - 3\texttt{X}^2 + \dfrac{1}{3}$. With solutions estimated at $x = 1$ and 2.5, the calculator shows that c may be either 1.244 or 2.727. These are the x-coordinates of points on the graph of $f(x)$ at which the tangents are parallel to the secant through points $(0,1)$ and $(3,0)$ on the curve.

89. A.   The line segment passes through $(1,-3)$ and $(2,-4)$.

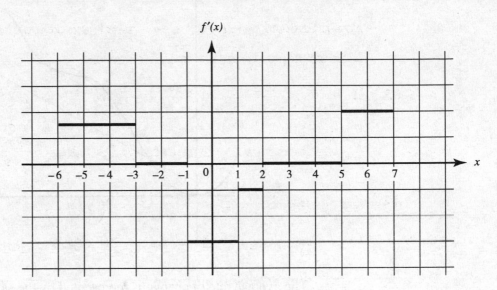

Use the graph of $f'(x)$, shown above, for Questions 90 through 93.

90. E. $f'(x) = 0$ when the slope of $f(x)$ is 0; that is, when the graph of f is a horizontal segment.

91. E.   The graph of $f'(x)$ jumps at each corner of the graph of $f(x)$, namely, at x equal to $-3, -1, 1, 2,$ and 5.

92. D.   On the interval $(-6,-3)$, $f(x) = \dfrac{3}{2}(x + 5)$.

93. B.   Verify that all choices but (B) are true. The graph of $f'(x)$ has five (not four) jump discontinuities.

94. B.   Since $x - 3 = 2 \sin t$ and $y + 1 = 2 \cos t$, we have

$$(x - 3)^2 + (y + 1)^2 = 4.$$

This is the equation of a circle with center at $(3,-1)$ and radius 2. In the domain given, $-\pi \leq t \leq \pi$, the entire circle is traced by a particle moving counterclockwise, starting from and returning to $(3, -3)$.

95. C.   Use L'Hôpital's rule; then

$$\lim_{x \to 0} \frac{\sec x - \cos x}{x^2} = \lim_{x \to 0} \frac{\sec x \tan x + \sin x}{2x}$$

$$= \lim_{x \to 0} \frac{\sec^3 x + \sec x \tan^2 x + \cos x}{2} = \frac{1 + 1 \cdot 0 + 1}{2} = 1.$$

96. C.   The best approximation to $f'(0.10)$ is $\dfrac{f(0.20) - f(0.10)}{0.20 - 0.10}$.

97. D.

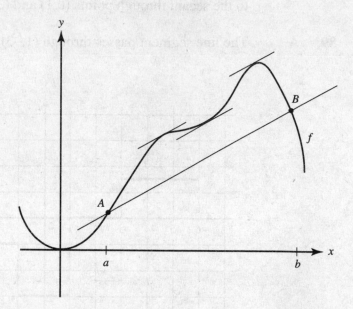

The average rate of change is represented by the slope of secant segment $\overline{AB}$.
There appear to be 3 points at which the tangent lines are parallel to $\overline{AB}$.

98. A. $\dfrac{dy}{dx} = \dfrac{\frac{dy}{dt}}{\frac{dx}{dt}} = \dfrac{\sin t}{1 - \cos t}$.

99. D. $\dfrac{dy}{dx} = \dfrac{\frac{dy}{d\theta}}{\frac{dx}{d\theta}} = \dfrac{3 \sin^2 \theta \cos \theta}{-3 \cos^2 \theta \sin \theta}$.

100. E. $\dfrac{dy}{dx} = \dfrac{\frac{dy}{dt}}{\frac{dx}{dt}} = \dfrac{1 - e^{-t}}{e^{-t}} = e^t - 1$.

101. C.   Since $\dfrac{dy}{dt} = \dfrac{1}{1-t}$ and $\dfrac{dx}{dt} = \dfrac{1}{(1-t)^2}$, then

$$\dfrac{dy}{dx} = 1 - t = \dfrac{1}{x}.$$

Chapter 4

Applications of Differential Calculus

Review of Definitions and Methods

A. Slope; Critical Points _____

If the derivative of $y = f(x)$ exists at $P(x_1, y_1)$, then the *slope* of the curve at P (which is defined to be the slope of the tangent to the curve at P) is $f'(x_1)$, the derivative of $f(x)$ at $x = x_1$.

Any c in the domain of f such that either $f'(c) = 0$ or $f'(c)$ is undefined is called a *critical point* or *critical value* of f. If f has a derivative everywhere, we find the critical points by solving the equation $f'(x) = 0$.

Example 1. If $f(x) = 4x^3 - 6x^2 - 8$, then

$$f'(x) = 12x^2 - 12x = 12x(x - 1),$$

which equals zero if x is 0 or 1. Thus, 0 and 1 are critical points.

Example 2. If $f(x) = 3x^3 + 2x$, then

$$f'(x) = 9x^2 + 2.$$

Since $f'(x)$ never equals zero (indeed, it is always positive), f has no critical values.

Example 3.   If $f(x) = (x - 1)^{1/3}$, then

$$f'(x) = \frac{1}{3(x-1)^{2/3}}.$$

Although f' is never zero, $x = 1$ is a critical value of f because f' does not exist at $x = 1$.

Average and Instantaneous Rate of Change.

Both average and instantaneous rates of change were defined in Chapter 3. If as x varies from a to $a + h$, the function f varies from $f(a)$ to $f(a + h)$, then we know that the difference quotient

$$\frac{f(a+h) - f(a)}{h}$$

is the average rate of change of f over the interval from a to $a + h$.

Thus, the *average velocity* of a moving object over some time interval is the change in distance divided by the change in time, the average rate of growth of a colony of fruit flies over some interval of time is the change in size of the colony divided by the time elapsed, the average rate of change in the profit of a company on some gadget with respect to production is the change in profit divided by the change in the number of gadgets produced.

The (instantaneous) rate of change of f at a, or the derivative of f at a, is the limit of the average rate of change as $h \to 0$:

$$f'(a) = \lim_{h \to 0} \frac{f(a+h) - f(a)}{h}.$$

On the graph of $y = f(x)$, the rate at which the y-coordinate changes with respect to the x-coordinate is $f'(x)$, the slope of the curve. The rate at which $s(t)$, the distance traveled by a particle in t seconds, changes with respect to time is $s'(t)$, the velocity of the particle; the rate at which a manufacturer's profit $P(x)$ changes relative to the production level x is $P'(x)$.

Example 4.   Let $G = 400(15 - t)^2$ be the number of gallons of water in a cistern t minutes after an outlet pipe is opened. Find the average rate of drainage during the first 5 minutes and the rate at which the water is running out at the end of 5 minutes.

The average rate of change during the first 5 min equals

$$\frac{G(5) - G(0)}{5} = \frac{400 \cdot 100 - 400 \cdot 225}{5} = -10,000 \text{ gal / min.}$$

The average rate of drainage during the first 5 min is 10,000 gal/min.

The instantaneous rate of change at $t = 5$ is $G'(5)$. Since

$$G'(t) = -800(15 - t),$$

$G'(5) = -800(10) = -8000$ gal/min. Thus the rate of drainage at the end of 5 min is 8000 gal/min.

B.   Tangents and Normals _____

The *equation of the tangent* to the curve $y = f(x)$ at point $P(x_1, y_1)$ is

$$y - y_1 = f'(x_1)(x - x_1).$$

The line through P that is perpendicular to the tangent, called the *normal* to the curve at P, has slope $-\dfrac{1}{f'(x_1)}$. Its equation is

$$y - y_1 = -\frac{1}{f'(x_1)}(x - x_1).$$

If the tangent to a curve is horizontal at a point, then the derivative at the point is 0. If the tangent is vertical at a point, then the derivative does not exist at the point.

*Tangents to Parametrically Defined Curves.

If the curve is defined parametrically, say in terms of t (as in Chapter 3, page 56), then we obtain the slope at any point from the parametric equations. We then evaluate the slope and the x- and y-coordinates by replacing t by the value specified in the question (see Example 9, page 96).

Example 5.   Find the equations of the tangent and normal to the curve of $f(x) = x^3 - 3x^2$ at the point $(1, -2)$.

Since $f'(x) = 3x^2 - 6x$ and $f'(1) = -3$, the equation of the tangent is

$$y + 2 = -3(x - 1) \qquad \text{or} \qquad y + 3x = 1,$$

and the equation of the normal is

$$y + 2 = \frac{1}{3}(x - 1) \qquad \text{or} \qquad 3y - x = -7.$$

Example 6.   Find the equation of the tangent to the curve of $x^2 y - x = y^3 - 8$ at the point where $x = 0$.

Here we differentiate implicitly to get $\dfrac{dy}{dx} = \dfrac{1 - 2xy}{x^2 - 3y^2}$. Since $y = 2$ when $x = 0$ and the slope at this point is $\dfrac{1 - 0}{0 - 12} = -\dfrac{1}{12}$, the equation of the tangent is

$$y - 2 = -\frac{1}{12}x \qquad \text{or} \qquad 12y + x = 24.$$

Example 7.   Find the coordinates of any point on the curve of $y^2 - 4xy = x^2 + 5$ for which the tangent is horizontal.

Since $\dfrac{dy}{dx} = \dfrac{x + 2y}{y - 2x}$ and the tangent is horizontal when $\dfrac{dy}{dx} = 0$, then $x = -2y$. If we substitute this in the equation of the curve, we get

$$y^2 - 4y(-2y) = 4y^2 + 5$$
$$5y^2 = 5.$$

Thus $y = \pm 1$ and $x = \pm 2$. The points, then, are $(2, -1)$ and $(-2, 1)$.

*An asterisk denotes a topic covered only in Calculus BC.

Example 8. Find the abscissa of any point on the curve of $y = \sin^2(x + 1)$ for which the tangent is parallel to the line $3x - 3y - 5 = 0$.

Since $\dfrac{dy}{dx} = 2\sin(x + 1)\cos(x + 1) = \sin 2(x + 1)$ and since the given line has slope 1, we seek x such that $\sin 2(x + 1) = 1$. Then

$$2(x + 1) = \frac{\pi}{2} + 2n\pi \qquad (n \text{ an integer})$$

or

$$(x + 1) = \frac{\pi}{4} + n\pi \qquad \text{and} \qquad x = \frac{\pi}{4} + n\pi - 1.$$

*Example 9. Find the equation of the tangent to $x = \cos t$, $y = 2\sin^2 t$ at the point where $t = \dfrac{\pi}{3}$.

Since $\dfrac{dx}{dt} = -\sin t$ and $\dfrac{dy}{dt} = 4\sin t \cos t$, we see that

$$\frac{dy}{dx} = \frac{4\sin t \cos t}{-\sin t} = -4\cos t .$$

At $t = \dfrac{\pi}{3}$, $x = \dfrac{1}{2}$, $y = 2\left(\dfrac{\sqrt{3}}{2}\right)^2 = \dfrac{3}{2}$, and $\dfrac{dy}{dx} = -2$. The equation of the tangent is

$$y - \frac{3}{2} = -2\left(x - \frac{1}{2}\right) \qquad \text{or} \qquad 4x + 2y = 5.$$

C.   Increasing and Decreasing Functions _____

Case I. Functions with Continuous Derivatives.

A function $y = f(x)$ is said to be $\begin{smallmatrix}increasing\\decreasing\end{smallmatrix}$ at $P(x_1, y_1)$ if its derivative $f'(x_1)$, the slope at P, is $\begin{smallmatrix}positive\\negative\end{smallmatrix}$. To find intervals over which $f(x)$ $\begin{smallmatrix}increases\\decreases\end{smallmatrix}$, that is, over which the curve $\begin{smallmatrix}rises\\falls\end{smallmatrix}$, compute $f'(x)$ and determine where it is $\begin{smallmatrix}positive\\negative\end{smallmatrix}$.

Example 10. If $f(x) = x^4 - 4x^3 + 4x^2$, then

$$f'(x) = 4x^3 - 12x^2 + 8x = 4x(x^2 - 3x + 2) = 4x(x - 1)(x - 2).$$

Since this derivative changes sign only at $x = 0$, 1, or 2, it has a constant sign in each of the intervals between these numbers. Thus:

$$\begin{array}{lll}
\text{if} & x < 0, & \text{then} & f'(x) < 0 \text{ and } f \text{ is decreasing;} \\
& 0 < x < 1, & & f'(x) > 0 \text{ and } f \text{ is increasing;} \\
& 1 < x < 2, & & f'(x) < 0 \text{ and } f \text{ is decreasing;} \\
& 2 < x, & & f'(x) > 0 \text{ and } f \text{ is increasing.}
\end{array}$$

Case II. Functions Whose Derivatives Have Discontinuities.

Here we proceed as in Case I, but also consider intervals bounded by any points of discontinuity of f or f'.

Example 11. If $f(x) = \dfrac{1}{x+1}$, then

$$f'(x) = -\frac{1}{(x+1)^2}.$$

We note that neither f nor f' is defined at $x = -1$; furthermore, $f'(x)$ never equals zero. We need therefore examine only the signs of $f'(x)$ when $x < -1$ and when $x > -1$.

When $x < -1$, $f'(x) < 0$; when $x > -1$, $f'(x) < 0$. So f decreases on both intervals. The curve is a hyperbola whose center is at the point $(-1, 0)$.

D. Maximum, Minimum, and Inflection Points: Definitions

The curve of $y = f(x)$ has a *local* (or *relative*) $\begin{smallmatrix}maximum\\minimum\end{smallmatrix}$ at a point where $x = c$ if $\begin{array}{l} f(c) \geqslant f(x) \\ f(c) \leqslant f(x) \end{array}$ for all x in the immediate neighborhood of c. If a curve has a local $\begin{smallmatrix}maximum\\minimum\end{smallmatrix}$ at $x = c$, then the curve changes from $\begin{smallmatrix}rising\\falling\end{smallmatrix}$ to $\begin{smallmatrix}falling\\rising\end{smallmatrix}$ as x increases through c. If a function is differentiable on the closed interval $[a, b]$ and has a local maximum or minimum at $x = c$ $(a < c < b)$, then $f'(c) = 0$. The converse of this statement is not true.

If $f(c)$ is either a local maximum or a local minimum, then $f(c)$ is called a *local extreme value* or *local extremum*. (The plural of *extremum* is *extrema*.)

The *global* or *absolute* $\begin{smallmatrix}maximum\\minimum\end{smallmatrix}$ of a function on $[a, b]$ occurs at $x = c$ if $\begin{array}{l} f(c) \geqslant f(x) \\ f(c) \leqslant f(x) \end{array}$ for all x on $[a, b]$.

We call $f(c)$ an *extreme value of f on $[a, b]$* if it is either the global maximum or the global minimum of f on $[a, b]$.

A curve is said to be *concave* $\begin{smallmatrix}upward\\downward\end{smallmatrix}$ at a point $P(x_1, y_1)$ if the curve lies $\begin{smallmatrix}above\\below\end{smallmatrix}$ its tangent. If $\begin{array}{l} y'' > 0 \\ y'' < 0 \end{array}$ at P, the curve is concave $\begin{smallmatrix}up\\down\end{smallmatrix}$. In Figure N4–1, the curves sketched in (*a*) and (*b*) are concave downward at *P* while in (*c*) and (*d*) they are concave upward at *P*.

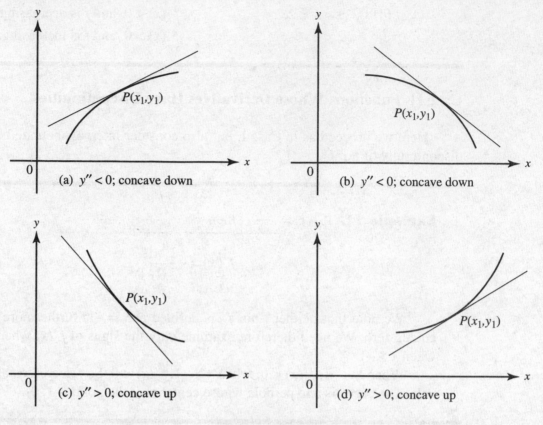

(a) $y'' < 0$; concave down

(b) $y'' < 0$; concave down

(c) $y'' > 0$; concave up

(d) $y'' > 0$; concave up

FIGURE N4–1

A *point of inflection* is a point where the curve changes its concavity from upward to downward or from downward to upward. See §I, page 110, for a table relating a function and its derivatives. It tells how to graph the derivatives of *f*, given the graph of *f*. On pages 112 and 197 we graph *f*, given the graph of *f'*.

E.   Maximum, Minimum, and Inflection Points: Curve Sketching

Case I. Functions That Are Everywhere Differentiable.

The following procedure is suggested to determine any maximum, minimum, or inflection point of a curve and to sketch the curve.

(1)   Find y' and y''.

(2)   Find all critical points of *y*, that is, all *x* for which $y' = 0$. At each of these *x*'s the tangent to the curve is horizontal.

(3)   Let *c* be a number for which y' is 0; investigate the sign of y'' at *c*. If $y''(c) > 0$, the curve is concave up and *c* yields a local minimum; if $y''(c) < 0$, the curve is concave down and *c* yields a local maximum. This procedure is known as the *second-derivative test* (for concavity). See Figure N4–2. If $y''(c) = 0$, the second-derivative test fails and we must use the test in step (4) below.

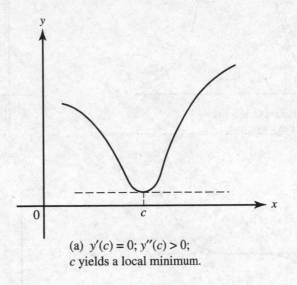

(a) $y'(c) = 0$; $y''(c) > 0$;
c yields a local minimum.

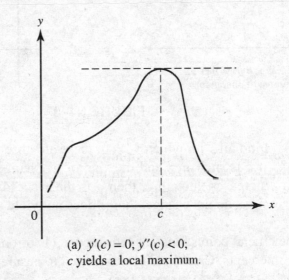

(a) $y'(c) = 0$; $y''(c) < 0$;
c yields a local maximum.

FIGURE N4–2

(4)   If $y'(c) = 0$ and $y''(c) = 0$, investigate the signs of y' as x increases through c. If $y'(x) > 0$ for x's (just) less than c but $y'(x) < 0$ for x's (just) greater than c, then the situation is that indicated in Figure N4–3a, where the tangent lines have been sketched as x increases through c; here c yields a local maximum. If the situation is reversed and the sign of y' changes from $-$ to $+$ as x increases through c, then c yields a local minimum. Figure N4–3b shows this case. The schematic sign pattern of y', $+ 0 -$ or $- 0 +$, describes each situation completely. If y' does not change sign as x increases through c, then c yields neither a local maximum nor a local minimum. Two examples of this appear in Figures N4–3c and N4–3d.

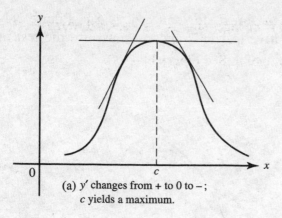

(a) y' changes from + to 0 to −;
c yields a maximum.

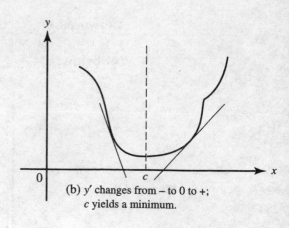

(b) y' changes from − to 0 to +;
c yields a minimum.

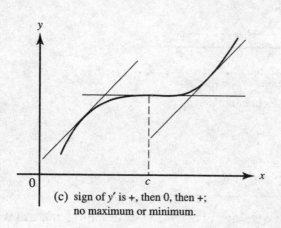

(c) sign of y' is +, then 0, then +;
no maximum or minimum.

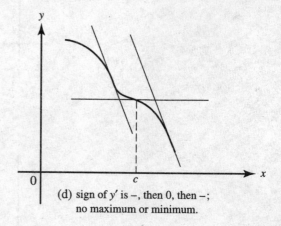

(d) sign of y' is −, then 0, then −;
no maximum or minimum.

FIGURE N4–3

(5) Find all x for which $y'' = 0$; these are abscissas of possible points of inflection. If c is such an x and the sign of y'' changes (from + to − or from − to +) as x increases through c, then c is the x-coordinate of a point of inflection. If the signs do not change, then c does not yield a point of inflection.

The crucial points found as indicated in (1) through (5) above should be plotted along with the intercepts. Care should be exercised to ensure that the tangent to the curve is horizontal whenever $\dfrac{dy}{dx} = 0$ and that the curve has the proper concavity.

Example 12. Find any maximum, minimum, or inflection points of $f(x) = x^3 - 5x^2 + 3x + 6$, and sketch the curve.
Steps:

(1) Here $f'(x) = 3x^2 - 10x + 3$ and $f''(x) = 6x - 10$.

(2) $f'(x) = (3x - 1)(x - 3)$, which is zero when $x = \dfrac{1}{3}$ or 3.

(3) Since $f''\left(\dfrac{1}{3}\right) < 0$, we know that the point $\left(\dfrac{1}{3}, f\left(\dfrac{1}{3}\right)\right)$ is a local maximum;

since $f''(3) > 0$, the point $(3, f(3))$ is a local minimum. Thus, $\left(\dfrac{1}{3}, \dfrac{175}{27}\right)$ is a local maximum and $(3, -3)$ a local minimum.

(4) is unnecessary for this problem.

(5) $f''(x) = 0$ when $x = \dfrac{5}{3}$, and f'' changes sign as x increases through $\dfrac{5}{3}$,

so this x does yield an inflection point. See Figure N4–4.

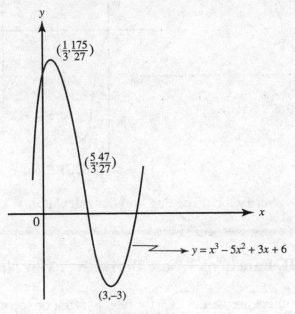

$$\left(\tfrac{1}{3}, \tfrac{175}{27}\right)$$

$$\left(\tfrac{5}{3}, \tfrac{47}{27}\right)$$

$$y = x^3 - 5x^2 + 3x + 6$$

$$(3, -3)$$

FIGURE N4–4

Verify the graph and information obtained above on your graphing calculator.

Example 13.   If we apply the procedure to $f(x) = x^4 - 4x^3$, we see that

(1) $f'(x) = 4x^3 - 12x^2$ and $f''(x) = 12x^2 - 24x$.

(2) $f'(x) = 4x^2(x - 3)$, which is zero when $x = 0$ or $x = 3$.

(3)   Since $f''(x) = 12x(x - 2)$ and $f''(3) > 0$, the point $(3, -27)$ is a relative minimum. Since $f''(0) = 0$, the second-derivative test fails to tell us whether 0 yields a maximum or a minimum.

(4)   Since $f'(x)$ does not change sign as x increases through 0, the point $(0, 0)$ yields neither a maximum nor a minimum.

(5) $f''(x) = 0$ when x is 0 or 2; f'' changes signs as x increases through 0 (+ 0 –), and also as x increases through 2 (– 0 +). Thus both $(0, 0)$ and $(2, -16)$ are inflection points.

The curve is sketched in Figure N4–5 on page 102.

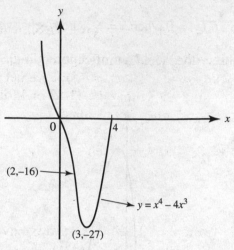

FIGURE N4–5

Verify the preceding on your calculator.

Case II. Functions Whose Derivatives May Not Exist Everywhere.

If there are values of x for which a first or second derivative does not exist, we consider those values separately, recalling that a local maximum or minimum point is one of transition between intervals of rise and fall and that an inflection point is one of transition between intervals of upward and downward concavity.

Example 14.   If $y = x^{2/3}$, then

$$\frac{dy}{dx} = \frac{2}{3x^{1/3}} \quad \text{and} \quad \frac{d^2y}{dx^2} = -\frac{2}{9x^{4/3}}.$$

Neither derivative is zero anywhere; both derivatives fail to exist when $x = 0$. As x increases through 0, $\frac{dy}{dx}$ changes from $-$ to $+$; $(0, 0)$ is therefore a minimum. Note that the tangent is vertical at the origin, and that since $\frac{d^2y}{dx^2}$ is negative everywhere except at 0, the curve is everywhere concave down. See Figure N4–6.

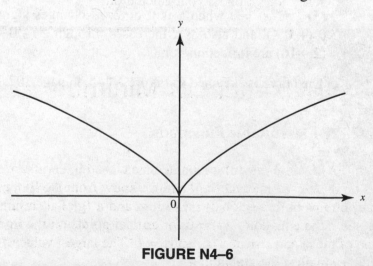

FIGURE N4–6

When we graph $y = x^{2/3}$ on a TI-82 using the exponent $\frac{2}{3}$, the calculator produces only the right half of Figure N4–6, which is incorrect. However, when we enter either $y = (x^{1/3})^2$ or $y = (x^2)^{1/3}$, we get the correct graph. There is no difficulty of this kind with either the TI-83 or TI-86.

Graph $y = x^{2/3}$ on your calculator.

Example 15.   If $y = x^{1/3}$, then

$$\frac{dy}{dx} = \frac{1}{3x^{2/3}} \quad \text{and} \quad \frac{d^2y}{dx^2} = -\frac{2}{9x^{5/3}}.$$

As in Example 14, neither derivative ever equals zero and both fail to exist when $x = 0$. Here, however, as x increases through 0, $\frac{dy}{dx}$ does not change sign. Since $\frac{dy}{dx}$ is positive for all x except 0, the curve rises for all x different from zero and can have neither maximum nor minimum points. The tangent is again vertical at the origin. Note here that $\frac{d^2y}{dx^2}$ does change sign (from + to –) as x increases through 0, so that (0, 0) is a point of inflection. See Figure N4–7.

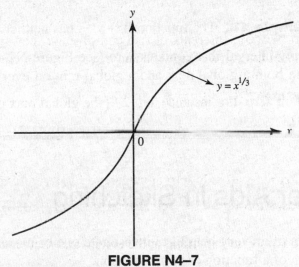

FIGURE N4–7

Verify the graph on your calculator.

F.   Global Maximum or Minimum

Case I. Differentiable Functions.

If a function f is differentiable on a closed interval $a \leqslant x \leqslant b$, then f is also continuous on the closed interval $[a,b]$ and we know from the Extreme Value Theorem (page 36) that f attains both a (global) maximum and a (global) minimum on $[a,b]$. To find these, we solve the equation $f'(x) = 0$ for critical points on the interval $[a,b]$, then evaluate f at each of those and also at $x = a$ and $x = b$. The largest value of f obtained is the global max, and the smallest the global min.

Example 16. Find the global max and global min of f on (a) $-2 \leqslant x \leqslant 3$, and (b) $0 \leqslant x \leqslant 3$, if $f(x) = 2x^3 - 3x^2 - 12x$.

(a) $f'(x) = 6x^2 - 6x - 12 = 6(x + 1)(x - 2)$, which equals zero if $x = -1$ or 2. Since $f(-2) = -4, f(-1) = 7, f(2) = -20$, and $f(3) = -9$, the global max of f occurs at $x = -1$ and equals 7, and the global min of f occurs at $x = 2$ and equals -20.

(b) Only the critical value 2 lies in $[0,3]$. We now evaluate f at 0, 2, and 3. Since $f(0) = 0, f(2) = -20$, and $f(3) = -9$, the global max of f equals 0 and the global min equals -20.

Case II. Functions That Are Not Everywhere Differentiable.

We proceed as for Case I but now evaluate f also at each point in a given interval for which f is defined but for which f' does not exist.

Example 17. The absolute-value function $f(x) = |x|$ is defined for all real x, but $f'(x)$ does not exist at $x = 0$. Since $f'(x) = -1$ if $x < 0$, but $f'(x) = 1$ if $x > 0$, we see that f has a global min at $x = 0$.

Example 18. The function $f(x) = \dfrac{1}{x}$ has neither a global max nor a global min on *any* interval that contains zero (see Figure N2–4, page 25). However, it does attain both a global max and a global min on every closed interval that does not contain zero. For instance, on $[2,5]$ the global max of f is $\dfrac{1}{2}$, the global min $\dfrac{1}{5}$.

G.   Further Aids in Sketching

It is often very helpful to investigate one or more of the following before sketching the graph of a function or of an equation:

(1) Intercepts. Set $x = 0$ and $y = 0$ to find any y- and x-intercepts respectively.

(2) Symmetry. Let the point (x, y) satisfy an equation. Then its graph is symmetric about
the x-axis if $(x, -y)$ also satisfies the equation;
the y-axis if $(-x, y)$ also satisfies the equation;
the origin if $(-x, -y)$ also satisfies the equation.

(3) Asymptotes. The line $y = b$ is a horizontal asymptote of the graph of a function f if either $\lim\limits_{x \to \infty} f(x) = b$ or $\lim\limits_{x \to -\infty} f(x) = b$. If $f(x) = \dfrac{P(x)}{Q(x)}$, inspect the degrees of $P(x)$ and $Q(x)$, then use the Rational Function Theorem, pages 30 and 31. The line $x = c$ is a vertical asymptote of the rational function $\dfrac{P(x)}{Q(x)}$ if $Q(c) = 0$ but $P(c) \neq 0$.

(4) Points of discontinuity. Identify points not in the domain of a function, particularly where the denominator equals zero.

Example 19. To sketch the graph of $y = \dfrac{2x+1}{x-1}$, we note that, if $x = 0$, then $y = -1$,

and that $y = 0$ when the numerator equals zero, which is when $x = -\dfrac{1}{2}$. A check

shows that the graph does not possess any of the symmetries described above. Since $y \to 2$ as $x \to \pm\infty$, $y = 2$ is a horizontal asymptote; also, $x = 1$ is a vertical asymptote. We see immediately that the function is defined for all reals except $x = 1$; the latter is the only point of discontinuity.

If we rewrite the function as follows:

$$y = \frac{2x+1}{x-1} = \frac{2x-2+3}{x-1} = \frac{2(x-1)+3}{x-1} = 2 + \frac{3}{x-1},$$

it is easy to find derivatives:

$$y' = -\frac{3}{(x-1)^2} \quad \text{and} \quad y'' = \frac{6}{(x-1)^3}.$$

From y' we see that the function decreases everywhere (except at $x = 1$), and from y'' that the curve is concave down if $x < 1$, up if $x > 1$. See Figure N4–8.

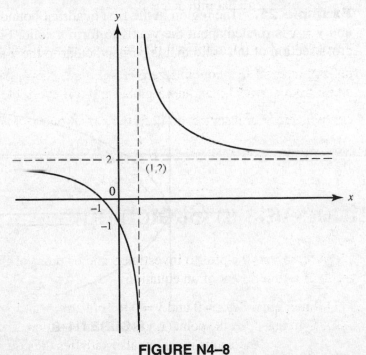

FIGURE N4–8

Verify the preceding on your calculator, using $[-4,4] \times [-4,8]$.

Example 20. Describe any symmetries of the graphs of (a) $3y^2 + x = 2$;

(b) $y = x + \dfrac{1}{x}$; (c) $x^2 - 3y^2 = 27$.

(a) Suppose point (x, y) is on this graph. Then so is point $(x, -y)$, since $3(-y)^2 + x = 2$ is equivalent to $3y^2 + x = 2$. So (a) is symmetric about the x-axis.

(b) Note that point $(-x, -y)$ satisfies the equation if point (x, y) does:

$$(-y) = (-x) + \frac{1}{(-x)} \quad \leftrightarrow \quad y = x + \frac{1}{x}.$$

Therefore the graph of this function is symmetric about the origin.

(c) This graph is symmetric about the x-axis, the y-axis, and the origin. It is easy to see that, if point (x, y) satisfies the equation, so do points $(x, -y)$, $(-x, y)$, and $(-x, -y)$.

H.   Optimization: Problems Involving Maxima and Minima

The techniques described above can be applied to problems in which a function is to be maximized (or minimized). Often it helps to draw a figure. If y, the quantity to be maximized (or minimized), can be expressed explicitly in terms of x, then the procedure outlined above can be used. If the domain of y is restricted to some closed interval, one should always check the endpoints of this interval so as not to overlook possible extrema. Often, implicit differentiation, sometimes of two or more equations, is indicated.

Example 21.   The region in the first quadrant bounded by the curves of $y^2 = x$ and $y = x$ is rotated about the y-axis to form a solid. Find the area of the largest cross section of this solid that is perpendicular to the y-axis.

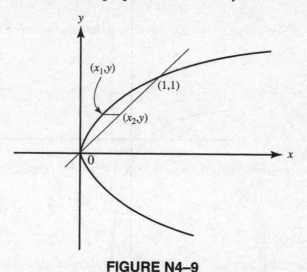

FIGURE N4–9

See Figure N4–9. The curves intersect at the origin and at $(1,1)$, so $0 < y < 1$. A cross section of the solid is a ring whose area A is the difference between the areas of two circles, one with radius x_2, the other with radius x_1. Thus

$$A = \pi x_2^{\,2} - \pi x_1^{\,2} = \pi(y^2 - y^4); \quad \frac{dA}{dy} = \pi(2y - 4y^3) = 2\pi y(1 - 2y^2).$$

The only relevant zero of the first derivative is $y = \dfrac{1}{\sqrt{2}}$. Thus for the maximum area A we have

$$A = \pi\left(\frac{1}{2} - \frac{1}{4}\right) = \frac{\pi}{4}.$$

Note that $\frac{d^2A}{dy^2} = \pi(2 - 12y^2)$ and that this is negative when $y = \frac{1}{\sqrt{2}}$, assuring a maximum there. Note further that A equals zero at each endpoint of the interval $[0,1]$ so that $\frac{\pi}{4}$ is the global maximum area.

Example 22.   The volume of a cylinder equals k cubic inches, where k is a constant. Find the proportions of the cylinder that minimize the total surface area.

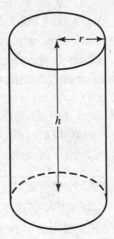

FIGURE N4–10

See Figure N4–10. We know that the volume is

$$V = \pi r^2 h = k, \tag{1}$$

where r is the radius and h the height. We seek to minimize S, the total surface area, where

$$S = 2\pi r^2 + 2\pi rh = 2\pi(r^2 + rh). \tag{2}$$

We can differentiate both (1) and (2) with respect to r, getting, from (1),

$$\pi\left(r^2 \cdot \frac{dh}{dr} + 2rh\right) = 0, \tag{3}$$

where we use the fact that $\frac{dk}{dr} = 0$; then, from (2),

$$2\pi\left(2r + r \cdot \frac{dh}{dr} + h\right) = 0, \tag{4}$$

where $\frac{dS}{dr}$ is set equal to zero because S is to be a minimum.

From (3) we see that $\frac{dh}{dr} = -\frac{2h}{r}$, and if we use this in (4) we get

$$2r + r\left(\frac{-2h}{r}\right) + h = 0 \quad \text{or} \quad h = 2r.$$

The total surface area of a cylinder of fixed volume is thus a minimum when its height equals its diameter.

(Note that we need not concern ourselves with the possibility that the value of r that renders $\dfrac{dS}{dr}$ equal to zero will produce a maximum surface area rather than a minimum one. With k fixed, we can choose r and h in such a way as to make S as large as we like.)

Example 23.   A charter bus company advertises a trip for a group as follows: At least 20 people must sign up. The cost when 20 participate is $80 per person. The price will drop by $2 per ticket for each member of the traveling group in excess of 20. If the bus can accommodate 28 people, how many participants will maximize the company's revenue?

Let x denote the number who sign up in excess of 20. Then $0 \leqq x \leqq 8$. The total number who agree to participate is $(20 + x)$, and the price per ticket is $(80 - 2x)$ dollars. So the revenue R, in dollars, is

$$R = (20 + x)(80 - 2x),$$
$$R'(x) = (20 + x)(-2) + (80 - 2x) \cdot 1$$
$$= 40 - 4x.$$

This is zero if $x = 10$. Although $x = 10$ yields maximum R—note that $R''(x) = -4$ and is always negative—this value of x is not within the restricted interval. We therefore evaluate R at the endpoints 0 and 8: $R(0) = 1600$ and $R(8) = 28 \cdot 64 = 1792$. So 28 participants will maximize revenue.

Example 24. A utilities company wants to deliver gas from a source S to a plant P located across a straight river 3 miles wide, then downstream 5 miles, as shown in Figure N4–11. It costs $4 per foot to lay the pipe in the river but only $2 per foot to lay it on land.

(a)   Express the cost of laying the pipe in terms of u.
(b)   How can the pipe be laid most economically?

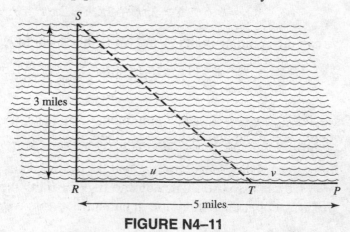

FIGURE N4–11

Note that the problem "allows" us to (1) lay all of the pipe in the river, along the line from S to P; (2) lay pipe along SR, in the river, then along RP on land; or (3) lay some pipe in the river, say, along ST, and lay the rest on land along TP. When T coincides with P, we have case (1), with $v = 0$; when T coincides with R, we have case (2), with $u = 0$. So case (3) includes both (1) and (2).

In any event, we need to find the lengths of pipe needed (that is, the distances involved); then we must figure out the cost.

In terms of u:

	In the River	On Land
Distances: miles	$ST = \sqrt{9 + u^2}$	$TP = v = 5 - u$
feet	$ST = 5280\sqrt{9 + u^2}$	$TP = 5280(5 - u)$
Costs (dollars):	$4(5280)\sqrt{9 + u^2}$	$2[5280(5 - u)]$

So, if $C(u)$ is the total cost,

$$C(u) = 21,120\sqrt{9 + u^2} + 10,560(5 - u)$$
$$= 10,560(2\sqrt{9 + u^2} + 5 - u). \qquad (*)$$

We now minimize $C(u)$:

$$C'(u) = 10,560\left(2 \cdot \frac{1}{2}\frac{2u}{\sqrt{9 + u^2}} - 1\right) = 10,560\left(\frac{2u}{\sqrt{9 + u^2}} - 1\right).$$

We now set $C'(u)$ equal to zero and solve for u:

$$\frac{2u}{\sqrt{9 + u^2}} - 1 = 0 \;\rightarrow\; \frac{2u}{\sqrt{9 + u^2}} = 1 \;\rightarrow\; \frac{4u^2}{9 + u^2} = 1,$$

where, in the last step, we squared both sides; then

$$4u^2 = 9 + u^2, \qquad 3u^2 = 9, \qquad u^2 = 3, \qquad u = \sqrt{3},$$

where we discard $u = -\sqrt{3}$ as meaningless for this problem.

The domain of $C(u)$ is $[0,5]$ and C is continuous on $[0,5]$. Since

$$C(0) = 10,560\left(2\sqrt{9} + 5\right) = \$116,160,$$
$$C(5) = 10,560\left(2\sqrt{34}\right) \approx \$123,150,$$
$$C\left(\sqrt{3}\right) = 10,560\left(2\sqrt{12} + 5 - \sqrt{3}\right) = \$107,671,$$

we see that $u = \sqrt{3}$ yields minimum cost. Thus, the pipe can be laid most economically if some of it is laid in the river from the source S to a point T that is $\sqrt{3}$ miles toward the plant P from R, and the rest is laid along the road from T to P.

We can also use a calculator to find the minimum of $C(u)$. Replacing u by X, we let Y_1 equal the expression in parentheses in the starred equation above:

$$Y_1 = 2\sqrt{9 + X^2} + 5 - X$$

We then graph Y_1 in $[0,5] \times [10,12]$. The [minimum] option yields, on the screen,

$$X = 1.732051, \qquad Y = 10.196152.$$

We evaluate C at its minimum by keying in $10,560Y$, which equals $107,671$, as before.

I. Relating a Function and Its Derivatives Graphically

The following table shows the characteristics of a function f and their implications for f's derivatives. These are crucial in obtaining one graph from another. The table can be used reading from left to right or from right to left.

Note that the slope at $x = c$ of any graph of a function is equal to the ordinate at c of the derivative of the function.

		f	f'	f''
ON AN INTERVAL		increasing	> 0	
		decreasing	< 0	

			$x < c$	$x = c$	$x > c$	
AT c	local maximum		$+$	0	$-$	$f''(c) < 0$
			(f' is decreasing)			
	local minimum		$-$	0	$+$	$f''(c) > 0$
			(f' is increasing)			
	neither local maximum nor local minimum		$+$	0	$+$	
			$-$	0	$-$	
			(f' does not change sign)			

			$x < c$	$x = c$	$x > c$
AT c	point of inflection	$f'(c)$ is a minimum; changes from decreasing to increasing	$-$	0	$+$
		$f'(c)$ is a maximum; changes from increasing to decreasing	$+$	0	$-$

		f'	f''
ON AN INTERVAL	concave up	f' is increasing	$f'' > 0$
	concave down	f' is decreasing	$f'' < 0$

If $f'(c)$ does not exist, check the signs of f' as x increases through c: plus-to-minus yields a local maximum; minus-to-plus yields a local minimum; no sign changes means no maximum or minimum, but check the possibility of a point of inflection.

You will find this table most helpful if you draw sketches, line by line, to illustrate the characteristics of f and its derivatives.

Example 25a.   Given the graph of $f(x)$ shown in Figure N4–12, sketch $f'(x)$.

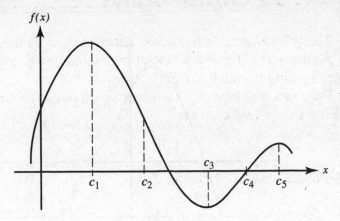

FIGURE N4–12

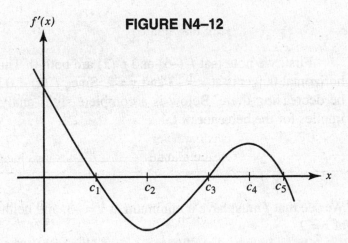

Point $x =$	Behavior of f	Behavior of f'
c_1	$f(c_1)$ is a local max	$f'(c_1) = 0$; f' changes sign from $+$ to $-$.
c_2	c_2 is an inflection point of f; f changes concavity from down to up	f' changes from decreasing to increasing; $f'(c_2)$ is a local minimum
c_3	$f(c_3)$ is a local minimum	$f'(c_3) = 0$; f' changes sign from $-$ to $+$
c_4	c_4 is an inflection point of f; f changes concavity from up to down	f' changes from increasing to decreasing; $f'(c_4)$ is a local maximum
c_5	$f(c_5)$ is a local maximum	$f'(c_5) = 0$; f' changes sign from $+$ to $-$

Example 25b. Given the graph of $f'(x)$ shown in Figure N4–13, sketch a possible graph of f.

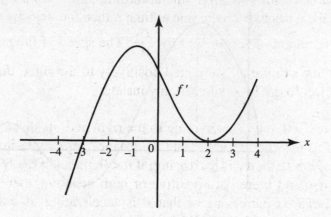

First, we note that $f'(-3)$ and $f'(2)$ are both 0. Thus the graph of f must have horizontal tangents at $x = -3$ and $x = 2$. Since $f'(x) < 0$ for $x - 3$, we see that f must be decreasing there. Below is a complete signs-analysis of f', showing what it implies for the behavior of f.

f	dec	-3	inc	2	inc
f'	$-$		$+$		$+$

We see that f must have a minimum at $x = -3$, and neither minimum nor maximum at $x = 2$.

We note next that f' is increasing for $x < -1$. This means that the derivative of f', f'', must be positive for $x < -1$ and that f is concave upward. Analyzing the signs of f'' yields the following:

f	conc. upward	-1	conc. down	2	conc. upward
f'	inc		dec		inc
f''	$+$		$-$		$+$

We conclude that f has points of inflection at $x = -1$ and $x = 2$. We use the information obtained to sketch a possible graph of f, shown in Figure N4-14. Note that other graphs are possible; in fact, any vertical translation of this f will do!

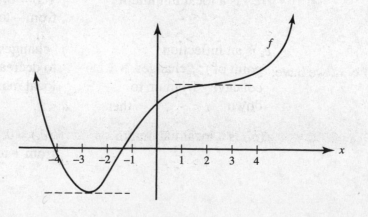

J.   Motion along a Line

If a particle moves along a line according to the law $s = f(t)$, where s represents the position of the particle P on the line at time t, then the velocity v of P at time t is given by $\frac{ds}{dt}$ and its acceleration a by $\frac{dv}{dt}$ or by $\frac{d^2s}{dt^2}$. The speed of the particle is $|v|$, the magnitude of v. If the line of motion is directed positively to the right, then the motion of the particle P is subject to the following: At any instant,

(1)   if $v > 0$, then P is moving to the right and its distance s is increasing; if $v < 0$, then P is moving to the left and its distance s is decreasing;

(2)   if $a > 0$, then v is increasing; if $a < 0$, then v is decreasing;

(3)   if a and v are both positive or both negative, then (1) and (2) imply that the speed of P is increasing or that P is accelerating; if a and v have opposite signs, then the speed of P is decreasing or P is decelerating;

(4)   if s is a continuous function of t, then P reverses direction whenever v is zero and a is different from zero; note that zero velocity does not necessarily imply a reversal in direction.

Example 26.   A particle moves along a line according to the law $s = 2t^3 - 9t^2 + 12t - 4$, where $t \geq 0$. (a) Find all t for which the distance s is increasing. (b) Find all t for which the velocity is increasing. (c) Find all t for which the speed of the particle is increasing. (d) Find the speed when $t = \frac{3}{2}$. (e) Find the total distance traveled between $t = 0$ and $t = 4$.

We have

$$v = \frac{ds}{dt} = 6t^2 - 18t + 12 = 6(t^2 - 3t + 2) = 6(t - 2)(t - 1)$$

and

$$a = \frac{dv}{dt} = \frac{d^2s}{dt^2} = 12t - 18 = 12\left(t - \frac{3}{2}\right).$$

The sign of v behaves as follows:

$$\begin{array}{lll} \text{if} & t < 1, & \text{then} \quad v > 0, \\ & 1 < t < 2, & \quad\quad\quad v < 0, \\ & t > 2, & \quad\quad\quad v > 0. \end{array}$$

For a, we have:

$$\begin{array}{lll} \text{if} & t < \frac{3}{2}, & \text{then} \quad a < 0, \\ \\ & t > \frac{3}{2}, & \quad\quad\quad a > 0. \end{array}$$

These signs of v and a immediately yield the answers, as follows:

(a) s increases when $t < 1$ or $t > 2$.

(b) v increases when $t > \dfrac{3}{2}$.

(c)   The speed $|v|$ is increasing when v and a are both positive, that is, for $t > 2$, and when v and a are both negative, that is, for $1 < t < \dfrac{3}{2}$.

(d)   The speed when $t = \dfrac{3}{2}$ equals $|v| = \left|-\dfrac{3}{2}\right| = \dfrac{3}{2}$.

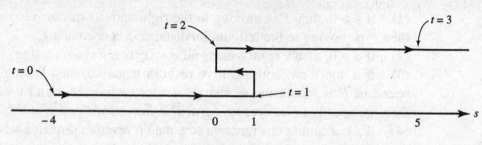

FIGURE N4–15

(e) P's motion can be indicated as shown in Figure N4–15. P moves to the right if $t < 1$, reverses its direction at $t = 1$, moves to the left when $1 < t < 2$, reverses again at $t = 2$, and continues to the right for all $t > 2$. The position of P at certain times t is shown in the following table:

t:	0	1	2	4
s:	–4	1	0	28

Thus P travels a total of 34 units between times $t = 0$ and $t = 4$.

Example 27.   Answer the questions of Example 26 if the law of motion is $s = t^4 - 4t^3$.

Since $v = 4t^3 - 12t^2 = 4t^2(t - 3)$ and $a = 12t^2 - 24t = 12t(t - 2)$, the signs of v and a are as follows:

	if		then	
	$t < 3$,		$v < 0$	
	$3 < t$,		$v > 0$;	
	if $t < 0$,		then $a > 0$	
	$0 < t < 2$,		$a < 0$	
	$2 < t$,		$a > 0$.	

Thus

(a) s increases if $t > 3$.

(b) v increases if $t < 0$ or $t > 2$.

(c)   Since v and a have the same sign if $0 < t < 2$ or if $t > 3$, the speed increases on these intervals.

(d)   The speed when $t = \dfrac{3}{2}$ equals $|v| = \left|-\dfrac{27}{2}\right| = \dfrac{27}{2}$.

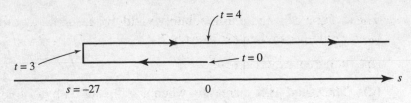

FIGURE N4–16

(e) The motion is shown in Figure N4–16. The particle moves to the left if $t < 3$ and to the right if $t > 3$, stopping instantaneously when $t = 0$ and $t = 3$, but reversing direction only when $t = 3$. Thus:

t:	0	3	4
s:	0	−27	0

The particle travels a total of 54 units between $t = 0$ and $t = 4$.

(Compare with Example 13, page 101, where the function $f(x) = x^4 - 4x^3$ is investigated for maximum and minimum values; also see the accompanying Figure N4–5 on page 102.)

*K   Motion along a Curve: Velocity and Acceleration Vectors

K1.   DERIVATIVE OF ARC LENGTH.

If the derivative of $y = f(x)$ is continuous, if Q is a fixed point on its curve and P any other point on it, and if s is the arc length from Q to P (see Figure N4–17), then the derivative of s is given by

$$\frac{ds}{dx} = \sqrt{1 + \left(\frac{dy}{dx}\right)^2} \tag{1}$$

or

$$\frac{ds}{dy} = \sqrt{1 + \left(\frac{dx}{dy}\right)^2}, \tag{2}$$

where it is assumed that s is increasing with x in (1), and with y in (2).

If the position of P is given parametrically by $x = f(t)$ and $y = g(t)$, then

$$\frac{ds}{dt} = \sqrt{\left(\frac{dx}{dt}\right)^2 + \left(\frac{dy}{dt}\right)^2}, \tag{3}$$

where s increases as t does.

Formulas (1), (2), and (3) above can all be derived easily from the very simple formula

$$ds^2 = dx^2 + dy^2. \tag{4}$$

*An asterisk denotes a topic covered only in Calculus BC.

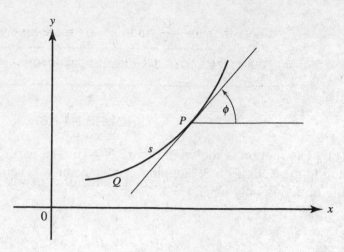

FIGURE N4–17

If, for instance, x is expressed in terms of y, making it convenient to use formula (2), note that it follows from (4) that

$$\frac{ds^2}{dy^2} = \frac{dx^2}{dy^2} + 1,$$

$$\left(\frac{ds}{dy}\right)^2 = \left(\frac{dx}{dy}\right)^2 + 1,$$

and

$$\frac{ds}{dy} = \sqrt{1 + \left(\frac{dx}{dy}\right)^2},$$

which is (2).

K2. VECTOR FUNCTIONS: VELOCITY AND ACCELERATION.

If a point P moves along a curve in accordance with the pair of parametric equations $x = f(t)$, $y = g(t)$, where t represents time, then the vector from the origin to P is called the *position vector*; often denoted by $[f(t), g(t)]$. Since a vector is defined for each t in the domain common to f and g, $[f(t), g(t)]$ is also called a *vector function*.

A vector is symbolized either by a boldface letter (thus: **R**, **i**, **j**) or by an italic letter with an arrow written over it (thus: $\vec{R}$, $\vec{i}$, $\vec{j}$). When writing on a blackboard or with pencil and paper the arrow notation is simpler, but in print the boldface notation is clearer, and will be used here.

The position vector is denoted by **R**, and

$$\mathbf{R} = x\mathbf{i} + y\mathbf{j},$$

where **i** is the (unit) vector from $(0, 0)$ to $(1, 0)$ and **j** is the (unit) vector from $(0, 0)$ to $(0, 1)$, and x and y are respectively the horizontal and vertical components of **R**.

The *velocity vector* is the derivative of the vector function (the position vector):

$$\mathbf{v} = \frac{d\mathbf{R}}{dt} = \frac{dx}{dt}\mathbf{i} + \frac{dy}{dt}\mathbf{j}.$$

Alternative notations for $\dfrac{dx}{dt}$ and $\dfrac{dy}{dt}$ are respectively v_x and v_y, or $\dot{x}$ and $\dot{y}$; these are the components of $\mathbf{v}$ in the horizontal and vertical directions respectively. The slope of $\mathbf{v}$ is

$$\frac{\dfrac{dy}{dt}}{\dfrac{dx}{dt}} = \frac{dy}{dx},$$

which is the slope of the curve; the magnitude of $\mathbf{v}$, denoted by $|\mathbf{v}|$, is

$$\sqrt{\left(\frac{dx}{dt}\right)^2 + \left(\frac{dy}{dt}\right)^2} = \sqrt{v_x^{\,2} + v_y^{\,2}},$$

which equals $\dfrac{ds}{dt}$, the derivative of arc length. Thus, if the vector $\mathbf{v}$ is drawn initiating at P, it will be tangent to the curve at P and its magnitude will be the *speed of the particle at P.*

The *acceleration vector* $\mathbf{a}$ is $\dfrac{dv}{dt}$ or $\dfrac{d^2\mathbf{R}}{dt^2}$, and can be obtained by a second differentiation of the components of $\mathbf{R}$. Thus

$$\mathbf{a} = \frac{d^2x}{dt^2}\mathbf{i} + \frac{d^2y}{dt^2}\mathbf{j};$$

the direction of $\mathbf{a}$ is

$$\tan^{-1}\frac{\dfrac{d^2y}{dt^2}}{\dfrac{d^2x}{dt^2}};$$

and its magnitude is

$$|\mathbf{a}| = \sqrt{\left(\frac{d^2x}{dt^2}\right)^2 + \left(\frac{d^2y}{dt^2}\right)^2} = \sqrt{a_x^{\,2} + a_y^{\,2}},$$

where we have used a_x and a_y for $\dfrac{d^2x}{dt^2}$ and $\dfrac{d^2y}{dt^2}$ respectively.

For both $\mathbf{v}$ and $\mathbf{a}$ the quadrant is determined by the signs of their components. Vectors $\mathbf{i}$ and $\mathbf{j}$ are shown in Figure N4–18a; $\mathbf{R}$, $\mathbf{v}$, and $\mathbf{a}$, and their components, are shown in Figure N4–18b. Note that v_x and v_y happen to be positive, so that ϕ is a first-quadrant angle, while $a_x < 0$ and $a_y > 0$ imply that θ is in the second quadrant.

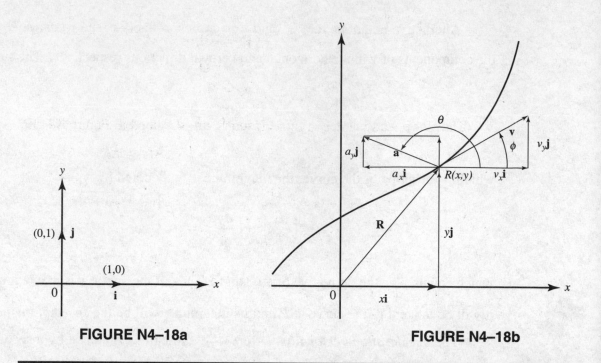

FIGURE N4–18a   **FIGURE N4–18b**

Example 28. A particle moves according to the equations $x = 3\cos t$, $y = 2\sin t$. (a) Find a single equation in x and y for the path of the particle and sketch the curve. (b) Find the velocity and acceleration vectors at any time t, and show that $\mathbf{a} = -\mathbf{R}$ at all times. (c) Find $\mathbf{R}$, $\mathbf{v}$, and $\mathbf{a}$ when (1) $t_1 = \frac{\pi}{6}$, (2) $t_2 = \pi$, and draw them on the sketch. (d) Find the speed of the particle and the magnitude of its acceleration at each instant in (c). (e) When is the speed a maximum? a minimum?

(a) Since $\frac{x^2}{9} = \cos^2 t$ and $\frac{y^2}{4} = \sin^2 t$, therefore

$$\frac{x^2}{9} + \frac{y^2}{4} = 1$$

and the particle moves in a counterclockwise direction along an ellipse, starting, when $t = 0$, at $(3, 0)$ and returning to this point when $t = 2\pi$.

(b) We have

$$\mathbf{R} = 3\cos t\mathbf{i} + 2\sin t\mathbf{j},$$

$$\mathbf{v} = -3\sin t\mathbf{i} + 2\cos t\mathbf{j},$$

$$\mathbf{a} = -3\cos t\mathbf{i} - 2\sin t\mathbf{j} = -\mathbf{R}.$$

The acceleration, then, is always directed toward the center of the ellipse.

(c) At $t_1 = \frac{\pi}{6}$,

$$\mathbf{R}_1 = \frac{3\sqrt{3}}{2}\mathbf{i} + \mathbf{j},$$

$$\mathbf{v}_1 = -\frac{3}{2}\mathbf{i} + \sqrt{3}\mathbf{j},$$

$$\mathbf{a}_1 = -\frac{3\sqrt{3}}{2}\mathbf{i} - \mathbf{j}.$$

At $t_2 = \pi$,

$$\mathbf{R}_2 = -3\mathbf{i},$$
$$\mathbf{v}_2 = -2\mathbf{j},$$
$$\mathbf{a}_2 = 3\mathbf{i}.$$

The curve, and $\mathbf{v}$ and $\mathbf{a}$ at those instants, are sketched in Figure N4–19.

(d) At $t_1 = \dfrac{\pi}{6}$,   At $t_2 = \pi$,

$$\left|\mathbf{v}_1\right| = \sqrt{\frac{9}{4} + 3} = \frac{\sqrt{21}}{2}, \qquad\qquad \left|\mathbf{v}_2\right| = \sqrt{0 + 4} = 2,$$

$$\left|\mathbf{a}_1\right| = \sqrt{\frac{27}{4} + 1} = \frac{\sqrt{31}}{2}. \qquad\qquad \left|\mathbf{a}_2\right| = \sqrt{9 + 0} = 3.$$

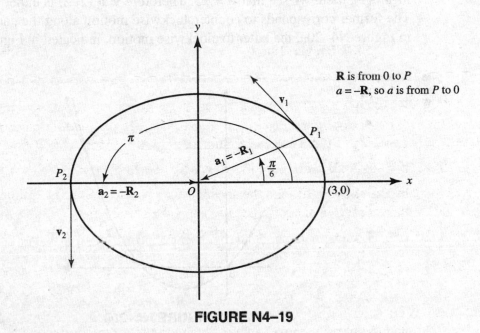

FIGURE N4–19

(e) For the speed $|\mathbf{v}|$ at any time t we have

$$\mathbf{v} = \sqrt{9\sin^2 t + 4\cos^2 t}$$
$$= \sqrt{4\sin^2 t + 4\cos^2 t + 5\sin^2 t}$$
$$= \sqrt{4 + 5\sin^2 t}.$$

We see immediately that the speed is a maximum when $t = \dfrac{\pi}{2}$ or $\dfrac{3\pi}{2}$, and a minimum when $t = 0$ or π. The particle goes fastest at the ends of the minor axis and most slowly at the ends of the major axis. Generally one can determine maximum or minimum speed by finding $\dfrac{d}{dt}|\mathbf{v}|$, setting it equal to zero, and applying the usual tests to sort out values of t that yield maximum or minimum speeds.

Example 29.   A particle moves along the parabola $y = x^2 - x$ with constant speed $\sqrt{10}$. Find **v** at $(2, 2)$.

Since

$$v_y = \frac{dy}{dt} = (2x-1)\frac{dx}{dt} = (2x-1)v_x \qquad (1)$$

and

$$v_x^2 + v_y^2 = 10, \qquad (2)$$

we have

$$v_x^2 + (2x-1)^2 v_x^2 = 10. \qquad (3)$$

Relation (3) holds at all times; specifically, at $(2, 2)$, $v_x^2 + 9v_x^2 = 10$ so that $v_x = \pm 1$. From (1), then, we see that $v_y = \pm 3$. Therefore **v** at $(2, 2)$ is either $\mathbf{i} + 3\mathbf{j}$ or $-\mathbf{i} - 3\mathbf{j}$. The former corresponds to counterclockwise motion along the parabola, as shown in Figure N4–20a; the latter to clockwise motion, indicated in Figure N4–20b.

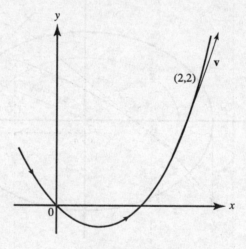

FIGURE N4–20a

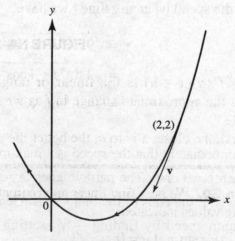

FIGURE N4–20b

L.   Local Linear Approximations _____

If $f'(a)$ exists, then the *local linear approximation of $f(x)$ at a* is

$$f(a) + f'(a)(x - a).$$

Since the equation of the tangent line to $y = f(x)$ at $x = a$ is

$$y - f(a) = f'(a)(x - a),$$

we see that the y value on the tangent line is an approximation for the actual or true value of f. Local linear approximation is therefore also called *tangent-line approximation*.[†] For values of x close to a, we have

$$f(x) \simeq f(a) + f'(a)(x - a), \tag{1}$$

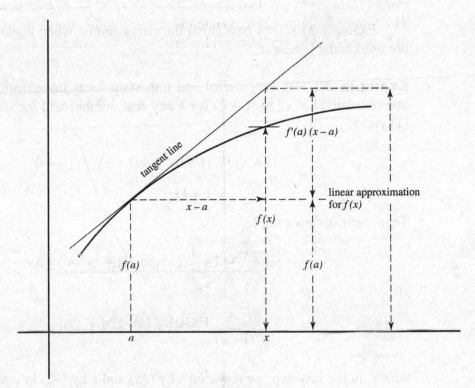

FIGURE N4–21

where $f(a) + f'(a)(x - a)$ is the linear or tangent-line approximation for $f(x)$, and $f'(a)(x - a)$ is the approximate change in f as we move along the curve from a to x. See Figure N4–21.

In general, the closer x is to a, the better the approximation is to $f(x)$.

Example 30. We now find linear approximations for each of the following functions at the values indicated:

(a) $\sin x$ at $a = 0$ (b) $\cos x$ at $a = \dfrac{\pi}{2}$

(c) $2x^3 - 3x$ at $a = 1$ (d) $\sqrt{1 + x}$ at $a = 8$

[†]Local linear approximation is also referred to as "linearization," "local linearization," "linear approximation," or even "best linear approximation" (the last because it is better than any other linear approximation).

(a) At $a = 0$, $\sin x \simeq \sin (0) + \cos (0)(x - 0) \simeq 0 + 1 \cdot x \simeq x$

(b) At $a = \dfrac{\pi}{2}$, $\cos x \simeq \cos \dfrac{\pi}{2} - \sin \left(\dfrac{\pi}{2} \right)(x - \dfrac{\pi}{2}) \simeq -x + \dfrac{\pi}{2}$

(c) At $a = 1$, $2x^3 - 3x \simeq -1 + 3(x - 1) \simeq 3x - 4$

(d) At $a = 8$, $\sqrt{1+x} = \sqrt{1+8} + \dfrac{1}{2\sqrt{1+8}} \ (x - 8) = 3 + \dfrac{1}{6}(x - 8)$

Example 31. Using the linearizations obtained in Example 30 and a calculator, we evaluate each function, then its linear approximation, at the indicated x-values:

Function	(a)	(b)	(c)	(d)
x-value	−0.80	2.00	1.10	5.50
True value	−0.72	−0.42	−0.64	2.55
Approximation	−0.80	−0.43	−0.70	2.58

Example 31 shows how small the errors can be when linear approximations are used and x is near a.

Example 32. A very useful and important local linearization enables us to approximate $(1 + x)^k$ by $1 + kx$ for k any real number and for x near 0. Equation (1) yields

$$(1 + x)^k \simeq (1 + 0)^k + k(1 + x)^{k-1}_{\text{at } 0} \cdot (x - 0)$$

$$\simeq 1 + kx. \tag{2}$$

Then, near 0, for example,

$$\sqrt{1+x} \simeq 1 + \frac{1}{2}x \quad \text{and} \quad (1 + x)^3 \simeq 1 + 3x;$$

also

$$\frac{3}{(1 - x)^2} = 3(1 - x)^{-2} \simeq 3(1 + 2x),$$

where, in the last step, we replaced x by $(-x)$ and k by (-2) in equation (2) above. Finally, $3(1 + 2x) = 3 + 6x$.

Example 33. The true value, to three decimal places, of $\dfrac{3}{(1 - x)^2}$ when $x = 0.05$ is 3.324; the local linear approximation yields 3.3. This tells us that the curve is concave up, lying above the tangent line to the curve near $x = 0$. Graph the curve and the tangent line on $[-1, 1] \times [-1, 6]$ to verify these statements.

Approximating the Change in a Function

Equation (1) above for a linear approximation also tells us by about how much *f changes* when we move along the curve from a to x: it is the quantity $f'(a)(x - a)$. (See Figure N4–21 on page 121)

Example 34.   To find by approximately how much the area of a circle changes when the radius increases from 3 to 3.01 inches, we use the formula $A = \pi r^2$. Then Equation (1) tells us that the linear approximation for $A(r)$, when A is near 3, is

$$A(3) + A'(3)(r-3).$$

Here we want only the change in area; that is,

$$A'(3)(r-3) \quad \text{when} \quad r = 3.01.$$

Since $A'(r) = 2\pi r$, therefore $A'(3) = 6\pi$; also, $(r-3) = 0.01$. So the approximate change is $(6\pi)(0.01) \simeq 0.1885$ in.^{2} The true increase in area, to four decimal places, is 0.1888 in.^{2}

Example 35.   Suppose the diameter of a cylinder is 8 centimeters. If its circumference is increased by 2 centimeters, how much larger, approximately, are (a) the diameter, and (b) the area of a cross section?

(a) Let D and C be respectively the diameter and circumference of the cylinder. Here, D plays the role of f, and C that of x, in the linear approximation equation (1) on page 121. The approximate increase in diameter, when $C = 8\pi$, is therefore equal to $D'(C)$ times (the change in C). Since $C = \pi D$, $D = \dfrac{C}{\pi}$ and $D'(C) = \dfrac{1}{\pi}$ (which is constant for *all* C). The change in C is given as 2 cm; so the increase in diameter is equal approximately to $\dfrac{1}{\pi} \cdot 2 \simeq 0.6366$ cm.^{2}

(b) The approximate increase in the area of a (circular) cross section is equal to

$$A'(C) \cdot (\text{change in } C),$$

where the area $A = \pi r^2 = \pi \left(\dfrac{C}{2\pi}\right)^2 = \dfrac{C^2}{4\pi}$. Therefore,

$$A'(C) = \frac{2C}{4\pi} = \frac{C}{2\pi} = \frac{8\pi}{2\pi} = 4.$$

Since the change in C is 2 cm, the area of a cross section increases by approximately $4 \cdot 2 = 8$ cm.^{2}

M.   Related Rates

If several variables that are functions of time t are related by an equation, we can obtain a relation involving their (time) rates of change by differentiating with respect to t.

Example 36.   If one leg AB of a right triangle increases at the rate of 2 inches per second, while the other leg AC decreases at 3 inches per second, find how fast the hypotenuse is changing when $AB = 6$ feet and $AC = 8$ feet.

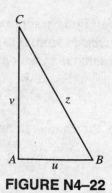

FIGURE N4–22

See Figure N4–22. Let u, v, and z denote the lengths respectively of AB, AC, and BC. We know that $\dfrac{du}{dt} = \dfrac{1}{6}$ (ft/sec) and $\dfrac{dv}{dt} = -\dfrac{1}{4}$. Since (at any time) $z^2 = u^2 + v^2$, then

$$2z\frac{dz}{dt} = 2u\frac{du}{dt} + 2v\frac{dv}{dt} \quad \text{and} \quad \frac{dz}{dt} = \frac{u\dfrac{du}{dt} + v\dfrac{dv}{dt}}{z}.$$

At the instant in question, $u = 6$, $v = 8$, and $z = 10$. So

$$\frac{dz}{dt} = \frac{6\left(\dfrac{1}{6}\right) + 8\left(-\dfrac{1}{4}\right)}{10} = -\frac{1}{10}\,\text{ft / sec}.$$

Example 37.   The diameter and height of a paper cup in the shape of a cone are both 4 inches, and water is leaking out at the rate of $\dfrac{1}{2}$ cubic inch per second. Find the rate at which the water level is dropping when the diameter of the surface is 2 inches.

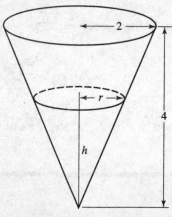

FIGURE N4–23

See Figure N4–23. We know that $\dfrac{dV}{dt} = -\dfrac{1}{2}$ and that $h = 2r$. Here, $V = \dfrac{1}{3}\pi r^2 h = \dfrac{\pi h^3}{12}$, so

$$\frac{dV}{dt} = \frac{\pi h^2}{4}\frac{dh}{dt} \quad \text{and} \quad \frac{dh}{dt} = -\frac{1}{2}\frac{4}{\pi h^2} = -\frac{2}{\pi h^2} \quad \text{at any time.}$$

When the diameter is 2 in., so is the height, and $\dfrac{dh}{dt} = -\dfrac{1}{2\pi}$. The water level is thus dropping at the rate of $\dfrac{1}{2\pi}$ in./sec.

Example 38.   Suppose liquid is flowing into a vessel at a constant rate. The vessel has the shape of a hemisphere capped by a cylinder, as shown in Figure N4–24. Graph $y = h(t)$, the height (= depth) of the liquid at time t, labeling and explaining any salient characteristics of the graph.

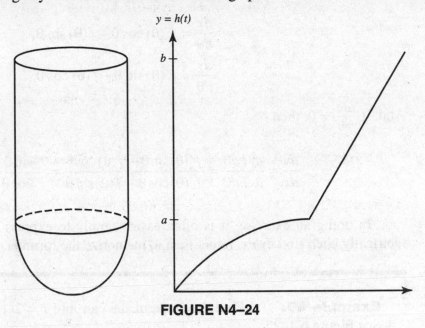

FIGURE N4–24

Liquid flowing in at a constant rate means the change in volume is constant per unit of time. Obviously, the depth of the liquid increases at t does, so $h'(t)$ is positive throughout. To maintain the constant increase in volume per unit of time, when the radius grows, $h'(t)$ must decrease. Thus, the rate of increase of h decreases as h increases from 0 to a (where the cross-sectional area of the vessel is largest). Therefore, since $h'(t)$ decreases, $h''(t) < 0$ from 0 to a and the curve is concave down.

As h increases from a to b, the radius of the vessel (here cylindrical) remains constant, as do the cross-sectional areas. Therefore $h'(t)$ is also constant, implying that $h(t)$ is linear from a to b.

Note that the inflection point at depth a does not exist, since $h''(t) < 0$ for all values less than a but is equal to 0 for all depths greater than or equal to a.

---

*N.   Slope of a Polar Curve

We know that, if a smooth curve is given by the parametric equations

$$x = f(t) \quad \text{and} \quad y = g(t),$$

then

$$\frac{dy}{dx} = \frac{g'(t)}{f'(t)} \qquad \text{provided } f'(t) \neq 0.$$

To find the slope of a polar curve $r = f(\theta)$, we must first express the curve in parametric form. Since

$$x = r \cos \theta \quad \text{and} \quad y = r \sin \theta,$$

*An asterisk denotes a topic covered only in Calculus BC.

therefore

$$x = f(\theta) \cos \theta \quad \text{and} \quad y = f(\theta) \sin \theta.$$

If $f(\theta)$ is differentiable, so are x and y; then

$$\frac{dx}{d\theta} = f'(\theta) \cos \theta - f(\theta) \sin \theta,$$

$$\frac{dy}{d\theta} = f'(\theta) \sin \theta + f(\theta) \cos \theta.$$

And, if $\frac{dx}{d\theta} \neq 0$, then

$$\frac{dy}{dx} = \frac{dy/d\theta}{dx/d\theta} = \frac{f'(\theta)\sin\theta + f(\theta)\cos\theta}{f'(\theta)\cos\theta - f(\theta)\sin\theta} = \frac{r'\sin\theta + r\cos\theta}{r'\cos\theta - r\sin\theta}.$$

In doing an exercise, it is often easier simply to express the polar equation parametrically, then find dy/dx, rather than to memorize the formula.

Example 39. (a) Find the slope of the cardioid $r = 2(1 + \cos \theta)$ at $\pi = \frac{\theta}{6}$. See Figure N4–25.

(b) Where is the tangent to the curve horizontal?

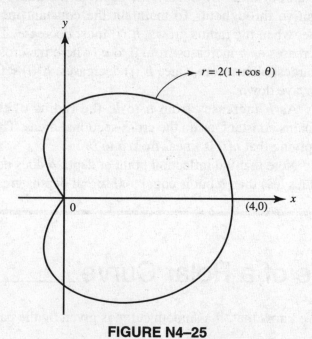

$r = 2(1 + \cos \theta)$

(4,0)

FIGURE N4–25

(a) Use $r = 2(1 + \cos \theta)$, $x = r \cos \theta$, $y = r \sin \theta$, and $r' = -2 \sin \theta$; then

$$\frac{dy}{dx} = \frac{dy/d\theta}{dx/d\theta} = \frac{(-2\sin\theta)\sin\theta + 2(1+\cos\theta)(\cos\theta)}{(-2\sin\theta)\cos\theta - 2(1+\cos\theta)(\sin\theta)}.$$

At $\theta = \frac{\pi}{6}$, $\frac{dy}{dx} = -1$. This can be verified on a grapher by finding $\frac{dy}{dx}$ at $\theta = \frac{\pi}{6}$.

(b)   Since the cardioid is symmetric to $\theta = 0$ we need consider only the upper half of the curve for part (b). The tangent is horizontal where $\frac{dy}{d\theta} = 0$ (provided $\frac{dx}{d\theta} \neq 0$).

Since $\frac{dy}{d\theta}$ factors into $2(2\cos\theta - 1)(\cos\theta + 1)$, which equals 0 for $\cos\theta = \frac{1}{2}$ or -1, $\theta = \frac{\pi}{3}$ or π. From part (a), $\frac{dx}{d\theta} \neq 0$ at $\frac{\pi}{3}$, but $\frac{dx}{d\theta}$ does equal 0 at π. Therefore, the tangent is horizontal only at $\frac{\pi}{3}$ (and, by symmetry, at $\frac{5\pi}{3}$).

It is obvious from Figure N4–25 that $f'(\theta)$—or $r'(\theta)$—does *not* give the slope of the cardioid. As θ varies from 0 to $\frac{2\pi}{3}$, the slope varies from $-\infty$ to 0 to $+\infty$ (with the tangent rotating counterclockwise), taking on every real value. However, $f'(\theta)$ equals $-2\sin\theta$, which takes on values only between -2 and 2!

Set 4: Multiple-Choice Questions on Applications of Differential Calculus

1. The slope of the curve $y^3 - xy^2 = 4$ at the point where $y = 2$ is

    (A) -2 **(B)** $\dfrac{1}{4}$ **(C)** $-\dfrac{1}{2}$ **(D)** $\dfrac{1}{2}$ **(E)** 2

2. The slope of the curve $y^2 - xy - 3x = 1$ at the point $(0, -1)$ is

    (A) -1 **(B)** -2 **(C)** $+1$ **(D)** 2 **(E)** -3

3. The equation of the tangent to the curve $y = x \sin x$ at the point $\left(\dfrac{\pi}{2}, \dfrac{\pi}{2} \right)$ is

    (A) $y = x - \pi$ **(B)** $y = \dfrac{\pi}{2}$ **(C)** $y = \pi - x$

    (D) $y = x + \dfrac{\pi}{2}$ **(E)** $y = x$

4. The tangent to the curve of $y = xe^{-x}$ is horizontal when x is equal to

    (A) 0 **(B)** 1 **(C)** -1 **(D)** $\dfrac{1}{e}$ **(E)** none of these

5. The point on the curve $y = \sqrt{2x+1}$ at which the normal is parallel to the line $y = -3x + 6$ is

    (A) $(4, 3)$ **(B)** $(0, 1)$ **(C)** $(1, \sqrt{3}\,)$
    (D) $(4, -3)$ **(E)** $(2, \sqrt{5}\,)$

6. The minimum value of the slope of the curve $y = x^5 + x^3 - 2x$ is

    (A) 0 **(B)** 2 **(C)** 6 **(D)** -2 **(E)** none of these

7. The equation of the tangent to the curve $x^2 = 4y$ at the point on the curve where $x = -2$ is

    (A) $x + y - 3 = 0$ **(B)** $y - 1 = 2x(x + 2)$ **(C)** $x - y + 3 = 0$
    (D) $x + y - 1 = 0$ **(E)** $x + y + 1 = 0$

8. The equation of the tangent to the hyperbola $x^2 - y^2 = 12$ at the point $(4, 2)$ on the curve is

    (A) $x - 2y + 6 = 0$ **(B)** $y = 2x$ **(C)** $y = 2x - 6$

    (D) $y = \dfrac{x}{2}$ **(E)** $x + 2y = 6$

9. The tangent to the curve $y^2 - xy + 9 = 0$ is vertical when

  (A) $y = 0$ (B) $y = \pm \sqrt{3}$ (C) $y = \dfrac{1}{2}$

  (D) $y = \pm 3$ (E)   none of these

10. The table shows the velocity at time t of an object moving along a line. Estimate the acceleration (in ft/sec^2) at $t = 6$ sec.

t (sec)	0	4	8	10
vel.	18	16	10	0

  (A)   −6   (B)   −1.8   (C)   −1.5   (D)   1.5   (E)   6

Use the graph shown, sketched on $[0, 7]$, for Questions 11–13.

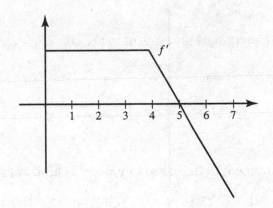

11. From the graph it follows that

  (A) f is discontinuous at $x = 4$
  (B) f is decreasing for $4 < x < 7$
  (C) f is constant for $0 < x < 4$
  (D) f has a local maximum at $x = 0$
  (E) f has a local minimum at $x = 7$

12. Which statement best describes f at $x = 5$?

  (A) f has a root.   (B) f is a maximum.   (C) f is a minimum.
  (D) f has a point of inflection.   (E)   none of these

13. The function is concave downward for which interval?

  (A) $(0,4)$ (B) $(4,5)$ (C) $(5,7)$
  (D) $(4,7)$ (E)   none of these

14. The best approximation, in cubic inches, to the increase in volume of a sphere when the radius is increased from 3 to 3.1 in. is

  (A)   11.3   (B)   11.7   (C)   12.1   (D)   33.9   (E)   39.7

15. When $x = 3$, the equation $2x^2 - y^3 = 10$ has the solution $y = 2$. When $x = 3.04$, $y \simeq$

    (A) 1.6   **(B)** 1.96   **(C)** 2.04   **(D)** 2.14   **(E)** 2.4

16. If the side e of a square is increased by 1%, then the area is increased approximately

    (A) $0.02e$ **(B)** $0.02e^2$ **(C)** $0.01e^2$ **(D)** 1%   **(E)** $0.01e$

17. The edge of a cube has length 10 in., with a possible error of 1%. The possible error, in cubic inches, in the volume of the cube is

    (A) 3   **(B)** 1   **(C)** 10   **(D)** 30   **(E)** none of these

Use the graph shown for Questions 18–24. It shows the velocity of an object moving along a straight line during the time interval $0 \leqslant t \leqslant 5$.

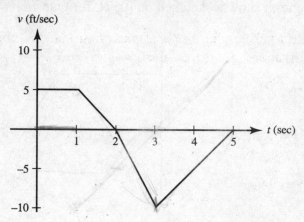

18. The object attains its maximum speed when $t =$

    (A) 0   **(B)** 1   **(C)** 2   **(D)** 3   **(E)** 5

19. The speed of the object is increasing during the time interval

    (A) (0,1)   **(B)** (1,2)   **(C)** (0,2)   **(D)** (2,3)   **(E)** (3,5)

20. The acceleration of the object is positive during the time interval

    (A) (0,1)   **(B)** (1,2)   **(C)** (0,2)   **(D)** (2,3)   **(E)** (3,5)

21. How many times on $0 < t < 5$ is the object's acceleration undefined?

    (A) none   **(B)** 1   **(C)** 2   **(D)** 3   **(E)** more than 3

22. During $2 < t < 3$ the object's acceleration (in ft/sec²) is

    (A) −10   **(B)** −5   **(C)** 0   **(D)** 5   **(E)** 10

23. The object is furthest to the right when $t =$

    (A) 0   **(B)** 1   **(C)** 2   **(D)** 3   **(E)** 5

24. The object's average acceleration (in ft/sec²) for the interval $0 \leqslant t \leqslant 3$ is

    (A) −15   **(B)** −5   **(C)** −3   **(D)** −1   **(E)** none of these

25. The function $f(x) = x^4 - 4x^2$ has

   (A)   one relative minimum and two relative maxima
   (B)   one relative minimum and one relative maximum
   (C)   two relative maxima and no relative minimum
   (D)   two relative minima and no relative maximum
   (E)   two relative minima and one relative maximum

26. The number of inflection points of the curve in Question 25 is

   (A) 0   (B) 1   (C) 2   (D) 3   (E) 4

27. The maximum value of the function $y = -4\sqrt{2-x}$ is

   (A) 0   (B) −4   (C) 2   (D) −2   (E) none of these

28. The total number of local maximum and minimum points of the function whose derivative, for all x, is given by $f'(x) = x(x-3)^2(x+1)^4$ is

   (A) 0   (B) 1   (C) 2   (D) 3   (E)   none of these

29. If $x \neq 0$, then the slope of $x \sin \dfrac{1}{x}$ equals zero whenever

   (A) $\tan \dfrac{1}{x} = x$ (B) $\tan \dfrac{1}{x} = -x$ (C) $\cos \dfrac{1}{x} = 0$

   (D) $\sin \dfrac{1}{x} = 0$ (E) $\tan \dfrac{1}{x} = \dfrac{1}{x}$

30. On the closed interval $[0, 2\pi]$, the maximum value of the function $f(x) = 4 \sin x - 3 \cos x$ is

   (A) 3   (B) 4   (C) $\dfrac{24}{5}$ (D) 5 .   (E)   none of these

31. If m_1 is the slope of the curve $xy = 2$ and m_2 is the slope of the curve $x^2 - y^2 = 3$, then at a point of intersection of the two curves

   (A) $m_1 = -m_2$ (B) $m_1 m_2 = -1$ (C) $m_1 = m_2$
   (D) $m_1 m_2 = 1$ (E) $m_1 m_2 = -2$

32. The line $y = 3x + k$ is tangent to the curve $y = x^3$ when k is equal to

   (A)   1 or −1   (B)   0   (C)   3 or −3   (D)   4 or −4   (E)   2 or −2

33. The two tangents that can be drawn from the point $(3,5)$ to the parabola $y = x^2$ have slopes

   (A)   1 and 5   (B)   0 and 4   (C)   2 and 10

   (D)   2 and $-\dfrac{1}{2}$ (E)   2 and 4

34. The table shows the velocity at various times of an object moving along a line. An estimate of its acceleration (in ft/sec²) at $t = 1$ is

t (sec)	1.0	1.5	2.0	2.5
v (ft/sec)	12.2	13.0	13.4	13.7

   (A) 0.8   **(B)** 1.0   **(C)** 1.2   **(D)** 1.4   **(E)** 1.6

In Questions 35–38, the position of a particle moving along a straight line is given by $s = t^3 - 6t^2 + 12t - 8$.

35. The distance s is increasing for

   (A) $t < 2$ **(B)** all t except $t = 2$ **(C)** $1 < t < 3$
   (D) $t < 1$ or $t > 3$ **(E)** $t > 2$

36. The minimum value of the speed is

   (A) 1   **(B)** 2   **(C)** 3   **(D)** 0   **(E)** none of these

37. The acceleration is positive

   (A) when $t > 2$ **(B)** for all t, $t \neq 2$ **(C)** when $t < 2$
   (D) for $1 < t < 3$ **(E)** for $1 < t < 2$

38. The speed of the particle is decreasing for

   (A) $t > 2$ **(B)** $t < 3$ **(C)** all t
   (D) $t < 1$ or $t > 2$ **(E)** none of these

In Questions 39–41, a particle moves along a horizontal line and its position at time t is $s = t^4 - 6t^3 + 12t^2 + 3$.

39. The particle is at rest when t is equal to

   (A) 1 or 2   **(B)** 0   **(C)** $\dfrac{9}{4}$ **(D)** 0, 2, or 3   **(E)** none of these

40. The velocity, v, is increasing when

   (A) $t > 1$ **(B)** $1 < t < 2$ **(C)** $t < 2$
   (D) $t < 1$ or $t > 2$ **(E)** $t > 0$

41. The speed of the particle is increasing for

   (A) $0 < t < 1$ or $t > 2$ **(B)** $1 < t < 2$ **(C)** $t < 2$
   (D) $t < 0$ or $t > 2$ **(E)** $t < 0$

For Questions 42 and 43, $f'(x) = x \sin x - \cos x$ for $0 < x < 4$.

42. f has a local maximum when x is approximately

(A) 0.9   (B) 1.2   (C) 2.3   (D) 3.4   (E) 3.7

43. f has a point of inflection when x is approximately

(A) 0.9   (B) 1.2   (C) 2.3   (D) 3.4   (E) 3.7

44. The displacement from the origin of a particle moving on a line is given by $s = t^4 - 4t^3$. The maximum displacement during the time interval $-2 \leqq t \leqq 4$ is

(A) 27   (B) 3   (C) $12\sqrt{3} + 3$
(D) 48   (E) none of these

45. If a particle moves along a line according to the law $s = t^5 + 5t^4$, then the number of times it reverses direction is

(A) 0   (B) 1   (C) 2   (D) 3   (E) 4

In Questions 46–49, the motion of a particle in a plane is given by the pair of equations $x = 2t$ and $y = 4t - t^2$.

*46. The particle moves along

(A) an ellipse   (B) a circle   (C) a hyperbola
(D) a line   (E) a parabola

*47. The speed of the particle at any time t is

(A) $\sqrt{6 - 2t}$ (B) $2\sqrt{t^2 - 4t + 5}$ (C) $2\sqrt{t^2 - 2t + 5}$

(D) $\sqrt{8}(|t - 2|)$ (E) $2(|3 - t|)$

*48. The minimum speed of the particle is

(A) 2   (B) $2\sqrt{2}$ (C) 0   (D) 1   (E) 4

*49. The acceleration of the particle

(A) depends on t
(B) is always directed upward
(C) is constant both in magnitude and in direction
(D) never exceeds 1 in magnitude
(E) is none of these

---

*Questions preceded by an asterisk are likely to appear only on the Calculus BC Examination.

In Questions 50–53, $\mathbf{R} = 3 \cos \frac{\pi}{3} t \mathbf{i} + 2 \sin \frac{\pi}{3} t \mathbf{j}$ is the (position) vector $x\mathbf{i} + y\mathbf{j}$ from the origin to a moving point $P(x, y)$ at time t.

*50.   A single equation in x and y for the path of the point is

   (A) $x^2 + y^2 = 13$ (B) $9x^2 + 4y^2 = 36$ (C) $2x^2 + 3y^2 = 13$
   (D) $4x^2 + 9y^2 = 1$ (E) $4x^2 + 9y^2 = 36$

*51.   When $t = 3$, the speed of the particle is

   (A) $\dfrac{2\pi}{3}$ (B) 2 (C) 3 (D) π (E) $\dfrac{\sqrt{13}}{3}\pi$

*52.   The magnitude of the acceleration when $t = 3$ is

   (A) 2 (B) $\dfrac{\pi^2}{3}$ (C) 3 (D) $\dfrac{2\pi^2}{9}$ (E) π

*53.   At the point where $t = \dfrac{1}{2}$, the slope of the curve along which the particle moves is

   (A) $-\dfrac{2\sqrt{3}}{9}$ (B) $-\dfrac{\sqrt{3}}{2}$ (C) $\dfrac{2}{\sqrt{3}}$

   (D) $-\dfrac{2\sqrt{3}}{3}$ (E)   none of these

*54.   If a particle moves along a curve with constant speed, then

   (A)   the magnitude of its acceleration must equal zero
   (B)   the direction of acceleration must be constant
   (C)   the curve along which the particle moves must be a straight line
   (D)   its velocity and acceleration vectors must be perpendicular
   (E)   the curve along which the particle moves must be a circle

*55.   A particle is moving on the curve of $y = 2x - \ln x$ so that $\dfrac{dx}{dt} = -2$ at all time t. At the point $(1,2)$, $\dfrac{dy}{dt}$ is

   (A) 4 (B) 2 (C) -4 (D) 1 (E) -2

---

*Questions preceded by an asterisk are likely to appear only on the Calculus BC Examination.

Use the graph of f' on [0,5], shown below, for Questions 56 and 57.

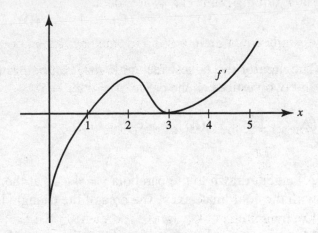

56. f has a local minimum at $x =$

    (A) 0   **(B)** 1   **(C)** 2   **(D)** 3   **(E)** 5

57. f has a point of inflection at $x =$

    (A) 1 only   **(B)** 2 only   **(C)** 3 only
    (D) 2 and 3 only   **(E)** none of these

58. It follows from the graph of f', shown at the right, that

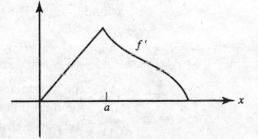

    (A) f is not continuous at $x = a$
    (B) f is continuous but not differentiable
        at $x = a$
    (C) f has a relative maximum at $x = a$
    (D) f has a point of inflection at $x = a$
    (E) none of these

59. A balloon is being filled with helium at the rate of 4 ft³/min. The rate, in square feet per minute, at which the surface area is increasing when the volume is $\dfrac{32\pi}{3}$ ft³ is

    (A) 4π **(B)** 2   **(C)** 4   **(D)** 1   **(E)** 2π

60. A circular conical reservoir, vertex down, has depth 20 ft and radius of the top 10 ft. Water is leaking out so that the surface is falling at the rate of $\dfrac{1}{2}$ ft/hr. The rate, in cubic feet per hour, at which the water is leaving the reservoir when the water is 8 ft deep is

    (A) 4π **(B)** 8π **(C)** 16π **(D)** $\dfrac{1}{4\pi}$ **(E)** $\dfrac{1}{8\pi}$

61. A vertical circular cylinder has radius r ft and height h ft. If the height and radius both increase at the constant rate of 2 ft/sec, then the rate, in square feet per second, at which the lateral surface area increases is

    (A) $4\pi r$ **(B)** $2\pi(r + h)$ **(C)** $4\pi(r + h)$ **(D)** $4\pi rh$ **(E)** $4\pi h$

62. A local minimum value of the function $y = \dfrac{e^x}{x}$ is

(A) $\dfrac{1}{e}$ (B) 1 (C) -1 (D) e (E) 0

63. The area of the largest rectangle that can be drawn with one side along the x-axis and two vertices on the curve of $y = e^{-x^2}$ is

(A) $\sqrt{\dfrac{2}{e}}$ (B) $\sqrt{2e}$ (C) $\dfrac{2}{e}$ (D) $\dfrac{1}{\sqrt{2e}}$ (E) $\dfrac{2}{e^2}$

64. A tangent drawn to the parabola $y = 4 - x^2$ at the point $(1, 3)$ forms a right triangle with the coordinate axes. The area of the triangle is

(A) $\dfrac{5}{4}$ (B) $\dfrac{5}{2}$ (C) $\dfrac{25}{2}$ (D) 1 (E) $\dfrac{25}{4}$

65. If the cylinder of largest possible volume is inscribed in a given sphere, the ratio of the volume of the sphere to that of the cylinder is

(A) $\sqrt{3} : 1$ (B) $\sqrt{3} : 3$ (C) $3 : 1$

(D) $2\sqrt{3} : 3$ (E) $3\sqrt{3} : 4$

66. A line is drawn through the point $(1, 2)$ forming a right triangle with the positive x- and y-axes. The slope of the line forming the triangle of least area is

(A) -1 (B) -2 (C) -4 (D) $-\dfrac{1}{2}$ (E) -3

67. The point(s) on the curve $x^2 - y^2 = 4$ closest to the point $(6, 0)$ is (are)

(A) $(2, 0)$ (B) $(\sqrt{5}, \pm 1)$ (C) $(3, \pm\sqrt{5})$

(D) $(\sqrt{13}, \pm\sqrt{3})$ (E) none of these

68. The sum of the squares of two positive numbers is 200; their minimum product is

(A) 100 (B) $25\sqrt{7}$ (C) 28

(D) $24\sqrt{14}$ (E) none of these

69. The first-quadrant point on the curve $y^2 x = 18$ which is closest to the point $(2, 0)$ is

(A) $(2, 3)$ (B) $(6, \sqrt{3})$ (C) $(3, \sqrt{6})$

(D) $(1, 3\sqrt{2})$ (E) none of these

70. Two cars are traveling along perpendicular roads, car A at 40 mph, car B at 60 mph. At noon, when car A reaches the intersection, car B is 90 mi away, and moving toward it. At 1 P.M. the rate, in miles per hour, at which the distance between the cars is changing is

(A) -40 (B) 68 (C) 4 (D) -4 (E) 40

71. For Question 70, if t is the number of hours of travel after noon, then the cars are closest together when t is

(A) 0   (B) $\dfrac{27}{26}$ (C) $\dfrac{9}{5}$ (D) $\dfrac{3}{2}$ (E) $\dfrac{14}{13}$

The graph for Questions 72 and 73 shows the velocity of an object moving along a straight line during the time interval $0 \leqslant t \leqslant 12$.

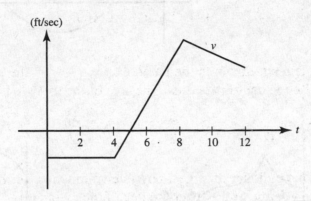

72. For what t does this object attain its maximum acceleration?

(A) $0 < t < 4$ (B) $4 < t < 8$ (C) $t = 5$ (D) $t = 8$ (E) $t = 12$

73. The object reverses direction at $t =$

(A) 4 only   (B) 5 only   (C) 8 only
(D) 5 and 8   (E) none of these

74. The graph of f' is shown below. If we know that $f(2) = 10$, then the local linearization of f at $x = 2$ is $f(x) \simeq$

(A) $\dfrac{x}{2} + 2$ (B) $\dfrac{x}{2} + 9$ (C) $3x - 3$

(D) $3x + 4$ (E) $10x - 17$

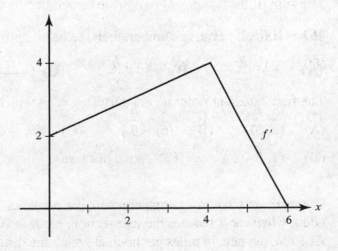

75. Given f' as graphed, which could be the graph of f?

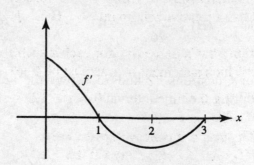

(A)

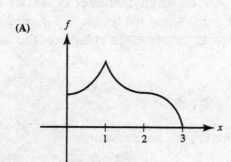

(B)

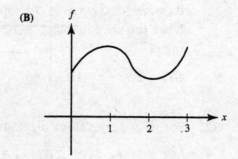

(C)

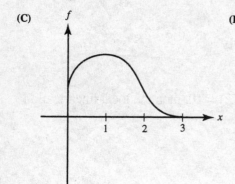

(D)

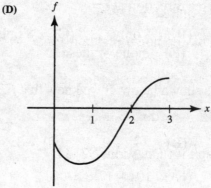

(E)

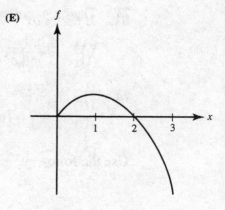

76. If h is a small negative number, then the best approximation for $\sqrt[3]{27 + h}$ is

(A) $\quad 3 + \dfrac{h}{27}$ **(B)** $\quad 3 - \dfrac{h}{27}$ **(C)** $\quad \dfrac{h}{27}$

(D) $\quad -\dfrac{h}{27}$ **(E)** $\quad 3 - \dfrac{h}{9}$

77. If $f(x) = xe^{-x}$, then at $x = 0$

(A) f is increasing   (B) f is decreasing   (C) f has a relative maximum
(D) f has a relative minimum   (E) f' does not exist

78. A function f has a derivative for each x such that $|x| < 2$ and has a local minimum at $(2, -5)$. Which statement below must be true?

(A) $f'(2) = 0$.
(B) f' exists at $x = 2$.
(C) The graph is concave up at $x = 2$.
(D) $f'(x) < 0$ if $x < 2$, $f'(x) > 0$ if $x > 2$.
(E) None of the preceding is necessarily true.

79. The height of a rectangular box is 10 in. Its length increases at the rate of 2 in./sec; its width decreases at the rate of 4 in./sec. When the length is 8 in. and the width is 6 in., the rate, in cubic inches per second, at which the volume of the box is changing is

(A) 200   (B) 80   (C) −80   (D) −200   (E) −20

80. The tangent to the curve $x^3 + x^2y + 4y = 1$ at the point $(3, -2)$ has slope

(A) −3   (B) $-\dfrac{23}{9}$ (C) $-\dfrac{27}{13}$ (D) $-\dfrac{11}{9}$ (E) $-\dfrac{15}{13}$

81. If $f(x) = ax^4 + bx^2$ and $ab > 0$, then

(A) the curve has no horizontal tangents
(B) the curve is concave up for all x
(C) the curve is concave down for all x
(D) the curve has no inflection point
(E) none of the preceding is necessarily true

Use the following graph for Questions 82–84.

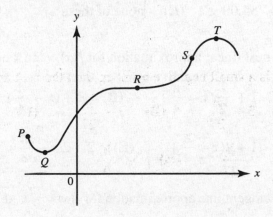

82. At which labeled point do both $\dfrac{dy}{dx}$ and $\dfrac{d^2y}{dx^2}$ equal zero?

(A) P (B) Q (C) R (D) S (E) T

83. At which labeled point is $\dfrac{dy}{dx}$ positive and $\dfrac{d^2y}{dx^2}$ equal to zero?

(A) P (B) Q (C) R (D) S (E) T

84. At which labeled point is $\dfrac{dy}{dx}$ equal to zero and $\dfrac{d^2y}{dx^2}$ negative?

(A) P (B) Q (C) R (D) S (E) T

85. Which statement below is true about the curve $y = \dfrac{2x^2 + 4}{2 + 7x - 4x^2}$?

(A)   The line $x = -\dfrac{1}{4}$ is a vertical asymptote.

(B)   The line $x = 1$ is a vertical asymptote.

(C)   The line $y = -\dfrac{1}{4}$ is a horizontal asymptote.

(D)   The graph has no vertical or horizontal asymptotes.

(E)   The line $x = 2$ is a horizontal asymptote.

*86. The equation of the tangent to the curve with parametric equations $x = 2t + 1$, $y = 3 - t^3$ at the point where $t = 1$ is

(A) $2x + 3y = 12$ (B) $3x + 2y = 13$ (C) $6x + y = 20$

(D) $3x - 2y = 5$ (E)   none of these

87. Approximately how much less than 4 is $\sqrt[3]{63}$?

(A) $\dfrac{1}{48}$ (B) $\dfrac{1}{16}$ (C) $\dfrac{1}{3}$ (D) $\dfrac{2}{3}$ (E)   1

88. If $f(6) = 30$ and $f'(x) = \dfrac{x^2}{x+3}$, then an estimate of $f(6.02)$, using the local linearization, is

(A)   29.92   (B)   30.02   (C)   30.08

(D)   34.00   (E)   none of these

89. The best linear approximation for $f(x) = \tan x$ near $x = \dfrac{\pi}{4}$ is

(A) $1 + \dfrac{1}{2}(x - \dfrac{\pi}{4})$ (B) $1 + (x - \dfrac{\pi}{4})$ (C) $1 + \sqrt{2}\,(x - \dfrac{\pi}{4})$

(D) $1 + 2(x - \dfrac{\pi}{4})$ (E) $2 + 2(x - \dfrac{\pi}{4})$

90. The tangent line approximation for $f(x) = \sqrt{x^2 + 16}$ near $x = -3$ is

(A) $5 - \dfrac{3}{5}(x - 3)$ (B) $5 + \dfrac{3}{5}(x - 3)$ (C) $5 - \dfrac{3}{5}(x + 3)$

(D) $3 - \dfrac{5}{3}(x - 3)$ (E) $3 + \dfrac{3}{5}(x + 3)$

*Questions preceded by an asterisk are likely to appear only on the Calculus BC Examination.

91. When h is near zero, e^{kh}, using local linearization, is approximately

  (A) k **(B)** kh **(C)**   1   **(D)** $1 + k$ **(E)** $1 + kh$

92. If $f(x) = cx^2 + dx + e$ for the function shown in the graph, then

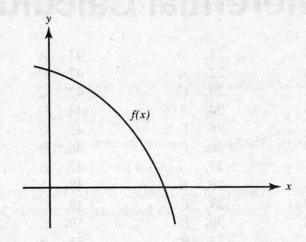

  (A) c, d, and e are all positive
  (B) $c > 0, d < 0, e < 0$
  (C) $c > 0, d < 0, e > 0$
  (D) $c < 0, d > 0, e > 0$
  (E) $c < 0, d < 0, e > 0$

Answers for Set 4: Applications of Differential Calculus

1.	D	21.	D	41.	A	61.	C	81.	D
2.	A	22.	A	42.	D	62.	D	82.	C
3.	E	23.	C	43.	C	63.	A	83.	D
4.	B	24.	B	44.	D	64.	E	84.	E
5.	A	25.	E	45.	C	65.	A	85.	A
6.	D	26.	C	46.	E	66.	B	86.	B
7.	E	27.	A	47.	B	67.	C	87.	A
8.	C	28.	B	48.	A	68.	E	88.	C
9.	D	29.	E	49.	C	69.	C	89.	D
10.	C	30.	D	50.	E	70.	D	90.	C
11.	E	31.	B	51.	A	71.	B	91.	E
12.	B	32.	E	52.	B	72.	B	92.	E
13.	D	33.	C	53.	D	73.	B		
14.	A	34.	E	54.	D	74.	D		
15.	C	35.	B	55.	E	75.	C		
16.	B	36.	D	56.	B	76.	A		
17.	D	37.	A	57.	D	77.	A		
18.	D	38.	E	58.	D	78.	E		
19.	D	39.	B	59.	C	79.	D		
20.	E	40.	D	60.	B	80.	E		

1. D. Substituting $y = 2$ yields $x = 1$. We find y' implicitly.

$$3y^2 y' - (2xyy' + y^2) = 0; \quad (3y^2 - 2xy)y' - y^2 = 0.$$

Replace x by 1 and y by 2; solve for y'.

2. A. $2yy' - (xy' + y) - 3 = 0$. Replace x by 0 and y by -1; solve for y'.

3. E. Find the slope of the curve at $x = \frac{\pi}{2}$: $y' = x \cos x + \sin x$. At $x = \frac{\pi}{2}$,

$y' = \frac{\pi}{2} \cdot 0 + 1$. The equation is $y - \frac{\pi}{2} = 1\left(x - \frac{\pi}{2}\right)$.

4. B. Since $y' = e^{-x}(1 - x)$ and $e^{-x} > 0$ for all x, $y' = 0$ when $x = 1$.

5. A. Since the slope of the tangent to the curve is $y' = \frac{1}{\sqrt{2x+1}}$, the slope of the normal is $-\sqrt{2x+1}$. So $-\sqrt{2x+1} = -3$ and $2x + 1 = 9$.

6. D. The slope $y' = 5x^4 + 3x^2 - 2$. Let $g = y'$. Since $g'(x) = 20x^3 + 6x = 2x(10x^2 + 3)$, $g'(x) = 0$ only if $x = 0$. Since $g''(x) = 60x^2 + 6$, g'' is always positive, assuring that $x = 0$ yields the minimum slope. Find y' when $x = 0$.

7. E. The slope $y' = \dfrac{2x}{4}$; at the given point $y' = -\dfrac{4}{4} = -1$ and $y = 1$. The equation is therefore

$$y - 1 = -1(x + 2) \qquad \text{or} \qquad x + y + 1 = 0.$$

8. C. Since $2x - 2yy' = 0$, $y' = \dfrac{x}{y}$. At $(4, 2)$, $y' = 2$. The equation of the tangent at $(4, 2)$ is $y - 2 = 2(x - 4)$.

9. D. Since $y' = \dfrac{y}{2y - x}$, the tangent is vertical for $x = 2y$. Substitute in the given equation.

10. C. $a \approx \dfrac{\Delta v}{\Delta t} = \dfrac{v(8) - v(4)}{8 - 4} = \dfrac{10 - 16}{4}$ ft/sec^2.

11. E. Since $f' < 0$ on $5 \leqslant x < 7$, the function decreases as it approaches the right endpoint.

12. B. For $x < 5$, $f' > 0$, so f is increasing; for $x > 5$, f is decreasing.

13. D. f being concave downward implies that $f'' < 0$, which implies that f' is decreasing.

14. A. Since $V = \dfrac{4}{3}\pi r^3$, therefore $V'(r) = 4\pi r^2$. The linearization equation gives the approximate increase in volume as $V'(3)(0.1)$ or 11.310 in.^{3}

15. C. Differentiating implicitly yields $4x - 3y^2 y' = 0$. So $y' = \dfrac{4x}{3y^2}$. The linear approximation for the true value of y when x changes from 3 to 3.04 is

$$\dot{y}_{\text{at } x=3} + y'_{\text{at point } (3,2)} \cdot (3.04 - 3).$$

Since it is given that, when $x = 3$, $y = 2$, the approximate value of y is

$$2 + \frac{4x}{3y^2}_{\text{at } (3,2)} \cdot (0.04)$$

or

$$2 + \frac{12}{12} \cdot (0.04) = 2.04 \,.$$

16. B. We want to approximate the change in area of the square when a side of length e increases by $0.01e$. The answer is

$$A'(e)(0.01e) \quad \text{or} \quad 2e(0.01e).$$

17. D. Linear approximation yields $V'(10)(\pm 0.1)$. Since $V = e^3$, we get $3e^2$ (at $e = 10$) times ± 0.1, which equals $300(\pm 0.1) = 30$ in.^{3}

18. D. Speed is the magnitude of velocity; at $t = 3$, speed $= 10$ ft/sec.

19. D.   Speed increases from 0 at $t = 2$ to 10 at $t = 3$; it is constant or decreasing elsewhere.

20. E.   Acceleration is positive when the *velocity* increases.

21. D.   Acceleration is undefined when velocity is not differentiable. Here that occurs at $t = 1, 2, 3$.

22. A.   Acceleration is the derivative of velocity. Since the velocity is linear, its derivative is its slope.

23. C.   Positive velocity implies motion to the right ($t < 2$); negative velocity ($t > 2$) implies motion to the left.

24. B.   The average rate of change of velocity is $\dfrac{v(3) - v(0)}{3 - 0} = \dfrac{-10 - 5}{-3}$ ft/sec².

25. E. $f'(x) = 4x^3 - 8x = 4x(x^2 - 2).\ f' = 0$ if $x = 0$ or $\pm\sqrt{2}$.
$f''(x) = 12x^2 - 8; f''$ is positive if $x = \pm\sqrt{2}$, negative if $x = 0$.

26. C.   Since $f''(x) = 4(3x^2 - 2)$, it equals 0 if $x = \pm\sqrt{\dfrac{2}{3}}$. Since f'' changes sign as x increases through each of these, both are inflection points.

27. A.   The domain of y is $\{x \mid x \leqq 2\}$. Note that y is negative for each x in the domain except 2, where $y = 0$.

28. B. $f'(x)$ changes sign only as x passes through zero.

29. E.   The derivative equals $x \cos \dfrac{1}{x}\left(-\dfrac{1}{x^2}\right) + \sin \dfrac{1}{x}$.

30. D.   Since $f'(x) = 4 \cos x + 3 \sin x$, the critical values of x are those for which $\tan x = -\dfrac{4}{3}$. For these values,

> when $\dfrac{\pi}{2} < x < \pi$, then $\sin x = \dfrac{4}{5}$ and $\cos x = -\dfrac{3}{5}$;
>
> but
>
> when $\dfrac{3\pi}{2} < x < 2\pi$, then $\sin x = -\dfrac{4}{5}$ and $\cos x = \dfrac{3}{5}$.

If these are used to determine the sign of $f''(x)$, we see that the second-quadrant x yield a negative second derivative, but the fourth-quadrant x give a positive second derivative. It follows that the function is a maximum for the former set, for which f has the value 5. Note that $f(0) = f(2\pi) = -3$.

31. B.   We find the points of intersection. From the first curve we get $y = \dfrac{2}{x}$, which yields, in the second curve,

$$x^2 - \frac{4}{x^2} = 3 \quad \text{or} \quad x^4 - 3x^2 - 4 = 0 .$$

$(x^2 - 4)(x^2 + 1) = 0$ if $x = \pm 2$. So the points of intersection are $(2, 1)$ and $(-2, -1)$. Since $m_1 = -\dfrac{2}{x^2}$ and $m_2 = \dfrac{x}{y}$, at each point $m_1 = -\dfrac{1}{2}$ and $m_2 = 2$.

32. E. The slope of $y = x^3$ is $3x^2$. It is equal to 3 when $x = \pm 1$. At $x = 1$, the equation of the tangent is

$$y - 1 = 3(x - 1) \qquad \text{or} \qquad y = 3x + 2.$$

At $x = -1$, the equation is

$$y + 1 = 3(x + 1) \qquad \text{or} \qquad y = 3x + 2.$$

33. C. Let the tangent to the parabola from $(3, 5)$ meet the curve at (x_1, y_1). Its equation is $y - 5 = 2x_1(x - 3)$. Since the point (x_1, y_1) is on both the tangent and the parabola, we solve simultaneously:

$$y_1 - 5 = 2x_1(x_1 - 3) \qquad \text{and} \qquad y_1 = x_1^2$$

The points of tangency are $(5, 25)$ and $(1, 1)$. The slopes, which equal $2x_1$, are 10 and 2.

34. E. $a \approx \dfrac{\Delta v}{\Delta t} = \dfrac{v(1.5) - v(1.0)}{0.5} = \dfrac{13.2 - 12.2}{0.5}$ ft/sec².

35. B. The distance is increasing when v is positive. Since $v = \dfrac{ds}{dt} = 3(t - 2)^2$,

$v > 0$ for all $t \neq 2$.

36. D. The speed $= |v|$. From Question 35, $|v| = v$. The least value of v is 0.

37. A. The acceleration $a = \dfrac{dv}{dt}$. From Question 35, $a = 6(t - 2)$.

38. E. The speed is decreasing when v and a have opposite signs. The answer is $t < 2$, since for all such t the velocity is positive while the acceleration is negative. For $t > 2$, both v and a are positive.

39. B. The particle is at rest when $v = 0$, $v = 2t(2t^2 - 9t + 12) = 0$ only if $t = 0$. Note that the discriminant of the quadratic factor $(b^2 - 4ac)$ is negative.

40. D. Since $a = 12(t - 1)(t - 2)$, we check the signs of a in the intervals $t < 1$, $1 < t < 2$, and $t > 2$. We choose those where $a > 0$.

41. A. From Questions 39 and 40 we see that $v > 0$ if $t > 0$ and that $a > 0$ if $t < 1$ or $t > 2$. So both v and a are positive if $0 < t < 1$ or $t > 2$. There are no values of t for which both v and a are negative.

42. D. The graph of $f'(x) = x \sin x - \cos x$ is drawn here in the window $[0,4] \times [-3,3]$:

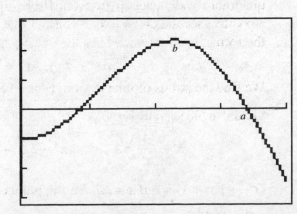

A local maximum exists where f' changes from positive to negative; use TRACE to approximate a.

43. **C.** f'' changes sign when f' changes from increasing to decreasing (or vice versa). Again, use TRACE to find the abscissa at b.

44. **D.** See the figure, which shows the motion of the particle during the time interval $-2 \leqslant t \leqslant 4$. The particle is at rest when $t = 0$ or 3, but reverses direction only at 3. The endpoints need to be checked here, of course. Indeed, the maximum displacement occurs at one of those, namely, when $t = -2$.

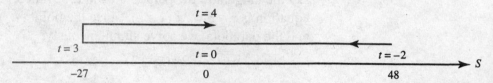

45. **C.** Since $v = 5t^3(t + 4)$, $v = 0$ when $t = -4$ or 0. Note that v does change sign at each of these times.

46. **E.** Eliminating t yields the equation $y = -\dfrac{1}{4}x^2 + 2x$.

47. **B.** $|\mathbf{v}| = \sqrt{\left(\dfrac{dx}{dt}\right)^2 + \left(\dfrac{dy}{dt}\right)^2} = \sqrt{2^2 + (4 - 2t)^2}$.

48. **A.** Since $|\mathbf{v}| = 2\sqrt{t^2 - 4t + 5}$, $\dfrac{d|\mathbf{v}|}{dt} = \dfrac{2t - 4}{\sqrt{t^2 - 4t + 5}} = 0$ if $t = 2$. We note that, as t increases through 2, the signs of $|\mathbf{v}|'$ are $-$, 0, $+$, assuring a minimum of $|\mathbf{v}|$ at $t = 2$. Evaluate $|\mathbf{v}|$ at $t = 2$.

49. **C.** The direction of $\mathbf{a}$ is $\tan^{-1} \dfrac{\frac{d^2y}{dt^2}}{\frac{d^2x}{dt^2}}$. Since $\dfrac{d^2x}{dt^2} = 0$ and $\dfrac{d^2y}{dt^2} = -2$, the acceleration is always directed downward. Its magnitude, $\sqrt{0^2 + (-2)^2}$, is 2 for all t.

50. **E.** Since $x = 3 \cos \dfrac{\pi}{3}t$ and $y = 2 \sin \dfrac{\pi}{3}t$, we note that $\left(\dfrac{x}{3}\right)^2 + \left(\dfrac{y}{2}\right)^2 = 1$.

51. **A.** Note that $\mathbf{v} = -\pi \sin \dfrac{\pi}{3}t\mathbf{i} + \dfrac{2\pi}{3} \cos \dfrac{\pi}{3}t\mathbf{j}$. At $t = 3$,

$$|\mathbf{v}| = \sqrt{(-\pi \cdot 0)^2 + \left(\dfrac{2\pi}{3} \cdot -1\right)^2}.$$

52. **B.** $\mathbf{a} = -\dfrac{\pi^2}{3} \cos \dfrac{\pi}{3}t\mathbf{i} - \dfrac{2\pi^2}{9} \sin \dfrac{\pi}{3}t\mathbf{j}$. At $t = 3$,

$$|\mathbf{a}| = \sqrt{\left(\dfrac{-\pi^2}{3} \cdot -1\right)^2 + \left(\dfrac{-2\pi^2}{9} \cdot 0\right)^2}.$$

53. D. The slope of the curve is the slope of **v**, namely, $\dfrac{\frac{dy}{dt}}{\frac{dx}{dt}}$. At $t = \dfrac{1}{2}$, the slope is equal to

$$\frac{\frac{2\pi}{3}\cdot\cos\frac{\pi}{6}}{-\pi\cdot\sin\frac{\pi}{6}} = -\frac{2}{3}\cot\frac{\pi}{6}.$$

54. D. Using the notations v_x, v_y, a_x, and a_y, we are given that $|\mathbf{v}| = \sqrt{v_x^2 + v_y^2} = k$, where k is a constant. Then

$$\frac{d|\mathbf{v}|}{dt} = \frac{v_x a_x + v_y a_y}{|\mathbf{v}|} = 0 \quad \text{or} \quad \frac{v_x}{v_y} = -\frac{a_y}{a_x}.$$

55. E. $\dfrac{dy}{dt} = \left(2 - \dfrac{1}{x}\right)\dfrac{dx}{dt} = \left(2 - \dfrac{1}{x}\right)(-2).$

56. B. A local minimum exists where f changes from decreasing ($f' < 0$) to increasing ($f' > 0$). Note that f has local maxima at both endpoints, $x = 0$ and $x = 5$.

57. D. See Answer 43.

58. D. f' changes from increasing ($f'' > 0$) to decreasing ($f'' < 0$). Note that f is differentiable at a (because $f'(a)$ exists) and therefore continuous at a.

59. C. Since $V = \dfrac{4}{3}\pi r^3$, $\dfrac{dV}{dt} = 4\pi r^2 \dfrac{dr}{dt}$. Since $\dfrac{dV}{dt} = 4$, $\dfrac{dr}{dt} = \dfrac{1}{\pi r^2}$. When $V = \dfrac{32\pi}{3}$, $r = 2$ and $\dfrac{dr}{dt} = \dfrac{1}{4\pi}$.

$$S = 4\pi r^2; \qquad \frac{dS}{dt} = 8\pi r\frac{dr}{dt};$$

when $r = 2$, $\dfrac{dS}{dt} = 8\pi(2)\left(\dfrac{1}{4\pi}\right) = 4.$

60. B. See Figure N4–23 on page 124. Replace the printed measurements of the radius and height by 10 and 20, respectively. We are given here that $r = \dfrac{h}{2}$ and that $\dfrac{dh}{dt} = -\dfrac{1}{2}$. Since $V = \dfrac{1}{3}\pi r^2 h$, we have $V = \dfrac{\pi}{3}\dfrac{h^3}{4}$, so

$$\frac{dV}{dt} = \frac{\pi h^2}{4}\frac{dh}{dt} = \frac{-\pi h^2}{8}.$$

Replace h by 8.

61. C. We know that $\dfrac{dh}{dt} = \dfrac{dr}{dt} = 2$. Since $S = 2\pi rh$,

$$\frac{dS}{dt} = 2\pi\left(r\frac{dh}{dt} + h\frac{dr}{dt}\right).$$

62. D. $y' = \dfrac{e^x(x-1)}{x^2}$ ($x \neq 0$). Since $y' = 0$ if $x = 1$ and changes from negative to positive as x increases through 1, $x = 1$ yields a minimum. Evaluate y at $x = 1$.

63.   **A.**   The domain of y is $-\infty < x < \infty$. The graph of y, which is nonnegative, is symmetric to the y-axis. The inscribed rectangle has area $A = 2xe^{-x^2}$. Thus $A' = \dfrac{2(1-2x^2)}{e^{x^2}}$, which is 0 when the positive value of x is $\dfrac{\sqrt{2}}{2}$. This value of x yields maximum area. Evaluate A.

64.   **E.**   The equation of the tangent is $y = -2x + 5$. Its intercepts are $\dfrac{5}{2}$ and 5.

65.   **A.**   See the figure, where V, the volume of the cylinder, equals $2\pi r^2 h$. Then $V = 2\pi(R^2h - h^3)$ is a maximum for $h^2 = \dfrac{R^2}{3}$, and the ratio of the volumes of sphere to cylinder is

$$\frac{4}{3}\pi R^3 : 2\pi \cdot \frac{2}{3}R^2 \cdot \frac{R}{\sqrt{3}} \quad \text{or} \quad \sqrt{3}:1.$$

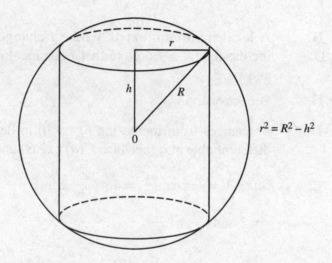

66.   **B.**   See the figure. If we let m be the slope of the line, then its equation is $y - 2 = m(x - 1)$ with intercepts as indicated in the figure.

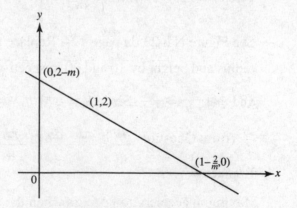

The area A of the triangle is given by

$$A = \frac{1}{2}(2 - m)\left(1 - \frac{2}{m}\right) = \frac{1}{2}\left(4 - \frac{4}{m} - m\right).$$

Then $\dfrac{dA}{dm} = \dfrac{1}{2}\left(\dfrac{4}{m^2} - 1\right)$ and equals 0 when $m = \pm 2$; m must be negative.

67. **C.**   Let $q = (x - 6)^2 + y^2$ be the quantity to be minimized. Then

$$q = (x - 6)^2 + (x^2 - 4);$$

$q' = 0$ when $x = 3$. Note that it suffices to minimize the square of the distance.

68. **E.**   Minimize, if possible, xy, where $x^2 + y^2 = 200$ $(x, y > 0)$. The derivative of the product is $\dfrac{2(100 - x^2)}{\sqrt{200 - x^2}}$, which equals 0 for $x = 10$. But the signs of the derivative as x increases through 10 show that 10 yields a maximum product. No minimum exists.

69. **C.**   Minimize $q = (x - 2)^2 + \dfrac{18}{x}$. Since

$$q' = 2(x - 2) - \frac{18}{x^2} = \frac{2(x^3 - 2x^2 - 9)}{x^2},$$

$q' = 0$ if $x = 3$. The signs of q' about $x = 3$ assure a minimum.

70. **D.**   See the figure. At noon, car A is at O, car B at N; the cars are shown t hours after noon. We know that $\dfrac{dx}{dt} = -60$ and that $\dfrac{dy}{dt} = 40$. Using $s^2 = x^2 + y^2$, we get

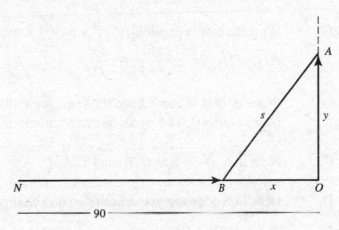

$$\frac{ds}{dt} = \frac{x \dfrac{dx}{dt} + y \dfrac{dy}{dt}}{s} = \frac{-60x + 40y}{s}.$$

At 1 P.M., $x = 30$, $y = 40$, and $s = 50$.

71. **B.** $\dfrac{ds}{dt}$ (from Question 70) is zero when $y = \dfrac{3}{2}x$. Note that $x = 90 - 60t$ and $y = 40t$.

72. **B.**   Maximum acceleration occurs when the derivative (slope) of velocity is greatest.

73. **B.**   The object changes direction only when velocity changes sign. Velocity changes sign from negative to positive at $t = 5$.

74. **D.**   From the graph, $f'(2) = 3$, and we are told the line passes through $(2,10)$. We therefore have $f(x) \simeq 10 + 3(x - 2) = 3x + 4$.

75. **C.** At $x = 1$ and 3, $f'(x) = 0$; therefore f has horizontal tangents.
For $x < 1$, $f' > 0$; therefore f is increasing.
For $x > 1$, $f' < 0$, so f is decreasing.
For $x < 2$, f' is decreasing, so $f'' < 0$ and f is concave downward.
For $x > 2$, f' is increasing, so $f'' > 0$ and f is concave upward.

76. **A.** The best approximation for $\sqrt[3]{27 + h}$ when h is small is the local linear (or tangent line) approximation. If we let $f(h) = \sqrt[3]{27 + h}$, then $f'(h) = \dfrac{1}{3(27 + h)^{2/3}}$ and $f'(0) = \dfrac{1}{3 \cdot 9}$. The approximation for $f(h)$ is $f(0) + f'(0) \cdot h$, which equals $3 + \dfrac{1}{27} h$.

77. **A.** Since $f'(x) = e^{-x}(1 - x)$, $f'(0) > 0$.

78. **E.** Cite counterexamples to show that no one of (A) through (D) is necessarily true.

79. **D.** Since $V = 10\ell w$, $V' = 10\left(\ell \dfrac{dw}{dt} + w \dfrac{d\ell}{dt} \right) = 10(8 \cdot -4 + 6 \cdot 2)$.

80. **E.** We differentiate implicitly: $3x^2 + x^2 y' + 2xy + 4y' = 0$. Then $y' = -\dfrac{3x^2 + 2xy}{x^2 + 4}$.
At $(3, -2)$, $y' = -\dfrac{27 - 12}{9 + 4} = -\dfrac{15}{13}$.

81. **D.** Since $ab > 0$, a and b have the same sign; therefore $f''(x) = 12ax^2 + 2b$ never equals 0. The curve has one horizontal tangent at $x = 0$.

82. **C.** Note that $\dfrac{dy}{dx} = 0$ at Q, R, and T. At Q, $\dfrac{d^2 y}{dx^2} > 0$; at T, $\dfrac{d^2 y}{dx^2} < 0$.

83. **D.** Only at S does the graph both rise and change concavity.

84. **E.** Only at T is the tangent horizontal and the curve concave down.

85. **A.** Note that, since $y = \dfrac{2x^2 + 4}{(2 - x)(1 + 4x)}$, both $x = 2$ and $x = -\dfrac{1}{4}$ are vertical asymptotes. Also, $y = -\dfrac{1}{2}$ is a horizontal asymptote. (See page 31.)

86. **B.** Since $\dfrac{dy}{dx} = -\dfrac{3t^2}{2}$, therefore, at $t = 1$, $\dfrac{dy}{dx} = -\dfrac{3}{2}$. Also, $x = 3$ and $y = 2$.

87. **A.** Let $f(x) = x^{1/3}$, and find the local linearization at $(64, 4)$. Since $f'(x) = \dfrac{1}{3} x^{-2/3}$, $f'(64) = \dfrac{1}{48}$. If we move one unit to the left of 64, the tangent line will drop approximately $\dfrac{1}{48}$ unit.

88. **C.** Since $f'(6) = 4$, the equation of the tangent at $(6, 30)$ is $y - 30 = 4(x - 6)$. Therefore $f(x) \simeq 4x + 6$ and $f(6.02) \simeq 30.08$.

In Questions 89–91, use $f(x) = f(a) + f'(a)(x - a)$ as the best linear (or tangent-line) approximation or local linearization for $f(x)$ near a.

89. **D.** $\quad \tan x \simeq \tan\left(\dfrac{\pi}{4}\right) + \sec^2\left(\dfrac{\pi}{4}\right)\left(x - \dfrac{\pi}{4}\right) = 1 + 2\left(x - \dfrac{\pi}{4}\right)$

90. **C.** $\quad \sqrt{x^2 + 16} \simeq \sqrt{9 + 16} + \dfrac{(-3)}{\sqrt{(-3)^2 + 16}}(x + 3) = 5 - \dfrac{3}{5}(x + 3)$

91. **E.** $\quad e^{kh} \simeq e^{k \cdot 0} + ke^{k \cdot 0}(h - 0) = 1 + kh$

92. **E.** $\quad$ Since the curve has a positive y-intercept, $e > 0$. Note that $f'(x) = 2cx + d$ and $f''(x) = 2c$. Since the curve is concave down, $f''(x) < 0$, implying that $c < 0$. Since the curve is decreasing at $x = 0$, $f'(0)$ must be negative, implying, since $f'(0) = d$, that $d < 0$. Therefore $c < 0$, $d < 0$, and $e > 0$.

Chapter 5

Integration

Review of Definitions and Methods

A. Antiderivatives _____

The *antiderivative* or *indefinite integral* of a function $f(x)$ is a function $F(x)$ whose derivative is $f(x)$. Since the derivative of a constant equals zero, the antiderivative of $f(x)$ is not unique; that is, if $F(x)$ is an integral of $f(x)$, then so is $F(x) + C$, where C is any constant. The arbitrary constant C is called the *constant of integration*. The indefinite integral of $f(x)$ is written as $\int f(x)\, dx$; thus

$$\int f(x)\, dx = F(x) + C \quad \text{if} \quad \frac{dF(x)}{dx} = f(x).$$

The function $f(x)$ is called the *integrand*. The Mean Value Theorem can be used to show that, if two functions have the same derivative on an interval, then they differ at most by a constant; that is, if $\dfrac{dF(x)}{dx} = \dfrac{dG(x)}{dx}$, then

$$F(x) - G(x) = C \qquad (C \text{ a constant}).$$

B. Basic Formulas _____

Familiarity with the following fundamental integration formulas is essential.

$$\int k f(x)\, dx = k \int f(x)\, dx \qquad (k \neq 0) \tag{1}$$

$$\int [f(x) + g(x)]\, dx = \int f(x)\, dx + \int g(x)\, dx \tag{2}$$

$$\int u^n \, du = \frac{u^{n+1}}{n+1} + C \qquad (n \neq -1) \tag{3}$$

$$\int \frac{du}{u} = \ln|u| + C \tag{4}$$

$$\int \cos u \, du = \sin u + C \tag{5}$$

$$\int \sin u \, du = -\cos u + C \tag{6}$$

$$\int \tan u \, du = \ln \left| \sec u \right| + C \tag{7}$$

$$\text{or} \qquad -\ln \left| \cos u \right| + C$$

$$\int \cot u \, du = \ln \left| \sin u \right| + C \tag{8}$$

$$\text{or} \qquad -\ln \left| \csc u \right| + C$$

$$\int \sec^2 u \, du = \tan u + C \tag{9}$$

$$\int \csc^2 u \, du = -\cot u + C \tag{10}$$

$$\int \sec u \tan u \, du = \sec u + C \tag{11}$$

$$\int \csc u \cot u \, du = -\csc u + C \tag{12}$$

$$\int \sec u \, du = \ln \left| \sec u + \tan u \right| + C \tag{13}$$

$$\int \csc u \, du = \ln \left| \csc u - \cot u \right| + C \tag{14}$$

$$\int e^u \, du = e^u + C \tag{15}$$

$$\int a^u \, du = \frac{a^u}{\ln a} + C \qquad (a > 0, \, a \neq 1) \tag{16}$$

$$\int \frac{du}{\sqrt{a^2 - u^2}} = \sin^{-1} \frac{u}{a} + C \tag{17}$$

$$\text{or} \qquad \arcsin \frac{u}{a} + C$$

$$\int \frac{du}{a^2 + u^2} = \frac{1}{a} \tan^{-1} \frac{u}{a} + C \tag{18}$$

$$\text{or} \qquad \frac{1}{a} \arctan \frac{u}{a} + C$$

$$\int \frac{du}{u\sqrt{u^2 - a^2}} = \frac{1}{a} \sec^{-1} \left| \frac{u}{a} \right| + C \tag{19}$$

$$\text{or} \qquad \frac{1}{a} \arcsec \left| \frac{u}{a} \right| + C$$

All the references in the following set of examples are to the preceding basic formulas. In all of these, whenever u is a function of x, we define du to be $u'(x)\ dx$; when u is a function of t, we define du to be $u'(t)\ dt$; and so on.

Example 1. $\displaystyle\int 5x\ dx = 5\int x\ dx$ by (1), $= 5\cdot\dfrac{x^2}{2} + C$ by (3).

Example 2. $\displaystyle\int\left(x^4 + \sqrt[3]{x^2} - \dfrac{2}{x^2} - \dfrac{1}{3\sqrt[3]{x}}\right)dx = \int\left(x^4 + x^{2/3} - 2x^{-2} - \dfrac{1}{3}x^{-1/3}\right)dx$

$\displaystyle = \int x^4\ dx + \int x^{2/3}\ dx - 2\int x^{-2}\ dx - \dfrac{1}{3}\int x^{-1/3}\ dx$ by (1) and (2)

$\displaystyle = \dfrac{x^5}{5} + \dfrac{x^{5/3}}{\frac{5}{3}} - \dfrac{2x^{-1}}{-1} - \dfrac{1}{3}\dfrac{x^{2/3}}{\frac{2}{3}} + C$ by (3) $= \dfrac{x^5}{5} + \dfrac{3}{5}x^{5/3} + \dfrac{2}{x} - \dfrac{1}{2}x^{2/3} + C.$

Example 3. Similarly, $\displaystyle\int(3 - 4x + 2x^3)\ dx = \int 3\ dx - 4\int x\ dx + 2\int x^3\ dx = $

$3x - \dfrac{4x^2}{2} + \dfrac{2x^4}{4} + C = 3x - 2x^2 + \dfrac{x^4}{2} + C.$

Example 4. $\displaystyle\int 2(1 - 3x)^2$ is integrated most efficiently by using formula (3) with $u = 1 - 3x$ and $du = u'(x)\ dx = -3\ dx.$ Since

$$2\int(1 - 3x)^2\ dx = \dfrac{2}{-3}\int(1 - 3x)^2(-3)\ dx,$$

we can now apply formula (3), getting

$$-\dfrac{2}{3}\int(1 - 3x)^2(-3)\ dx = -\dfrac{2}{3}\int u^2\ du = -\dfrac{2}{9}u^3 + C \text{ by (3)}$$

$$= -\dfrac{2}{9}(1 - 3x)^3 + C.$$

If this last function is expanded, note that the two answers to the given integral differ only by a constant.

Example 5. $\displaystyle\int(2x^3 - 1)^5 \cdot x^2\ dx = \dfrac{1}{6}\int(2x^3 - 1)^5 \cdot 6x^2\ dx = \dfrac{1}{6}\int u^5\ du,$ where $u = $

$2x^3 - 1$ and $du = u'(x)\ dx = 6x^2\ dx$; this, by formula (3), equals $\dfrac{1}{6}\cdot\dfrac{u^6}{6} + C = $

$\dfrac{1}{36}(2x^3 - 1)^6 + C.$

Example 6. $\displaystyle\int\sqrt[3]{1 - x}\ dx = \int(1 - x)^{1/3}\ dx = -\int(1 - x)^{1/3}(-1)\ dx = -\int u^{1/3}\ du,$ where

$u = 1 - x$ and $du = -1\ dx$; this, by formula (3) yields $-\dfrac{u^{4/3}}{\frac{4}{3}} + C = -\dfrac{3}{4}(1 - x)^{4/3} + C.$

Example 7. $\int \dfrac{x}{\sqrt{3-4x^2}}\, dx = \int (3-4x^2)^{-1/2} \cdot x\, dx = -\dfrac{1}{8}\int (3-4x^2)^{-1/2}(-8x)\, dx =$

$-\dfrac{1}{8}\int u^{-1/2}\, du = -\dfrac{1}{8}\dfrac{u^{1/2}}{\frac{1}{2}} + C$ by (3) $= -\dfrac{1}{4}\sqrt{3-4x^2} + C.$

Example 8. $\int \dfrac{4x^2}{(x^3-1)^3}\, dx = 4\int (x^3-1)^{-3} \cdot x^2\, dx = \dfrac{4}{3}\int (x^3-1)^{-3}(3x^2)\, dx =$

$\dfrac{4}{3}\dfrac{(x^3-1)^{-2}}{-2} + C = -\dfrac{2}{3}\dfrac{1}{(x^3-1)^2} + C.$

Example 9. $\int \dfrac{(1+\sqrt{x})^4}{\sqrt{x}}\, dx = \int (1+x^{1/2})^4 \cdot \dfrac{1}{x^{1/2}}\, dx.$ Now let $u = 1 + x^{1/2}$, and note

that $du = \dfrac{1}{2}x^{1/2}\, dx$; this gives $2\int (1+x^{1/2})^4\, \dfrac{1}{2x^{1/2}}\, dx = \dfrac{2}{5}(1+\sqrt{x})^5 + C.$

Example 10. $\int (2-y)^2 \cdot \sqrt{y}\, dy = \int (4-4y+y^2) \cdot y^{1/2}\, dy = \int (4y^{1/2} - 4y^{3/2} + y^{5/2})\, dy =$

$4 \cdot \dfrac{2}{3}y^{3/2} - 4 \cdot \dfrac{2}{5} \cdot y^{5/2} + \dfrac{2}{7} \cdot y^{7/2} + C$ by (2) $= \dfrac{8}{3}y^{3/2} + -\dfrac{8}{5}y^{5/2} + \dfrac{2}{7}y^{7/2} + C.$

Example 11. $\int \dfrac{x^3 - x - 4}{2x^2}\, dx = \dfrac{1}{2}\int \left(x - \dfrac{1}{x} - \dfrac{4}{x^2}\right)\, dx = \dfrac{1}{2}\left(\dfrac{x^2}{2} - \ln|x| + \dfrac{4}{x}\right) + C.$

Example 12. $\int \dfrac{3x-1}{\sqrt[3]{1-2x+3x^2}}\, dx = \int (1-2x+3x^2)^{-1/3}(3x-1)\, dx$

$= \dfrac{1}{2}\int (1-2x+3x^2)^{-1/3}(6x-2)\, dx$ or $\dfrac{1}{2} \cdot \dfrac{3}{2}(1-2x+3x^2)^{2/3} + C$ by (3)

$= \dfrac{3}{4}(1-2x+3x^2)^{2/3} + C.$

Example 13. $\int \dfrac{2x^2 - 4x + 3}{(x-1)^2}\, dx = \int \dfrac{2x^2 - 4x + 3}{x^2 - 2x + 1}\, dx = \int \left(2 + \dfrac{1}{(x-1)^2}\right)\, dx =$

$\int 2\, dx + \dfrac{dx}{(x-1)^2} = 2x - \dfrac{1}{x-1} + C.$ This example illustrates the following principle:

If the degree of the numerator of a rational function is not less than that of the denominator, divide until a remainder of lower degree is obtained.

Example 14. $\int \dfrac{du}{u-3} = \ln|u-3| + C$ by (4).

Example 15. $\int \dfrac{z\, dz}{1-4z^2} = -\dfrac{1}{8}\int \dfrac{-8z\, dz}{1-4z^2} = -\dfrac{1}{8}\ln|1-4z^2| + C$ by (4).

Example 16. $\int \dfrac{\cos x}{5+2\sin x}\, dx = \dfrac{1}{2}\int \dfrac{2\cos x}{5+2\sin x}\, dx = \dfrac{1}{2}\ln(5+2\sin x) + C$ by (4) with $u = 5 + 2\sin x$. The absolute-value sign is not necessary here since $(5+2\sin x) > 0$ for all x.

Example 17. $\int \dfrac{e^x}{1-2e^x}\, dx = -\dfrac{1}{2}\int \dfrac{-2e^x}{1-2e^x}\, dx = -\dfrac{1}{2}\ln|1-2e^x| + C$ by (4).

Example 18. $\int \frac{x}{1-x}\, dx = \int \left(-1 + \frac{1}{1-x}\right) dx$ (by long division) $= -x - \ln|1-x| + C$.

Example 19. $\int \sin(1-2y)\, dy = -\frac{1}{2} \int \sin(1-2y)(-2\,dy) =$

$-\frac{1}{2}[-\cos(1-2y)] + C$ by (6) $= \frac{1}{2}\cos(1-2y) + C$.

Example 20. $\int \sin^2 \frac{x}{2} \cos \frac{x}{2}\, dx = 2\int \sin^2 \frac{x}{2} \cos \frac{x}{2} \frac{dx}{2} = \frac{2}{3} \sin^3 \frac{x}{2} + C$ by (3).

Example 21. $\int \frac{\sin x}{1 + 3\cos x}\, dx = -\frac{1}{3} \int \frac{-3\sin x}{1 + 3\cos x}\, dx = -\frac{1}{3} \ln|1 + 3\cos x| + C$.

Example 22. $\int e^{\tan y} \sec^2 y\, dy = e^{\tan y} + C$ by (15) with $u = \tan y$.

Example 23. $\int e^x \tan e^x\, dx = -\ln|\cos e^x| + C$ by (7) with $u = e^x$.

Example 24. $\int \frac{\cos z}{\sin^2 z}\, dz = \int \csc z \cot z\, dz = -\csc z + C$ by (12).

Example 25. $\int \tan t \sec^2 t\, dt = \frac{\tan^2 t}{2} + C$ by (3) with $u = \tan t$ and $du = u'(t)\, dt = \sec^2 t\, dt$.

Example 26. $\int \sec^4 x\, dx = \int \sec^2 x \sec^2 x\, dx = \int (\tan^2 x + 1) \sec^2 x\, dx =$

$\int \tan^2 x \sec^2 x\, dx + \int \sec^2 x\, dx = \frac{\tan^3 x}{3} + \tan x + C$ by (3) and (9).

Example 27. (a) $\int \frac{dz}{\sqrt{9 - z^2}} = \sin^{-1} \frac{z}{3} + C$ by (17) with $u = z$ and $a = 3$.

(b) $\int \frac{z\,dz}{\sqrt{9 - z^2}} = -\frac{1}{2} \int (9 - z^2)^{-1/2}(-2z\,dz) = -\frac{1}{2} \frac{(9 - z^2)^{1/2}}{1/2} + C$ by (3)

$\left(\text{with } u = 9 - z^2, n = -\frac{1}{2}\right) = -\sqrt{9 - z^2} + C$.

(c) $\int \frac{z\,dz}{9 - z^2} = -\frac{1}{2} \ln|9 - z^2| + C$ by (4) with $u = 9 - z^2$.

(d) $\int \frac{z\,dz}{(9 - z^2)^2} = \frac{1}{2(9 - z^2)} + C$ by (3).

(e) $\int \frac{dz}{9 + z^2} = \frac{1}{3} \tan^{-1} \frac{z}{3} + C$ by (18) with $u = z$ and $a = 3$.

Example 28. $\int \frac{dx}{\sqrt{x}(1 + 2\sqrt{x})} = \ln(1 + 2\sqrt{x}) + C$ by (4) with $u = 1 + 2\sqrt{x}$ and

$du = \frac{dx}{\sqrt{x}}$.

Example 29. $\int \sin x \cos x \, dx = \dfrac{1}{2} \sin^2 x + C$ by (3) with $u = \sin x$; OR $=$

$-\dfrac{1}{2} \cos^2 x + C$ by (3) with $u = \cos x$; OR $= -\dfrac{1}{4} \cos 2x + C$ by (6), where we use the trigonometric identity $\sin 2x = 2 \sin x \cos x$.

Example 30. $\int \dfrac{\cos \sqrt{x}}{\sqrt{x}} \, dx = 2 \sin \sqrt{x} + C$ by (5) with $u = \sqrt{x}$.

Example 31. $\int \sin^2 y \, dy = \int \left(\dfrac{1}{2} - \dfrac{\cos 2y}{2} \right) dy = \dfrac{y}{2} - \dfrac{\sin 2y}{4} + C.$

Example 32. $\int \cos^2 z \sin^3 z \, dz = \int \cos^2 z \sin^2 z \sin z \, dz = \int \cos^2 z \, (1 - \cos^2 z) \sin z \, dz =$

$\int \cos^2 z \sin z \, dz - \int \cos^4 z \sin z \, dz = -\dfrac{\cos^3 z}{3} + \dfrac{\cos^5 z}{5} + C$, by repeated application of (3).

Example 33. $\int \dfrac{x \, dx}{x^4 + 1} = \dfrac{1}{2} \tan^{-1} x^2 + C$ by (18) with $u = x^2$ and $a = 1$.

Example 34. $\int \dfrac{dy}{\sqrt{6y - y^2}} = \int \dfrac{dy}{\sqrt{9 - (y^2 - 6y + 9)}} = \sin^{-1} \dfrac{y - 3}{3} + C$ by (17).

Example 35. $\int \dfrac{e^x}{3 + e^{2x}} \, dx = \dfrac{1}{\sqrt{3}} \tan^{-1} \dfrac{e^x}{\sqrt{3}} + C$ by (18) with $u = e^x$ and $a = \sqrt{3}$.

Example 36. $\int \dfrac{e^x - e^{-x}}{e^x + e^{-x}} \, dx = \ln (e^x + e^{-x}) + C$ by (4) with $u = e^x + e^{-x}$.

Example 37. To evaluate $\int \dfrac{x + 1}{x^2 + 4x + 13} \, dx$, we let $u = x^2 + 4x + 13$, note that du is $(2x + 4) \, dx$, and rewrite the integral as

$$\dfrac{1}{2} \int \dfrac{2x + 2}{x^2 + 4x + 13} \, dx = \dfrac{1}{2} \int \dfrac{2x + 2 + 2 - 2}{x^2 + 4x + 13} \, dx$$

$$= \dfrac{1}{2} \int \dfrac{2x + 4}{x^2 + 4x + 13} \, dx - \int \dfrac{dx}{x^2 + 4x + 13}$$

$$= \dfrac{1}{2} \ln (x^2 + 4x + 13) - \int \dfrac{dx}{(x + 2)^2 + 3^2}$$

$$= \dfrac{1}{2} \ln (x^2 + 4x + 13) - \dfrac{1}{3} \tan^{-1} \dfrac{x + 2}{3} + C.$$

Example 38. $\displaystyle\int \frac{2x-1}{\sqrt{8-2x-x^2}}\,dx = -\int \frac{-2x+1}{\sqrt{8-2x-x^2}}\,dx$

$\displaystyle = -\int \frac{-2x-2+1+2}{\sqrt{8-2x-x^2}}\,dx = -\int \frac{-2x-2}{\sqrt{8-2x-x^2}}\,dx - 3\int \frac{dx}{\sqrt{8-2x-x^2}}$

$\displaystyle = -\int (8-2x-x^2)^{-1/2}(-2x-2)\,dx - 3\int \frac{dx}{\sqrt{9-(x^2+2x+1)}}$

$\displaystyle = -2\sqrt{8-2x-x^2} \;\;\text{(by (3))} - 3\sin^{-1}\frac{x+1}{3} \;\;\text{(by (17))} + C.$

Example 39. $\displaystyle\int \frac{dt}{\sin^2 2t} = \int \csc^2 2t\,dt = \frac{1}{2}\int \csc^2 2t\,(2\,dt) = -\frac{1}{2}\cot 2t + C \text{ by (10)}.$

Example 40. $\displaystyle\int \cos^2 4z\,dz = \int \left(\frac{1}{2}+\frac{\cos 8z}{2}\right)dz = \frac{z}{2}+\frac{\sin 8z}{16}+C.$

Example 41. $\displaystyle\int \frac{\sin 2x}{1+\sin^2 x}\,dx = \ln(1+\sin^2 x) + C \text{ by (4)}.$

Example 42. $\displaystyle\int x^2 e^{-x^3}\,dx = -\frac{1}{3}e^{-x^3}+C \text{ by (15) with } u = -x^3.$

Example 43. $\displaystyle\int \frac{dy}{y\sqrt{1+\ln y}} = 2\sqrt{1+\ln y}+C \text{ by (3)}.$

Example 44. $\displaystyle\int \frac{\sqrt{x-1}}{x}\,dx.$ We let $u = \sqrt{x-1}$, $u^2 = x-1$. Then $2u\,du = dx$.
Then we substitute, as follows:

$$\int \frac{\sqrt{x-1}}{x}\,dx = 2\int \frac{u^2\,du}{u^2+1} = 2\int \frac{u^2+1-1}{u^2+1}\,du = 2\int \left(1-\frac{1}{u^2+1}\right)du$$

$$= 2(u-\tan^{-1} u) + C = 2\left(\sqrt{x-1}-\tan^{-1}\sqrt{x-1}\right)+C.$$

---

†C.   Trigonometric Substitutions

Examples 1–44 illustrate the technique of integration by substitution. Trigonometric substitutions are also effective for certain integrals, as we now show.

(1)   Integrals involving $\sqrt{a^2-u^2}$. Let $u = a\sin\theta$ and replace a^2-u^2 by $a^2\cos^2\theta$.

(2)   Integrals involving $\sqrt{a^2+u^2}$. Let $u = a\tan\theta$ and replace a^2+u^2 by $a^2\sec^2\theta$.

(3)   Integrals involving $\sqrt{u^2-a^2}$. Let $u = a\sec\theta$ and replace u^2-a^2 by $a^2\tan^2\theta$.

†Topic not included in 2002–2003 Topical Outline.

We thus obtain a new integrand, in each case involving trigonometric functions of θ. This technique is illustrated in the following examples.

Example 45. $\int \dfrac{x^2}{\sqrt{4-x^2}}\,dx$. Let $x = 2 \sin \theta \ \left(-\dfrac{\pi}{2} < \theta < \dfrac{\pi}{2}\right)$. Then

$$dx = 2 \cos \theta \, d\theta \qquad \text{and} \qquad \sqrt{4-x^2} = 2 \cos \theta,$$

where we have noted that $\cos \theta > 0$ on the prescribed interval. Then

$$\int \frac{x^2\,dx}{\sqrt{4-x^2}} = \int \frac{4 \sin^2 \theta}{2 \cos \theta}\, 2 \cos \theta \, d\theta = 4 \int \sin^2 \theta \, d\theta$$

$$= 4 \int \left(\frac{1}{2} - \frac{\cos 2\theta}{2}\right) d\theta = 4 \left(\frac{\theta}{2} - \frac{\sin 2\theta}{4}\right) + C$$

$$= 2\theta - \sin 2\theta + C \qquad \text{or} \qquad 2\theta - 2 \sin \theta \cos \theta + C.$$

If we draw an appropriate right triangle involving θ and x (see Figure N5–1), we can easily obtain the answer in terms of x.

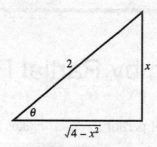

FIGURE N5–1

Thus

$$\int \frac{x^2\,dx}{\sqrt{4-x^2}} = 2 \sin^{-1} \frac{x}{2} - 2 \frac{x}{2}\, \frac{\sqrt{4-x^2}}{2} + C$$

$$= 2 \sin^{-1} \frac{x}{2} - \frac{x\sqrt{4-x^2}}{2} + C.$$

Example 46. $\int \dfrac{dx}{x\sqrt{4x^2+9}}$. Let $2x = 3 \tan \theta \ \left(-\dfrac{\pi}{2} < \theta < \dfrac{\pi}{2}\right)$. Then

$$2\,dx = 3 \sec^2 \theta \, d\theta \qquad \text{and} \qquad \sqrt{4x^2+9} = 3 \sec \theta,$$

where $\sec \theta > 0$ on the above interval. Then

$$\int \frac{dx}{x\sqrt{4x^2+9}} = \int \frac{3 \sec^2 \theta \, d\theta}{2 \cdot \dfrac{3 \tan \theta}{2} \cdot 3 \sec \theta}$$

$$= \frac{1}{3} \int \frac{\sec \theta}{\tan \theta}\, d\theta = \frac{1}{3} \int \csc \theta \, d\theta$$

$$= \frac{1}{3} \ln |\csc \theta - \cot \theta| + C$$

by (14), page 153. We use Figure N5–2 to express the answer in terms of x:

$$\int \frac{dx}{x\sqrt{4x^2+9}} = \frac{1}{3} \ln \left| \frac{\sqrt{4x^2+9}}{2x} - \frac{3}{2x} \right| + C$$

$$= \frac{1}{3} \ln \left| \frac{\sqrt{4x^2+9}-3}{x} \right| + C',$$

where we have replaced $C - \frac{1}{3} \ln 2$ by C' in the final answer.

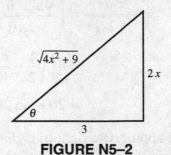

FIGURE N5–2

†*D.   Integration by Partial Fractions_____

The method of partial fractions makes it possible to express a rational function $\frac{f(x)}{g(x)}$ as a sum of simpler fractions. Here $f(x)$ and $g(x)$ are real polynomials in x and it is assumed that $\frac{f(x)}{g(x)}$ is a proper fraction; that is, that $f(x)$ is of lower degree than $g(x)$. If not, we divide $f(x)$ by $g(x)$ to express the given rational function as the sum of a polynomial and a proper rational function. Thus,

$$\frac{x^3 - x^2 - 2}{x(x+1)} = x - \frac{2}{x(x-1)},$$

where the fraction on the right is proper.

Theoretically, every real polynomial can be expressed as a product of (powers of) real linear factors and (powers of) real quadratic factors.†

In the following, the capital letters denote constants to be determined. We consider only nonrepeating linear factors. For each distinct linear factor $(x - a)$ of $g(x)$ we set up one partial fraction of the type $\frac{A}{x - a}$. The techniques for determining the unknown constants are illustrated in the following examples.

---

*An asterisk denotes a topic covered only in Calculus BC.
†In the 2002–2003 Topical Outline for Calculus BC, integration by partial fractions is restricted to "simple partial fractions (nonrepeating linear factors only)."

Example 47. Find $\displaystyle\int \frac{x^2 - x + 4}{x^3 - 3x^2 + 2x}\, dx$.

We factor the denominator and then set

$$\frac{x^2 - x + 4}{x(x-1)(x-2)} = \frac{A}{x} + \frac{B}{x-1} + \frac{C}{x-2}, \tag{1}$$

where the constants A, B, and C are to be determined. It follows that

$$x^2 - x + 4 = A(x-1)(x-2) + Bx(x-2) + Cx(x-1). \tag{2}$$

Since the polynomial on the right in (2) is to be identical to the one on the left, we can find the constants by either of the following methods:

METHOD ONE. We expand and combine the terms on the right in (2), getting

$$x^2 - x + 4 = (A + B + C)x^2 - (3A + 2B + C)x + 2A.$$

We then *equate coefficients of like powers in x* and solve simultaneously. Thus:

using the coefficients of x^2, we get $1 = A + B + C$;

using the coefficients of x, we get $-1 = -(3A + 2B + C)$;

using the constant coefficient, $4 = 2A$

These equations yield $A = 2$, $B = -4$, $C = 3$.

METHOD TWO. Although equation (1) above is meaningless for $x = 0$, $x = 1$, or $x = 2$, it is still true that equation (2) must hold even for these special values. We see, in (2), that:

if $x = 0$,   then $4 = 2A$ and $A = 2$;

if $x = 1$,   then $4 = -B$ and $B = -4$;

if $x = 2$,   then $6 = 2C$ and $C = 3$.

The second method is shorter than the first and more convenient when the denominator of the given fraction can be decomposed into nonrepeating linear factors.

Finally, then, the original integral equals

$$\int \left(\frac{2}{x} - \frac{4}{x-1} + \frac{3}{x-2} \right) dx = 2\ln|x| - 4\ln|x-1| + 3\ln|x-2| + C'$$

$$= \ln \frac{x^2|x-2|^3}{(x-1)^4} + C'.$$

[The symbol "C'" appears here for the constant of integration because C was used in simplifying the original rational function.]

*E. Integration by Parts

The Parts Formula stems from the equation for the derivative of a product:

$$\frac{d}{dx}(uv) = uv' + vu' \qquad \text{or} \qquad uv' = \frac{d}{dx}(uv) - vu'.$$

Integrating, we get

$$\int uv'\, dx = \int \frac{d}{dx}(uv)\, dx - \int vu'\, dx$$

$$= uv - \int vu'\, dx\,,$$

the Parts Formula. Success in using this important technique depends on being able to separate a given integral into parts u and v' so that (a) v' can be integrated, and (b) $\int vu'\, dx$ is no more difficult to calculate than the original integral.

Example 48. To integrate $\int x \cos x\, dx$, we let $u = x$ and $v' = \cos x$. Then $u' = 1$ and $v = \sin x$. Thus, the Parts Formula yields

$$\int x \cos x\, dx = x \sin x - \int \sin x\, dx = x \sin x + \cos x + C.$$

Example 49. To integrate $\int x^2 e^x\, dx$, we let $u = x^2$ and $v' = e^x\, dx$. Then $u' = 2x$ and $v = e^x$, so $\int x^2 e^x\, dx = x^2 e^x - \int 2\, x\, e^x\, dx$. We use the Parts Formula, this time letting $u = x$ and $v' = e^x$ so that $u' = 1$ and $v = e^x$. Thus,

$$\int x^2 e^x\, dx = x^2 e^x - 2\left(xe^x - \int e^x\, dx\right) = x^2 e^x - 2xe^x + 2e^x + C.$$

Example 50. Let $I = e^x \cos x\, dx$. To integrate, we can let $u = e^x$ and $v' = \cos x$; then $u' = e^x$, $v = \sin x$. Thus,

$$I = e^x \sin x - \int e^x \sin x\, dx.$$

To evaluate the integral on the right we again let $u = e^x$, $v' = \sin x$, so that $u' = e^x$ and $v = -\cos x$. Then

$$I = e^x \sin x - \left(-e^x \cos x + \int e^x \cos x\, dx\right)$$

$$= e^x \sin x + e^x \cos x - I.$$

$$2I = e^x (\sin x + \cos x),$$

$$I = \frac{1}{2} e^x (\sin x + \cos x) + C.$$

*An asterisk denotes a topic covered only in Calculus BC.

Example 51. To integrate $\int x^4 \ln x\, dx$, we let $u = \ln x$ and $v' = x^4$. Then, $u' = \dfrac{1}{x}$ and $v = \dfrac{x^5}{5}$. Thus,

$$\int x^4 \ln x\, dx = \frac{x^5}{5} \ln x - \frac{1}{5} \int x^4 dx = \frac{x^5}{5} \ln x - \frac{x^5}{25} + C.$$

F. Applications of Antiderivatives; Differential Equations

The following examples show how we use given conditions to determine constants of integration.

Example 52. If $f'(x) = 3x^2$ and $f(1) = 6$, then

$$f(x) = \int 3x^2\, dx = x^3 + C.$$

Since $f(1) = 6$, $1^3 + C$ must equal 6; so C must equal $6 - 1$ or 5, and $f(x) = x^3 + 5$.

Example 53. If the slope of a curve at each point (x, y) equals the reciprocal of the abscissa and if the curve contains the point $(e, -3)$, we are given that $\dfrac{dy}{dx} = \dfrac{1}{x}$ and that $y = -3$ when $x = e$. This equation is also solved by integration. Since $\dfrac{dy}{dx} = \dfrac{1}{x}$, $dy = \dfrac{1}{x}\, dx$. Thus, $y = \ln x + C$. We now use the given condition, by substituting the point $(e, -3)$, to determine C. Since $-3 = \ln e + C$, we have $-3 = 1 + C$, and $C = -4$. Then, the solution of the given equation subject to the given condition is

$$y = \ln x - 4.$$

Differential Equations: Motion Problems.

An equation involving a derivative is called a *differential equation*. In Examples 52 and 53 we solved two simple differential equations. In each one we were given the derivative of a function and the value of the function at a particular point. The problem of finding the function is called an *initial-value problem* and the given condition is called the *initial condition*.

In Examples 54 and 55, we use the velocity (or the acceleration) of a particle moving on a line to find the position of the particle. Note especially how the initial conditions are used to evaluate constants of integration.

Example 54. The velocity of a particle moving along a line is given by $v(t) = 4t^3 - 3t^2$ at time t. If initially the particle is at $x = 3$ on the line, find its position when $t = 2$.

Since

$$v(t) = \frac{dx}{dt} = 4t^3 - 3t^2,$$

$$x = \int (4t^3 - 3t^2)\, dt = t^4 - t^3 + C.$$

Since $x(0) = 0^4 - 0^3 + C = 3$, we see that $C = 3$, and that the position function is $x(t) = t^4 - t^3 + 3$. When $t = 2$, we see that

$$x(2) = 2^4 - 2^3 + 3 = 16 - 8 + 3 = 11.$$

Example 55. Suppose that $a(t)$, the acceleration of a particle at time t, is given by $a(t) = 4t - 3$, that $v(1) = 6$, and that $f(2) = 5$, where $f(t)$ is the position function. (a) Find $v(t)$ and $f(t)$. (b) Find the position of the particle when $t = 1$.

(a) $$a(t) = v'(t) = \frac{dv}{dt} = 4t - 3,$$

$$v = \int (4t - 3)\, dt = 2t^2 - 3t + C_1.$$

Using $v(1) = 6$, we get $6 = 2(1)^2 - 3(1) + C_1$, and $C_1 = 7$, from which it follows that $v(t) = 2t^2 - 3t + 7$. Since,

$$v(t) = f'(t) = \frac{df}{dt},$$

$$f(t) = \int (2t^2 - 3t + 7)\, dt = \frac{2t^3}{3} - \frac{3t^2}{2} + 7t + C_2.$$

Using $f(2) = 5$, we get $5 = \frac{2}{3}(2)^3 - \frac{3}{2}(2)^2 + 7(2) + C_2$, $5 = \frac{16}{3} - 6 + 14 + C_2$, so $C_2 = -\frac{25}{3}$. Thus,

$$f(t) = \frac{2}{3}t^3 - \frac{3}{2}t^2 + 7t - \frac{25}{3}.$$

(b) When $t = 1$,

$$f(1) = \frac{2}{3} - \frac{3}{2} + 7 - \frac{25}{3} = -\frac{13}{6}.$$

For more examples of motion along a line see Chapter 8, Further Applications of Integration, and Chapter 9 Differential Equations.

Set 5: Multiple-Choice Questions on Integration

1. $\int (3x^2 - 2x + 3)\, dx =$

   (A) $x^3 - x^2 + C$ (B) $3x^3 - x^2 + 3x + C$ (C) $x^3 - x^2 + 3x + C$

   (D) $\dfrac{1}{2}(3x^2 - 2x + 3)^2 + C$ (E) none of these

2. $\int \left(x - \dfrac{1}{2x}\right)^2 dx =$

   (A) $\dfrac{1}{3}\left(x - \dfrac{1}{2x}\right)^3 + C$ (B) $x^2 - 1 + \dfrac{1}{4x^2} + C$ (C) $\dfrac{x^3}{3} - 2x - \dfrac{1}{4x} + C$

   (D) $\dfrac{x^3}{3} - x - \dfrac{4}{x} + C$ (E) none of these

3. $\int \sqrt{4 - 2t}\ dt =$

   (A) $-\dfrac{1}{3}(4 - 2t)^{3/2} + C$ (B) $\dfrac{2}{3}(4 - 2t)^{3/2} + C$ (C) $-\dfrac{1}{6}(4 - 2t)^3 + C$

   (D) $+\dfrac{1}{2}(4 - 2t)^2 + C$ (E) $\dfrac{4}{3}(4 - 2t)^{3/2} + C$

4. $\int (2 - 3x)^5\, dx =$

   (A) $\dfrac{1}{6}(2 - 3x)^6 + C$ (B) $-\dfrac{1}{2}(2 - 3x)^6 + C$ (C) $\dfrac{1}{2}(2 - 3x)^6 + C$

   (D) $-\dfrac{1}{18}(2 - 3x)^6 + C$ (E) none of these

5. $\int \dfrac{1 - 3y}{\sqrt{2y - 3y^2}}\, dy =$

   (A) $4\sqrt{2y - 3y^2} + C$ (B) $\dfrac{1}{4}(2y - 3y^2)^2 + C$ (C) $\dfrac{1}{2}\ln\sqrt{2y - 3y^2} + C$

   (D) $\dfrac{1}{4}(2y - 3y^2)^{1/2} + C$ (E) $\sqrt{2y - 3y^2} + C$

6. $\int \dfrac{dx}{3(2x-1)^2} =$

    (A) $\dfrac{-3}{2x-1} + C$ (B) $\dfrac{1}{6-12x} + C$ (C) $+\dfrac{6}{2x-1} + C$

    (D) $\dfrac{2}{3\sqrt{2x-1}} + C$ (E) $\dfrac{1}{3}\ln|2x-1| + C$

7. $\int \dfrac{2\,du}{1+3u} =$

    (A) $\dfrac{2}{3}\ln|1+3u| + C$ (B) $-\dfrac{1}{3(1+3u)^2} + C$ (C) $2\ln|1+3u| + C$

    (D) $\dfrac{3}{(1+3u)^2} + C$ (E) none of these

8. $\int \dfrac{t}{\sqrt{2t^2-1}}\,dt =$

    (A) $\dfrac{1}{2}\ln\sqrt{2t^2-1} + C$ (B) $4\ln\sqrt{2t^2-1} + C$ (C) $8\sqrt{2t^2-1} + C$

    (D) $-\dfrac{1}{4(2t^2-1)} + C$ (E) $\dfrac{1}{2}\sqrt{2t^2-1} + C$

9. $\int \cos 3x\,dx =$

    (A) $3\sin 3x + C$ (B) $-\sin 3x + C$ (C) $-\dfrac{1}{3}\sin 3x + C$

    (D) $\dfrac{1}{3}\sin 3x + C$ (E) $\dfrac{1}{2}\cos^2 3x + C$

10. $\int \dfrac{x\,dx}{1+4x^2} =$

    (A) $\dfrac{1}{8}\ln(1+4x^2) + C$ (B) $\dfrac{1}{8(1+4x^2)^2} + C$ (C) $\dfrac{1}{4}\sqrt{1+4x^2} + C$

    (D) $\dfrac{1}{2}\ln|1+4x^2| + C$ (E) $\dfrac{1}{2}\tan^{-1} 2x + C$

11. $\int \dfrac{dx}{1+4x^2} =$

    (A) $\tan^{-1}(2x) + C$ (B) $\dfrac{1}{8}\ln(1+4x^2) + C$ (C) $\dfrac{1}{8(1+4x^2)^2} + C$

    (D) $\dfrac{1}{2}\tan^{-1}(2x) + C$ (E) $\dfrac{1}{8x}\ln|1+4x^2| + C$

12. $\int \dfrac{x}{(1 + 4x^2)^2}\, dx =$

(A) $\dfrac{1}{8}\ln(1 + 4x^2)^2 + C$ (B) $\dfrac{1}{4}\sqrt{1 + 4x^2} + C$ (C) $-\dfrac{1}{8(1 + 4x^2)} + C$

(D) $-\dfrac{1}{3(1 + 4x^2)^3} + C$ (E) $-\dfrac{1}{(1 + 4x^2)} + C$

13. $\int \dfrac{x\, dx}{\sqrt{1 + 4x^2}} =$

(A) $\dfrac{1}{8}\sqrt{1 + 4x^2} + C$ (B) $\dfrac{\sqrt{1 + 4x^2}}{4} + C$ (C) $\dfrac{1}{2}\sin^{-1} 2x + C$

(D) $\dfrac{1}{2}\tan^{-1} 2x + C$ (E) $\dfrac{1}{8}\ln\sqrt{1 + 4x^2} + C$

14. $\int \dfrac{dy}{\sqrt{4 - y^2}} =$

(A) $\dfrac{1}{2}\sin^{-1}\dfrac{y}{2} + C$ (B) $-\sqrt{4 - y^2} + C$ (C) $\sin^{-1}\dfrac{y}{2} + C$

(D) $-\dfrac{1}{2}\ln\sqrt{4 - y^2} + C$ (E) $-\dfrac{1}{3(4 - y^2)^{3/2}} + C$

15. $\int \dfrac{y\, dy}{\sqrt{4 - y^2}} =$

(A) $\dfrac{1}{2}\sin^{-1}\dfrac{y}{2} + C$ (B) $-\sqrt{4 - y^2} + C$ (C) $\sin^{-1}\dfrac{y}{2} + C$

(D) $-\dfrac{1}{2}\ln\sqrt{4 - y^2} + C$ (E) $2\sqrt{4 - y^2} + C$

16. $\int \dfrac{2x + 1}{2x}\, dx =$

(A) $x + \dfrac{1}{2}\ln|x| + C$ (B) $1 + \dfrac{1}{2}x^{-1} + C$ (C) $x + 2\ln|x| + C$

(D) $x + \ln|2x| + C$ (E) $\dfrac{1}{2}\left(2x - \dfrac{1}{x^2}\right) + C$

17. $\int \dfrac{(x - 2)^3}{x^2}\, dx =$

(A) $\dfrac{(x - 2)^4}{4x^2} + C$ (B) $\dfrac{x^2}{2} - 6x + 6\ln|x| - \dfrac{8}{x} + C$

(C) $\dfrac{x^2}{2} - 3x + 6\ln|x| + \dfrac{4}{x} + C$ (D) $-\dfrac{(x - 2)^4}{4x} + C$

(E)   none of these

18. $\int \left(\sqrt{t} - \frac{1}{\sqrt{t}} \right)^2 dt =$

(A) $\quad t - 2 + \frac{1}{t} + C$

(B) $\quad \frac{t^3}{3} - 2t - \frac{1}{t} + C$

(C) $\quad \frac{t^2}{2} + \ln|t| + C$

(D) $\quad \frac{t^2}{2} - 2t + \ln|t| + C$

(E) $\quad \frac{t^2}{2} - t - \frac{1}{t^2} + C$

19. $\int (4x^{1/3} - 5x^{3/2} - x^{-1/2})\, dx =$

(A) $\quad 3x^{4/3} - 2x^{5/2} - 2x^{1/2} + C$

(B) $\quad 3x^{4/3} - 2x^{5/2} + 2x^{1/2} + C$

(C) $\quad 6x^{2/3} - 2x^{5/2} - \frac{1}{2}x^2 + C$

(D) $\quad \frac{4}{3}x^{-2/3} - \frac{15}{2}x^{1/2} + \frac{1}{2}x^{-3/2} + C$

(E) $\quad$ none of these

20. $\int \frac{x^3 - x - 1}{(x+1)^2}\, dx =$

(A) $\quad (x-2) + \frac{2x+1}{(x+1)^2} + C$

(B) $\quad x^2 - 2x + \frac{1}{2}\ln(x^2 + 2x + 1) + C$

(C) $\quad \frac{1}{2}x^2 - 2x + \ln|x+1|^2 - \frac{1}{x+1} + C$

(D) $\quad \frac{1}{2}x^2 - 2x + 2\ln|x+1| + \frac{1}{x+1} + C$

(E) $\quad$ none of these

21. $\int \frac{dy}{\sqrt{y}(1 - \sqrt{y})} =$

(A) $\quad 4\sqrt{1 - \sqrt{y}} + C$

(B) $\quad \frac{1}{2}\ln\left|1 - \sqrt{y}\right| + C$

(C) $\quad 2\ln(1 - \sqrt{y}) + C$

(D) $\quad 2\sqrt{y} - \ln|y| + C$

(E) $\quad -2\ln\left|1 - \sqrt{y}\right| + C$

22. $\int \frac{u\, du}{\sqrt{4 - 9u^2}} =$

(A) $\quad \frac{1}{3}\sin^{-1}\frac{3u}{2} + C$

(B) $\quad -\frac{1}{18}\ln\sqrt{4 - 9u^2} + C$

(C) $\quad 2\sqrt{4 - 9u^2} + C$

(D) $\quad \frac{1}{6}\sin^{-1}\frac{3}{2}u + C$

(E) $\quad -\frac{1}{9}\sqrt{4 - 9u^2} + C$

23. $\displaystyle\int \sin\theta \cos\theta \, d\theta =$

    (A) $\quad -\dfrac{\sin^2\theta}{2} + C$
    (B) $\quad -\dfrac{1}{4}\cos 2\theta + C$
    (C) $\quad \dfrac{\cos^2\theta}{2} + C$

    (D) $\quad \dfrac{1}{2}\sin 2\theta + C$
    (E) $\quad \cos 2\theta + C$

24. $\displaystyle\int \dfrac{\sin\sqrt{x}}{\sqrt{x}} \, dx =$

    (A) $\quad -2\cos^{1/2} x + C$
    (B) $\quad -\cos\sqrt{x} + C$
    (C) $\quad -2\cos\sqrt{x} + C$

    (D) $\quad \dfrac{3}{2}\sin^{3/2} x + C$
    (E) $\quad \dfrac{1}{2}\cos\sqrt{x} + C$

25. $\displaystyle\int t\cos(2t)^2 \, dt =$

    (A) $\quad \dfrac{1}{8}\sin(4t^2) + C$
    (B) $\quad \dfrac{1}{2}\cos^2(2t) + C$
    (C) $\quad -\dfrac{1}{8}\sin(4t^2) + C$

    (D) $\quad \dfrac{1}{4}\sin(2t)^2 + C$
    (E) $\quad$ none of these

26. $\displaystyle\int \cos^2 2x \, dx =$

    (A) $\quad \dfrac{x}{2} + \dfrac{\sin 4x}{8} + C$
    (B) $\quad \dfrac{x}{2} - \dfrac{\sin 4x}{8} + C$
    (C) $\quad \dfrac{x}{4} + \dfrac{\sin 4x}{4} + C$

    (D) $\quad \dfrac{x}{4} + \dfrac{\sin 4x}{16} + C$
    (E) $\quad \dfrac{1}{4}(x + \sin 4x) + C$

27. $\displaystyle\int \sin 2\theta \, d\theta =$

    (A) $\quad \dfrac{1}{2}\cos 2\theta + C$
    (B) $\quad -2\cos 2\theta + C$
    (C) $\quad -\sin^2\theta + C$

    (D) $\quad \cos^2\theta + C$
    (E) $\quad -\dfrac{1}{2}\cos 2\theta + C$

*28. $\displaystyle\int x\cos x \, dx =$

    (A) $\quad x\sin x + C$
    (B) $\quad x\sin x + \cos x + C$
    (C) $\quad x\sin x - \cos x + C$

    (D) $\quad \cos x - x\sin x + C$
    (E) $\quad \dfrac{x^2}{2}\sin x + C$

29. $\displaystyle\int \dfrac{du}{\cos^2 3u} =$

    (A) $\quad -\dfrac{\sec 3u}{3} + C$
    (B) $\quad \tan 3u + C$
    (C) $\quad u + \dfrac{\sec 3u}{3} + C$

    (D) $\quad \dfrac{1}{3}\tan 3u + C$
    (E) $\quad \dfrac{1}{3\cos 3u} + C$

*An asterisk denotes a topic covered only in Calculus BC.

30. $\displaystyle\int \frac{\cos x \, dx}{\sqrt{1 + \sin x}} =$

    (A) $\displaystyle -\frac{1}{2}(1 + \sin x)^{1/2} + C$

    (B) $\ln \sqrt{1 + \sin x} + C$

    (C) $2\sqrt{1 + \sin x} + C$

    (D) $\ln |1 + \sin x| + C$

    (E) $\displaystyle \frac{2}{3(1 + \sin x)^{3/2}} + C$

31. $\displaystyle\int \frac{\cos (\theta - 1) \, d\theta}{\sin^2 (\theta - 1)} =$

    (A) $2 \ln \sin |\theta - 1| + C$ (B) $-\csc (\theta - 1) + C$ (C) $-\dfrac{1}{3} \sin^{-3} (\theta - 1) + C$

    (D) $-\cot (\theta - 1) + C$ (E) $\csc (\theta - 1) + C$

32. $\displaystyle\int \sec \frac{t}{2} \, dt =$

    (A) $\ln \left| \sec \dfrac{t}{2} + \tan \dfrac{t}{2} \right| + C$ (B) $2 \tan^2 \dfrac{t}{2} + C$ (C) $2 \ln \cos \dfrac{t}{2} + C$

    (D) $\ln |\sec t + \tan t| + C$ (E) $2 \ln \left| \sec \dfrac{t}{2} + \tan \dfrac{t}{2} \right| + C$

33. $\displaystyle\int \frac{\sin 2x \, dx}{\sqrt{1 + \cos^2 x}} =$

    (A) $-2\sqrt{1 + \cos^2 x} + C$ (B) $\dfrac{1}{2} \ln (1 + \cos^2 x) + C$

    (C) $\sqrt{1 + \cos^2 x} + C$ (D) $-\ln \sqrt{1 + \cos^2 x} + C$

    (E) $2 \ln |\sin x| + C$

34. $\displaystyle\int \sec^{3/2} x \tan x \, dx =$

    (A) $\dfrac{2}{5} \sec^{5/2} x + C$ (B) $-\dfrac{2}{3} \cos^{-3/2} x + C$ (C) $\sec^{3/2} x + C$

    (D) $\dfrac{2}{3} \sec^{3/2} x + C$ (E) none of these

35. $\displaystyle\int \tan \theta \, d\theta =$

    (A) $-\ln |\sec \theta| + C$ (B) $\sec^2 \theta + C$ (C) $\ln |\sin \theta| + C$

    (D) $\sec \theta + C$ (E) $-\ln |\cos \theta| + C$

36. $\displaystyle\int \frac{dx}{\sin^2 2x} =$

(A) $\displaystyle\frac{1}{2}\csc 2x \cot 2x + C$ (B) $\displaystyle -\frac{2}{\sin 2x} + C$ (C) $\displaystyle -\frac{1}{2}\cot 2x + C$

(D) $-\cot x + C$ (E) $-\csc 2x + C$

37. $\displaystyle\int \frac{\tan^{-1} y}{1 + y^2} \, dy =$

(A) $\sec^{-1} y + C$ (B) $(\tan^{-1} y)^2 + C$ (C) $\ln(1 + y^2) + C$

(D) $\ln(\tan^{-1} y) + C$ (E) none of these

38. $\displaystyle\int \sin 2\theta \cos \theta \, d\theta =$

(A) $\displaystyle -\frac{2}{3}\cos^3 \theta + C$ (B) $\displaystyle \frac{2}{3}\cos^3 \theta + C$ (C) $\sin^2 \theta \cos \theta + C$

(D) $\cos^3 \theta + C$ (E) none of these

39. $\displaystyle\int \frac{\sin 2t}{1 - \cos 2t} \, dt =$

(A) $\displaystyle \frac{2}{(1 - \cos 2t)^2} + C$ (B) $-\ln|1 - \cos 2t| + C$ (C) $\ln \sqrt{|1 - \cos 2t|} + C$

(D) $\sqrt{1 - \cos 2t} + C$ (E) $2\ln|1 - \cos 2t| + C$

40. $\displaystyle\int \cot 2u \, du =$

(A) $\ln|\sin u| + C$ (B) $\displaystyle \frac{1}{2}\ln|\sin 2u| + C$ (C) $\displaystyle -\frac{1}{2}\csc^2 2u + C$

(D) $-\sec 2u + C$ (E) $2\ln|\sin 2u| + C$

41. $\displaystyle\int \frac{e^x}{e^x - 1} \, dx =$

(A) $x + \ln|e^x - 1| + C$ (B) $x - e^x + C$ (C) $\displaystyle x - \frac{1}{(e^x - 1)^2} + C$

(D) $\displaystyle 1 + \frac{1}{e^x - 1} + C$ (E) $\ln|e^x - 1| + C$

*42. $\displaystyle\int \frac{x - 1}{x(x - 2)} \, dx =$

(A) $\displaystyle \frac{1}{2}\ln|x| + \ln|x - 2| + C$ (B) $\displaystyle \frac{1}{2}\ln\left|\frac{x - 2}{x}\right| + C$

(C) $\ln|x - 2| + \ln|x| + C$ (D) $\displaystyle \frac{1}{2}\ln|x(x - 2)| + C$

(E) none of these

*An asterisk denotes a topic covered only in Calculus BC.

43. $\int xe^{x^2}\,dx =$

(A) $\dfrac{1}{2}e^{x^2} + C$ (B) $e^{x^2}(2x^2 + 1) + C$ (C) $2e^{x^2} + C$

(D) $e^{x^2} + C$ (E) $\dfrac{1}{2}e^{x^2+1} + C$

44. $\int \cos\theta\, e^{\sin\theta}\,d\theta =$

(A) $e^{\sin\theta+1} + C$ (B) $e^{\sin\theta} + C$ (C) $-e^{\sin\theta} + C$

(D) $e^{\cos\theta} + C$ (E) $e^{\sin\theta}(\cos\theta - \sin\theta) + C$

45. $\int e^{2\theta} \sin e^{2\theta}\,d\theta =$

(A) $\cos e^{2\theta} + C$ (B) $2e^{4\theta}(\cos e^{2\theta} + \sin e^{2\theta}) + C$ (C) $-\dfrac{1}{2}\cos e^{2\theta} + C$

(D) $-2\cos e^{2\theta} + C$ (E) none of these

46. $\int \dfrac{e^{\sqrt{x}}}{\sqrt{x}}\,dx =$

(A) $2\sqrt{x}(e^{\sqrt{x}} - 1) + C$ (B) $2e^{\sqrt{x}} + C$ (C) $\dfrac{e^{\sqrt{x}}}{2}\left(\dfrac{1}{x} + \dfrac{1}{x\sqrt{x}}\right) + C$

(D) $\dfrac{1}{2}e^{\sqrt{x}} + C$ (E) none of these

*47. $\int xe^{-x}\,dx =$

(A) $e^{-x}(1 - x) + C$ (B) $\dfrac{e^{1-x}}{1-x} + C$ (C) $-e^{-x}(x + 1) + C$

(D) $-\dfrac{x^2}{2}e^{-x} + C$ (E) $e^{-x}(x + 1) + C$

*48. $\int x^2 e^x\,dx =$

(A) $e^x(x^2 + 2x) + C$ (B) $e^x(x^2 - 2x - 2) + C$ (C) $e^x(x^2 - 2x + 2) + C$

(D) $e^x(x - 1)^2 + C$ (E) $e^x(x + 1)^2 + C$

49. $\int \dfrac{e^x + e^{-x}}{e^x - e^{-x}}\,dx =$

(A) $x - \ln\left|e^x - e^{-x}\right| + C$ (B) $x + 2\ln\left|e^x - e^{-x}\right| + C$

(C) $-\dfrac{1}{2}(e^x - e^{-x})^{-2} + C$ (D) $\ln\left|e^x - e^{-x}\right| + C$

(E) $\ln(e^x + e^{-x}) + C$

*An asterisk denotes a topic covered only in Calculus BC.

50. $\int \dfrac{e^x}{1+e^{2x}}\, dx =$

    (A) $\tan^{-1} e^x + C$ **(B)** $\dfrac{1}{2}\ln(1+e^{2x}) + C$ **(C)** $\ln(1+e^{2x}) + C$

    (D) $\dfrac{1}{2}\tan^{-1} e^x + C$ **(E)** $2\tan^{-1} e^x + C$

51. $\int \dfrac{\ln v\, dv}{v} =$

    (A) $\ln|\ln v| + C$ **(B)** $\ln\dfrac{v^2}{2} + C$ **(C)** $\dfrac{1}{2}(\ln v)^2 + C$

    (D) $2\ln v + C$ **(E)** $\dfrac{1}{2}\ln v^2 + C$

52. $\int \dfrac{\ln \sqrt{x}}{x}\, dx =$

    (A) $\dfrac{\ln^2 \sqrt{x}}{\sqrt{x}} + C$ **(B)** $\ln^2 x + C$ **(C)** $\dfrac{1}{2}\ln|\ln x| + C$

    (D) $\dfrac{(\ln \sqrt{x})^2}{2} + C$ **(E)** $\dfrac{1}{4}\ln^2 x + C$

53. $\int x^3 \ln x\, dx =$

    (A) $x^2(3\ln x + 1) + C$ **(B)** $\dfrac{x^4}{16}(4\ln x - 1) + C$ **(C)** $\dfrac{x^4}{4}(\ln x - 1) + C$

    (D) $3x^2\left(\ln x - \dfrac{1}{2}\right) + C$ **(E)** none of these

*54. $\int \ln \eta\, d\eta =$

    (A) $\dfrac{1}{2}\ln^2 \eta + C$ **(B)** $\eta(\ln \eta - 1) + C$ **(C)** $\dfrac{1}{2}\ln \eta^2 + C$

    (D) $\ln \eta(\eta - 1) + C$ **(E)** $\eta \ln \eta + \eta + C$

*55. $\int \ln x^3\, dx =$

    (A) $\dfrac{3}{2}\ln^2 x + C$ **(B)** $3x(\ln x - 1) + C$ **(C)** $3\ln x(x - 1) + C$

    (D) $\dfrac{3x \ln^2 x}{2} + C$ **(E)** none of these

*An asterisk denotes a topic covered only in Calculus BC.

*56. $\int \dfrac{\ln y}{y^2}\, dy =$

(A) $\dfrac{1}{y}(1 - \ln y) + C$ (B) $\dfrac{1}{2y}\ln^2 y + C$ (C) $-\dfrac{1}{3y^3}(4\ln y + 1) + C$

(D) $-\dfrac{1}{y}(\ln y + 1) + C$ (E) $\dfrac{\ln y}{y} - \dfrac{1}{y} + C$

57. $\int \dfrac{dv}{v \ln v} =$

(A) $\dfrac{1}{\ln v^2} + C$ (B) $-\dfrac{1}{\ln^2 v} + C$ (C) $-\ln|\ln v| + C$

(D) $\ln\dfrac{1}{v} + C$ (E) $\ln|\ln v| + C$

58. $\int \dfrac{y-1}{y+1}\, dy =$

(A) $y - 2\ln|y+1| + C$ (B) $1 - \dfrac{2}{y+1} + C$ (C) $\ln\dfrac{|y|}{(y+1)^2} + C$

(D) $1 - 2\ln|y+1| + C$ (E) $\ln\left|\dfrac{e^y}{y+1}\right| + C$

59. $\int t\sqrt{t+1}\, dt =$

(A) $\dfrac{2}{3}(t+1)^{3/2} + C$ (B) $\dfrac{2}{15}(3t-2)(t+1)^{3/2} + C$

(C) $2\left[\dfrac{(t+1)^{5/2}}{5} + \dfrac{(t+1)^{3/2}}{5}\right] + C$ (D) $2t(t+1) + C$

(E) none of these

60. $\int \sqrt{x}(\sqrt{x} - 1)\, dx =$

(A) $2(x^{3/2} - x) + C$ (B) $\dfrac{x^2}{2} - x + C$ (C) $\dfrac{1}{2}(\sqrt{x} - 1)^2 + C$

(D) $\dfrac{1}{2}x^2 - \dfrac{2}{3}x^{3/2} + C$ (E) $x - 2\sqrt{x} + C$

*61. $\int e^\theta \cos\theta\, d\theta =$

(A) $e^\theta(\cos\theta - \sin\theta) + C$
(B) $e^\theta \sin\theta + C$
(C) $\dfrac{1}{2}e^\theta(\sin\theta + \cos\theta) + C$
(D) $2e^\theta(\sin\theta + \cos\theta) + C$
(E) $\dfrac{1}{2}e^\theta(\sin\theta - \cos\theta) + C$

*An asterisk denotes a topic covered only in Calculus BC.

62. $\int \dfrac{(1-\ln t)^2}{t}\, dt =$

(A) $\dfrac{1}{3}(1-\ln t)^3 + C$ (B) $\ln t - 2\ln^2 t + \ln^3 t + C$ (C) $-2(1-\ln t) + C$

(D) $\ln t - \ln^2 t + \dfrac{\ln t^3}{3} + C$ (E) $-\dfrac{(1-\ln t)^3}{3} + C$

*63. $\int u\,\sec^2 u\, du =$

(A) $u\tan u + \ln|\cos u| + C$ (B) $\dfrac{u^2}{2}\tan u + C$ (C) $\dfrac{1}{2}\sec u\,\tan u + C$

(D) $u\tan u - \ln|\sin u| + C$ (E) $u\sec u - \ln|\sec u + \tan u| + C$

64. $\int \dfrac{2x+1}{4+x^2}\, dx =$

(A) $\ln(x^2+4) + C$ (B) $\ln(x^2+4) + \tan^{-1}\dfrac{x}{2} + C$ (C) $\dfrac{1}{2}\tan^{-1}\dfrac{x}{2} + C$

(D) $\ln(x^2+4) + \dfrac{1}{2}\tan^{-1}\dfrac{x}{2} + C$ (E) none of these

65. $\int \dfrac{x+2}{x^2+2x+10}\, dx =$

(A) $\dfrac{1}{2}\ln(x^2+2x+10) + C$ (B) $\dfrac{1}{3}\tan^{-1}\dfrac{x+1}{3} + C$

(C) $\dfrac{1}{2}\ln(x^2+2x+10) + \dfrac{1}{3}\tan^{-1}\dfrac{x+1}{3} + C$ (D) $\tan^{-1}\dfrac{x+1}{3} + C$

(E) $\dfrac{1}{2}\ln|x^2+2x+10| - 2\tan^{-1}\dfrac{x+1}{3} + C$

66. $\int \dfrac{2x-1}{\sqrt{4x-4x^2}}\, dx =$

(A) $4\ln\sqrt{4x-4x^2} + C$ (B) $\sin^{-1}(1-2x) + C$

(C) $\dfrac{1}{2}\sqrt{4x-4x^2} + C$ (D) $-\dfrac{1}{4}\ln(4x-4x^2) + C$

(E) $-\dfrac{1}{2}\sqrt{4x-4x^2} + C$

67. $\int \dfrac{e^{2x}}{1+e^x}\, dx =$

(A) $\tan^{-1}e^x + C$ (B) $e^x - \ln(1+e^x) + C$ (C) $e^x - x + \ln|1+e^x| + C$

(D) $e^x + \dfrac{1}{(e^x+1)^2} + C$ (E) none of these

*An asterisk denotes a topic covered only in Calculus BC.

68. $\int \dfrac{\cos \theta}{1 + \sin^2 \theta}\, d\theta =$

(A) $\sec \theta \tan \theta + C$ (B) $\sin \theta - \csc \theta + C$ (C) $\ln (1 + \sin^2 \theta) + C$

(D) $\tan^{-1} (\sin \theta) + C$ (E) $-\dfrac{1}{(1 + \sin^2 \theta)^2} + C$

*69. $\int \arctan x\, dx =$

(A) $\arctan x + C$

(B) $x \arctan x - \ln (1 + x^2) + C$

(C) $x \arctan x + \ln (1 + x^2) + C$

(D) $x \arctan x + \dfrac{1}{2} \ln (1 + x^2) + C$

(E) $x \arctan x - \dfrac{1}{2} \ln (1 + x^2) + C$

70. $\int \dfrac{dx}{1 - e^x} =$

(A) $-\ln|1 - e^x| + C$ (B) $x - \ln|1 - e^x| + C$ (C) $\dfrac{1}{(1 - e^x)^2} + C$

(D) $e^{-x} \ln|1 + e^x| + C$ (E) none of these

71. $\int \dfrac{(2 - y)^2}{4\sqrt{y}}\, dy =$

(A) $\dfrac{1}{6}(2 - y)^3 \sqrt{y} + C$

(B) $2\sqrt{y} - \dfrac{2}{3} y^{3/2} + \dfrac{8}{5} y^{5/2} + C$

(C) $\ln|y| - y + 2y^2 + C$

(D) $2y^{1/2} - \dfrac{2}{3} y^{3/2} + \dfrac{1}{10} y^{5/2} + C$

(E) none of these

72. $\int e^{2 \ln u}\, du =$

(A) $\dfrac{1}{3} e^{u^3} + C$ (B) $e^{u^3/3} + C$ (C) $\dfrac{1}{3} u^3 + C$

(D) $\dfrac{2}{u} e^{2 \ln u} + C$ (E) $e^{1 + 2 \ln u} + C$

73. $\int \dfrac{dy}{y(1 + \ln y^2)} =$

(A) $\dfrac{1}{2} \ln|1 + \ln y^2| + C$ (B) $-\dfrac{1}{(1 + \ln y^2)^2} + C$

(C) $\ln|y| + \dfrac{1}{2} \ln|\ln y| + C$ (D) $\tan^{-1}(\ln|y|) + C$ (E) none of these

*An asterisk denotes a topic covered only in Calculus BC.

74. $\int (\tan \theta - 1)^2 \, d\theta =$

(A) $\sec \theta + \theta + 2 \ln |\cos \theta| + C$ (B) $\tan \theta + 2 \ln |\cos \theta| + C$

(C) $\tan \theta - 2 \sec^2 \theta + C$ (D) $\sec \theta + \theta - \tan^2 \theta + C$

(E) $\tan \theta - 2 \ln |\cos \theta| + C$

75. $\int \dfrac{d\theta}{1 + \sin \theta} =$

(A) $\sec \theta - \tan \theta + C$ (B) $\ln (1 + \sin \theta) + C$

(C) $\ln |\sec \theta + \tan \theta| + C$ (D) $\theta + \ln |\csc \theta - \cot \theta| + C$

(E) none of these

76. A particle starting at rest at $t = 0$ moves along a line so that its acceleration at time t is $12t$ ft/sec². How much distance does the particle cover during the first 3 sec?

(A) 16 ft   (B) 32 ft   (C) 48 ft   (D) 54 ft   (E) 108 ft

77. The equation of the curve whose slope at point (x, y) is $x^2 - 2$ and which contains the point $(1, -3)$ is

(A) $y = \dfrac{1}{3} x^3 - 2x$ (B) $y = 2x - 1$ (C) $y = \dfrac{1}{3} x^3 - \dfrac{10}{3}$

(D) $y = \dfrac{1}{3} x^3 - 2x - \dfrac{4}{3}$ (E) $3y = x^3 - 10$

78. A particle moves along a line with acceleration $2 + 6t$ at time t. When $t = 0$, its velocity equals 3 and it is at position $s = 2$. When $t = 1$, it is at position $s =$

(A) 2   (B) 5   (C) 6   (D) 7   (E) 8

79. If a deceleration of k ft/sec² is needed to bring a particle moving with a velocity of 75 ft/sec to a stop in 5 sec, then $k =$

(A) 25   (B) 15   (C) –6   (D) –15   (E) –25

*80. $\int \dfrac{x^2}{x^2 - 1} \, dx =$

(A) $x + \dfrac{1}{2} \ln \left| \dfrac{x - 1}{x + 1} \right| + C$ (B) $\ln |x^2 - 1| + C$ (C) $x + \tan^{-1} x + C$

(D) $x + \dfrac{1}{2} \ln \left| \dfrac{x + 1}{x - 1} \right| + C$ (E) $1 + \dfrac{1}{2} \ln \left| \dfrac{x + 1}{x - 1} \right| + C$

*An asterisk denotes a topic covered only in Calculus BC.

Answers for Set 5: Integration

1.	C	17.	E	33.	A	49.	D	65.	C
2.	E	18.	D	34.	D	50.	A	66.	E
3.	A	19.	A	35.	E	51.	C	67.	B
4.	D	20.	D	36.	C	52.	E	68.	D
5.	E	21.	E	37.	E	53.	B	69.	E
6.	B	22.	E	38.	A	54.	B	70.	B
7.	A	23.	B	39.	C	55.	B	71.	D
8.	E	24.	C	40.	B	56.	D	72.	C
9.	D	25.	A	41.	E	57.	E	73.	A
10.	A	26.	A	42.	D	58.	A	74.	B
11.	D	27.	E	43.	A	59.	B	75.	E
12.	C	28.	B	44.	B	60.	D	76.	D
13.	B	29.	D	45.	C	61.	C	77.	D
14.	C	30.	C	46.	B	62.	E	78.	D
15.	B	31.	B	47.	C	63.	A	79.	B
16.	A	32.	E	48.	C	64.	D	80.	A

All the references in parentheses below are to the basic integration formulas on pages 152 and 153. In general, if u is a function of x, then $du = u'(x)\, dx$.

1. C.   Use, first, formula (2), then (3), replacing u by x.

2. E.   Hint: Expand. The correct answer is $\dfrac{x^3}{3} - x - \dfrac{1}{4x} + C$.

3. A.   By formula (3), with $u = 4 - 2t$ and $n = \dfrac{1}{2}$,

$$\int \sqrt{4 - 2t}\, dx = -\frac{1}{2} \int \sqrt{4 - 2t} \cdot (-2)\, dt = -\frac{1}{2} \frac{(4 - 2t)^{3/2}}{3/2} + C.$$

4. D.   Use (3) with $u = 2 - 3x$, noting that $du = -3\, dx$.

5. E.   Rewrite:

$$\int (2y - 3y^2)^{-1/2}(1 - 3y)\, dy = \frac{1}{2} \int (2y - 3y^2)^{-1/2}(2 - 6y)\, dy.$$

Use (3).

6. B.   Rewrite:

$$\frac{1}{3} \int (2x - 1)^{-2}\, dx = \frac{1}{3} \cdot \frac{1}{2} \int (2x - 1)^{-2} \cdot 2\, dx.$$

Using (3) yields $-\dfrac{1}{6(2x - 1)} + C$.

7. A.   This is equivalent to $\frac{2}{3}\int \frac{dv}{v}$. Use (4).

8. E.   Rewrite and modify to $\frac{1}{4}\int (2t^2 - 1)^{-1/2} \cdot 4t\, dt$. Use (3).

9. D.   Use (5) with $u = 3x$; $du = 3\, dx$.

10. A.   Use (4). If $u = 1 + 4x^2$, $du = 8x\, dx$.

11. D.   Use (18). Let $u = 2x$; then $du = 2\, dx$.

12. C.   Rewrite and modify to $\frac{1}{8}\int (1 + 4x^3)^{-2} \cdot 8x\, dx$. Use (3) with $n = -2$.

13. B.   Rewrite and modify to $\frac{1}{8}\int (1 + 4x^2)^{-1/2} \cdot 8x\, dx$. Use (3) with $n = -\frac{1}{2}$.
Note carefully the differences in the integrands in questions 10 through 13.

14. C.   Use (17). Note that $u = y$, $a = 2$.

15. B.   Rewrite and modify to $-\frac{1}{2}\int (4 - y^2)^{-1/2} \cdot -2y\, dy$. Use (3).
Compare the integrands in questions 14 and 15, noting the difference.

16. A.   Divide to obtain $\int \left(1 + \frac{1}{2} \cdot \frac{1}{x}\right) dx$. Use (2), (3), and (4). Remember that $\int k\, dx = kx + C$ whenever $k \neq 0$.

17. E. $\int \frac{(x-2)^3}{x^2}\, dx = \int \left(x - 6 + \frac{12}{x} - \frac{8}{x^2}\right) dx = \frac{x^2}{2} - 6x + 12\ln|x| + \frac{8}{x} + C$. We used the Binomial Theorem on page 623 with $n = 3$ to expand $(x-2)^3$.

18. D.   The integral is equivalent to $\int \left(t - 2 + \frac{1}{t}\right) dt$. Integrate term by term.

19. A.   Integrate term by term.

20. D.   Long division yields
$$\int \left((x-2) + \frac{2x+1}{x^2+2x+1}\right) dx = \frac{1}{2}x^2 - 2x + \int \frac{2x+2-1}{x^2+2x+1}\, dx.$$

The integral equals $\int \frac{2x+2}{x^2+2x+1}\, dx - \int \frac{1}{(x+1)^2}\, dx$. Use formula (4) for the first integral, (3) for the second with $u = x + 1$ and $n = -2$. Note that $\ln|x+1|^2 = 2\ln|x+1|$.

21. E.   Use formula (4) with $u = 1 - \sqrt{y} = 1 - y^{1/2}$. Then $du = -\frac{1}{2\sqrt{y}}\, dy$. Note that the integral can be written as $-2\int \frac{1}{(1-\sqrt{y})}\left(-\frac{1}{2\sqrt{y}}\right) dy$.

22. E.   Use formula (3) after multiplying outside the integral by $-\dfrac{1}{18}$, inside by -18.

23. B.   The integral is equal to $\dfrac{1}{2}\displaystyle\int \sin 2\theta \, d\theta$. Use formula (6) with $u = 2\theta$; $du = 2\, d\theta$.

24. C.   Use formula (6) with $u = \sqrt{x}$; $du = \dfrac{1}{2\sqrt{x}}\, dx$.

25. A.   Use formula (5) with $u = 4t^2$; $du = 8t\, dt$.

26. A.   Using the Half-Angle Formula (23) on page 625 with $\alpha = 2x$ yields
$$\int \left(\frac{1}{2} + \frac{1}{2}\cos 4x \right) dx.$$

27. E.   Use formula (6).

28. B.   Integrate by parts (page 162). Let $u = x$ and $v' = \cos x$. Then $u' = 1$ and $v = \sin x$. The given integral equals $x \sin x - \int \sin x \, dx$.

29. D.   Replace $\dfrac{1}{\cos^2 3u}$ by $\sec^2 3u$; then use formula (9).

30. C.   The integral is of the form $\displaystyle\int \dfrac{du}{\sqrt{u}}$, where $u = 1 + \sin x$ and $du = \cos x \, dx$. Use formula (3) with $n = -\dfrac{1}{2}$.

31. B.   The integral is equivalent to $\displaystyle\int \csc(\theta - 1)\cot(\theta - 1)\, d\theta$. Use formula (12).

32. E.   Use formula (13) with $u = \dfrac{t}{2}$; $du = \dfrac{1}{2}\, dt$.

33. A.   Replace $\sin 2x$ by $2 \sin x \cos x$; then the integral is equivalent to
$$-\int \frac{-2 \sin x \cos x}{\sqrt{1 + \cos^2 x}}\, dx = -\int \frac{du}{\sqrt{u}},$$
where $u = 1 + \cos^2 x$ and $du = -2 \sin x \cos x \, dx$. Use formula (3).

34. D.   Rewriting in terms of sines and cosines yields
$$\int \frac{\sin x}{\cos^{5/2} x}\, dx = -\int \cos^{-5/2} x(-\sin x)dx = -\left(-\frac{2}{3}\right)\cos^{-3/2} x + C.$$

35. E.   Use formula (7).

36. C.   Replace $\dfrac{1}{\sin^2 2x}$ by $\csc^2 2x$ and use formula (10).

37. E.   Let $u = \tan^{-1} y$; then integrate $\displaystyle\int u \, du$. The correct answer is
$\dfrac{1}{2}(\tan^{-1} y)^2 + C.$

38. A.   Replacing $\sin 2\theta$ by $2 \sin \theta \cos \theta$ yields
$$\int 2 \sin \theta \cos^2 \theta \, d\theta = -\frac{2}{3}\cos^3 \theta + C.$$

39. C. The answer is equivalent to $\frac{1}{2}\ln|1-\cos 2t| + C$.

40. B. To use formula (8), multiply outside the integral by $\frac{1}{2}$, inside by 2.

41. E. Use formula (4) with $u = e^x - 1$; $du = e^x\, dx$.

42. D. Use partial fractions; find A and B such that

$$\frac{x-1}{x(x-2)} = \frac{A}{x} + \frac{B}{x-2}.$$

Then $x - 1 = A(x - 2) + Bx$.

Set $x = 0$: $-1 = -2A$ and $A = \frac{1}{2}$.

Set $x = 2$: $1 = 2B$ and $B = \frac{1}{2}$.

So the given integral equals

$$\int\left(\frac{1}{2x} + \frac{1}{2(x-2)}\right)dx = \frac{1}{2}\ln|x| + \frac{1}{2}\ln|x-2| + C$$

$$= \frac{1}{2}\ln|x(x-2)| + C.$$

43. A. Use formula (15) with $u = x^2$; $du = 2x\, dx$.

44. B. Use formula (15) with $u = \sin\theta$; $du = \cos\theta\, d\theta$.

45. C. Use formula (6) with $u = e^{2\theta}$; $du = 2e^{2\theta}\, d\theta$.

46. B. Use formula (15) with $u = \sqrt{x} = x^{1/2}$; $du = \frac{1}{2\sqrt{x}}\, dx$.

47. C. Use the Parts Formula. Let $u = x$, $v' = e^{-x}$; $u' = 1$, $v = -e^{-x}$. Then

$$-xe^{-x} + \int e^{-x}\, dx = -xe^{-x} - e^{-x} + C.$$

48. C. See Example 49, page 162.

49. D. The integral is of the form $\int \frac{du}{u}$; use (4).

50. A. The integral has the form $\int \frac{du}{1+u^2}$. Use formula (18), with $u = e^x$, $du = e^x\, dx$, and $a = 1$.

51. C. Let $u = \ln v$; then $du = \frac{dv}{v}$. Use formula (3).

52. E. Hint: $\ln\sqrt{x} = \frac{1}{2}\ln x$.

53. B. Use parts, letting $u = \ln x$ and $v' = x^3$. Then $u' = \frac{1}{x}$ and $v = \frac{x^4}{4}$. The integral equals $\frac{x^4}{4}\ln x - \frac{1}{4}\int x^3\, dx$.

54. B. Use parts, letting $u = \ln\eta$ and $v' = 1$. Then $u' = \frac{1}{\eta}$ and $v = \eta$. The integral equals $\eta\ln\eta - \int d\eta$.

55. B. Rewrite $\ln x^3$ as $3\ln x$, and use the method of Answer 54.

56. D.   Use parts, letting $u = \ln y$ and $v' = y^{-2}$. Then $u' = \dfrac{1}{y}$ and $v = -\dfrac{1}{y}$. The Parts Formula yields $\dfrac{-\ln y}{y} + \displaystyle\int \dfrac{1}{y^2}\,dy$.

57. E.   The integral has the form $\displaystyle\int \dfrac{du}{u}$, where $u = \ln v$.

58. A.   Hint: The integrand is equivalent to $1 - \dfrac{2}{y+1}$.

59. B.   Hint: Let $u = \sqrt{t+1}$. Then
$$u^2 = t + 1, \qquad 2u\,du = dt, \qquad \text{and} \qquad t = u^2 - 1.$$

60. D.   Hint: Multiply.

61. C.   See Example 50, page 162. Replace x by θ.

62. E.   The integral equals $-\displaystyle\int (1 - \ln t)^2 \left(-\dfrac{1}{t}\,dt\right)$; it is equivalent to $-\displaystyle\int u^2\,du$, where $u = 1 - \ln t$.

63. A.   Replace u by x in the given integral to avoid confusion in applying the Parts Formula. To integrate $\displaystyle\int x \sec^2 x\,dx$, let the variable u in the Parts Formula be x, and let v' be $\sec^2 x\,dx$. Then $u' = 1$ and $v = \tan x$, so
$$\int x \sec^2 x\,dx = x \tan x - \int \tan x\,dx$$
$$= x \tan x + \ln |\cos x| + C.$$

64. D.   The integral is equivalent to $\displaystyle\int \dfrac{2x}{4+x^2}\,dx + \int \dfrac{1}{4+x^2}\,dx$. Use formula (4) on the first integral and (18) on the second.

65. C.   Rewrite:
$$\frac{1}{2}\int \frac{2x+4}{x^2+2x+10}\,dx = \frac{1}{2}\int \frac{2x+2+2}{x^2+2x+10}\,dx$$
$$= \frac{1}{2}\int \frac{2x+2}{x^2+2x+10}\,dx + \int \frac{dx}{(x+1)^2 + 3^2}\,dx.$$
Use formulas (4) and (18).

66. E.   Hint: Let $u = 4x - 4x^2$; then you are integrating $-\dfrac{1}{4}\displaystyle\int u^{-1/2}\,du$.

67. B.   Hint: Divide, getting $\displaystyle\int \left[e^x - \dfrac{e^x}{1+e^x}\right]dx$.

68. D.   Letting $u = \sin \theta$ yields the integral $\displaystyle\int \dfrac{du}{1+u^2}$. Use formula (18).

69. E.   Use integration by parts, letting $u = \arctan x$ and $v' = dx$. Then
$$du = \frac{dx}{1+x^2} \quad \text{and} \quad v = x.$$

The Parts Formula yields

$$x \arctan x - \int \frac{x\,dx}{1+x^2} \qquad \text{or} \qquad x \arctan x - \frac{1}{2}\ln(1+x^2) + C.$$

70. B.   Hint: Note that

$$\frac{1}{1-e^x} = \frac{1-e^x+e^x}{1-e^x} = 1 + \frac{e^x}{1-e^x}.$$

Or multiply the integrand by $\frac{e^{-x}}{e^{-x}}$, recognizing that the correct answer is equivalent to $-\ln\left|e^{-x}-1\right|$.

71. D.   Hint: Expand the numerator and divide. Then integrate term by term.

72. C.   Hint: Observe that $e^{2\ln u} = u^2$.

73. A.   If you let $u = 1 + \ln y^2 = 1 + 2\ln|y|$, you want to integrate $\frac{1}{2}\int \frac{du}{u}$.

74. B.   Hint: Expand and note that

$$\int(\tan^2\theta - 2\tan\theta + 1)\,d\theta = \int\sec^2\theta\,d\theta - 2\int\tan\theta\,d\theta.$$

Use formulas (9) and (7).

75. E.   Multiply by $\frac{1-\sin\theta}{1-\sin\theta}$. The correct answer is $\tan\theta - \sec\theta + C$.

76. D.   Note the initial conditions: when $t = 0$, $v = 0$ and $s = 0$. Integrate twice: $v = 6t^2$ and $s = 2t^3$. Let $t = 3$.

77. D.   Since $y' = x^2 - 2$, $y = \frac{1}{3}x^3 - 2x + C$. Replacing x by 1 and y by -3 yields

$$C = -\frac{4}{3}.$$

78. D.   When $t = 0$, $v = 3$ and $s = 2$. So

$$v = 2t + 3t^2 + 3 \quad\text{and}\quad s = t^2 + t^3 + 3t + 2.$$

Let $t = 1$.

79. B.   Let $a = \frac{dv}{dt} = -k$; then

$$v = -kt + C. \qquad\qquad (*)$$

Since $v = 75$ when $t = 0$, therefore $C = 75$. Then (*) becomes

$$v = -kt + 75$$

so

$$0 = -5k + 75 \quad\text{and}\quad k = 15.$$

80. A.   Divide to obtain $\int\left(1 + \frac{1}{x^2-1}\right)dx$. Use partial fractions to get

$$\frac{1}{x^2-1} = \frac{1}{2(x-1)} - \frac{1}{2(x+1)}.$$

Chapter 6

Definite Integrals

Review of Definitions and Methods

A. Fundamental Theorem of Calculus (FTC); Definition of Definite Integral _____

If f is continuous on the closed interval $[a,b]$ and $F' = f$, then, according to the Fundamental Theorem of Calculus,

$$\int_a^b f(x)\ dx = F(b) - F(a).$$

Here $\int_a^b f(x)$ is the *definite integral of f from a to b*; $f(x)$ is called the *integrand*; and a and b are called respectively the *lower* and *upper limits of integration*.

This important theorem says that if f is the derivative of F then the definite integral of f gives the net change in F as x varies from a to b. It also says that we can evaluate any definite integral for which we can find an antiderivative of f.

B. Properties of Definite Integrals _____

The following theorems about definite integrals are important. (The first is simply a restatement of the Fundamental Theorem.)

$$\frac{d}{dx} \int_a^x f(t)\, dt = f(x) \tag{1}$$

$$\int_a^b kf(x)\, dx = k \int_a^b f(x)\, dx \qquad (k \text{ a constant}) \tag{2}$$

$$\int_a^a f(x)\, dx = 0 \tag{3}$$

$$\int_a^b f(x)\, dx = - \int_b^a f(x)\, dx \tag{4}$$

$$\int_a^c f(x)\, dx + \int_c^b f(x)\, dx = \int_a^b f(x)\, dx \qquad (a < c < b) \tag{5}$$

If f and g are both integrable functions of x on $[a,b]$, then

$$\int_a^b [f(x) \pm g(x)]\, dx = \int_a^b f(x)\, dx \pm \int_a^b g(x)\, dx \tag{6}$$

The Mean Value Theorem for Integrals: If f is continuous on $[a,b]$ there exists at least one number c, $a < c < b$, such that

$$\int_a^b f(x)\, dx = f(c)(b - a) \tag{7}$$

By the comparison property, if f and g are integrable on $[a,b]$ and if $f(x) \leqslant g(x)$ for all x in $[a,b]$, then

$$\int_a^b f(x)\, dx \leqslant \int_a^b g(x)\, dx \tag{8}$$

The evaluation of a definite integral is illustrated in the following examples. A calculator will be helpful for some numerical calculations.

Example 1. $\displaystyle \int_{-1}^{2} (3x^2 - 2x)\, dx = x^3 - x^2 \Big|_{-1}^{2} = (8 - 4) - (-1 - 1) = 6.$

Example 2. $\displaystyle \int_{1}^{2} \frac{x^2 + x - 2}{2x^2}\, dx = \frac{1}{2} \int_{1}^{2} \left(1 + \frac{1}{x} - \frac{2}{x^2}\right) dx = \frac{1}{2} \left(x + \ln x + \frac{2}{x}\right) \Big|_{1}^{2} =$

$\frac{1}{2} [(2 + \ln 2 + 1) - (1 + 2)] = \frac{1}{2} \ln 2,$ or $\ln \sqrt{2}$.

Example 3. $\displaystyle \int_{5}^{8} \frac{dy}{\sqrt{9 - y}} = - \int_{5}^{8} (9 - y)^{-1/2} (-dy) = -2 \sqrt{9 - y} \Big|_{5}^{8} = -2(1 - 2) = 2.$

Example 4. $\displaystyle \int_{0}^{1} \frac{x\, dx}{(2 - x^2)^3} = -\frac{1}{2} \int_{0}^{1} (2 - x^2)^{-3}(-2x\, dx) = -\frac{1}{2} \frac{(2 - x^2)^{-2}}{-2} \Big|_{0}^{1} =$

$\frac{1}{4} \left(1 - \frac{1}{4}\right) = \frac{3}{16}.$

Example 5. $\int_0^3 \frac{dt}{9+t^2} = \frac{1}{3} \tan^{-1} \frac{t}{3} \Big|_0^3 = \frac{1}{3} (\tan^{-1} 1 - \tan^{-1} 0) = \frac{1}{3} \left(\frac{\pi}{4} - 0 \right) = \frac{\pi}{12}$.

Example 6. $\int_0^1 (3x-2)^3 \, dx = \frac{1}{3} \int_0^1 (3x-2)^3(3\,dx) = \frac{(3x-2)^4}{12} \Big|_0^1 = \frac{1}{12} (1-16) = -\frac{5}{4}$.

Example 7. $\int_0^1 xe^{-x^2} \, dx = -\frac{1}{2} e^{-x^2} \Big|_0^1 = -\frac{1}{2} \left(\frac{1}{e} - 1 \right) = \frac{e-1}{2e}$.

Example 8. $\int_{-\pi/4}^{\pi/4} \cos 2x \, dx = \frac{1}{2} \sin 2x \Big|_{-\pi/4}^{\pi/4} = \frac{1}{2} (1+1) = 1$.

Example 9. $\int_{-1}^1 xe^x \, dx = (xe^x - e^x) \Big|_{-1}^1 = e - e - \left(-\frac{1}{e} - \frac{1}{e} \right) = \frac{2}{e}$.

Example 10. $\int_0^{1/2} \frac{dx}{\sqrt{1-x^2}} = \sin^{-1} x \Big|_0^{1/2} = \frac{\pi}{6}$.

*Example 11. $\int_0^{e-1} \ln (x+1) \, dx = \left[(x+1) \ln (x+1) - x \right]_0^{e-1}$ (where the Parts Formula has been used) $= e \ln e - (e-1) - 0 = 1$.

*Example 12. To evaluate $\int_{-1}^1 \frac{dy}{y^2-4}$ we use the method of partial fractions and set

$$\frac{1}{y^2-4} = \frac{A}{y+2} + \frac{B}{y-2}.$$

Solving for A and B yields $A = -\frac{1}{4}$, $B = \frac{1}{4}$. Thus

$$\int_{-1}^1 \frac{dy}{y^2-4} = \frac{1}{4} \ln \left| \frac{y-2}{y+2} \right|_{-1}^1 = \frac{1}{4} \left(\ln \frac{1}{3} - \ln 3 \right) = -\frac{1}{2} \ln 3.$$

Example 13. $\int_{\pi/3}^{\pi/2} \tan \frac{\theta}{2} \sec^2 \frac{\theta}{2} \, d\theta = \frac{2}{2} \tan^2 \frac{\theta}{2} \Big|_{\pi/3}^{\pi/2} = 1 - \frac{1}{3} = \frac{2}{3}$.

Example 14. $\int_0^{\pi/2} \sin^2 \frac{1}{2} x \, dx = \int_0^{\pi/2} \left(\frac{1}{2} - \frac{\cos x}{2} \right) dx = \frac{x}{2} - \frac{\sin x}{2} \Big|_0^{\pi/2} = \frac{\pi}{4} - \frac{1}{2}$.

*An asterisk denotes a topic covered only in Calculus BC.

Example 15. $\dfrac{d}{dx} \displaystyle\int_{-1}^{x} \sqrt{1+\sin^2 t}\, dt = \sqrt{1+\sin^2 x}$ by theorem (1), page 185.

Example 16. $\dfrac{d}{dx} \displaystyle\int_{x}^{1} e^{-t^2} dt = \dfrac{d}{dx}\left(-\displaystyle\int_{1}^{x} e^{-t^2} dt\right)$ by theorem (4), page 185,

$= -\dfrac{d}{dx} \displaystyle\int_{1}^{x} e^{-t^2} dt = -e^{-x^2}$ by theorem (1).

Example 17. If $F(x) = \displaystyle\int_{1}^{x^2} \dfrac{dt}{3+t}$, then

$$F'(x) = \dfrac{d}{dx} \int_{1}^{x^2} \dfrac{dt}{3+t}$$

$$= \dfrac{d}{dx} \int_{1}^{u} \dfrac{dt}{3+t} \qquad (\text{where } u = x^2)$$

$$= \dfrac{d}{du} \int_{1}^{u} \dfrac{dt}{3+t} \cdot \dfrac{du}{dx} \qquad \text{by the chain rule}$$

$$= \left(\dfrac{1}{3+u}\right)(2x) = \dfrac{2x}{3+x^2}.$$

Example 18. If $F(x) = \displaystyle\int_{0}^{\cos x} \sqrt{1-t^3}\, dt$, then to find $F'(x)$ we let $u = \cos x$. Thus

$$\dfrac{dF}{dx} = \dfrac{dF}{du} \cdot \dfrac{du}{dx} = \sqrt{1-u^3}\,(-\sin x) = -\sin x \sqrt{1-\cos^3 x}.$$

Example 19. $\displaystyle\lim_{h \to 0} \dfrac{1}{h} \int_{x}^{x+h} \sqrt{e^t - 1}\, dt = \sqrt{e^x - 1}$. Here we have let $f(t) = \sqrt{e^t - 1}$
and noted that

$$\lim_{h \to 0} \dfrac{1}{h} \int_{x}^{x+h} f(t)\, dt = \lim_{h \to 0} \dfrac{F(x+h) - F(x)}{h} \qquad (*)$$

where

$$\dfrac{dF(x)}{dx} = f(x) = \sqrt{e^x - 1}.$$

The limit on the right in the starred equation is, by definition, the derivative of $F(x)$, that is, $f(x)$.

Example 20. Evaluate $\displaystyle\int_{3}^{6} x\sqrt{x-2}\, dx$.

Here we let $u = \sqrt{x-2}$, $u^2 = x - 2$, and $2u\, du = dx$. The limits of the given integral are values of x. When we write the new integral in terms of the variable u, then the limits, if written, must be the values of u that correspond to the given limits. Thus, when $x = 3$, $u = 1$, and when $x = 6$, $u = 2$. Then

$$\int_3^6 x\sqrt{x-2}\,dx = 2\int_1^2 (u^2+2)u^2\,du = 2\int_1^2 (u^4+2u^2)\,du$$

$$= 2\left(\frac{u^5}{5}+\frac{2u^3}{3}\right)\Big|_1^2 = 2\left[\left(\frac{32}{5}+\frac{16}{3}\right)-\left(\frac{1}{5}+\frac{2}{3}\right)\right]$$

$$= \frac{326}{15}.$$

Example 21.   If g' is continuous, then

$$\lim_{h\to 0}\frac{1}{h}\int_c^{c+h} g'(x)\,dx = g'(c)$$

because

$$\lim_{h\to 0}\frac{1}{h}\int_c^{c+h} g'(x)\,dx = \lim_{h\to 0}\frac{g(c+h)-g(c)}{h},$$

where the limit on the right is, by definition, the derivative of $g(x)$ at c, namely, $g'(c)$.

*C.   Integrals Involving Parametrically Defined Functions

The techniques are illustrated in Examples 22 and 23.

Example 22.   To evaluate $\int_{-2}^{2} y\,dx$, where $x = 2\sin\theta$ and $y = 2\cos\theta$, we note that $dx = 2\cos\theta\,d\theta$ and that

$$\theta = -\frac{\pi}{2} \quad \text{when } x = -2,$$

$$= \frac{\pi}{2} \quad \text{when } x = 2.$$

Then

$$\int_{-2}^{2} y\,dx = \int_{-\pi/2}^{\pi/2} 2\cos\theta\,(2\cos\theta)\,d\theta = 4\int_{-\pi/2}^{\pi/2}\frac{1+\cos 2\theta}{2}\,d\theta$$

$$= 2\left(\theta + \frac{\sin 2\theta}{2}\right)\Big|_{-\pi/2}^{\pi/2} = 2\pi.$$

*An asterisk denotes a topic covered only in Calculus BC.

When using parametric equations we must be sure to express everything in terms of the parameter. In Example 22 we replaced in terms of θ: (1) the integrand, (2) dx, and (3) both limits. Remember that we have defined dx as $x'(\theta)\,d\theta$.

Example 23. Evaluate $\displaystyle\int_0^{2\pi} \sqrt{1+\left(\frac{dy}{dx}\right)^2}\ dx$, where $\begin{aligned} x &= t - \sin t. \\ y &= 1 - \cos t. \end{aligned}$

We see that the integral can be rewritten as $\displaystyle\int_0^{2\pi} \sqrt{dx^2 + dy^2}$. Since $dx =$ $(1 - \cos t)\,dt$, $dy = \sin t\,dt$, $t = 0$ when $x = 0$ and $t = 2\pi$ when $x = 2\pi$, it follows that

$$\int_0^{2\pi} \sqrt{dx^2 + dy^2} = \int_0^{2\pi} \sqrt{(1-\cos t)^2 + \sin^2 t}\ dt$$

$$= \sqrt{2} \int_0^{2\pi} \sqrt{1 - \cos t}\ dt = 2 \int_0^{2\pi} \sin\frac{t}{2}\,dt$$

$$= 4\left(-\cos\frac{t}{2}\right)\bigg|_0^{2\pi} = 8.$$

D. Definition of Definite Integral as the Limit of a Sum: The Fundamental Theorem Again____

Most applications of integration are based on the FTC. This theorem provides the tool for evaluating an infinite sum by means of a definite integral. Suppose that a function $f(x)$ is continuous on the closed interval $[a,b]$. Divide the interval into n equal* subintervals, of length $\Delta x = \dfrac{b-a}{n}$. Choose numbers, one in each subinterval, as follows: x_1 in the first, x_2 in the second, . . . , x_k in the kth, . . . , x_n in the nth. Then

$$\lim_{n\to\infty} \sum_{k=1}^{n} f(x_k)\,\Delta x = \int_a^b f(x)\,dx = F(b) - F(a), \text{ where } \frac{dF(x)}{dx} = f(x).$$

Any sum of the form $\displaystyle\sum_{k=1}^{n} f(x_k)\,\Delta x$ is called a *Riemann sum.*

Area

If $f(x)$ is nonnegative on $[a,b]$, we see (Figure N6–1) that $f(x_k)\,\Delta x$ can be regarded as the area of a typical approximating rectangle. As the number of rectangles increases, or, equivalently, as the width Δx of the rectangles approaches zero, the rectangles become an increasingly better fit to the curve. The sum of their areas gets closer and closer to the exact area under the curve. Finally, the area bounded by the x-axis, the curve, and the vertical lines $x = a$ and $x = b$ is given exactly by

---

*It is not in fact necessary that the subintervals be of equal length, but the formulation is simpler if one assumes they are.

$$\lim_{n \to \infty} \sum_{k=1}^{n} f(x_k)\, \Delta x \qquad \text{and hence by} \qquad \int_a^b f(x)\, dx.$$

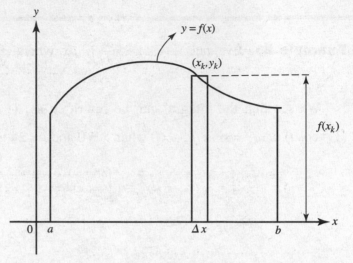

FIGURE N6–1

What if $f(x)$ is negative? Then *any area above the graph and below the x-axis is counted as negative* (Figure N6–2).

The shaded area above the curve and below the x-axis equals

$$-\int_a^b f(x)\, dx,$$

where the integral yields a negative number. Note that every product $f(x_k)\, \Delta x$ in the shaded region is negative, since $f(x_k)$ is negative for all x between a and b.

We see from Figure N6–3 that the graph of f crosses the x-axis at c, that area A_1 lies above the x-axis, and that area A_2 lies below the x-axis. Since, by property (5) on page 185,

$$\int_a^b f(x)\, dx = \int_a^c f(x)\, dx + \int_c^b f(x)\, dx,$$

therefore

$$\int_a^b f(x)\, dx = A_1 - A_2.$$

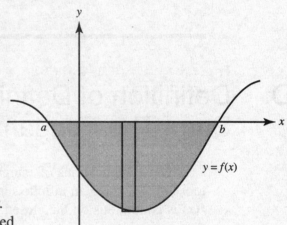

FIGURE N6–2

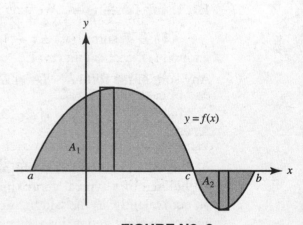

FIGURE N6–3

Note that if f is continuous then the area between the graph of f on $[a,b]$ and the x-axis is given by

$$\int_a^b |f(x)| dx.$$

This implies that, over any interval within $[a,b]$ for which $f(x) < 0$ (for which its graph dips below the x-axis), $|f(x)| = -f(x)$. The area between the graph of f and the x-axis in Figure N6–3 equals

$$\int_a^b |f(x)| dx = \int_a^c f(x)\, dx - \int_c^b f(x)\, dx.$$

We can find the area between the graph of f and the x-axis on $[a,b]$ on a calculator. If $a = 1$ and $b = 4$, for example, and we let $Y_1 = f(x)$, we key in

$$\text{fnInt(abs } Y_1, X, 1, 4).$$

This topic is discussed further in Chapter 7.

E.   Approximations to the Definite Integral; Riemann Sums

It is always possible to approximate the value of a definite integral, even when an integrand cannot be expressed in terms of elementary functions. If f is nonnegative on $[a,b]$, we interpret $\int_a^b f(x)\, dx$ as the area bounded above by $y = f(x)$, below by the x-axis, and vertically by the lines $x = a$ and $x = b$. The value of the definite integral is then approximated by dividing the area into n strips, approximating the area of each strip by a rectangle or other geometric figure, then summing these approximations. In this section we will divide the interval from a to b into n strips of equal width, Δx.

E1. Using Rectangles. We may approximate $\int_a^b f(x)\, dx$ by any of the following sums:

(1) Left sum: $f(x_0)\, \Delta x + f(x_1)\, \Delta x + \cdots + f(x_{n-1})\, \Delta x$, using the value of f at the left endpoint of each subinterval.

(2) Right sum: $f(x_1)\, \Delta x + f(x_2)\, \Delta x + \cdots + f(x_n)\, \Delta x$, using the value of f at the right end of each subinterval.

(3) Midpoint sum: $f\left(\dfrac{x_0 + x_1}{2}\right) \Delta x + f\left(\dfrac{x_1 + x_2}{2}\right) \Delta x + \cdots + f\left(\dfrac{x_{n-1} + x_n}{2}\right) \Delta x$, using the value of f at the midpoint of each subinterval.

These approximations are illustrated in Figures N6–4 and N6–5, which accompany Example 24.

Example 24. Approximate $\displaystyle\int_0^2 x^3\,dx$ by using four subintervals and calculating (a) the left sum, (b) the right sum, and (c) the midpoint sum. Evaluate the integral exactly.

Here $\Delta x = \dfrac{2-0}{4} = \dfrac{1}{2}$.

(a) For a left sum we use the left-hand altitudes at $x = 0,\ \dfrac{1}{2},\ 1,$ and $\dfrac{3}{2}$. The approximating sum is

$$(0)^3 \cdot \frac{1}{2} + \left(\frac{1}{2}\right)^3 \cdot \frac{1}{2} + (1)^3 \cdot \frac{1}{2} + \left(\frac{3}{2}\right)^3 \cdot \frac{1}{2} = \frac{9}{4}\ .$$

The dashed lines in Figure N6–4 show the inscribed rectangles used.

(b) For the right sum we use right-hand altitudes at $x = \dfrac{1}{2},\ 1,\ \dfrac{3}{2},$ and 2. The approximating sum is

$$\left(\frac{1}{2}\right)^3 \cdot \frac{1}{2} + (1)^3 \cdot \frac{1}{2} + \left(\frac{3}{2}\right)^3 \cdot \frac{1}{2} + (2)^3 \cdot \frac{1}{2} = \frac{25}{4}\ .$$

This sum uses the circumscribed rectangles shown in Figure N6–4.

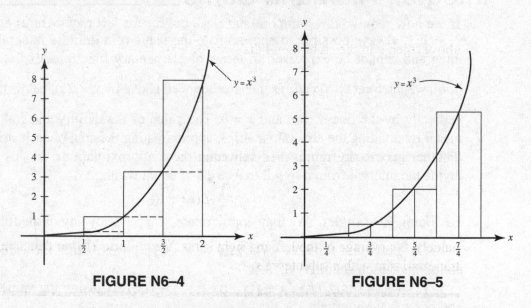

FIGURE N6–4   **FIGURE N6–5**

(c) The midpoint sum uses the heights at the midpoints of the subintervals, as shown in Figure N6–5. The approximating sum is

$$\left(\frac{1}{4}\right)^3 \cdot \frac{1}{2} + \left(\frac{3}{4}\right)^3 \cdot \frac{1}{2} + \left(\frac{5}{4}\right)^3 \cdot \frac{1}{2} + \left(\frac{7}{4}\right)^3 \cdot \frac{1}{2} = \frac{31}{8}\ \text{or}\ \frac{15.5}{4}\ .$$

Since the exact value of $\displaystyle\int_0^2 x^3\,dx$ is $\left.\dfrac{x^4}{4}\right|_0^2$ or 4, the midpoint sum is the best of the three approximations. This is usually the case.

We will denote the three Riemann sums, with n subintervals, by $L(n)$, $R(n)$, and $M(n)$. (These sums are also sometimes called "rules.")

E2. Using Trapezoids. If we represent the area of a strip in Figure N6–6 by that of a trapezoid, then the following sum approximates the area between f and the x-axis from a to b:

$$\frac{(f(x_0) + f(x_1))}{2} \Delta x + \frac{(f(x_1) + f(x_2))}{2} \Delta x + \cdots + \frac{(f(x_{n-1}) + f(x_n))}{2} \Delta x$$

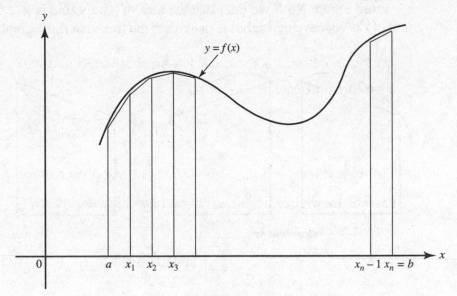

FIGURE N6–6

If we now remove the common factor $\frac{1}{2}$, collect the left members from the parentheses above, then add the right members, we get

$$\frac{1}{2}[(f(x_0) \Delta x + f(x_1) \Delta x + \cdots + f(x_{n-1}) \Delta x)$$

$$+ (f(x_1) \Delta x + f(x_2) \Delta x + \cdots + f(x_n) \Delta x)].$$

This approximating sum is precisely equal to

$$\frac{1}{2}[L(n) + R(n)],$$

namely, the average of the left and right sums. We will use $T(n)$ to denote the approximating trapezoid sum with n subintervals.

Example 25. Use $T(4)$ to approximate $\displaystyle\int_0^2 x^3 \, dx$.

From Example 24, $\Delta x = \frac{1}{2}$, $L(4) = \frac{9}{4}$, and $R(4) = \frac{25}{4}$. Then,

$$T(4) = \frac{1}{2}\left(\frac{9}{4} + \frac{25}{4}\right) = \frac{34}{8} = \frac{17}{4}.$$

This is better than either $L(4)$ or $R(4)$, but $M(4)$ is the best approximation here.

COMPARING APPROXIMATING SUMS

If f is an increasing function on $[a,b]$, then $L(n) \leqslant \int_a^b f(x)\,dx \leqslant R(n)$, while if f is

decreasing, then $R(n) \leqslant \int_a^b f(x)\,dx \leqslant L(n)$.

From Figure N6–7 we infer that the area of a trapezoid is less than the true area if the graph of f is concave down, but is more than the true area if the graph of f is concave up.

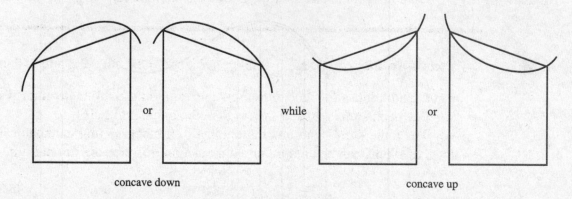

concave down   concave up

FIGURE N6–7

Figure N6–8 is helpful in showing how the area of a midpoint rectangle compares with that of a trapezoid and with the true area. Our graph here is concave down. If M is the midpoint of AB, then the midpoint rectangle is AM_1M_2B. We've drawn T_1T_2 tangent to the curve at T (where the midpoint ordinate intersects the curve). Since the shaded triangles have equal areas, we see that area AM_1M_2B = area AT_1T_2B.[†] But area AT_1T_2B clearly exceeds the true area, as does the area of the midpoint rectangle. This fact justifies the right half of the inequality below; Figure N6–7 verifies the left half.

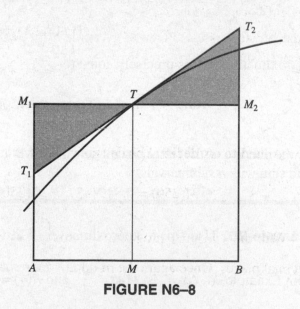

FIGURE N6–8

Generalizing to n subintervals, we conclude:
If the graph of f is concave down, then

$$T(n) \leqslant \int_a^b f(x)\,dx \leqslant M(n).$$

[†]Note that the trapezoid AT_1T_2B is *different* from the trapezoids in Figure N6–7, which are like the ones we use in applying the trapezoid rule.

If the graph of f is concave up, then

$$M(n) \leq \int_a^b f(x)\, dx \leq T(n).$$

USING A GRAPHING CALCULATOR

Obviously an approximating sum for n subintervals improves as n increases.

Example 26.   Estimate $\int_0^{\pi/2} \cos(x^2)\, dx$ for $n = 10$, 50, 100, and 250, using left, right, midpoint, and trapezoid sums and rounding to four decimal places. Tell whether each sum is greater or less than the given integral.

The table shows the outputs obtained on our graphing calculator. Programs for these sums can be entered in your calculator if they are not built in.

n	LEFT	RIGHT	MID.	TRAP.
10	0.9850	0.7052	0.8512	0.8451
50	0.8770	0.8210	0.8492	0.8490
100	0.8631	0.8351	0.8492	0.8491
250	0.8547	0.8435	0.8491	0.8491

Since $\cos(x^2)$ decreases on $[0, \frac{\pi}{2}]$, $R(n) \leq \int_0^{\pi/2} \cos(x^2)\, dx \leq L(n)$ for every n.

Since the graph is concave down on $[0, \frac{\pi}{2}]$ (as can be seen from the calculator), the trapezoid sum is less than the given definite integral, while the midpoint sum is greater. Thus

$$R(n) \leq T(n) \leq \int_0^{\pi/2} \cos(x^2)\, dx \leq M(n) \leq L(n).$$

If we round to six decimal places for $n = 250$, we get

$$M(250) = 0.849143; \qquad T(250) = 0.849133.$$

With the [fnInt] program, $\int_0^{\pi/2} \cos(x^2)\, dx$ equals 0.849139, rounded to six decimal places. Once again the midpoint approximation is best.

Example 27.   Write an inequality including $L(n)$, $R(n)$, $M(n)$, $T(n)$, and $\int_a^b f(t)\, dt$ for the graph of f shown in Figure N6–9.

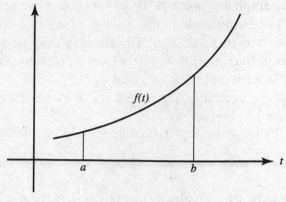

FIGURE N6–9

Since f increases on $[a,b]$ and is concave up, the inequality is

$$L(n) \leq M(n) \leq \int_a^b f(x)\,dx \leq T(n) \leq R(n).$$

GRAPHING A FUNCTION FROM ITS DERIVATIVE; ANOTHER LOOK

Example 28. Figure N6–10 is the graph of function $f'(x)$; it consists of two line segments and a semicircle. If $f(0) = 1$, sketch the graph of $f(x)$. Identify any critical or inflection points of f and give their coordinates.

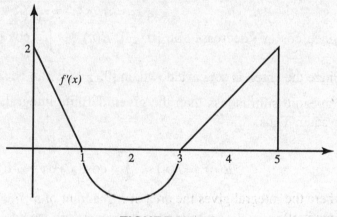

FIGURE N6–10

We know that if $f' > 0$ on an interval then f increases on the interval, while if $f' < 0$ then f decreases; also, if f' is increasing on an interval then f is concave up on the interval, while if f' is decreasing then f is concave down. These statements lead to the following conclusions:

f increases on $[0,1]$ and $[3,5]$,

but f decreases on $[1,3]$;

also f is concave down on $[0,2]$;

but f is concave up on $[2,5]$.

Additionally, since $f'(1) = f'(3) = 0$, f has critical points at $x = 1$ and $x = 3$. As x passes through 1, the sign of f' changes from positive to negative; as x passes through 3, the sign of f' changes from negative to positive. Therefore $f(1)$ is a local maximum and $f(3)$ a local minimum. Since f changes concavity at $x = 2$, the latter is an inflection point of f.

These conclusions enable us to get the general shape of the curve, as displayed in Figure N6–11a.

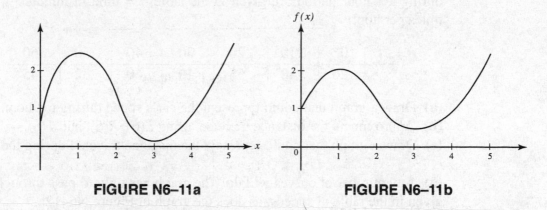

FIGURE N6–11a   **FIGURE N6–11b**

All that remains is to evaluate $f(x)$ at $x = 1$, 2, and 3. We use the Fundamental Theorem of Calculus to accomplish this, finding f also at $x = 4$ and 5 for completeness.

We are given that $f(0) = 1$. Then

$$f(1) = f(0) + \int_0^1 f'(x)\, dx$$

$$= 1 \quad + \quad 1 \qquad\qquad = 2,$$

where the integral yields the area of the triangle with height 2 and base 1;

$$f(2) = f(1) + \int_1^2 f'(x)\, dx$$

$$= 2 \quad - \quad \frac{\pi}{4} \qquad\qquad \simeq 1.2,$$

where the integral gives the area of a quadrant of a circle of radius 1 (this integral is negative!);

$$f(3) = f(2) + \int_2^3 f'(x)\, dx$$

$$1.2 \quad - \quad \frac{\pi}{4} \quad \text{(why?)} \qquad \simeq 0.4,$$

$$f(4) = f(3) + \int_3^4 f'(x)\, dx$$

$$\simeq 0.4 \quad + \quad \frac{1}{2} \qquad\qquad \simeq 0.9,$$

where the integral is the area of the triangle with height 1 and base 1;

$$f(5) = f(4) + \int_4^5 f'(x)\, dx$$

$$\simeq 0.9 \quad + \quad 1.5 \quad \text{(why?)} \qquad \simeq 2.4.$$

So the function $f(x)$ has a local maximum at (1,2), a point of inflection at (2,1.2), and a local minimum at (3,0.4) where we have rounded to one decimal place when necessary.

In Figure N6–11b, the graph of f is shown again, but now it incorporates the information just obtained using the FTC.

Example 29. Readings from a car's speedometer at 10-minute intervals during a 1-hour period are given in the table; t = time in minutes, v = speed in miles per hour:

t	0	10	20	30	40	50	60
v	26	40	55	10	60	32	45

(a) Draw a graph that could represent the car's speed during the hour.
(b) Approximate the distance traveled, using $L(6)$, $R(6)$, and $T(6)$.
(c) Draw a graph that could represent the distance traveled during the hour.

(a) Any number of curves will do. The graph has only to pass through the points given in the table of speeds, as does the graph in Figure N6–12a.

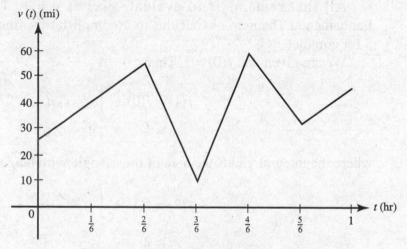

FIGURE N6–12a

(b) $L(6) = (26 + 40 + 55 + 10 + 60 + 32) \cdot \dfrac{1}{6} = 37\dfrac{1}{6}$ mi;

$R(6) = (40 + 55 + 10 + 60 + 32 + 45) \dfrac{1}{6} = 40\dfrac{1}{3}$ mi;

$T(6) = 38\dfrac{3}{4}$ mi.

(c) To calculate the distance traveled during the hour, we use the methods demonstrated in Example 28. We know that, since $v(t) > 0$, $s = \displaystyle\int_a^b v(t)\, dt$ is the distance covered from time a to time b, where $v(t)$ is the speed (or velocity). Thus

$s(0) = 0,$

$s\left(\dfrac{1}{6}\right) = 0 + \displaystyle\int_0^{1/6} v(t)\, dt = 0 + (26 + 40) \cdot \dfrac{1}{12} = \dfrac{66}{12},$

$$s\left(\frac{2}{6}\right) = \frac{66}{12} + \int_{1/6}^{2/6} v(t) \, dt = \frac{66}{12} + (40 + 55) \cdot \frac{1}{12} = \frac{161}{12},$$

$$s\left(\frac{6}{6}\right) = \frac{388}{12} + \int_{5/6}^{6/6} v(t) \, dt = \frac{388}{12} + (32 + 45) \cdot \frac{1}{12} = \frac{465}{12}.$$

It is left to the student to complete the missing steps above and to verify the distances in the following table (t = time in minutes, s = distance in miles):

t	0	10	20	30	40	50	60
s	0	5.5	13.4	18.8	24.7	32.3	38.8

Figure N6–12b is one possible graph for the distance covered during the hour.

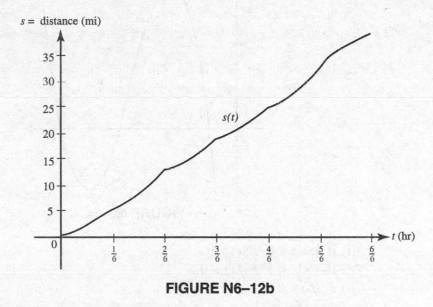

FIGURE N6–12b

Example 30.   The graph of $f(t)$ is given in Figure N6–13. If $F(x) = \int_{0}^{x} f(t) \, dt$, fill in the values for $F(x)$ in the table:

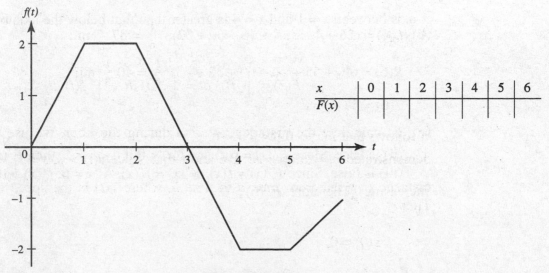

x	0	1	2	3	4	5	6
$F(x)$							

FIGURE N6–13

Here is the completed table:

x	0	1	2	3	4	5	6
$F(x)$	0	1	3	4	3	1	–0.5

Example 31. The graph of the function $f(t)$ is shown in Figure N6–14. Let

$$F(x) = \int_1^x f(t)\, dt.$$

Decide whether each statement is true or false; justify your answers:

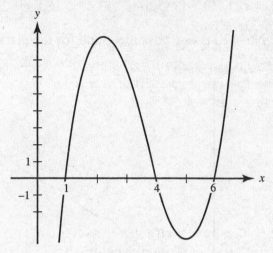

FIGURE N6–14

   (i) If $4 < x < 6$, $F(x) > 0$.
   (ii) If $4 < x < 5$, $F'(x) > 0$.
   (iii) $F''(6) < 0$.

(i) is true. We know that, if a function g is positive on (a,b), then $\int_a^b g(x)\, dx > 0$,

whereas if g is negative on (a,b), then $\int_a^b g(x)\, dx < 0$. However, the area above the x-axis between $x = 1$ and $x = 4$ is greater than that below the axis between 4 and 6. Since

$$F(x) = \int_1^x f(t)\, dt = \int_1^4 f(t)\, dt + \int_4^x f(t)\, dt,$$

it follows that $F(x) > 0$ if $4 < x < 6$.
   (ii) is false. Since $F'(x) = f(x)$ and $f(x) < 0$ if $4 < x < 5$, then $F'(x) < 0$.
   (iii) is false. Since $F'(x) = f(x)$, $F''(x) = f'(x)$. At $x = 6$, $f'(x) > 0$ (because f is increasing). Therefore, $F''(6) > 0$.

Example 32.   Graphs of functions $f(x)$, $g(x)$, and $h(x)$ are given in Figures N6–15a, N6–15b, and N6–15c. Consider the following statements:

$$\text{(I)} \ f(x) = g'(x) \qquad \text{(II)} \ h(x) = f'(x) \qquad \text{(III)} \ g(x) = \int_{-2.5}^{x} f(t) \, dt$$

Which of these statements is (are) true?

   (A)   I only
   (B)   II only
   (C)   III only
   (D)   all three
   (E)   none of them

The correct answer is D.

I is true since, for example, $f(x) = 0$ for the critical values of g: f is positive where g increases, negative where g decreases, and so on.

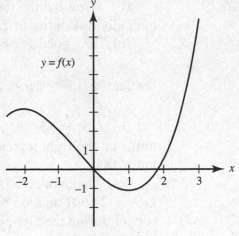

FIGURE N6–15a

II is true for similar reasons.

III is also true. Verify that the value of the integral $g(x)$ increases on the interval $-2.5 < x < 0$ (where $f > 0$), decreases between the zeros of f (where $f < 0$), then increases again when f becomes positive.

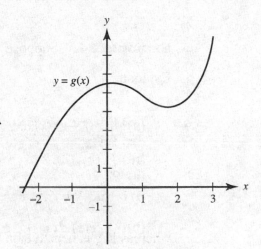

FIGURE N6–15b

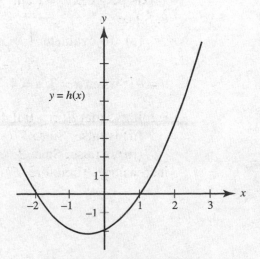

FIGURE N6–15c

Example 33. The world use of copper has been increasing at a rate given by $f(t)$, where t is measured in years, with $t = 0$ the beginning of 1995, and $f(t)$ is measured in millions of tons per year. Assume $f(t) = 15e^{0.015t}$.

(a) What definite integral gives the total amount of copper that has been used for the 5-year period from $t = 0$ to the beginning of the year 2000?

(b) Write out the terms in the left sum $L(5)$ for the integral in (a). What do the individual terms of $L(5)$ mean in terms of the world use of copper?

(c) How good an approximation is $L(5)$ for the definite integral in (a)?

(a) $\displaystyle\int_0^5 15e^{0.015t}\, dt.$

(b) $L(5) = 15e^{0.015 \cdot 0} + 15e^{0.015 \cdot 1} + 15e^{0.015 \cdot 2} + 15e^{0.015 \cdot 3} + 15e^{0.015 \cdot 4}$. The five terms on the right represent the world's use of copper for the 5 years from 1995 until 2000.

(c) The answer to (a) (using our [fnInt] program) is 77.884 million tons. $L(5) = 77.301$ million tons, so $L(5)$ underestimates the projected world use of copper during the 5-yr period by approximately 583,000 tons.

Example 33 is an excellent instance of the FTC: if $f = F'$, then $\displaystyle\int_a^b f(x)\, dx$ gives the total change in F as x varies from a to b.

Example 34. Suppose $\displaystyle\int_{-1}^4 f(x)\, dx = 6$, $\displaystyle\int_{-1}^4 g(x)\, dx = -3$, and $\displaystyle\int_{-1}^0 g(x)\, dx = -1$.
Evaluate

(a) $\displaystyle\int_{-1}^4 (f - g)(x)\, dx;$ (b) $\displaystyle\int_0^4 g(x)\, dx;$ (c) $\displaystyle\int_2^7 f(x - 3)\, dx.$

(a) 9.

(b) $\displaystyle\int_0^4 g(x)\, dx = \int_0^{-1} g(x)\, dx + \int_{-1}^4 g(x)\, dx = -\int_{-1}^0 g(x)\, dx + \int_{-1}^4 g(x)\, dx$

$= +1 + (-3) = -2.$

(c) To evaluate $\displaystyle\int_2^7 f(x - 3)\, dx$, let $u = x - 3$. Then $du = dx$ and, when $x = 2$,

$u = -1$; when $x = 7$, $u = 4$. Therefore $\displaystyle\int_2^7 f(x - 3)\, dx = \int_{-1}^4 f(u)\, du = 6.$

F.   Interpreting In x as an Area _____

It is quite common to define ln x, the natural logarithm of x, as a definite integral, as follows:

$$\ln x = \int_1^x \frac{1}{t}\, dt \qquad (x > 0).$$

This integral can be interpreted as the area bounded above by the curve $y = \frac{1}{t}\,(t > 0)$, below by the t-axis, at the left by $t = 1$, and at the right by $t = x\,(x > 1)$. See Figure N6–16.

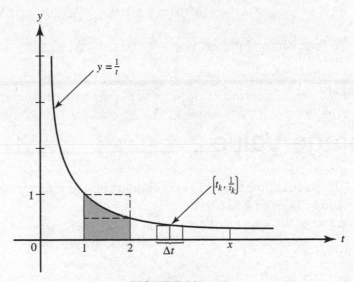

FIGURE N6–16

Note that if $x = 1$ the above definition yields ln $1 = 0$, and if $0 < x < 1$ we can rewrite as follows:

$$\ln x = -\int_x^1 \frac{1}{t}\, dt,$$

showing that ln $x < 0$ if $0 < x < 1$.

With this definition of ln x we can approximate ln x using rectangles or trapezoids.

_____

Example 35.   Show that $\frac{1}{2} < \ln 2 < 1$.

Using the definition of ln x above yields $\ln 2 = \int_1^2 \frac{1}{t}\, dt$, which we interpret as the area under $y = \frac{1}{t}$, above the t-axis, and bounded at the left by $t = 1$ and at the right by $t = 2$ (the shaded region in Figure N6–16). Since $y = \frac{1}{t}$ is strictly decreasing, the area of the inscribed rectangle (height $\frac{1}{2}$, width 1) is less than ln 2, which, in turn, is less than the area of the circumscribed rectangle (height 1, width 1). Thus $\frac{1}{2} \cdot 1 < \ln 2 < 1 \cdot 1$ or $\frac{1}{2} < \ln 2 < 1$.

Example 36. Find $L(5)$, $R(5)$, $M(5)$ and $L(20)$, $R(20)$, and $M(20)$ for

$$\int_1^2 \frac{1}{t} \, dt.$$

All of the above sums will approximate the given definite integral for $f(t) = \frac{1}{t}$, with $a = x_1 = 1$ and $b = x_2 = 2$. First we let $n = 5$, then let $n = 20$, getting, to four decimal places,

$$L(5) = 0.7456, \qquad L(20) = 0.7058;$$
$$R(5) = 0.6456, \qquad R(20) = 0.6808;$$
$$M(5) = 0.6919, \qquad M(20) = 0.6931.$$

To four decimal places, if $\text{Y}_1 = \frac{1}{\text{X}}$, $\texttt{fnInt(Y}_1\texttt{,X,1,2)}$ yields 0.6931.

G.   Average Value

If the function $y = f(x)$ is continuous on the interval $a \leqslant x \leqslant b$, then we define *average value of f from a to b* to be

$$\frac{1}{b-a} \int_a^b f(x) \, dx. \tag{1}$$

Note that (1) is equivalent to

$$(\text{average value of } f) \cdot (b - a) = \int_a^b f(x) \, dx. \tag{2}$$

If $f(x) \geqslant 0$ for all x on $[a,b]$, we can interpret (2) in terms of areas as follows: The right-hand member represents the area under the curve of $y = f(x)$, above the x-axis, and bounded by the vertical lines $x = a$ and $x = b$. The left-hand member of (2) represents the area of a rectangle with the same base $(b - a)$ and with the average value of f as its height. See Figure N6–17.

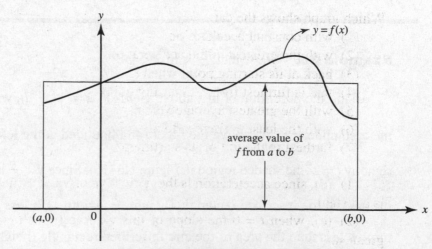

FIGURE N6–17

Example 37.   The average value of $f(x) = \ln x$ on the interval $[1,4]$ is

$$\frac{1}{4-1} \int_1^4 \ln x \, dx = \frac{1}{3} (x \ln x - x) \Big|_1^4 = \frac{4 \ln 4 - 3}{3}.$$

Example 38.   The average value of ordinates of the semicircle $y = \sqrt{4 - x^2}$ on $[-2,2]$ is given by

$$\frac{1}{2-(-2)} \int_{-2}^2 \sqrt{4 - x^2} \, dx = \frac{1}{4} \frac{\pi(2^2)}{2} = \frac{\pi}{2}. \qquad (3)$$

In (3) we have used the fact that the definite integral equals exactly the area of a semicircle of radius 2.

Example 39.   The graphs (a) through (e) in Figure N6–18 show the velocities of five cars moving along an east-west road (the x-axis) at time t, where $0 \leqslant t \leqslant 6$. In each graph the scales on the two axes are the same.

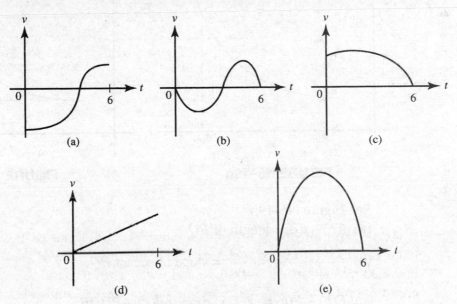

FIGURE N6–18

Which graph shows the car
    (1) with constant acceleration?
    (2) with the greatest initial acceleration?
    (3) back at its starting point when $t = 6$?
    (4) that is furthest from its starting point at $t = 6$?
    (5) with the greatest average velocity?
    (6) with the least average velocity?
    (7) farthest to the left of its starting point when $t = 6$?

    (1) (d), since acceleration is the derivative of velocity and in (d) v', the slope, is constant.
    (2) (e): when $t = 0$ the slope of this v-curve (which equals acceleration) is greatest.

(3) (b), since for this car the net distance traveled (given by the net area) equals zero.

(4) (e): since the area under the v-curve is greatest, this car is farthest east.

(5) (e): the average velocity equals the total distance divided by 6, which is the net area divided by 6 (see (4)).

(6) (a), since only for this car is the net area negative.

(7) (a) again, since net area is negative only for this car.

Example 40. Identify each of the following quantities for the function $f(x)$, whose graph is shown in Figure N6–19a (note: $F'(x) = f(x)$):

(a) $f(b) - f(a)$

(b) $\dfrac{f(b) - f(a)}{b - a}$

(c) $F(b) - F(a)$

(d) $\dfrac{F(b) - F(a)}{b - a}$

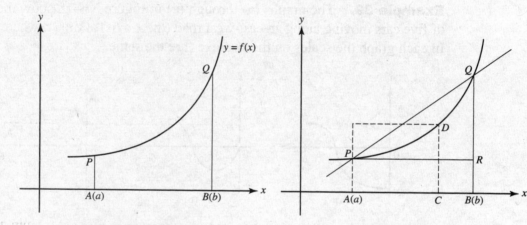

FIGURE N6–19a   **FIGURE N6–19b**

See Figure N6–19b.

(a) $f(b) - f(a)$ = length of RQ.

(b) $\dfrac{f(b) - f(a)}{b - a} = \dfrac{RQ}{PR}$ = slope of secant PQ.

(c) $F(b) - F(a) = \displaystyle\int_a^b f(x)\,dx$ = area of $APDQB$.

(d) $\dfrac{F(b) - F(a)}{b - a}$ = average value of f over $[a,b]$ = length of CD, where $CD \cdot AB$ or $CD \cdot (b - a)$ is equal to the area $F(b) - F(a)$.

Example 41.   The graph in Figure N6–20 shows the speed $v(t)$ of a car, in miles per hour, at 10-minute intervals during a 1-hour period.

   (a) Give an upper and a lower estimate of the total distance traveled.
   (b) When does the acceleration appear greatest?
   (c) Estimate the acceleration when $t = 20$.
   (d) Estimate the average speed of the car during the interval $30 \leqslant t \leqslant 50$.

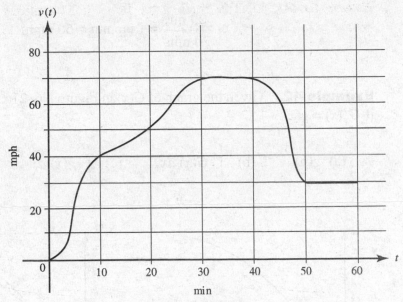

FIGURE N6–20

   (a) A lower estimate, using minimum speeds and $\dfrac{1}{6}$ hr for 10 min, is

$$0\left(\frac{1}{6}\right) + 40\left(\frac{1}{6}\right) + 50\left(\frac{1}{6}\right) + 70\left(\frac{1}{6}\right) + 30\left(\frac{1}{6}\right) + 30\left(\frac{1}{6}\right).$$

This yields $36\dfrac{2}{3}$ mi for the total distance traveled during the hour. An upper estimate uses maximum speeds; it equals

$$40\left(\frac{1}{6}\right) + 50\left(\frac{1}{6}\right) + 70\left(\frac{1}{6}\right) + 70\left(\frac{1}{6}\right) + 70\left(\frac{1}{6}\right) + 30\left(\frac{1}{6}\right)$$

or 55 mi for the total distance.
   (b) The acceleration, which is the slope of $v(t)$, appears greatest at $t = 5$ min, when the curve is steepest.
   (c) To estimate the acceleration $v'(t)$ at $t = 20$, we approximate the slope of the curve at $t = 20$. The slope of the tangent at $t = 20$ appears to be equal to

(10 mph)/(10 min) = (10 mph)/($\frac{1}{6}$ hr) = 60 mi/hr².

(d) The average speed equals the distance traveled divided by the time. We can approximate the distance from $t = 30$ to $t = 50$ by the area under the curve, or, roughly, by the sum of the areas of a rectangle and a trapezoid:

$$70\left(\frac{1}{6}\right) + \frac{70 + 30}{2}\left(\frac{1}{6}\right) = 20 \text{ mi.}$$

Thus the average speed from $t = 30$ to $t = 50$ is

$$\frac{20 \text{ mi}}{20 \text{ min}} = 1 \text{ mpmin} = 60 \text{ mph.}$$

Example 42.   Given the graph of $G(x)$ in Figure N6–21a, identify the following if $G'(x) = g(x)$:

(a) $g(b)$ (b) $\displaystyle\int_a^b G(x)\, dx$ (c) $\displaystyle\int_a^b g(x)\, dx$ (d) $\dfrac{\displaystyle\int_a^b g(x)\, dx}{b - a}$.

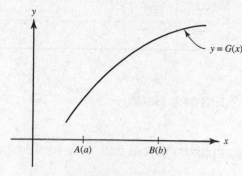

FIGURE N6–21a

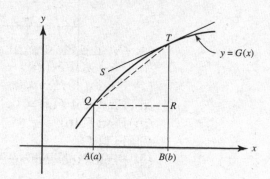

FIGURE N6–21b

See Figure N6–21b.

(a) $g(b)$ is the slope of $G(x)$ at b, the slope of line ST.

(b) $\displaystyle\int_a^b G(x)\, dx$ is equal to the area under $G(x)$ from a to b.

(c) $\displaystyle\int_a^b g(x)\, dx = G(b) - G(a) = $ length of $BT -$ length of $BR = $ length of RT.

(d) $\dfrac{\displaystyle\int_a^b g(x)\, dx}{b - a} = \dfrac{\text{length of } RT}{\text{length of } QR} = $ slope of QT.

Example 43.   The function $f(t)$ is graphed in Figure N6–22a. Let

$$F(x) = \int_{4}^{x/2} f(t)\, dt.$$

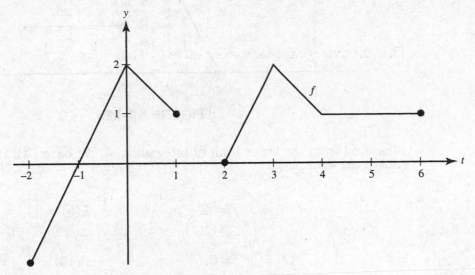

FIGURE N6–22a

(a) What is the domain of F?
(b) Find x, if $F'(x) = 0$.
(c) Find x, if $F(x) = 0$.
(d) Find x, if $F(x) = 1$.
(e) Find $F'(6)$.
(f) Find $F(6)$.
(g) Sketch the complete graph of F.

(a) The domain of f is $[-2,1]$ and $[2,6]$. We choose the portion of this domain that contains the lower limit of integration, 4. Thus the domain of F is $2 \leqslant \frac{x}{2} < 6$, or $4 \leqslant x < 12$.

(b) Since $F'(x) = f\left(\frac{x}{2}\right) \cdot \frac{1}{2}$, $F'(x) = 0$ if $f\left(\frac{x}{2}\right) = 0$. Then $\frac{x}{2} = 2$ and $x = 4$.

(c) $F(x) = 0$ when $\frac{x}{2} = 4$ or $x = 8$. $F(8) = \int_{4}^{8/2} f(t)\, dt = 0$.

(d) For $F(x)$ to equal 1, we need a region under f whose left endpoint is 4 with area equal to 1. The region from 4 to 5 works nicely; so $\frac{x}{2} = 5$ and $x = 10$.

(e) $F'(6) = f\left(\frac{6}{2}\right) \cdot \frac{1}{2} = f(3) \cdot \left(\frac{1}{2}\right) = 2 \cdot \left(\frac{1}{2}\right) = 1$.

(f) $F(6) = \int_{4}^{3} f(t)\, dt = -\text{(area of trapezoid)} = -\frac{2+1}{2} \cdot 1 = -\frac{3}{2}$.

(g) In Figure N6–22b we evaluate the areas in the original graph.

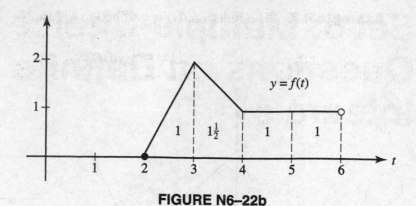

FIGURE N6–22b

Measured from the lower limit of integration, 4, we have (with "f" as an abbreviation for "$f(t)\,dt$"):

$$F(4) = \int_4^2 f = -2\frac{1}{2}, \qquad\qquad F(6) = \int_4^3 f = -1\frac{1}{2},$$

$$F(8) = \int_4^4 f = 0, \qquad\qquad F(10) = \int_4^5 f = 1,$$

$$F(12) = \int_4^6 f = 2.$$

We note that, since $F'(= f)$ is linear on $(2,4)$, F is quadratic on $(4,8)$; also, since F' is positive and increasing on $(2,3)$, F is increasing and concave up on $(4,6)$, while since F' is positive and decreasing on $(3,4)$, F is increasing but concave down on $(6,8)$. Finally, since F' is constant on $(4,6)$, F is linear on $(8,12)$. (See Figure N6–22c.)

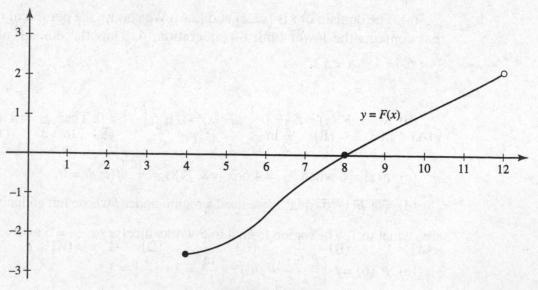

FIGURE N6–22c

Set 6: Multiple-Choice Questions on Definite Integrals

1. $\displaystyle\int_{-1}^{1} (x^2 - x - 1)\, dx =$

(A) $\dfrac{2}{3}$ (B) 0 (C) $-\dfrac{4}{3}$ (D) -2 (E) -1

2. $\displaystyle\int_{1}^{2} \dfrac{3x - 1}{3x}\, dx =$

(A) $\dfrac{3}{4}$ (B) $1 - \dfrac{1}{3}\ln 2$ (C) $1 - \ln 2$ (D) $-\dfrac{1}{3}\ln 2$ (E) 1

3. $\displaystyle\int_{0}^{3} \dfrac{dt}{\sqrt{4 - t}} =$

(A) 1 (B) -2 (C) 4 (D) -1 (E) 2

4. $\displaystyle\int_{-1}^{0} \sqrt{3u + 4}\, du =$

(A) 2 (B) $\dfrac{14}{9}$ (C) $\dfrac{14}{3}$ (D) 6 (E) $\dfrac{7}{2}$

5. $\displaystyle\int_{2}^{3} \dfrac{dy}{2y - 3} =$

(A) $\ln 3$ (B) $\dfrac{1}{2}\ln\dfrac{3}{2}$ (C) $\dfrac{16}{9}$ (D) $\ln\sqrt{3}$ (E) $\sqrt{3} - 1$

6. $\displaystyle\int_{0}^{\sqrt{3}} \dfrac{x}{\sqrt{4 - x^2}}\, dx =$

(A) 1 (B) $\dfrac{\pi}{6}$ (C) $\dfrac{\pi}{3}$ (D) -1 (E) 2

7. $\displaystyle\int_{0}^{1} (2t - 1)^3\, dt =$

(A) $\dfrac{1}{4}$ (B) 6 (C) $\dfrac{1}{2}$ (D) 0 (E) 4

8. To three decimal places, $\displaystyle\int_0^1 \frac{dx}{\sqrt{4-x^2}} =$

    (A) 0.262   (B) 0.268   (C) 0.524   (D) 0.536   (E) 1.047

9. $\displaystyle\int_4^9 \frac{2+x}{2\sqrt{x}}\, dx =$

    (A) $\dfrac{25}{3}$ (B) $\dfrac{41}{3}$ (C) $\dfrac{100}{3}$ (D) $\dfrac{5}{3}$ (E) $\dfrac{1}{3}$

10. $\displaystyle\int_{-3}^3 \frac{dx}{9+x^2} =$

    (A) $\dfrac{\pi}{2}$ (B) 0   (C) $\dfrac{\pi}{6}$ (D) $-\dfrac{\pi}{2}$ (E) $\dfrac{\pi}{3}$

11. $\displaystyle\int_0^1 e^{-x}\, dx =$

    (A) $\dfrac{1}{e}-1$ (B) $1-e$ (C) $-\dfrac{1}{e}$ (D) $1-\dfrac{1}{e}$ (E) $\dfrac{1}{e}$

12. $\displaystyle\int_0^1 xe^{x^2}\, dx =$

    (A) $e-1$ (B) $\dfrac{1}{2}(e-1)$ (C) $2(e-1)$ (D) $\dfrac{e}{2}$ (E) $\dfrac{e}{2}-1$

13. $\displaystyle\int_0^{\pi/4} \sin 2\theta\, d\theta =$

    (A) 2   (B) $\dfrac{1}{2}$ (C) -1 (D) $-\dfrac{1}{2}$ (E) -2

14. $\displaystyle\int_1^2 \frac{dz}{3-z} =$

    (A) $-\ln 2$ (B) $\dfrac{3}{4}$ (C) $2(\sqrt{2}-1)$ (D) $\dfrac{1}{2}\ln 2$ (E) $\ln 2$

*15. If we let $x = 2\sin\theta$, then $\displaystyle\int_1^2 \frac{\sqrt{4-x^2}}{x}\, dx$ is equivalent to

    (A) $2\displaystyle\int_0^2 \frac{\cos^2\theta}{\sin\theta}\, d\theta$ (B) $\displaystyle\int_{\pi/6}^{\pi/2} \frac{\cos\theta}{\sin\theta}\, d\theta$ (C) $2\displaystyle\int_{\pi/6}^{\pi/2} \frac{\cos^2\theta}{\sin\theta}\, d\theta$

    (D) $\displaystyle\int_1^2 \frac{\cos\theta}{\sin\theta}\, d\theta$ (E) none of these

*An asterisk denotes a topic covered only in Calculus BC.

16. The integral $\int_{-4}^{4} \sqrt{16 - x^2}\ dx$ gives the area of

    (A)   a circle of radius 4
    (B)   a semicircle of radius 4
    (C)   a quadrant of a circle of radius 4
    (D)   an ellipse whose semimajor axis is 4
    (E)   none of these

17. $\int_{0}^{\pi} \cos^2 \theta \sin \theta\ d\theta =$

    (A) $-\dfrac{2}{3}$ **(B)** $\dfrac{1}{3}$ **(C)** 1 **(D)** $\dfrac{2}{3}$ **(E)** 0

18. $\int_{1}^{e} \dfrac{\ln x}{x}\ dx =$

    (A) $\dfrac{1}{2}$ **(B)** $\dfrac{1}{2}(e^2 - 1)$ **(C)** 0 **(D)** 1 **(E)** $e - 1$

*19. $\int_{0}^{1} x e^x\ dx =$

    (A) -1 **(B)** $e + 1$ **(C)** 1 **(D)** $e - 1$ **(E)** $\dfrac{1}{2}(e - 1)$

20. $\int_{0}^{\pi/6} \dfrac{\cos \theta}{1 + 2 \sin \theta}\ d\theta =$

    (A) $\ln 2$ **(B)** $\dfrac{3}{8}$ **(C)** $-\dfrac{1}{2}\ln 2$ **(D)** $\dfrac{3}{2}$ **(E)** $\ln \sqrt{2}$

21. $\int_{0}^{\pi/4} \sqrt{1 - \cos 2\alpha}\ d\alpha =$

    (A) 0.25 **(B)** 0.414 **(C)** 1.000 **(D)** 1.414 **(E)** 2.000

22. $\int_{\sqrt{2}}^{2} \dfrac{u}{u^2 - 1}\ du =$

    (A) $\ln \sqrt{3}$ **(B)** $\dfrac{8}{9}$ **(C)** $\ln \dfrac{3}{2}$ **(D)** $\ln 3$ **(E)** $1 - \sqrt{3}$

*An asterisk denotes a topic covered only in Calculus BC.

23. $\displaystyle\int_{\sqrt{2}}^{2} \frac{u\,du}{(u^2-1)^2} =$

(A) $-\dfrac{1}{3}$ (B) $-\dfrac{2}{3}$ (C) $\dfrac{2}{3}$ (D) -1 (E) $\dfrac{1}{3}$

24. $\displaystyle\int_{0}^{\pi/4} \cos^2\theta\,d\theta =$

(A) $\dfrac{1}{2}$ (B) $\dfrac{\pi}{8}$ (C) $\dfrac{\pi}{8}+\dfrac{1}{4}$ (D) $\dfrac{\pi}{8}+\dfrac{1}{2}$ (E) $\dfrac{\pi}{8}-\dfrac{1}{4}$

25. $\displaystyle\int_{\pi/12}^{\pi/4} \frac{\cos 2x\,dx}{\sin^2 2x} =$

(A) $-\dfrac{1}{4}$ (B) 1 (C) $\dfrac{1}{2}$ (D) $-\dfrac{1}{2}$ (E) -1

26. $\displaystyle\int_{0}^{1} \frac{e^{-x}+1}{e^{-x}}\,dx =$

(A) e (B) $2+e$ (C) $\dfrac{1}{e}$ (D) $1+e$ (E) $e-1$

27. $\displaystyle\int_{0}^{1} \frac{e^x}{e^x+1}\,dx =$

(A) $\ln 2$ (B) e (C) $1+e$ (D) $-\ln 2$ (E) $\ln\dfrac{e+1}{2}$

28. If $f(x)$ is continuous on the interval $a \le x \le b$ and $a < c < b$, then $\displaystyle\int_{c}^{b} f(x)\,dx$ is equal to

(A) $\displaystyle\int_{a}^{c} f(x)\,dx + \int_{c}^{b} f(x)\,dx$ (B) $\displaystyle\int_{a}^{c} f(x)\,dx - \int_{a}^{b} f(x)\,dx$

(C) $\displaystyle\int_{c}^{a} f(x)\,dx + \int_{b}^{a} f(x)\,dx$ (D) $\displaystyle\int_{a}^{b} f(x)\,dx - \int_{a}^{c} f(x)\,dx$

(E) $\displaystyle\int_{a}^{c} f(x)\,dx - \int_{b}^{c} f(x)\,dx$

29. If $f(x)$ is continuous on $a \le x \le b$, then

(A) $\displaystyle\int_{a}^{b} f(x)\,dx = f(b)-f(a)$ (B) $\displaystyle\int_{a}^{b} f(x)\,dx = -\int_{b}^{a} f(x)\,dx$

(C) $\displaystyle\int_{a}^{b} f(x)\,dx \geqq 0$ (D) $\dfrac{d}{dx}\displaystyle\int_{a}^{x} f(t)\,dt = f'(x)$

(E) $\dfrac{d}{dx}\displaystyle\int_{a}^{x} f(t)\,dt = f(x)-f(a)$

30. If $f(x)$ is continuous on the interval $a \leqq x \leqq b$, if this interval is partitioned into n equal subintervals of length Δx, and if x_k is a number in the kth subinterval, then

$$\lim_{n \to \infty} \sum_{1}^{n} f(x_k) \, \Delta x \text{ is equal to}$$

(A) $f(b) - f(a)$

(B) $F(x) + C$, where $\dfrac{dF(x)}{dx} = f(x)$ and C is an arbitrary constant

(C) $\displaystyle\int_a^b f(x) \, dx$

(D) $F(b - a)$, where $\dfrac{dF(x)}{dx} = f(x)$

(E) none of these

31. If $F'(x) = G'(x)$ for all x, then

(A) $\displaystyle\int_a^b F'(x) \, dx = \int_a^b G'(x) \, dx$ (B) $\displaystyle\int F(x) \, dx = \int G(x) \, dx$

(C) $\displaystyle\int_a^b F(x) \, dx = \int_a^b G(x) \, dx$ (D) $\displaystyle\int F(x) \, dx = \int G(x) \, dx + C$

(E) None of the preceding is necessarily true

32. If $f(x)$ is continuous on the closed interval $[a,b]$, then there exists at least one number c, $a < c < b$, such that $\displaystyle\int_a^b f(x) \, dx$ is equal to

(A) $\dfrac{f(c)}{b - a}$ (B) $f'(c)(b - a)$ (C) $f(c)(b - a)$

(D) $\dfrac{f'(c)}{b - a}$ (E) $f(c)[f(b) - f(a)]$

33. If $f(x)$ is continuous on the closed interval $[a,b]$ and k is a constant, then

$$\int_a^b kf(x) \, dx \text{ is equal to}$$

(A) $k(b - a)$ (B) $k[f(b) - f(a)]$ (C) $kF(b - a)$, where $\dfrac{dF(x)}{dx} = f(x)$

(D) $k \displaystyle\int_a^b f(x) \, dx$ (E) $\dfrac{[kf(x)]^2}{2} \Big]_a^b$

34. $\dfrac{d}{dt} \displaystyle\int_0^t \sqrt{x^3 + 1} \, dx =$

(A) $\sqrt{t^3 + 1}$ (B) $\dfrac{\sqrt{t^3 + 1}}{3t^2}$ (C) $\dfrac{2}{3}(t^3 + 1)(\sqrt{t^3 + 1} - 1)$

(D) $3x^2 \sqrt{x^3 + 1}$ (E) none of these

35. If $F(u) = \displaystyle\int_{1}^{u} (2 - x^2)^3 \, dx$, then $F'(u)$ is equal to

(A) $-6u(2 - u^2)^2$ (B) $\dfrac{(2 - u^2)^4}{4} - \dfrac{1}{4}$ (C) $(2 - u^2)^3 - 1$

(D) $(2 - u^2)^3$ (E) $-2u(2 - u^2)^3$

36. $\dfrac{d}{dx} \displaystyle\int_{\pi/2}^{x^2} \sqrt{\sin t} \, dt =$

(A) $\sqrt{\sin t^2}$ (B) $2x\sqrt{\sin x^2} - 1$ (C) $\dfrac{2}{3}(\sin^{3/2} x^2 - 1)$

(D) $\sqrt{\sin x^2} - 1$ (E) $2x\sqrt{\sin x^2}$

37. If we let $x = \tan \theta$, then $\displaystyle\int_{1}^{\sqrt{3}} \sqrt{1 + x^2} \, dx$ is equivalent to

(A) $\displaystyle\int_{\pi/4}^{\pi/3} \sec \theta \, d\theta$ (B) $\displaystyle\int_{1}^{\sqrt{3}} \sec^3 \theta \, d\theta$ (C) $\displaystyle\int_{\pi/4}^{\pi/3} \sec^3 \theta \, d\theta$

(D) $\displaystyle\int_{\pi/4}^{\pi/3} \sec^2 \theta \tan \theta \, d\theta$ (E) $\displaystyle\int_{1}^{\sqrt{3}} \sec \theta \, d\theta$

38. If the substitution $u = \sqrt{x + 1}$ is used, then $\displaystyle\int_{0}^{3} \dfrac{dx}{x\sqrt{x + 1}}$ is equivalent to

(A) $\displaystyle\int_{1}^{2} \dfrac{du}{u^2 - 1}$ (B) $\displaystyle\int_{1}^{2} \dfrac{2 \, du}{u^2 - 1}$ (C) $2\displaystyle\int_{0}^{3} \dfrac{du}{(u - 1)(u + 1)}$

(D) $2\displaystyle\int_{1}^{2} \dfrac{du}{u(u^2 - 1)}$ (E) $2\displaystyle\int_{0}^{3} \dfrac{du}{u(u - 1)}$

39. If $x = 4 \cos \theta$ and $y = 3 \sin \theta$, then $\displaystyle\int_{2}^{4} xy \, dx$ is equivalent to

(A) $48\displaystyle\int_{\pi/3}^{0} \sin \theta \cos^2 \theta \, d\theta$ (B) $48\displaystyle\int_{2}^{4} \sin^2 \theta \cos \theta \, d\theta$

(C) $36\displaystyle\int_{2}^{4} \sin \theta \cos^2 \theta \, d\theta$ (D) $-48\displaystyle\int_{0}^{\pi/3} \sin \theta \cos^2 \theta \, d\theta$

(E) $48\displaystyle\int_{0}^{\pi/3} \sin^2 \theta \cos \theta \, d\theta$

*40.   A curve is defined by the parametric equations $y = 2a \cos^2 \theta$ and $x = 2a \tan \theta$, where $0 \leqq \theta \leqq \pi$. Then the definite integral $\pi \displaystyle\int_0^{2a} y^2\, dx$ is equivalent to

(A) $4\pi a^2 \displaystyle\int_0^{\pi/4} \cos^4 \theta\, d\theta$ (B) $8\pi a^3 \displaystyle\int_{\pi/2}^{\pi} \cos^2 \theta\, d\theta$ (C) $8\pi a^3 \displaystyle\int_0^{\pi/4} \cos^2 \theta\, d\theta$

(D) $8\pi a^3 \displaystyle\int_0^{2a} \cos^2 \theta\, d\theta$ (E) $8\pi a^3 \displaystyle\int_0^{\pi/4} \sin \theta \cos^2 \theta\, d\theta$

*41.   A curve is given parametrically by $x = 1 - \cos t$ and $y = t - \sin t$, where $0 \leqq t \leqq \pi$. Then $\displaystyle\int_0^{3/2} y\, dx$ is equivalent to

(A) $\displaystyle\int_0^{3/2} \sin t(t - \sin t)\, dt$ (B) $\displaystyle\int_{2\pi/3}^{\pi} \sin t(t - \sin t)\, dt$

(C) $\displaystyle\int_0^{2\pi/3} (t - \sin t)\, dt$ (D) $\displaystyle\int_0^{2\pi/3} \sin t(t - \sin t)\, dt$

(E) $\displaystyle\int_0^{3/2} (t - \sin t)\, dt$

42.   If we approximate the area of the shaded region below by $M(20)$ (that is, the midpoint sum with 20 subintervals), then the difference $M(20) - \displaystyle\int_0^6 f(x)\, dx$ is equal to

(A)   0.004   (B)   0.008   (C)   0.010
(D)   0.045   (E)   none of these

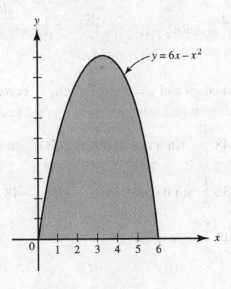

*An asterisk denotes a topic covered only in Calculus BC.

43.   The area of the shaded region in the figure is equal exactly to ln 3. If we approximate ln 3 using $L(2)$ and $R(2)$, which inequality follows?

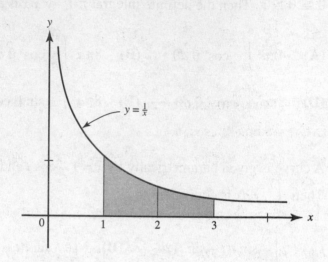

$y = \frac{1}{x}$

(A) $\dfrac{1}{2} < \displaystyle\int_1^2 \dfrac{1}{x}\,dx < 1$ (B) $\dfrac{1}{3} < \displaystyle\int_1^3 \dfrac{1}{x}\,dx < 2$ (C) $\dfrac{1}{2} < \displaystyle\int_0^2 \dfrac{1}{x}\,dx < 2$

(D) $\dfrac{1}{3} < \displaystyle\int_2^3 \dfrac{1}{x}\,dx < \dfrac{1}{2}$ (E) $\dfrac{5}{6} < \displaystyle\int_1^3 \dfrac{1}{x}\,dx < \dfrac{3}{2}$

44. $\displaystyle\int_0^2 \sqrt{3x+1}\;dx$ is equal approximately to

(A)   3.646   (B)   3.893   (C)   4.116   (D)   4.646   (E)   11.680

45. $\displaystyle\int_1^2 \dfrac{dx}{\sqrt{4-x^2}}$ is equal, to three decimal places, approximately to

(A)   0.322   (B)   0.524   (C)   1.047
(D)   1.570   (E)   none of these

46.   If the trapezoidal rule is used with $n = 5$, then $\displaystyle\int_0^1 \dfrac{dx}{1+x^2}$ is equal, to three decimal places, to

(A)   0.784   (B)   1.567   (C)   1.959   (D)   3.142   (E)   7.837

47. $\displaystyle\int_{-1}^3 |x|\;dx =$

(A) $\dfrac{7}{2}$ (B)   4   (C) $\dfrac{9}{2}$ (D)   5   (E) $\dfrac{11}{2}$

48. $\int_{-3}^{2} |x + 1| \, dx =$

(A) $\dfrac{5}{2}$ (B) $\dfrac{7}{2}$ (C) 5   (D) $\dfrac{11}{2}$ (E) $\dfrac{13}{2}$

49.   If $M(4)$ is used to approximate $\int_{0}^{1} \sqrt{1 + x^3} \, dx$, then the definite integral is equal, to two decimal places, to

(A) 1.00   (B) 1.11   (C) 1.20   (D) 2.22   (E) 3.33

50. $\int_{2}^{5} \dfrac{1}{3x} \, dx$ is best approximated, to three decimal places, by

(A) 0.268   (B) 0.286   (C) 0.305   (D) 0.916   (E) 2.749

51.   The average value of $\cos x$ over the interval $\dfrac{\pi}{3} \leq x \leq \dfrac{\pi}{2}$ is

(A) $\dfrac{3}{\pi}$ (B) $\dfrac{1}{2}$ (C) $\dfrac{3(2 - \sqrt{3})}{\pi}$ (D) $\dfrac{3}{2\pi}$ (E) $\dfrac{2}{3\pi}$

52.   The average value of $\csc^2 x$ over the interval from $x = \dfrac{\pi}{6}$ to $x = \dfrac{\pi}{4}$ is

(A) $\dfrac{3\sqrt{3}}{\pi}$ (B) $\dfrac{\sqrt{3}}{\pi}$ (C) $\dfrac{12}{\pi}(\sqrt{3} - 1)$

(D) $3\sqrt{3}$ (E) $3(\sqrt{3} - 1)$

Answers for Set 6: Definite Integrals

1.	C	12.	B	23.	E	34.	A	45.	C
2.	B	13.	B	24.	C	35.	D	46.	A
3.	E	14.	E	25.	C	36.	E	47.	D
4.	B	15.	C	26.	A	37.	C	48.	E
5.	D	16.	B	27.	E	38.	B	49.	B
6.	A	17.	D	28.	D	39.	E	50.	C
7.	D	18.	A	29.	B	40.	C	51.	C
8.	C	19.	C	30.	C	41.	D	52.	C
9.	A	20.	E	31.	A	42.	D		
10.	C	21.	B	32.	C	43.	E		
11.	D	22.	A	33.	D	44.	B		

1. C. The integral is equal to

$$\left(\frac{1}{3}x^3 - \frac{1}{2}x^2 - x\right)\Bigg|_{-1}^{1} = -\frac{7}{6} - \frac{1}{6}.$$

2. B. Rewrite as $\int_{1}^{2}\left(1 - \frac{1}{3}\cdot\frac{1}{x}\right)dx$. This equals

$$\left(x - \frac{1}{3}\ln x\right)\Bigg|_{1}^{2} = 2 - \frac{1}{3}\ln 2 - 1.$$

3. E. Rewrite as

$$-\int_{0}^{3}(4-t)^{-1/2}(-1\,dt) = -2\sqrt{4-t}\,\Bigg|_{0}^{3} = -2(1-2).$$

4. B. This integral equals

$$\frac{1}{3}\int_{-1}^{0}(3u+4)^{1/2}\cdot 3\,du = \frac{1}{3}\cdot\frac{2}{3}(3u+4)^{3/2}\,\Bigg|_{-1}^{0}$$

$$= \frac{2}{9}(4^{3/2} - 1^{3/2}).$$

5. D. You have:

$$\frac{1}{2}\int_{2}^{3}\frac{2\,dy}{2y-3} = \frac{1}{2}\ln(2y-3)\,\Bigg|_{2}^{3} = \frac{1}{2}(\ln 3 - \ln 1).$$

6. A. Rewrite as

$$-\frac{1}{2}\int_0^{\sqrt{3}}(4-x^2)^{-1/2}(-2x\,dx)=-\frac{1}{2}\cdot 2\sqrt{4-x^2}\,\Big|_0^{\sqrt{3}}=-(1-2).$$

7. D. Expand, then integrate term by term:

$$(8t^3-12t^2+6t-1)\Big|_0^1=(2t^4-4t^3+3t^2-t)\Big|_0^1=(2-4+3-1)=0.$$

8. C. Use a graphing calculator. Compare with Question 6.

9. A. Divide:

$$\int_4^9\left(x^{-1/2}+\frac{1}{2}x^{1/2}\right)dx=\left(2x^{1/2}+\frac{1}{2}\cdot\frac{2}{3}x^{3/2}\right)\Big|_4^9$$

$$=\left(2\cdot 3+\frac{1}{3}\cdot 27\right)-\left(2\cdot 2+\frac{1}{3}\cdot 8\right).$$

10. C. The integral equals

$$\frac{1}{3}\tan^{-1}\frac{x}{3}\,\Big|_{-3}^3=\frac{1}{3}\left(\frac{\pi}{4}-\left(-\frac{\pi}{4}\right)\right).$$

11. D. You get $-e^{-x}\Big|_0^1=-(e^{-1}-1).$

12. B. The integral equals $\frac{1}{2}e^x\Big|_0^1=\frac{1}{2}(e^1-1).$

13. B. Evaluate $-\frac{1}{2}\cos 2\theta\,\Big|_0^{\pi/4}$, which equals $-\frac{1}{2}(0-1).$

14. E. Evaluate $-\ln(3-z)\,\Big|_1^2$ and get $-(\ln 1-\ln 2).$

15. C. If $x=2\sin\theta$, $\sqrt{4-x^2}=2\cos\theta$, $dx=2\cos\theta\,d\theta$. When $x=1$, $\theta=\frac{\pi}{6}$;

when $x=2$, $\theta=\frac{\pi}{2}$. The integral is equivalent to $\displaystyle\int_{\pi/6}^{\pi/2}\frac{(2\cos\theta)(2\cos\theta)\,d\theta}{2\sin\theta}$.

16. B. The given integral equals the area of a semicircle of radius 4.

17. D. Evaluate $-\displaystyle\int_0^\pi\cos^2\theta(-\sin\theta)\,d\theta$. This equals $-\frac{1}{3}\cos^3\theta\,\Big|_0^\pi=-\frac{1}{3}(-1-1).$

18. A. The integral equals $\frac{1}{2}\ln^2 x\,\Big|_1^e=\frac{1}{2}(1-0).$

19. **C.**   Use the Parts Formula with $u = x$ and $dv = e^x\,dx$. Then $du = dx$ and $v = e^x$. The result is

$$\left(xe^x - \int e^x\,dx\right)\Bigg|_0^1 = (xe^x - e^x)\Bigg|_0^1 = (e - e) - (0 - 1).$$

20. **E.**   Evaluate $\dfrac{1}{2}\ln(1 + 2\sin\theta)\Bigg|_0^{\pi/6}$ and get $\dfrac{1}{2}(\ln(1+1) - \ln 1)$.

21. **B.**   Use a graphing calculator.

22. **A.**   Evaluate the integral $\dfrac{1}{2}\displaystyle\int_{\sqrt{2}}^{2} \dfrac{2u}{u^2 - 1}\,du$. It equals

$$\frac{1}{2}\ln(u^2 - 1)\Bigg|_{\sqrt{2}}^{2} \qquad \text{or} \qquad \frac{1}{2}(\ln 3 - \ln 1).$$

23. **E.**   Evaluate $\dfrac{1}{2}\displaystyle\int_{\sqrt{2}}^{2} (u^2 - 1)^{-2}\cdot 2u\,du$ and get

$$-\frac{1}{2(u^2 - 1)}\Bigg|_{\sqrt{2}}^{2} \qquad \text{or} \qquad -\frac{1}{2}\left(\frac{1}{3} - \frac{1}{1}\right).$$

Compare with Question 22.

24. **C.**   Use this formula: $\cos^2\theta = \dfrac{1}{2}(1 + \cos 2\theta)$. Then

$$\frac{1}{2}\int_0^{\pi/4}(1 + \cos 2\theta)\,d\theta = \frac{1}{2}\left(\theta + \frac{1}{2}\sin 2\theta\right)\Bigg|_0^{\pi/4} = \frac{1}{2}\left(\left(\frac{\pi}{4} + \frac{1}{2}\right) - 0\right).$$

25. **C.**   Rewrite:

$$\frac{1}{2}\int_{\pi/12}^{\pi/4}\sin^{-2} 2x \cos 2x\,(2\,dx) = -\frac{1}{2}\cdot\frac{1}{\sin 2x}\Bigg|_{\pi/12}^{\pi/4}$$

$$= -\frac{1}{2}\left(\frac{1}{1} - \frac{1}{1/2}\right).$$

26. **A.**   The integral is equivalent to

$$\int_0^1 (1 + e^x)\,dx = (x + e^x)\Bigg|_0^1 = (1 + e) - 1.$$

27. **E.**   Evaluate $\ln(e^x + 1)\Bigg|_0^1$, getting $\ln(e + 1) - \ln 2$.

28. **D.**   Note that the integral from a to b is the sum of the two integrals from a to c and from c to b.

29. **B.**   Find the errors in (A), (C), (D), and (E).

30. **C.**   This is the definition of the definite integral given on page 184.

31. A.   Find examples of functions F and G that show that (B), (C), and (D) are false.

32. C.   This is the Mean Value Theorem for Integrals (page 185). Draw some sketches that illustrate the theorem.

33. D.   This is theorem (2) on page 185. Prove by counterexamples that (A), (B), (C), and (D) are false.

34. A.   This is a restatement of the Fundamental Theorem. In theorem (1) on page 185, interchange t and x.

35. D.   Apply theorem (1) on page 185, noting that

$$F'(u) = \frac{d}{du} \int_a^u f(x)\,dx = f(u).$$

36. E.   Let $y = \int_{\pi/2}^{x^2} \sqrt{\sin t}\,dt$ and $u = x^2$; then

$$y = \int_{\pi/2}^u \sqrt{\sin t}\,dt$$

By the chain rule, $\dfrac{dy}{dx} = \dfrac{dy}{du} \cdot \dfrac{du}{dx} = \sqrt{\sin u} \cdot 2x$, where theorem (1) on page 185 is used to find $\dfrac{dy}{du}$. Replace u by x^2.

37. C.   Note that $dx = \sec^2 \theta\,d\theta$ and that $\sqrt{1 + \tan^2 \theta} = \sec \theta$. Be sure to express the limits as values of θ: $1 = \tan \theta$ yields $\theta = \dfrac{\pi}{4}$; $\sqrt{3} = \tan \theta$ yields $\theta = \dfrac{\pi}{3}$.

38. B.   If $u = \sqrt{x+1}$, then $u^2 = x + 1$, and $2u\,du = dx$. When you substitute for the limits, you get $2 \int_1^2 \dfrac{u\,du}{u(u^2 - 1)}$. Since $u \neq 0$ on its interval of integration, you may divide numerator and denominator by it.

39. E.   Since $dx = -4 \sin \theta\,d\theta$, you get the new integral $-48 \int_{\pi/3}^0 \sin^2 \theta \cos \theta\,d\theta$. Use theorem (4) on page 185 to get the correct answer.

40. C.   Since $dx = 2a \sec^2 \theta\,d\theta$, you get $8\pi a^3 \int_0^{\pi/4} \cos^4 \theta \sec^2 \theta\,d\theta$. Use the fact that $\cos^2 \theta \sec^2 \theta = 1$.

41. D.   Use the facts that $dx = \sin t\,dt$, that $t = 0$ when $x = 0$, and that $t = \dfrac{2\pi}{3}$ when $x = \dfrac{3}{2}$.

42. D. $M(20) - \int_0^6 (6x - x^2)\,dx = 36.045 - 36 = 0.045.$

43.   E.   For $L(2)$ use the circumscribed rectangles:

$$1 \cdot 1 + \frac{1}{2} \cdot 1 = \frac{3}{2};$$

for $R(2)$ use the inscribed rectangles:

$$\frac{1}{2} \cdot 1 + \frac{1}{3} \cdot 1 = \frac{5}{6}.$$

44.   B.   Using the [fnInt] program gives choice (B) for the integral.

45.   C.   This is the answer given by our calculator.

46.   A. $\int_0^1 \frac{dx}{1 + x^2}$, calculated approximately by $T(5)$ on our graphing calculator, yields choice A.

47.   D   Rewrite the integral to evaluate it, using the fact that x changes sign at 0. The result is

$$\int_{-1}^0 (-x)\, dx + \int_0^3 x\, dx = -\frac{x^2}{2}\bigg|_{-1}^0 + \frac{x^2}{2}\bigg|_0^3.$$

Draw a sketch of $y = |x|$, and verify that the area over $-1 \leqq x \leqq 3$ equals 5.

48.   E.   Since $x + 1$ changes sign at $x = -1$, $|x + 1| = -(x + 1)$ if $x < -1$ but equals $x + 1$ if $x \geqq -1$. The given integral is therefore equivalent to

$$\int_{-3}^{-1} -(x + 1)\, dx + \int_{-1}^2 (x + 1)\, dx = -\frac{(x+1)^2}{2}\bigg|_{-3}^{-1} + \frac{(x+1)^2}{2}\bigg|_{-1}^2$$

$$= -\frac{1}{2}(0 - 4) + \frac{1}{2}(9 - 0).$$

Draw a sketch of $y = |x + 1|$, and verify that the area over $-3 \leqq x \leqq 2$ is $\frac{13}{2}$.

49.   B.   Our calculator gives 1.11 when we round to two decimal places.

50.   C. $\int_2^5 \frac{1}{3x}\, dx = 0.305$, using our [fnInt] program.

51.   C.   Equation (1) on page 204 yields

$$(y_{av})_x = \frac{1}{\pi/2 - \pi/3} \int_{\pi/3}^{\pi/2} \cos x\, dx.$$

52.   C.   The average value is equal to $\frac{1}{\pi/4 - \pi/6} \int_{\pi/6}^{\pi/4} \csc^2 x\, dx.$

Chapter 7

Applications of Integration to Geometry

In this chapter we will use the definite integral to compute areas, volumes, and arc lengths. In each of these applications, we will set up a Riemann sum whose limit is a definite integral. In the last section we will review the topic of improper integrals.

A.   Area

To find an area, we
    (1) draw a sketch of the given region and of a typical element;
    (2) write the expression for the area of a typical rectangle; and
    (3) set up the definite integral that is the limit of the Riemann sum of n areas as $n \to \infty$.

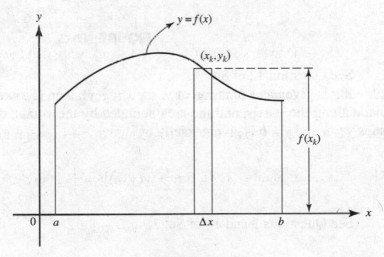

FIGURE N7–1

If $f(x)$ is nonnegative on $[a,b]$, as in Figure N7–1, then $f(x_k) \Delta x$ can be regarded as the area of a typical approximating rectangle, and the area bounded by the x-axis, the curve, and the vertical lines $x = a$ and $x = b$ is given exactly by

$$\lim_{n \to \infty} \sum_{k=1}^{n} f(x_k) \Delta x \qquad \text{and hence by} \qquad \int_{a}^{b} f(x)\, dx.$$

See Questions 1, 6, and 11 of Set 7.

If $f(x)$ changes sign on the interval (Figure N7–2), we find the values of x for which $f(x) = 0$ and note where the function is positive, where it is negative. The total area bounded by the x-axis, the curve, $x = a$, and $x = b$ is here given exactly by

$$\int_{a}^{c} f(x)\, dx - \int_{c}^{d} f(x)\, dx + \int_{d}^{b} f(x)\, dx,$$

where we have taken into account that $f(x_k) \Delta x$ is a negative number if $c < x < d$.

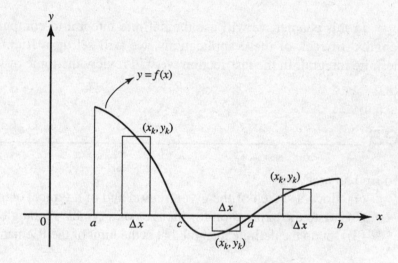

FIGURE N7–2

See Question 12 of Set 7.

If x is given as a function of y, say $x = g(y)$, then (Figure N7–3) the subdivisions are made along the y-axis, and the area bounded by the y-axis, the curve, and the horizontal lines $y = a$ and $y = b$ is given exactly by

$$\lim_{n \to \infty} \sum_{k=1}^{n} g(y_k) \Delta y = \int_{a}^{b} g(y)\, dy.$$

See Questions 3 and 14 of Set 7.

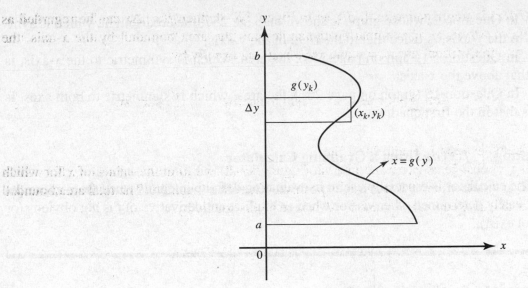

FIGURE N7–3

A1. Area Between Curves

To find the area between curves (Figure N7–4), we first find where they intersect and then write the area of a typical element for each region between the points of intersection. For the total area bounded by the curves $y = f(x)$ and $y = g(x)$ between $x = a$ and $x = e$, we see that, if they intersect at $[c,d]$, the total area is given exactly by

$$\int_a^c [f(x) - g(x)] \, dx + \int_c^e [g(x) - f(x)] \, dx.$$

See Questions 4, 7, 8, and 10 of Set 7.

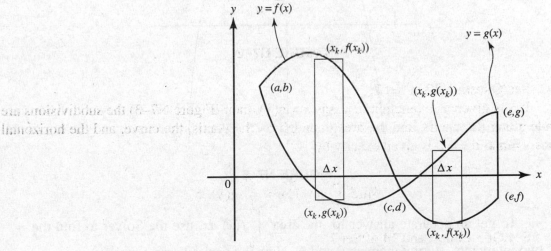

FIGURE N7–4

A2. Using Symmetry

Frequently we seek the area of a region that is symmetric to the x- or y-axis (or both) or to the origin. In such cases it is almost always simpler to make use of this symmetry when integrating. For example:

- In Question 2, whose graph is shown on page 257, the area sought, which is symmetric to the y-axis, is twice that to the right of the y-axis;
- In Question 5 (graph on page 258), the area, which is symmetric to the x-axis, is twice that above the x-axis;
- In Question 15 (graph on page 261), the area, which is symmetric to both axes, is 4 times that in the first quadrant.

Evaluating $\int_a^b f(x)\,dx$ Using a Graphing Calculator

The calculator is especially useful in evaluating definite integrals when the x-intercepts are not easily determined otherwise or when an explicit antiderivative of f is not obvious (or does not exist).

Example 1.   Evaluate $\int_0^1 e^{-x^2}\,dx$.

We enter `fnInt (e`$^{-x^2}$`,X,0,1)`, getting 0.7468 to four decimal places.

Example 2.   In Figure N7–5, find the area under $f(x) = -x^4 + x^2 + x + 10$ and above the x-axis.

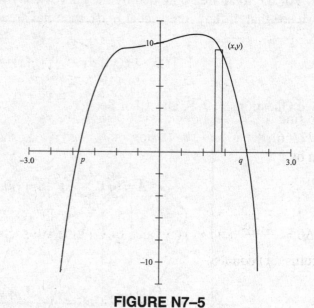

FIGURE N7–5

To get an accurate answer for the area $\int_p^q f(x)\,dx$, use the Solver to find the two intercepts, storing them as P and Q. Now the desired area equals

$$\texttt{fnInt(Y}_1\texttt{,X,P,Q)} \qquad \text{or} \qquad 32.8324,$$

which is accurate to four decimal places.

*Region Bounded by Parametric Curve

If x and y are given parametrically, say by $x = f(\theta)$, $y = g(\theta)$, then to evaluate $\int_a^b y\, dx$, we express y, dx, and the limits a and b in terms of θ and $d\theta$, then integrate. Remember that we define dx to be $x'(\theta)\, d\theta$, or $f'(\theta)\, d\theta$.

See Questions 15, 16, and 19 of Set 7.

*Region Bounded by Polar Curve

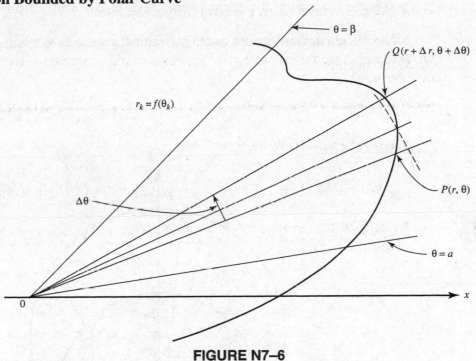

FIGURE N7–6

To find the area A bounded by the polar curve $r = f(\theta)$ and the rays $\theta = \alpha$, $\theta = \beta$ (see Figure N7–6), we divide the region into n sectors as shown. Then the area of the region is given by

$$A = \lim_{n \to \infty} \sum_{k=1}^{n} \frac{1}{2} f^2(\theta_k)\, \Delta\theta, \tag{1}$$

where $\Delta\theta = \dfrac{\beta - \alpha}{n}$ and θ_k is a value of θ in the kth sector. By the Fundamental Theorem of Calculus, (1) equals

$$\int_\alpha^\beta \frac{1}{2} f^2(\theta)\, d\theta = \int_\alpha^\beta \frac{1}{2} r^2\, d\theta. \tag{2}$$

Note that, if we think of dA as an element of area, then $dA = \dfrac{1}{2} r^2\, d\theta$, and this is the area of a circular sector of central angle $d\theta$ and radius r.

We have assumed above that $f(\theta) \geq 0$ on $[\alpha, \beta]$. We must be careful in determining the limits α and β in (2); often it helps to think of the required area as that "swept out" (or generated) as the radius vector (from the pole) rotates from $\theta = \alpha$ to $\theta = \beta$. It is also useful to exploit symmetry of the curve wherever possible.

The relations between rectangular and polar coordinates, some common polar equations, and graphs of polar curves are given in the Appendix, starting on page 628.

*An asterisk denotes a topic covered only in Calculus BC.

Example 3.   Find the area inside both the circle $r = 3 \sin \theta$ and the cardioid $r = 1 + \sin \theta$.

Choosing an appropriate window, graph the curves on your calculator.

See Figure N7–7, where one half of the required area is shaded. Since $3 \sin \theta = 1 + \sin \theta$ when $\theta = \dfrac{\pi}{6}$ or $\dfrac{5\pi}{6}$, we see that the desired area is twice the sum of two parts: the area of the circle swept out by θ as it varies from 0 to $\dfrac{\pi}{6}$ plus the area of the cardioid swept out by a radius vector as θ varies from $\dfrac{\pi}{6}$ to $\dfrac{\pi}{2}$. Consequently

$$A = 2 \left[\int_0^{\pi/6} \frac{9}{2} \sin^2 \theta \, d\theta + \int_{\pi/6}^{\pi/2} \frac{1}{2} (1 + \sin \theta)^2 \, d\theta \right] = \frac{5\pi}{4} .$$

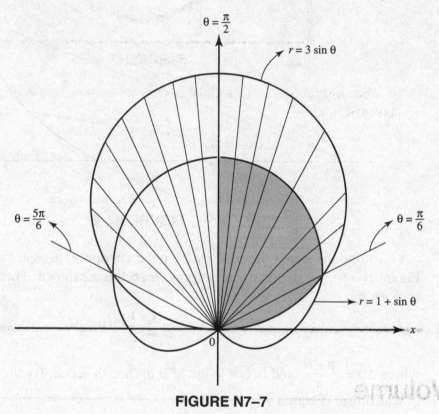

FIGURE N7–7

The integrals that comprise A above can be evaluated on a graphing calculator. Ours, using [fnInt], yielded $\dfrac{5\pi}{4}$ to ten decimal places.

See also Questions 22 and 23 of Set 7.

NOTE: Computation of areas involving improper integrals are discussed in §D of this chapter.

Example 4. Find the area enclosed by the cardioid $r = 2(1 + \cos \theta)$.

We graphed the cardioid on our calculator, using polar mode, in the window $[-2, 5] \times [-3, 3]$ with θ in $[0, 2\pi]$.

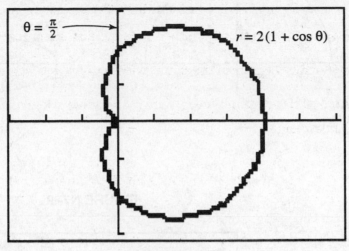

FIGURE N7–8

See Figure N7–8. We use the symmetry of the curve with respect to the polar axis and write

$$A = 2 \cdot \frac{1}{2} \int_0^\pi r^2 \, d\theta = 4 \int_0^\pi (1 + \cos \theta)^2 \, d\theta$$

$$= 4 \int_0^\pi (1 + 2 \cos \theta + \cos^2 \theta) \, d\theta$$

$$= 4 \int_0^\pi \left(1 + 2 \cos \theta + \frac{1}{2} + \frac{\cos 2\theta}{2} \right) d\theta$$

$$= 4 \left[\theta + 2 \sin \theta + \frac{\theta}{2} + \frac{\sin 2\theta}{4} \right] \Big|_0^\pi = 6\pi.$$

B. Volume

B1. Solids of Revolution

A *solid of revolution* is obtained when a plane region is revolved about a fixed line, called the *axis of revolution*. There are two major methods of obtaining the volume of a solid of revolution.

DISKS. For example, if the region bounded by a semicircle (with center O and radius r) and its diameter is revolved about the x-axis, the solid of revolution obtained is a sphere of radius r. We think of the "rectangular strip" of the semicircular region at the left in Figure N7–9 as generating the solid disk or slice of the sphere shown at the right.

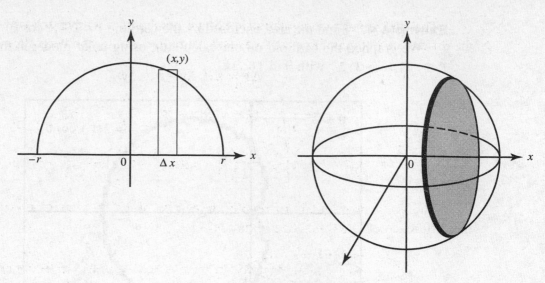

FIGURE N7–9

The volume of a disk (cylinder) is $\pi R^2 H$,
where R is the radius and H is the height.

In our example, the volume ΔV of a typical disk is given by $\Delta V = \pi y^2 \, \Delta x$. The equation of the circle is $x^2 + y^2 = r^2$. To find the volume of the sphere, we form a Riemann sum whose limit as n becomes infinite is a definite integral. Then,

$$V = \int_{-r}^{r} \pi y^2 \, dx = \pi \int_{-r}^{r} (r^2 - x^2) \, dx = \pi \left[r^2 x - \frac{x^3}{3} \right] \Big|_{-r}^{r} = \frac{4}{3} \pi r^3.$$

See Questions 24, 30, 34, 35, and 36 of Set 7 for examples of finding volumes by disks.

WASHERS. A washer is a disk with a hole in it. The volume may be regarded as the difference in the volumes of two concentric disks. As an example, we find the volume obtained when the region bounded by $y = x^2$ and $y = 2x$ is revolved about the y-axis. The curves intersect at the origin and at (2, 4). Note that we distinguish between the functions by letting (x_1, y) be a point on the line, (x_2, y) a point on the parabola. See Figure N7–10.

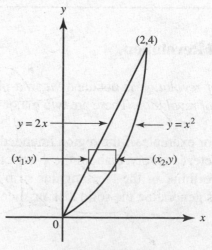

FIGURE N7–10

The horizontal rectangular strip shown generates a washer whose volume, ΔV, is the difference in the volumes of two disks, of radii x_1 and x_2 and height Δy. Then

$$\Delta V = \pi x_2^2 \, \Delta y - \pi x_1^2 \, \Delta y;$$

$$V = \pi \int_0^4 \left(x_2^2 - x_1^2 \right) dy = \pi \int_0^4 \left(y - \frac{y^2}{4} \right) dy$$

$$= \pi \left[\frac{y^2}{2} - \frac{y^3}{12} \right] \Bigg|_0^4 = \pi \left(\frac{16}{2} - \frac{64}{12} \right) = \frac{8}{3} \pi .$$

The volume of a solid of revolution can be found by viewing a typical washer as the difference of two disks. See Figure N7–11.

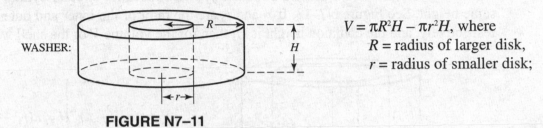

WASHER:

$V = \pi R^2 H - \pi r^2 H$, where
R = radius of larger disk,
r = radius of smaller disk;

FIGURE N7–11

See Questions 25, 27, 29, 31, and 40 of Set 7 for examples in which washers are regarded as the differences of two disks.

Occasionally when more than one method is satisfactory we try to use the most efficient. In the answers for Set 7, a sketch is shown for each question and for each the type and volume of a typical element are given. The required volume is then found by letting the number of elements become infinite and applying the Fundamental Theorem.

B2. Solids with Known Cross Sections

If the *area of a cross section* of a solid *is known* and can be expressed in terms of x, then the volume of a typical slice, ΔV, can be determined. The volume of the solid is obtained, as usual, by letting the number of slices increase indefinitely. In Figure N7–12, the slices are taken perpendicular to the x-axis so that $\Delta V = A(x) \, \Delta x$, where $A(x)$ is the area of a cross section and Δx is the thickness of the slice.

Questions 37, 38, and 39 of Set 7 illustrate solids with known cross sections.

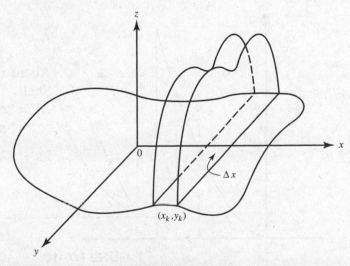

FIGURE N7–12

When the cross section of a solid is a circle, a typical slice, as we saw above, is a disk. When the cross section is the region between two circles, a typical slice, as also noted above, is a washer—a disk with a hole in it. Both of these solids, which are special cases of solids with known cross sections, can be generated by revolving a plane area about a fixed line.

Shells

The following examples involve finding volumes by the method of shells. Although shells are not included in the Topic Outline, we include this method here because it is often the most efficient (and elegant) way to find a volume.

A cylindrical shell is a solid bounded by two concentric circular cylinders with the same height. See Figure N7–13. If r_1 and r_2 are the radii of the inner and outer cylinders, respectively, and the common height is H, then for the volume V of the shell we have

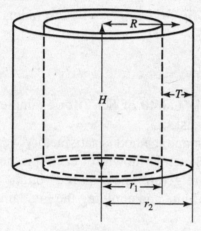

$$V = \pi r_2^2 H - \pi r_1^2 H = \pi(r_2^2 - r_1^2)H$$
$$= \pi(r_2 + r_1)(r_2 - r_1)H$$
$$= 2\pi \frac{(r_2 + r_1)}{2}(r_2 - r_1)H.$$

FIGURE N7–13

With $r_2 - r_1 = T$, the *thickness* of the shell, and $\frac{r_2 + r_1}{2} = R$, the *midradius*, and H as previously defined, we see that the volume of a shell is $2\pi RHT$.

Example 5.   Find the volume of the solid generated when the region bounded by $y = x^2$, $x = 2$, and $y = 0$ is rotated about the line $x = 2$. See Figure N7–14.

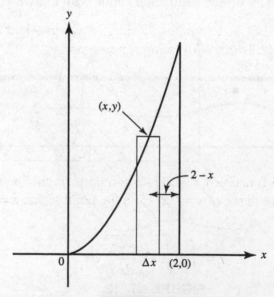

About $x = 2$.
Shell.
$$\Delta V = 2\pi(2 - x)y\,\Delta x$$
$$= 2\pi(2 - x)x^2\,\Delta x.$$
$$2\pi \int_0^2 (2 - x)x^2\,dx = \frac{8\pi}{3}.$$

FIGURE N7–14

Example 6.   The region bounded by $y = 3x - x^2$ and $y = x$ is rotated about the
y-axis. Find the volume of the solid obtained. See Figure N7–15.

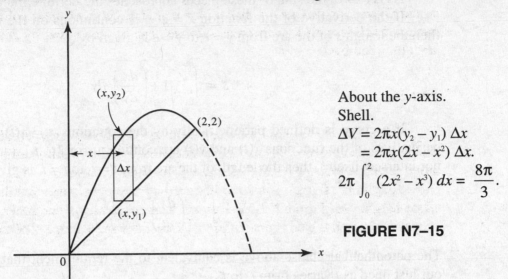

About the y-axis.
Shell.
$$\Delta V = 2\pi x (y_2 - y_1)\, \Delta x$$
$$= 2\pi x (2x - x^2)\, \Delta x.$$
$$2\pi \int_0^2 (2x^2 - x^3)\, dx = \frac{8\pi}{3}.$$

FIGURE N7–15

Example 7.   Write an integral in terms of y for the volume of the solid obtained
when the region bounded by $y = \ln x$, $y = 0$, $x = e$ is rotated about the line $x = e$. See
Figure N7–16.

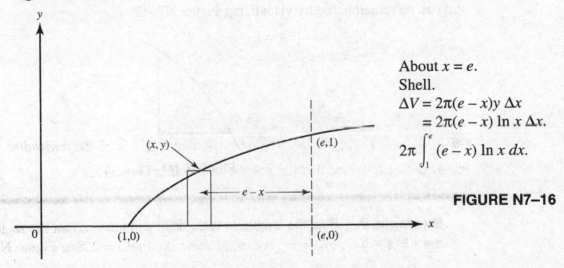

About $x = e$.
Shell.
$$\Delta V = 2\pi (e - x)y\, \Delta x$$
$$= 2\pi (e - x)\ln x\, \Delta x.$$
$$2\pi \int_1^e (e - x)\ln x\, dx.$$

FIGURE N7–16

NOTE: On page 245 in Examples 27 and 28 we consider finding the volumes of solids using shells that lead to improper integrals.

*C.   Arc Length

If the derivative of a function $y = f(x)$ is continuous on the interval $a \leqq x \leqq b$, then
the length s of the arc of the curve of $y = f(x)$ from the point where $x = a$ to the point where
$x = b$ is given by

$$s = \int_a^b \sqrt{1 + \left(\frac{dy}{dx}\right)^2}\, dx .$$ (1)

*An asterisk denotes a topic covered only in Calculus BC.

Here a small piece of the curve is equal approximately to $\sqrt{1 + (f'(x))^2}\,\Delta x$.
As $\Delta x \to 0$, the sum of these pieces approaches the definite integral above.

If the derivative of the function $x = g(y)$ is continuous on the interval $c \leqslant y \leqslant d$, then the length s of the arc from $y = c$ to $y = d$ is given by

$$s = \int_c^d \sqrt{1 + \left(\frac{dx}{dy}\right)^2}\,dy. \tag{2}$$

If a curve is defined parametrically by the equations $x = x(t)$ and $y = y(t)$, if the derivatives of the functions $x(t)$ and $y(t)$ are continuous on $[t_a, t_b]$, (and if the curve does not intersect itself), then the length of the arc from $t = t_a$ to $t = t_b$ is given by

$$s = \int_{t_a}^{t_b} \sqrt{\left(\frac{dx}{dt}\right)^2 + \left(\frac{dy}{dt}\right)^2}\,dt. \tag{3}$$

The parenthetical clause above is equivalent to the requirement that the curve is traced out just once as t varies from t_a to t_b.

As indicated in Equation (4), formulas (1), (2), and (3) can all be derived easily from the very simple relation

$$ds^2 = dx^2 + dy^2 \tag{4}$$

and can be remembered by visualizing Figure N7–17.

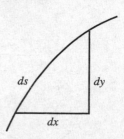

FIGURE N7–17

Example 8.   Find the length, to three decimal places, of the arc of $y = x^{3/2}$ from $x = 1$ to $x = 8$.

Here $\dfrac{dy}{dx} = \dfrac{3}{2}\,x^{1/2}$, so, by (1),

$$s = \int_1^8 \sqrt{1 + \frac{9}{4}\,x}\,dx = 22.803,$$

where we used the [fnInt] program on our graphing calculator.

Example 9.   Find the length, to three decimal places, of the curve $(x - 2)^2 = 4y^3$ from $y = 0$ to $y = 1$.

Since

$$x - 2 = 2y^{3/2} \qquad \text{and} \qquad \frac{dx}{dy} = 3y^{1/2},$$

Equation (2) above yields

$$s = \int_0^1 \sqrt{1+9y}\, dy = 2.268,$$

again using the calculator to evaluate the definite integral.

Example 10.   The position (x, y) of a particle at time t is given parametrically by $x = t^2$ and $y = \dfrac{t^3}{3} - t$. Find the distance the particle travels between $t = 1$ and $t = 2$.

We can use (4): $ds^2 = dx^2 + dy^2$, where $dx = 2t\, dt$ and $dy = (t^2 - 1)\, dt$. Thus,

$$ds = \sqrt{4t^2 + t^4 - 2t^2 + 1}\ dt\,,$$

and

$$s = \int_1^2 \sqrt{(t^2 + 1)^2}\ dt = \int_1^2 (t^2 + 1)\, dt$$

$$= \frac{t^3}{3} + t\ \Big|_1^2 = \frac{10}{3}\,.$$

Example 11.   Find the length of the arc of $y = \ln \sec x$ from $x = 0$ to $x = \dfrac{\pi}{3}$. Here

$$\frac{dy}{dx} = \frac{\sec x \tan x}{\sec x}\,,$$

so

$$s = \int_0^{\pi/3} \sqrt{1 + \tan^2 x}\ dx = \int_0^{\pi/3} \sec x\ dx$$

$$= \ln(\sec x + \tan x)\ \Big|_0^{\pi/3} = \ln(2 + \sqrt{3})\,.$$

*D.   Improper Integrals

There are two classes of improper integrals: (1) those in which at least one of the limits of integration is infinite; and (2) those of the type $\int_a^b f(x)\ dx$, where $f(x)$ has a point of discontinuity (becoming infinite) at $x = c$, $a \leqq c \leqq b$.

Illustrations of improper integrals of class (1) are:

$$\int_0^\infty \frac{dx}{\sqrt[3]{x+1}}\ ; \qquad \int_1^\infty \frac{dx}{x}\ ; \qquad \int_{-\infty}^\infty \frac{dx}{a^2 + x^2}\ ; \qquad \int_{-\infty}^0 e^{-x}\, dx\ ;$$

$$\int_{-\infty}^{-1} \frac{dx}{x^n} \qquad (n \text{ a real number}); \qquad \int_{-\infty}^0 \frac{dx}{e^x + e^{-x}}\ ;$$

$$\int_0^\infty \frac{dx}{(4 + x)^2}\ ; \qquad \int_{-\infty}^0 e^{-x^2}\, dx\ ; \qquad \int_1^\infty \frac{e^{-x^2}}{x^2}\, dx\,.$$

*An asterisk denotes a topic covered only in Calculus BC.

The following improper integrals of type $\int_a^b f(x)\,dx$ are of class (2):

$$\int_0^1 \frac{dx}{x} \ ; \qquad \int_1^2 \frac{dx}{(x-1)^n} \qquad (n \text{ a real number}); \qquad \int_{-1}^1 \frac{dx}{1-x^2} \ ;$$

$$\int_0^2 \frac{x}{\sqrt{4-x^2}}\,dx \ ; \qquad \int_\pi^{2\pi} \frac{dx}{1+\sin x} \ ; \qquad \int_{-1}^2 \frac{dx}{x(x-1)^2} \ ;$$

$$\int_0^{2\pi} \frac{\sin x \, dx}{\cos x + 1} \ ; \qquad \int_a^b \frac{dx}{(x-c)^n} \qquad (n \text{ real}; \ a \leqq c \leqq b);$$

$$\int_0^1 \frac{dx}{\sqrt{x+x^4}} \ ; \qquad \int_{-2}^2 \sqrt{\frac{2+x}{2-x}}\,dx \ .$$

Sometimes an improper integral belongs to both classes. Consider, for example:

$$\int_0^\infty \frac{dx}{x} \ ; \qquad \int_0^\infty \frac{dx}{\sqrt{x+x^4}} \ ; \qquad \int_{-\infty}^1 \frac{dx}{\sqrt{1-x}} \ .$$

Each of the integrands in this set fails to exist at some point on the interval of integration. Note, however, that each integral of the following set is *proper*:

$$\int_{-1}^3 \frac{dx}{\sqrt{x+2}} \ ; \qquad \int_{-2}^2 \frac{dx}{x^2+4} \ ; \qquad \int_0^{\pi/6} \frac{dx}{\cos x} \ ;$$

$$\int_0^e \ln(x+1)\,dx; \qquad \int_{-3}^3 \frac{dx}{e^x+1} \ .$$

In each of the above the integrand is defined at each number on the interval of integration. Improper integrals of class (1) are handled as follows:

$$\int_a^\infty f(x)\,dx = \lim_{b\to\infty} \int_a^b f(x)\,dx \ ,$$

where f is continuous on $[a,b]$. If the limit on the right exists, the improper integral on the left is said to *converge* to this limit; if the limit on the right fails to exist, we say that the improper integral *diverges* (or is *meaningless*).

The evaluation of improper integrals of class (1) is illustrated in Examples 12–18.

Example 12. $\displaystyle\int_1^\infty \frac{dx}{x^2} = \lim_{b\to\infty} \int_1^b x^{-2}\,dx = \lim_{b\to\infty} -\frac{1}{x} \ \Big|_1^b = \lim_{b\to\infty} -\left(\frac{1}{b}-1\right) = 1.$ The

given integral thus converges to 1. In Figure N7–18 we interpret $\displaystyle\int_1^\infty \frac{dx}{x^2}$ as the

area above the x-axis, under the curve of $y = \dfrac{1}{x^2}$, and bounded at the left by the

vertical line $x = 1$.

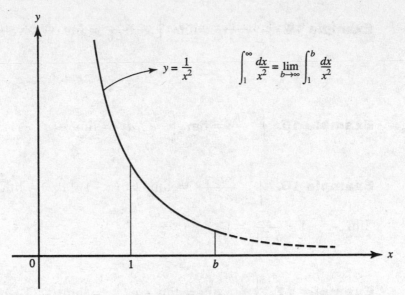

$$y = \frac{1}{x^2} \qquad \int_1^\infty \frac{dx}{x^2} = \lim_{b\to\infty} \int_1^b \frac{dx}{x^2}$$

FIGURE N7–18

Example 13. $\displaystyle \int_1^\infty \frac{dx}{\sqrt{x}} = \lim_{b\to\infty} \int_1^b x^{-1/2}\,dx = \lim_{b\to\infty} 2\sqrt{x}\ \bigg|_1^b = \lim_{b\to\infty} 2(\sqrt{b}-1) = +\infty.$

Then $\displaystyle \int_1^\infty \frac{dx}{\sqrt{x}}$ diverges. It can be proved that $\displaystyle \int_1^\infty \frac{dx}{x^p}$ converges if $p > 1$ but

diverges if $p \leqq 1$. Figure N7–19 gives a geometric interpretation in terms of area of

$\displaystyle \int_1^\infty \frac{dx}{x^p}$ for $p = \frac{1}{2}, 1, 2$. Only the first-quadrant area under $y = \frac{1}{x^2}$ bounded at the

left by $x = 1$ exists. Note that

$$\int_1^\infty \frac{dx}{x} = \lim_{b\to\infty} \ln x\ \bigg|_1^b = +\infty.$$

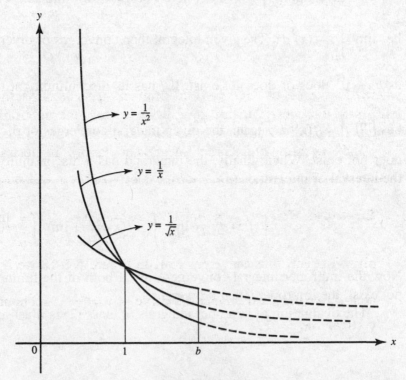

$$y = \frac{1}{x^2}$$

$$y = \frac{1}{x}$$

$$y = \frac{1}{\sqrt{x}}$$

FIGURE N7–19

Example 14. $\displaystyle\int_0^\infty \frac{dx}{x^2+9} = \lim_{b\to\infty}\int_0^b \frac{dx}{x^2+9} = \lim_{b\to\infty}\frac{1}{3}\tan^{-1}\frac{x}{3}\Big|_0^b = \lim_{b\to\infty}\frac{1}{3}\tan^{-1}\frac{b}{3}$

$= \dfrac{1}{3}\cdot\dfrac{\pi}{2} = \dfrac{\pi}{6}$.

Example 15. $\displaystyle\int_0^\infty \frac{dy}{e^y} = \lim_{b\to\infty}\int_0^b e^{-y}\,dy = \lim_{b\to\infty} -(e^{-b}-1) = 1$.

Example 16. $\displaystyle\int_{-\infty}^0 \frac{dz}{(z-1)^2} = \lim_{b\to-\infty}\int_b^0 (z-1)^{-2}\,dz = \lim_{b\to-\infty} -\frac{1}{z-1}\Big|_b^0 =$

$\displaystyle\lim_{b\to-\infty} -\left(-1-\frac{1}{b-1}\right) = 1$.

Example 17. $\displaystyle\int_{-\infty}^0 e^{-x}\,dx = \lim_{b\to-\infty} - e^{-x}\Big|_b^0 = \lim_{b\to-\infty} -(1-e^{-b}) = +\infty$. Thus, this improper integral diverges.

Example 18. $\displaystyle\int_0^\infty \cos x\,dx = \lim_{b\to\infty}\sin x\Big|_b^0 = \lim_{b\to\infty}\sin b$. Since this limit does not exist ($\sin b$ takes on values between -1 and 1 as $b\to\infty$), it follows that the given integral diverges. Note, however, that it does not become infinite; rather, it diverges by oscillation.

Improper integrals of class (2) are handled as follows.

To investigate $\displaystyle\int_a^b f(x)\,dx$, where f becomes infinite at $x=a$, we define $\displaystyle\int_a^b f(x)\,dx$ to

be $\displaystyle\lim_{h\to 0^+}\int_{a+h}^b f(x)\,dx$. The given integral then converges or diverges according as the limit

as $h\to 0^+$ does or does not exist. If f has its discontinuity at b, we define $\displaystyle\int_a^b f(x)\,dx$ to

be $\displaystyle\lim_{h\to 0^+}\int_a^{b-h} f(x)\,dx$; again, the given integral converges or diverges as the limit does or

does not exist. When, finally, the integrand has a discontinuity at an interior point c on the interval of integration ($a<c<b$), we let

$$\int_a^b f(x)\,dx = \lim_{h\to 0^+}\int_a^{c-h} f(x)\,dx + \lim_{h\to 0^+}\int_{c+h}^b f(x)\,dx.$$

Now the improper integral converges only if both of the limits exist. If *either* limit does not exist, the improper integral diverges.

The evaluation of improper integrals of class (2) is illustrated in Examples 19–26.

Example 19. $\displaystyle\int_0^1 \frac{dx}{\sqrt[3]{x}} = \lim_{h\to 0^+} \int_h^1 x^{-1/3}\,dx = \lim_{h\to 0^+} \frac{3}{2} x^{2/3}\Big|_h^1 = \lim_{h\to 0^+} \frac{3}{2}(1-h^{2/3}) = \frac{3}{2}.$

In Figure N7–20 we interpret this integral as the first-quadrant area under $y = \dfrac{1}{\sqrt[3]{x}}$ and to the left of $x = 1$.

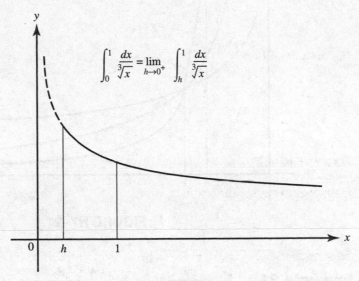

FIGURE N7–20

Example 20. $\displaystyle\int_0^1 \frac{dx}{x^3} = \lim_{h\to 0^+} \int_h^1 x^{-3}\,dx = \lim_{h\to 0^+} -\frac{1}{2x^2}\Big|_h^1 = \lim_{h\to 0^+} -\frac{1}{2}\left(1-\frac{1}{h^2}\right) = \infty.$

So this integral diverges.

It can be shown that $\displaystyle\int_0^a \frac{dx}{x^p}\ (a > 0)$ converges if $p < 1$ but diverges if $p \geqq 1$.

Figure N7–21 shows an interpretation of $\displaystyle\int_0^1 \frac{dx}{x^p}$ in terms of areas where $p = \dfrac{1}{3}$,

1, and 3. Only the first-quadrant area under $y = \dfrac{1}{\sqrt[3]{x}}$ to the left of $x = 1$ exists.

Note that

$$\int_0^1 \frac{dx}{x} = \lim_{h\to 0^+} \ln x \Big|_h^1 = \lim_{h\to 0^+} (\ln 1 - \ln h) = +\infty.$$

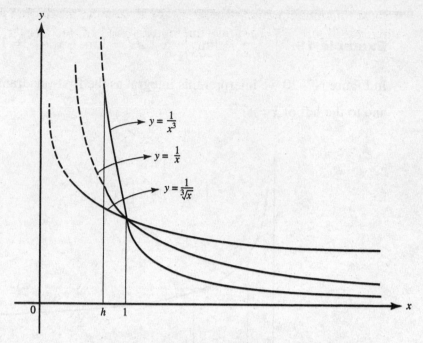

FIGURE N7–21

Example 21. $\displaystyle\int_0^2 \frac{dy}{\sqrt{4-y^2}} = \lim_{h\to 0^+} \int_0^{2-h} \frac{dy}{\sqrt{4-y^2}} = \lim_{h\to 0^+} \sin^{-1} \frac{y}{2} \Big|_0^{2-h} =$

$\sin^{-1} 1 - \sin^{-1} 0 = \dfrac{\pi}{2}$.

Example 22. $\displaystyle\int_2^3 \frac{dt}{(3-t)^2} = \lim_{h\to 0^+} -\int_2^{3-h} (3-t)^{-2}(-dt) = \lim_{h\to 0^+} \frac{1}{3-t} \Big|_2^{3-h} = +\infty.$
This integral diverges.

Example 23. $\displaystyle\int_0^2 \frac{dx}{(x-1)^{2/3}} = \lim_{h\to 0^+} \int_0^{1-h} (x-1)^{-2/3}\, dx + \lim_{h\to 0^+} \int_{1+h}^2 (x-1)^{-2/3}\, dx =$

$\displaystyle\lim_{h\to 0^+} 3(x-1)^{1/3} \Big|_0^{1-h} + \lim_{h\to 0^+} 3(x-1)^{1/3} \Big|_{1+h}^2 = 3(0+1) + 3(1-0) = 6.$

Example 24. $\displaystyle\int_{-2}^2 \frac{dx}{x^2} = \lim_{h\to 0^+} \int_{-2}^{0-h} x^{-2}\, dx + \lim_{h\to 0^+} \int_{0+h}^2 x^{-2}\, dx =$

$\displaystyle\lim_{h\to 0^+} -\frac{1}{x} \Big|_{-2}^{-h} + \lim_{h\to 0^+} -\frac{1}{x} \Big|_h^2$. Neither limit exists; the integral diverges.

    This example demonstrates how careful one must be to notice a discontinuity at an interior point. If it were overlooked, one might proceed as follows:

$$\int_{-2}^2 \frac{dx}{x^2} = -\frac{1}{x} \Big|_{-2}^2 = -\left(\frac{1}{2} + \frac{1}{2}\right) = -1.$$

Since this integrand is positive except at zero the result obtained is clearly meaningless. Figure N7–22 shows the impossibility of this answer.

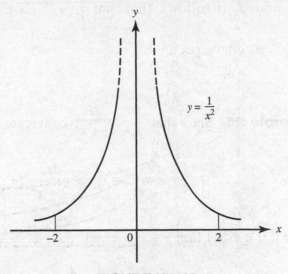

$$y = \frac{1}{x^2}$$

FIGURE N7–22

Example 25. Determine whether or not $\displaystyle\int_1^\infty e^{-x^2}\,dx$ converges.

Although there is no elementary function whose derivative is e^{-x^2}, we can still show that the given improper integral converges. Note, first, that if $x \geqq 1$ then $x^2 \geqq x$, so that $-x^2 \leqq -x$ and $e^{-x^2} \leqq e^{-x}$. Figure N7–23 enables us to compare the areas under the curves of $y = e^{-x^2}$ and $y = e^{-x}$ from $x = 1$ to $x = b$. Thus, if $b > 1$, then

$$\int_1^b e^{-x^2}\,dx \leqq \int_1^b e^{-x}\,dx \leqq \int_1^\infty e^{-x}\,dx = \frac{1}{e}.$$

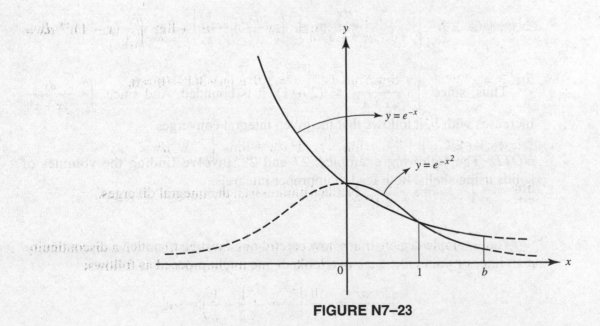

$y = e^{-x}$

$y = e^{-x^2}$

FIGURE N7–23

Note also that $\int_1^b e^{-x^2} \, dx$ increases with b. Since this is true and since $\int_1^b e^{-x^2} \, dx$ is bounded, it follows that $\lim\limits_{b \to \infty} \int_1^b e^{-x^2} \, dx$ exists $\left(\text{and is} \leq \frac{1}{e}\right)$. Therefore $\int_1^\infty e^{-x^2} \, dx$ converges.

Example 26. Show that $\int_0^\infty \dfrac{dx}{\sqrt{x + x^4}}$ converges.

$$\int_0^\infty \frac{dx}{\sqrt{x + x^4}} = \int_0^1 \frac{dx}{\sqrt{x + x^4}} + \int_1^\infty \frac{dx}{\sqrt{x + x^4}}.$$

Since if $0 < x \leq 1$ then $x + x^4 > x$ and $\sqrt{x + x^4} > \sqrt{x}$, it follows that

$$\frac{1}{\sqrt{x + x^4}} < \frac{1}{\sqrt{x}} \quad (0 < x \leq 1).$$

Thus

$$\int_0^1 \frac{dx}{\sqrt{x + x^4}} \leq \int_0^1 \frac{dx}{\sqrt{x}} = 2.$$

Further, if $x \geq 1$ then $x + x^4 \geq x^4$ and $\sqrt{x + x^4} \geq \sqrt{x^4} = x^2$, so

$$\frac{1}{\sqrt{x + x^4}} \leq \frac{1}{x^2} \quad (x \geq 1)$$

and

$$\int_1^\infty \frac{dx}{\sqrt{x + x^4}} \leq \int_1^\infty \frac{dx}{x^2} = -\frac{1}{x} \Big|_1^\infty = 1.$$

Thus, since $\int_0^\infty \dfrac{dx}{\sqrt{x + x^4}} \leq (2 + 1)$, it is bounded. And since $\int_0^b \dfrac{dx}{\sqrt{x + x^4}}$ increases with b, it follows that the given integral converges.

NOTE: The following examples, 27 and 28, involve finding the volumes of solids using shells. Both lead to improper integrals.

Example 27. Find the volume, if it exists, of the solid generated by rotating the region in the first quadrant bounded above by $y = \dfrac{1}{x}$, at the left by $x = 1$, and below by $y = 0$, about the y-axis. See Figure N7–24.

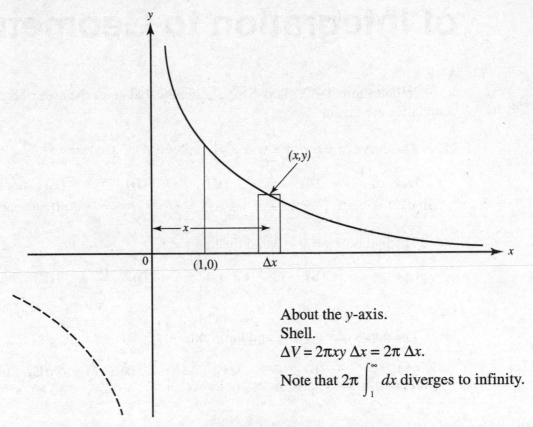

About the y-axis.
Shell.
$\Delta V = 2\pi xy\, \Delta x = 2\pi\, \Delta x.$

Note that $2\pi \displaystyle\int_{1}^{\infty} dx$ diverges to infinity.

FIGURE N7–24

Example 28. The region under $y = e^{-x}$ in the first quadrant is rotated about the y-axis. Find the volume, if it exists, of the solid generated. See Figure N7–25.

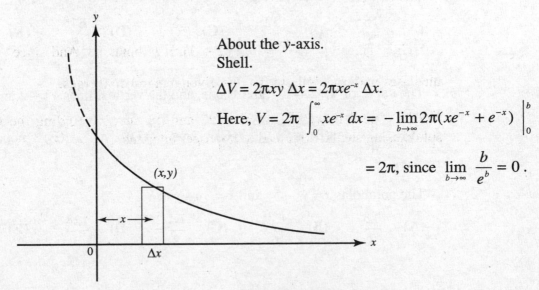

About the y-axis.
Shell.

$\Delta V = 2\pi xy\, \Delta x = 2\pi x e^{-x}\, \Delta x.$

Here, $V = 2\pi \displaystyle\int_{0}^{\infty} x e^{-x}\, dx = -\lim_{b\to\infty} 2\pi (x e^{-x} + e^{-x}) \Big|_{0}^{b}$

$= 2\pi,$ since $\displaystyle\lim_{b\to\infty} \frac{b}{e^{b}} = 0.$

FIGURE N7–25

Set 7: Multiple-Choice Questions on Applications of Integration to Geometry

AREA

In Questions 1–12, choose the alternative that gives the area of the region whose boundaries are given.

1. The curve of $y = x^2$, $y = 0$, $x = -1$, and $x = 2$.

    (A) $\dfrac{11}{3}$ (B) $\dfrac{7}{3}$ (C) 3 (D) 5 (E) none of these

2. The parabola $y = x^2 - 3$ and the line $y = 1$.

    (A) $\dfrac{8}{3}$ (B) 32 (C) $\dfrac{32}{3}$ (D) $\dfrac{16}{3}$ (E) none of these

3. The curve of $x = y^2 - 1$ and the y-axis.

    (A) $\dfrac{4}{3}$ (B) $\dfrac{2}{3}$ (C) $\dfrac{8}{3}$ (D) $\dfrac{1}{2}$ (E) none of these

4. The parabola $y^2 = x$ and the line $x + y = 2$.

    (A) $\dfrac{5}{2}$ (B) $\dfrac{3}{2}$ (C) $\dfrac{11}{6}$ (D) $\dfrac{9}{2}$ (E) $\dfrac{29}{6}$

5. The curve of $x^2 = y^2(4 - y^2)$.

    (A) $\dfrac{16}{3}$ (B) $\dfrac{32}{3}$ (C) $\dfrac{8}{3}$ (D) $\dfrac{64}{15}$ (E) 4

6. The curve of $y = \dfrac{4}{x^2 + 4}$, the x-axis, and the vertical lines $x = -2$ and $x = 2$.

    (A) $\dfrac{\pi}{4}$ (B) $\dfrac{\pi}{2}$ (C) 2π (D) π (E) none of these

7. The parabolas $x = y^2 - 5y$ and $x = 3y - y^2$.

    (A) $\dfrac{32}{3}$ (B) $\dfrac{139}{6}$ (C) $\dfrac{64}{3}$ (D) $\dfrac{128}{3}$ (E) none of these

8.   The curve of $y = \dfrac{2}{x}$ and $x + y = 3$.

(A) $\dfrac{1}{2} - 2\ln 2$ (B) $\dfrac{3}{2}$ (C) $\dfrac{1}{2} - \ln 4$

(D) $\dfrac{5}{2}$ (E) $\dfrac{3}{2} - \ln 4$

9.   In the first quadrant, bounded below by the x-axis and above by the curves of $y = \sin x$ and $y = \cos x$.

(A) $2 - \sqrt{2}$ (B) $2 + \sqrt{2}$ (C) 2 (D) $\sqrt{2}$ (E) $2\sqrt{2}$

10.   Bounded above by the curve $y = \sin x$ and below by $y = \cos x$ from $x = \dfrac{\pi}{4}$ to $x = \dfrac{5\pi}{4}$.

(A) $2\sqrt{2}$ (B) $\dfrac{2}{\sqrt{2}}$ (C) $\dfrac{1}{2\sqrt{2}}$

(D) $2(\sqrt{2} - 1)$ (E) $2(\sqrt{2} + 1)$

11.   The curve $y = \cot x$, the line $x = \dfrac{\pi}{4}$ and the x-axis.

(A) $\ln 2$ (B) $\dfrac{1}{2}\ln\dfrac{1}{2}$ (C) 1 (D) $\dfrac{1}{2}\ln 2$ (E) 2

12.   The curve of $y = x^3 - 2x^2 - 3x$ and the x-axis.

(A) $\dfrac{28}{3}$ (B) $\dfrac{79}{6}$ (C) $\dfrac{45}{4}$ (D) $\dfrac{71}{6}$ (E)   none of these

13.   The total area bounded by the cubic $x = y^3 - y$ and the line $x = 3y$ is equal to

(A) 4 (B) $\dfrac{16}{3}$ (C) 8 (D) $\dfrac{32}{3}$ (E) 16

14.   The area bounded by $y = e^x$, $y = 1$, $y = 2$, and $x = 3$ is equal to

(A) $3 + \ln 2$ (B) $3 - 3\ln 3$ (C) $4 + \ln 2$

(D) $3 - \dfrac{1}{2}\ln^2 2$ (E) $4 - \ln 4$

*15.   The area enclosed by the ellipse with parametric equations $x = 2\cos\theta$ and $y = 3\sin\theta$ equals

(A) 6π (B) $\dfrac{9}{2}\pi$ (C) 3π (D) $\dfrac{3}{2}\pi$ (E)   none of these

*An asterisk denotes a topic covered only in Calculus BC.

*16.   The area enclosed by one loop of the cycloid with parametric equations
$x = \theta - \sin\theta$ and $y = 1 - \cos\theta$ equals

(A) $\dfrac{3\pi}{2}$ (B) 3π (C) 2π (D) 6π (E)   none of these

17.   The area enclosed by the curve $y^2 = x(1-x)$ is given by

(A) $2\displaystyle\int_0^1 x\sqrt{1-x}\,dx$ (B) $2\displaystyle\int_0^1 \sqrt{x-x^2}\,dx$ (C) $4\displaystyle\int_0^1 \sqrt{x-x^2}\,dx$

(D) π (E) 2π

18.   The area bounded by the parabola $y = 2 - x^2$ and the line $y = x - 4$ is given by

(A) $\displaystyle\int_{-2}^3 (6 - x - x^2)\,dx$ (B) $\displaystyle\int_{-2}^1 (2 + x + x^2)\,dx$ (C) $\displaystyle\int_{-3}^2 (6 - x - x^2)\,dx$

(D) $2\displaystyle\int_0^{\sqrt{2}} (2 - x^2)\,dx + \displaystyle\int_{-3}^2 (4 - x)\,dx$ (E)   none of these

*19.   The area enclosed by the hypocycloid with parametric equations $x = \cos^3 t$ and
$y = \sin^3 t$ is given by

(A) $3\displaystyle\int_{\pi/2}^0 \sin^4 t \cos^2 t\,dt$ (B) $4\displaystyle\int_0^1 \sin^3 t\,dt$ (C) $-4\displaystyle\int_{\pi/2}^0 \sin^6 t\,dt$

(D) $12\displaystyle\int_0^{\pi/2} \sin^4 t \cos^2 t\,dt$ (E)   none of these

20.   The figure below shows part of the curve of $y = x^3$ and a rectangle with two vertices
at $(0, 0)$ and $(c, 0)$. What is the ratio of the area of the rectangle to the shaded part of
it above the cubic?

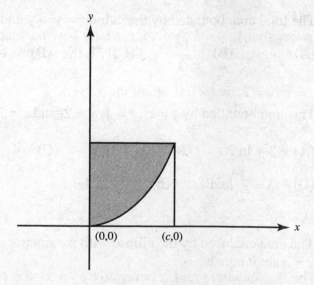

(A) $3:4$ (B) $5:4$ (C) $4:3$ (D) $3:1$ (E) $2:1$

---

*An asterisk denotes a topic covered only in Calculus BC.

21.   Suppose the following is a table of ordinates for $y = f(x)$, given that f is continuous on [1, 5]:

x	1	2	3	4	5
y	1.62	4.15	7.5	9.0	12.13

If a trapezoid sum in used, with $n = 4$, then the area under the curve, from $x = 1$ to $x = 5$, is equal, to two decimal places, to

(A) 6.88   (B) 13.76   (C) 20.30   (D) 25.73   (E) 27.53

*22.   The area A enclosed by the four-leaved rose $r = \cos 2\theta$ equals, to three decimal places,

(A) 0.785   (B) 1.571   (C) 2.071   (D) 3.142   (E) 6.283

*23.   The area bounded by the small loop of the limaçon $r = 1 - 2 \sin \theta$ is given by the definite integral

(A) $\displaystyle\int_{\pi/3}^{5\pi/3} \left[\frac{1}{2}(1 - 2\sin\theta)\right]^2 d\theta$

(B) $\displaystyle\int_{7\pi/6}^{3\pi/2} (1 - 2\sin\theta)^2\, d\theta$

(C) $\displaystyle\int_{\pi/6}^{\pi/2} (1 - 2\sin\theta)^2\, d\theta$

(D) $\displaystyle\int_{0}^{\pi/6} \left[\frac{1}{2}(1 - 2\sin\theta)\right]^2 d\theta + \int_{5\pi/6}^{\pi} \left[\frac{1}{2}(1 - 2\sin\theta)\right]^2 d\theta$

(E) $\displaystyle\int_{0}^{\pi/3} (1 - 2\sin\theta)^2\, d\theta$

VOLUME

In Questions 24–35 the region whose boundaries are given is rotated about the line indicated. Choose the alternative that gives the volume of the solid generated.

24. $y = x^2$, $x = 2$, and $y = 0$; about the x-axis.

(A) $\dfrac{64\pi}{3}$ (B) 8π (C) $\dfrac{8\pi}{3}$ (D) $\dfrac{128\pi}{5}$ (E) $\dfrac{32\pi}{5}$

25. $y = x^2$, $x = 2$, and $y = 0$; about the y-axis.

(A) $\dfrac{16\pi}{3}$ (B) 4π (C) $\dfrac{32\pi}{5}$ (D) 8π (E) $\dfrac{8\pi}{3}$

26.   The first quadrant region bounded by $y = x^2$, the y-axis, and $y = 4$; about the y-axis.

(A) 8π (B) 4π (C) $\dfrac{64\pi}{3}$ (D) $\dfrac{32\pi}{3}$ (E) $\dfrac{16\pi}{3}$

---

*An asterisk denotes a topic covered only in Calculus BC.

27. $y = x^2$ and $y = 4$; about the x-axis.

(A) $\dfrac{64\pi}{5}$ (B) $\dfrac{512\pi}{15}$ (C) $\dfrac{256\pi}{5}$

(D) $\dfrac{128\pi}{5}$ (E) none of these

28. $y = x^2$ and $y = 4$; about the line $y = 4$.

(A) $\dfrac{256\pi}{15}$ (B) $\dfrac{256\pi}{5}$ (C) $\dfrac{512\pi}{5}$ (D) $\dfrac{512\pi}{15}$ (E) $\dfrac{64\pi}{3}$

29. $y = x^2$ and $y = 4$; about the line $y = -1$.

(A) $4\pi \displaystyle\int_{-1}^{4} (y + 1) \sqrt{y}\, dy$ (B) $2\pi \displaystyle\int_{0}^{2} (4 - x^2)^2\, dx$ (C) $\pi \displaystyle\int_{-2}^{2} (16 - x^4)\, dx$

(D) $2\pi \displaystyle\int_{0}^{2} (24 - 2x^2 - x^4)\, dx$ (E) none of these

30. $y = 3x - x^2$ and $y = 0$; about the x-axis.

(A) $\pi \displaystyle\int_{0}^{3} (9x^2 + x^4)\, dx$ (B) $\pi \displaystyle\int_{0}^{3} (3x - x^2)^2\, dx$ (C) $\pi \displaystyle\int_{0}^{\sqrt{3}} (3x - x^2)\, dx$

(D) $2\pi \displaystyle\int_{0}^{3} y \sqrt{9 - 4y}\, dy$ (E) $\pi \displaystyle\int_{0}^{9/4} y^2\, dy$

31. $y = 3x - x^2$ and $y = x$; about the x-axis.

(A) $\pi \displaystyle\int_{0}^{3/2} [(3x - x^2)^2 - x^2]\, dx$ (B) $\pi \displaystyle\int_{0}^{2} (9x^2 - 6x^3)\, dx$

(C) $\pi \displaystyle\int_{0}^{2} [(3x - x^2)^2 - x^2]\, dx$ (D) $\pi \displaystyle\int_{0}^{3} [(3x - x^2)^2 - x^4]\, dx$

(E) $\pi \displaystyle\int_{0}^{3} (2x - x^2)^2\, dx$

32.   An arch of $y = \sin x$ and the x-axis; about the x-axis.

(A) $\dfrac{\pi}{2}\left(\pi - \dfrac{1}{2}\right)$ (B) $\dfrac{\pi^2}{2}$ (C) $\dfrac{\pi^2}{4}$ (D) π^2 (E) $\pi(\pi - 1)$

33.   A trapezoid with vertices at $(2, 0)$, $(2, 2)$, $(4, 0)$, and $(4, 4)$; about the x-axis.

(A) $\dfrac{56\pi}{3}$ (B) $\dfrac{128\pi}{3}$ (C) $\dfrac{92\pi}{3}$

(D) $\dfrac{112\pi}{3}$ (E) none of these

34. $y = \ln x$, $y = 0$, $x = e$; about the line $x = e$.

(A) $\pi \displaystyle\int_{1}^{e} (e - x) \ln x \, dx$ (B) $\pi \displaystyle\int_{0}^{1} (e - e^{y})^2 \, dy$ (C) $2\pi \displaystyle\int_{1}^{e} (e - \ln x) \, dx$

(D) $\pi \displaystyle\int_{0}^{e} (e^2 - 2e^{y+1} + e^{2y}) \, dy$ (E) none of these

*35. The curve with parametric equations $x = \tan \theta$, $y = \cos^2 \theta$, and the lines $x = 0$, $x = 1$, and $y = 0$; about the x-axis.

(A) $\pi \displaystyle\int_{0}^{\pi/4} \cos^4 \theta \, d\theta$ (B) $\pi \displaystyle\int_{0}^{\pi/4} \cos^2 \theta \sin \theta \, d\theta$ (C) $\pi \displaystyle\int_{0}^{\pi/4} \cos^2 \theta \, d\theta$

(D) $\pi \displaystyle\int_{0}^{1} \cos^2 \theta \, d\theta$ (E) $\pi \displaystyle\int_{0}^{1} \cos^4 \theta \, d\theta$

36. A sphere of radius r is divided into two parts by a plane at distance h ($0 < h < r$) from the center. The volume of the smaller part equals

(A) $\dfrac{\pi}{3}(2r^3 + h^3 - 3r^2 h)$ (B) $\dfrac{\pi h}{3}(3r^2 - h^2)$ (C) $\dfrac{4}{3}\pi r^3 + \dfrac{h^3}{3} - r^2 h$

(D) $\dfrac{\pi}{3}(2r^3 + 3r^2 h - h^3)$ (E) none of these

37. The base of a solid is a circle of radius a, and every plane section perpendicular to a diameter is a square. The solid has volume

(A) $\dfrac{8}{3}a^3$ (B) $2\pi a^3$ (C) $4\pi a^3$ (D) $\dfrac{16}{3}a^3$ (E) $\dfrac{8\pi}{3}a^3$

38. The base of a solid is the region bounded by the parabola $x^2 = 8y$ and the line $y = 4$, and each plane section perpendicular to the y-axis is an equilateral triangle. The volume of the solid is

(A) $\dfrac{64\sqrt{3}}{3}$ (B) $64\sqrt{3}$ (C) $32\sqrt{3}$

(D) 32 (E) none of these

39. The base of a solid is the region bounded by $y = e^{-x}$, the x-axis, the y-axis, and the line $x = 1$. Each cross section perpendicular to the x-axis is a square. The volume of the solid is

(A) $\dfrac{e^2}{2}$ (B) $e^2 - 1$ (C) $1 - \dfrac{1}{e^2}$

(D) $\dfrac{e^2 - 1}{2}$ (E) $\dfrac{1}{2}\left(1 - \dfrac{1}{e^2}\right)$

*An asterisk denotes a topic covered only in Calculus BC.

40. If the curves of $f(x)$ and $g(x)$ intersect for $x = a$ and $x = b$ and if $f(x) > g(x) > 0$ for all x on (a, b), then the volume obtained when the region bounded by the curves is rotated about the x-axis is equal to

(A) $\pi \int_a^b f^2(x)\, dx - \int_a^b g^2(x)\, dx$

(B) $\pi \int_a^b [f(x) - g(x)]^2\, dx$

(C) $2\pi \int_a^b x[f(x) - g(x)]\, dx$

(D) $\pi \int_a^b [f^2(x) - g^2(x)]\, dx$

(E) none of these

ARC LENGTH

*41. The length of the arc of the curve $y^2 = x^3$ cut off by the line $x = 4$ is

(A) $\dfrac{4}{3}(10\sqrt{10} - 1)$ (B) $\dfrac{8}{27}(10^{3/2} - 1)$ (C) $\dfrac{16}{27}(10^{3/2} - 1)$

(D) $\dfrac{16}{27}10\sqrt{10}$ (E) none of these

*42. The length of the arc of $y = \ln \cos x$ from $x = \dfrac{\pi}{4}$ to $x = \dfrac{\pi}{3}$ equals

(A) $\ln \dfrac{\sqrt{3} + 2}{\sqrt{2} + 1}$ (B) 2 (C) $\ln(1 + \sqrt{3} - \sqrt{2})$

(D) $\sqrt{3} - 2$ (E) $\dfrac{\ln(\sqrt{3} + 2)}{\ln(\sqrt{2} + 1)}$

*43. The length of one arch of the cycloid $\begin{aligned} x &= t - \sin t \\ y &= 1 - \cos t \end{aligned}$ equals

(A) $\displaystyle\int_0^\pi \sqrt{1 - \cos t}\, dt$ (B) $\displaystyle\int_0^{2\pi} \sqrt{\dfrac{1 - \cos t}{2}}\, dt$ (C) $\displaystyle\int_0^\pi \sqrt{2 - 2\cos t}\, dt$

(D) $\displaystyle\int_0^{2\pi} \sqrt{2 - 2\cos t}\, dt$ (E) $2\displaystyle\int_0^\pi \sqrt{\dfrac{1 - \cos t}{2}}\, dt$

*44. The length of the arc of the parabola $4x = y^2$ cut off by the line $x = 2$ is given by the integral

(A) $\displaystyle\int_{-1}^1 \sqrt{x^2 + 1}\, dx$ (B) $\dfrac{1}{2}\displaystyle\int_0^2 \sqrt{4 + y^2}\, dy$ (C) $\displaystyle\int_{-1}^1 \sqrt{1 + x}\, dx$

(D) $\displaystyle\int_0^{2\sqrt{2}} \sqrt{4 + y^2}\, dy$ (E) none of these

*An asterisk denotes a topic covered only in Calculus BC.

*45. The length of $x = e^t \cos t$, $y = e^t \sin t$ from $t = 2$ to $t = 3$ is equal to

(A) $\sqrt{2}e^2\sqrt{e^2 - 1}$ (B) $\sqrt{2}(e^3 - e^2)$ (C) $2(e^3 - e^2)$

(D) $e^3(\cos 3 + \sin 3) - e^2(\cos 2 + \sin 2)$ (E) none of these

IMPROPER INTEGRALS

*46. Which one of the following is an improper integral?

(A) $\displaystyle\int_0^2 \frac{dx}{\sqrt{x+1}}$ (B) $\displaystyle\int_{-1}^1 \frac{dx}{1+x^2}$ (C) $\displaystyle\int_0^2 \frac{x\,dx}{1-x^2}$

(D) $\displaystyle\int_0^{\pi/3} \frac{\sin x\,dx}{\cos^2 x}$ (E) none of these

*47. $\displaystyle\int_0^\infty e^{-x}\,dx =$

(A) 1   (B) $\dfrac{1}{e}$ (C) -1 (D) $-\dfrac{1}{e}$ (E) none of these

*48. $\displaystyle\int_0^e \frac{du}{u} =$

(A) 1   (B) $\dfrac{1}{e}$ (C) $-\dfrac{1}{e^2}$ (D) -1 (E) none of these

*49. $\displaystyle\int_1^2 \frac{dt}{\sqrt[3]{t-1}} =$

(A) $\dfrac{2}{3}$ (B) $\dfrac{3}{2}$ (C) 3   (D) 1   (E) none of these

*50. $\displaystyle\int_2^4 \frac{dx}{(x-3)^{2/3}} =$

(A) 6   (B) $\dfrac{6}{5}$ (C) $\dfrac{2}{3}$ (D) 0   (E) none of these

*51. $\displaystyle\int_2^4 \frac{dx}{(x-3)^2} =$

(A) 2   (B) -2 (C) 0   (D) $\dfrac{2}{3}$ (E) none of these

*52. $\displaystyle\int_0^{\pi/2} \frac{\sin x}{\sqrt{1-\cos x}}\,dx$

(A) -2 (B) $\dfrac{2}{3}$ (C) 2   (D) $\dfrac{1}{2}$ (E) none of these

*An asterisk denotes a topic covered only in Calculus BC.

*53.   Which one of the following improper integrals diverges?

   (A) $\displaystyle\int_1^\infty \frac{dx}{x^2}$ **(B)** $\displaystyle\int_0^\infty \frac{dx}{e^x}$ **(C)** $\displaystyle\int_{-1}^1 \frac{dx}{\sqrt[3]{x}}$

   (D) $\displaystyle\int_{-1}^1 \frac{dx}{x^2}$ **(E)** none of these

*54.   Which one of the following improper integrals diverges?

   (A) $\displaystyle\int_0^\infty \frac{dx}{1+x^2}$ **(B)** $\displaystyle\int_0^1 \frac{dx}{x^{1/3}}$ **(C)** $\displaystyle\int_0^\infty \frac{dx}{x^3+1}$

   (D) $\displaystyle\int_0^\infty \frac{dx}{e^x+2}$ **(E)** $\displaystyle\int_1^\infty \frac{dx}{x^{1/3}}$

In Questions 55–59, choose the alternative that gives the area, if it exists, of the region described.

*55.   In the first quadrant under the curve of $y = e^{-x}$.

   (A) 1   **(B)** e **(C)** $\dfrac{1}{e}$ **(D)** 2   **(E)** none of these

*56.   In the first quadrant under the curve of $y = xe^{-x^2}$.

   (A) 2   **(B)** $\dfrac{2}{e}$ **(C)** $\dfrac{1}{2}$ **(D)** $\dfrac{1}{2e}$ **(E)** none of these

*57.   In the first quadrant above $y = 1$, bounded by the y-axis and the curve $xy = 1$.

   (A) 1   **(B)** 2   **(C)** $\dfrac{1}{2}$ **(D)** 4   **(E)** none of these

*58.   Between the curve $y = \dfrac{4}{1+x^2}$ and its asymptote.

   (A) 2π **(B)** 4π **(C)** 8π **(D)** π **(E)** none of these

*59.   Between the curve $y = \dfrac{4}{\sqrt{1-x^2}}$ and its asymptotes.

   (A) $\dfrac{\pi}{2}$ **(B)** π **(C)** 2π **(D)** 4π **(E)** none of these

*An asterisk denotes a topic covered only in Calculus BC.

In Questions 60 and 61, choose the alternative that gives the volume, if it exists, of the solid generated.

*60. $y = \dfrac{1}{x}$, at the left by $x = 1$, and below by $y = 0$; about the x-axis.

    (A) $\dfrac{\pi}{2}$ (B) π (C) 2π (D) 4π (E) none of these

*61. The first-quadrant region under $y = e^{-x}$; about the x-axis.

    (A) $\dfrac{\pi}{2}$ (B) π (C) 2π (D) 4π (E) none of these

*An asterisk denotes a topic covered only in Calculus BC.

Answers for Set 7: Applications of Integration to Geometry

1.	C	14.	E	27.	C	40.	D	53.	D
2.	C	15.	A	28.	D	41.	C	54.	E
3.	A	16.	B	29.	D	42.	A	55.	A
4.	D	17.	B	30.	B	43.	D	56.	C
5.	B	18.	C	31.	C	44.	D	57.	E
6.	D	19.	D	32.	B	45.	B	58.	B
7.	C	20.	C	33.	A	46.	C	59.	D
8.	E	21.	E	34.	B	47.	A	60.	B
9.	A	22.	B	35.	C	48.	E	61.	A
10.	A	23.	C	36.	A	49.	B		
11.	D	24.	E	37.	D	50.	A		
12.	D	25.	D	38.	B	51.	E		
13.	C	26.	A	39.	E	52.	C		

AREA

We give below, for each of Questions 1–20, a sketch of the region, and indicate a typical element of area. The area of the region is given by the definite integral. We exploit symmetry wherever possible.

1. C.

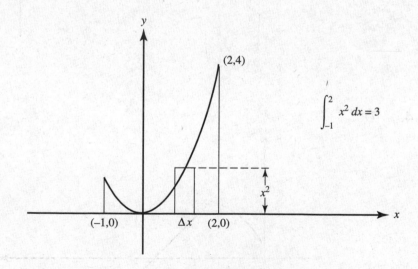

$$\int_{-1}^{2} x^2 \, dx = 3$$

2.   C.

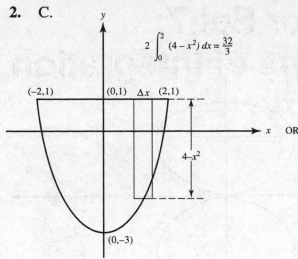

$$2 \int_0^2 (4 - x^2)\, dx = \frac{32}{3}$$

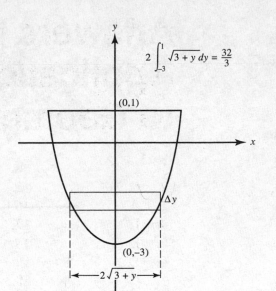

$$2 \int_{-3}^1 \sqrt{3 + y}\, dy = \frac{32}{3}$$

3.   A.

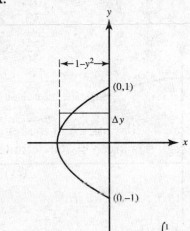

OR

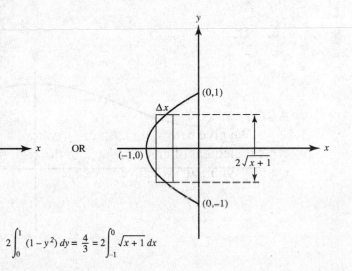

$$2 \int_0^1 (1 - y^2)\, dy = \frac{4}{3} = 2 \int_{-1}^0 \sqrt{x + 1}\, dx$$

4.   D.

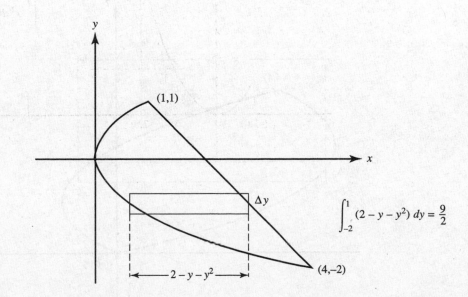

$$\int_{-2}^1 (2 - y - y^2)\, dy = \frac{9}{2}$$

5. B.

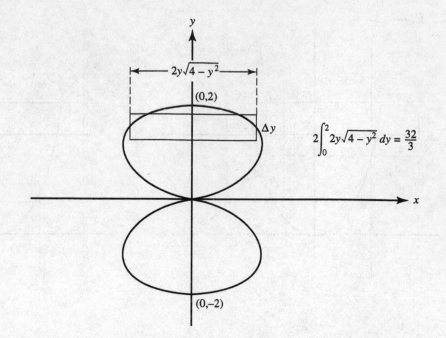

$$2\int_0^2 2y\sqrt{4-y^2}\,dy = \frac{32}{3}$$

6. D.

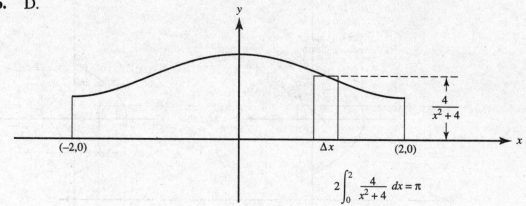

$$2\int_0^2 \frac{4}{x^2+4}\,dx = \pi$$

7. C.

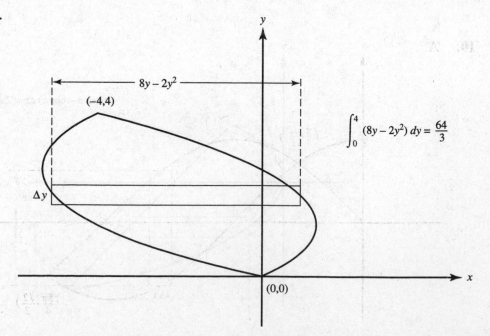

$$\int_0^4 (8y - 2y^2)\,dy = \frac{64}{3}$$

8. E.

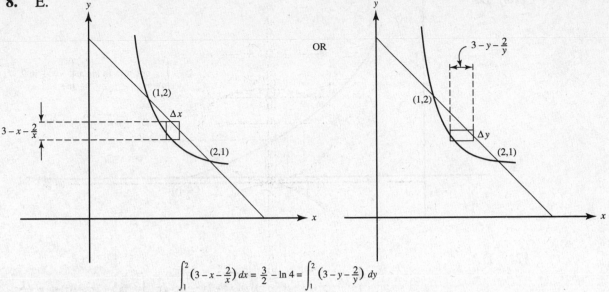

OR

$$\int_1^2 \left(3 - x - \frac{2}{x}\right) dx = \frac{3}{2} - \ln 4 = \int_1^2 \left(3 - y - \frac{2}{y}\right) dy$$

9. A.

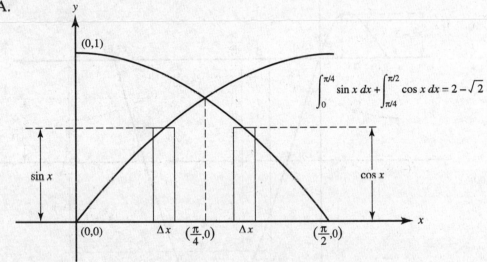

$$\int_0^{\pi/4} \sin x \, dx + \int_{\pi/4}^{\pi/2} \cos x \, dx = 2 - \sqrt{2}$$

10. A.

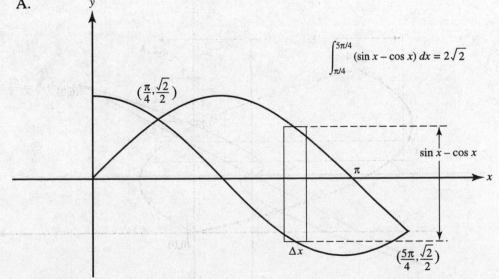

$$\int_{\pi/4}^{5\pi/4} (\sin x - \cos x) \, dx = 2\sqrt{2}$$

11. D.

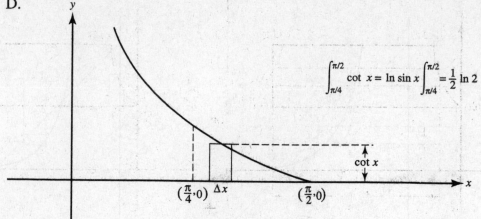

$$\int_{\pi/4}^{\pi/2} \cot x = \ln \sin x \Big|_{\pi/4}^{\pi/2} = \frac{1}{2} \ln 2$$

12. D.

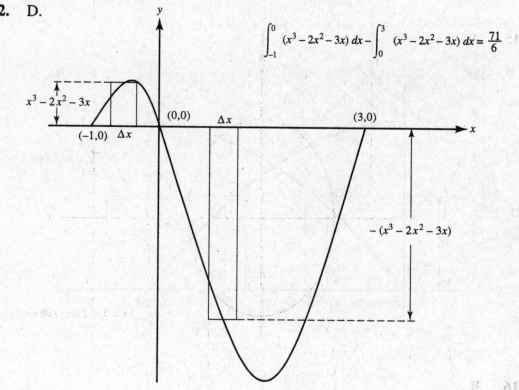

$$\int_{-1}^{0} (x^3 - 2x^2 - 3x)\, dx - \int_{0}^{3} (x^3 - 2x^2 - 3x)\, dx = \frac{71}{6}$$

13. C.

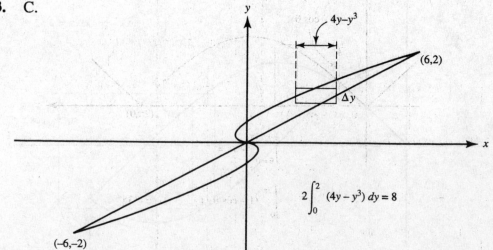

$$2\int_{0}^{2} (4y - y^3)\, dy = 8$$

14.   E.

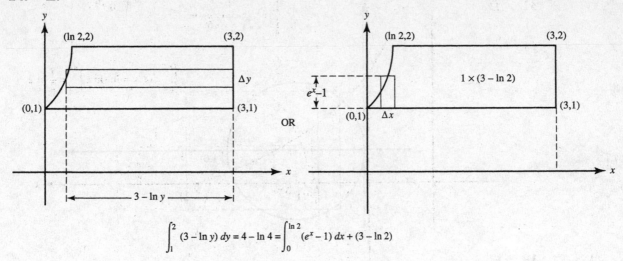

$$\int_1^2 (3 - \ln y)\, dy = 4 - \ln 4 = \int_0^{\ln 2} (e^x - 1)\, dx + (3 - \ln 2)$$

15.   A.

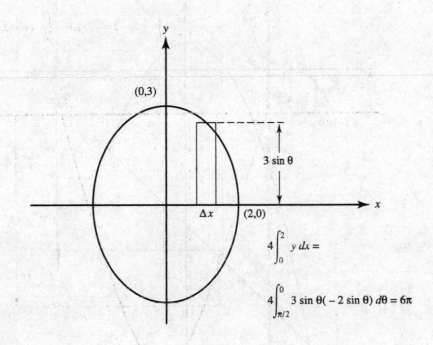

$$4 \int_0^2 y\, dx =$$

$$4 \int_{\pi/2}^0 3 \sin \theta (-2 \sin \theta)\, d\theta = 6\pi$$

16.   B.

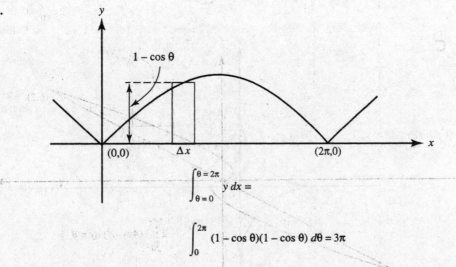

$$\int_{\theta = 0}^{\theta = 2\pi} y\, dx =$$

$$\int_0^{2\pi} (1 - \cos \theta)(1 - \cos \theta)\, d\theta = 3\pi$$

17. **B.**

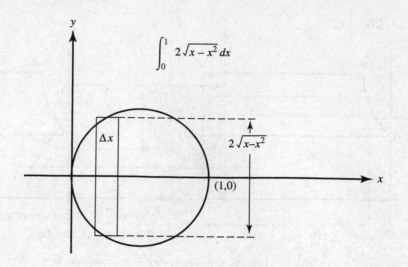

$$\int_0^1 2\sqrt{x-x^2}\,dx$$

$2\sqrt{x-x^2}$

Δx

$(1,0)$

18. **C.**

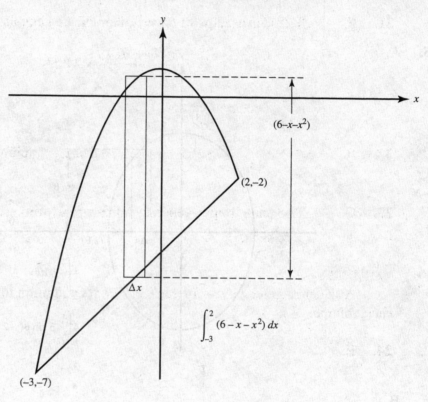

$(6-x-x^2)$

$(2,-2)$

Δx

$$\int_{-3}^{2}(6-x-x^2)\,dx$$

$(-3,-7)$

19. **D.**

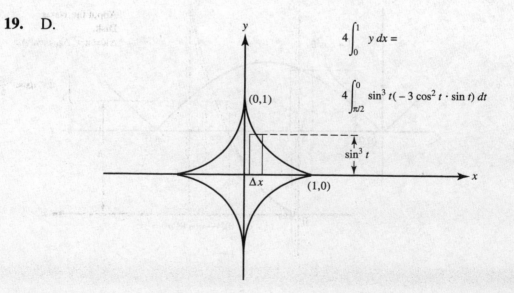

$$4\int_0^1 y\,dx =$$

$$4\int_{\pi/2}^{0} \sin^3 t\,(-3\cos^2 t \cdot \sin t)\,dt$$

$(0,1)$

$\sin^3 t$

Δx

$(1,0)$

20.   C.

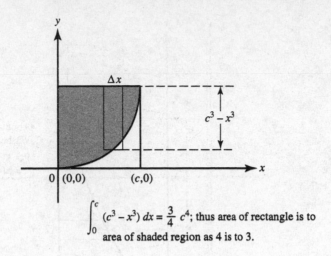

$$\int_0^c (c^3 - x^3)\, dx = \frac{3}{4}\, c^4;\ \text{thus area of rectangle is to}$$
area of shaded region as 4 is to 3.

21.   E.   If $T(4)$ is used, with $\Delta x = 1$, then the area equals

$$\frac{L(4) + R(4)}{2} = 27.53$$

to two decimal places.

22.   B. $A = 8 \displaystyle\int_0^{\pi/4} \frac{1}{2}\, \cos^2 2\theta\, d\theta = 1.571$, using a graphing calculator.

23.   C.   The small loop is generated as θ varies from $\dfrac{\pi}{6}$ to $\dfrac{5\pi}{6}$. (C) uses the loop's symmetry.

VOLUME

A sketch is given below for each question, in addition to the definite integral for each volume.

24.   E.

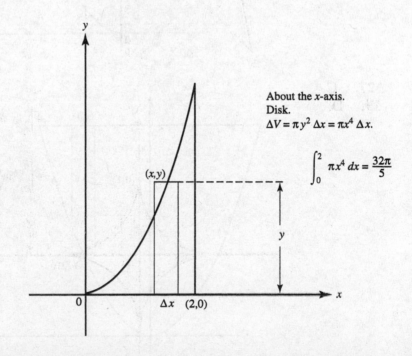

About the x-axis.
Disk.
$\Delta V = \pi y^2\, \Delta x = \pi x^4\, \Delta x.$

$$\int_0^2 \pi x^4\, dx = \frac{32\pi}{5}$$

25. D.

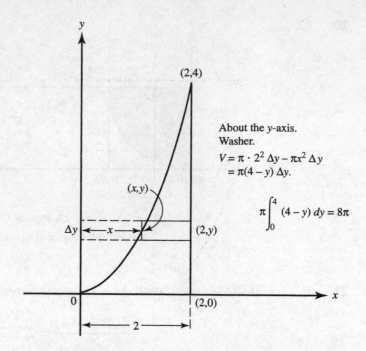

About the y-axis.
Washer.
$V = \pi \cdot 2^2 \, \Delta y - \pi x^2 \, \Delta y$
   $= \pi(4 - y) \, \Delta y.$

$\pi \displaystyle\int_0^4 (4 - y) \, dy = 8\pi$

26. A.

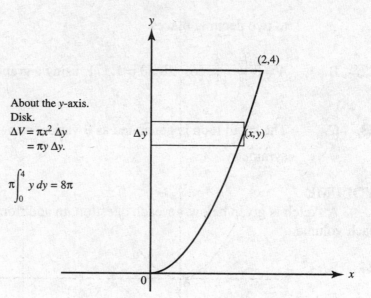

About the y-axis.
Disk.
$\Delta V = \pi x^2 \, \Delta y$
   $= \pi y \, \Delta y.$

$\pi \displaystyle\int_0^4 y \, dy = 8\pi$

27. C.

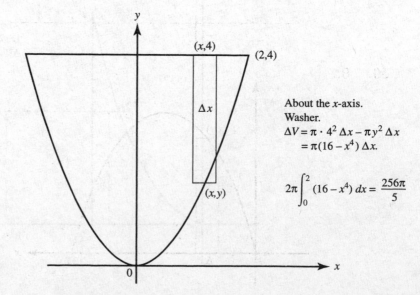

About the x-axis.
Washer.
$\Delta V = \pi \cdot 4^2 \, \Delta x - \pi y^2 \, \Delta x$
   $= \pi(16 - x^4) \, \Delta x.$

$2\pi \displaystyle\int_0^2 (16 - x^4) \, dx = \dfrac{256\pi}{5}$

28. D.

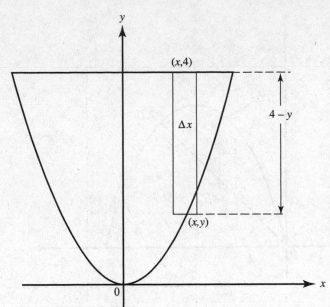

About $y = 4$.
Disk.
$$\Delta V = \pi(4 - y)^2 \, \Delta x$$
$$= \pi(4 - x^2)^2 \, \Delta x.$$

$$2\pi \int_0^2 (4 - x^2)^2 \, dx = \frac{512\pi}{15}$$

29. D.

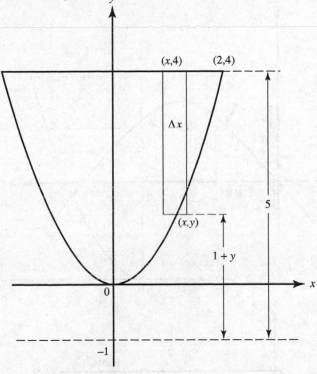

About $y = -1$.
Washer.
$$\Delta V = \pi \cdot 5^2 \, \Delta x - \pi(1 + y)^2 \, \Delta x$$
$$= \pi[25 - (1 + y)^2] \, \Delta x.$$

$$2\pi \int_0^2 (24 - 2x^2 - x^4) \, dx$$

30. B.

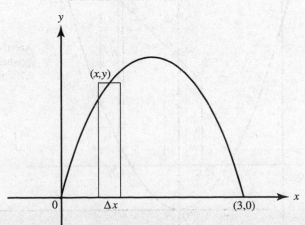

About the x-axis.
Disk.
$$\Delta V = \pi y^2 \, \Delta x$$
$$= \pi(3x - x^2)^2 \, \Delta x.$$

$$\pi \int_0^3 (3x - x^2)^2 \, dx$$

31. C.

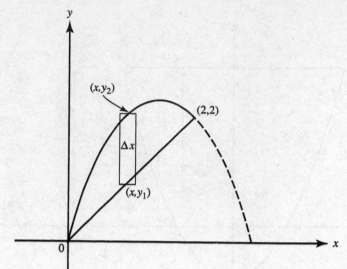

About the x-axis.
Washer.
$$\Delta V = \pi y_2^2\, \Delta x - \pi y_1^2\, \Delta x$$
$$= \pi\left[(3x - x^2)^2 - x^2\right]\Delta x.$$

$$\pi \int_0^2 \left[(3x - x)^2 - x^2\right] dx$$

32. B.

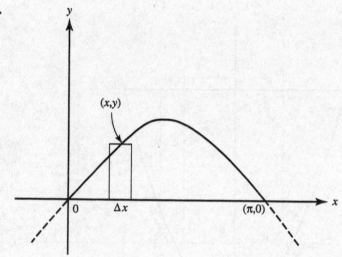

About the x-axis.
Disk.
$$\Delta V = \pi y^2\, \Delta x$$
$$= \pi \sin^2 x\, \Delta x.$$

$$\pi \int_0^\pi \sin^2 x\; dx = \frac{\pi^2}{2}$$

33. A.

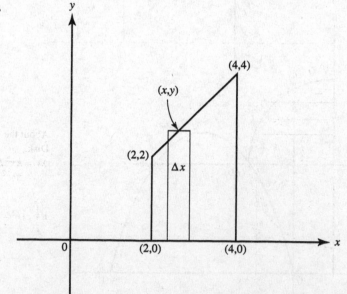

About the x-axis.
Disk.
$$\Delta V = \pi y^2\, \Delta x$$

$$\pi \int_2^4 x^2\, dx = \frac{56\pi}{3}$$

34. B.

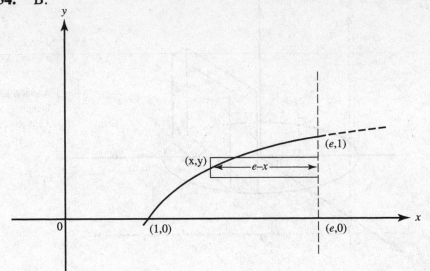

About $x = e$.
Disk.
$$\Delta V = \pi(e - x)^2 \, \Delta y$$
$$\quad = \pi(e - e^y)^2 \, \Delta y.$$

$$\pi \int_0^1 (e - e^y)^2 \, dy$$

35. C.

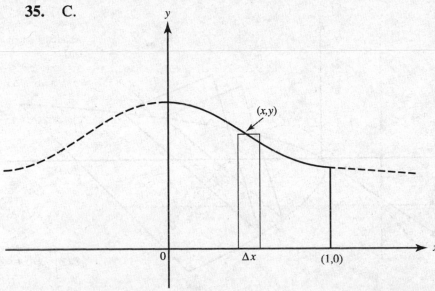

About the x-axis.
Disk.
$$\Delta V = \pi y^2 \, \Delta x.$$

$$\pi \int_0^1 y^2 \, dx = \pi \int_0^{\pi/4} \cos^2 \theta \, d\theta$$

36. A.

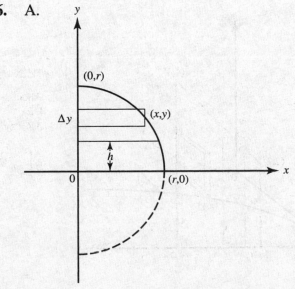

About the y-axis.
Disk.
$$\Delta V = \pi x^2 \, \Delta y$$
$$\quad = \pi(r^2 - y^2) \, \Delta y.$$

$$\pi \int_h^r (r^2 - y^2) \, dy = \frac{\pi}{3} (2r^3 + h^3 - 3r^2 h)$$

37. D.

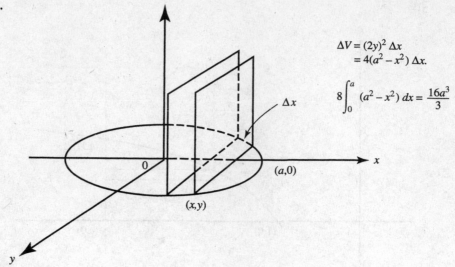

$\Delta V = (2y)^2 \, \Delta x$
$\quad = 4(a^2 - x^2) \, \Delta x.$

$8 \int_0^a (a^2 - x^2) \, dx = \dfrac{16a^3}{3}$

38. B.

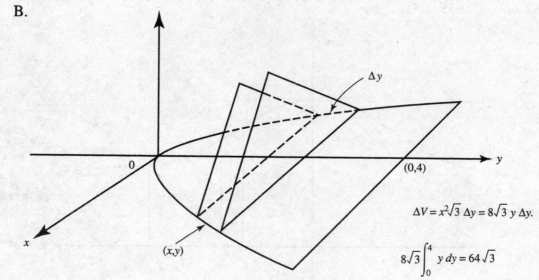

$\Delta V = x^2 \sqrt{3} \, \Delta y = 8\sqrt{3} \, y \, \Delta y.$

$8\sqrt{3} \int_0^4 y \, dy = 64\sqrt{3}$

39. E.

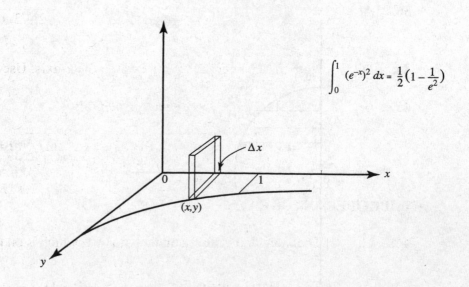

$\int_0^1 (e^{-x})^2 \, dx = \dfrac{1}{2}\left(1 - \dfrac{1}{e^2}\right)$

40. D.

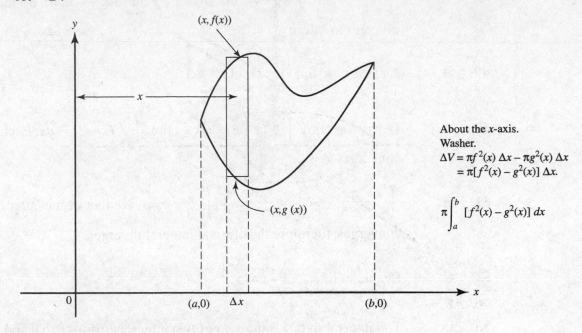

About the x-axis.
Washer.
$$\Delta V = \pi f^2(x)\,\Delta x - \pi g^2(x)\,\Delta x$$
$$= \pi[f^2(x) - g^2(x)]\,\Delta x.$$

$$\pi \int_a^b [f^2(x) - g^2(x)]\,dx$$

ARC LENGTH

41. C.   Note that the curve is symmetric to the x-axis. The arc length equals
$$2\int_0^4 \sqrt{1 + \frac{9}{4}x}\;dx\,.$$

42. A.   Integrate $\displaystyle\int_{\pi/4}^{\pi/3} \sqrt{1 + \tan^2 x}\;dx$. Replace the integrand by $\sec x$, and use

formula (13) on page 153 to get $\left.\ln|\sec x + \tan x|\,\right|_{\pi/4}^{\pi/3}$

43. D.   From (3) on page 236 we obtain the length:
$$\int_0^{2\pi} \sqrt{(1 - \cos t)^2 + (\sin t)^2}\;dt = \int_0^{2\pi} \sqrt{2 - 2\cos t}\;dt$$

44. D.   Note that the curve is symmetric to the x-axis. Use (2) on page 236.

45. B.   Use (3) on page 236 to get the integral:
$$\int_2^3 \sqrt{(-e^t \sin t + e^t \cos t)^2 + (e^t \cos t + e^t \sin t)^2}\;dt = \left.\sqrt{2}e^t\,\right|_2^3\,.$$

IMPROPER INTEGRALS

46. C.   The integrand is discontinuous at $x = 1$, which is on the interval of integration.

47. A.   The integral equals $\displaystyle\lim_{b\to\infty} \left.-\frac{1}{e^x}\,\right|_0^b = -(0 - 1).$

48. **E.** $\displaystyle\int_0^e \frac{du}{u} = \lim_{h\to 0^+}\int_h^e \frac{du}{u} = \lim_{h\to 0^+} \ln|u|\ \Big|_h^e = \lim_{h\to 0^+}(\ln e - \ln h)$. So the integral diverges to infinity.

49. **B.**   Redefine as $\displaystyle\lim_{h\to 0^+}\int_{1+h}^2 (t-1)^{-1/3}\,dt$.

50. **A.**   Rewrite as $\displaystyle\lim_{h\to 0^+}\int_2^{3-h}(x-3)^{-2/3}\,dx + \lim_{h\to 0^+}\int_{3+h}^4 (x-3)^{-2/3}\,dx$. Each integral converges to 3.

51. **E.** $\displaystyle\int_2^4 \frac{dx}{(x-3)^2} = \int_2^3 \frac{dx}{(x-3)^2} + \int_3^4 \frac{dx}{(x-3)^2}$. Neither of the latter integrals converges; therefore the original integral diverges.

52. **C.**   Evaluate $\displaystyle\lim_{h\to 0^+} 2\sqrt{1-\cos x}\ \Big|_0^{(\pi/2)-h}$.

53. **D.**   The integral in (D) is the sum of two integrals from -1 to 0 and from 0 to 1. Both diverge (see Example 24, page 242). Note that (A), (B), and (C) all converge.

54. **E.**   Note, first, that

$$\int_0^\infty \frac{dx}{1+x^2} = \lim_{b\to\infty} \tan^{-1} x\ \Big|_0^b = \frac{\pi}{2}.$$

Since $\displaystyle Q(b) = \int_0^b \frac{dx}{x^3+1}$ increases with b and, further,

$$Q(b) \leq \int_0^b \frac{dx}{x^2+1},$$

$Q(b)$ converges as $b \to \infty$. Similarly, since $\displaystyle\int_0^\infty \frac{dx}{e^x} = 1$ and since, further,

$S(b) = \displaystyle\int_0^b \frac{dx}{e^x+2}$ increases with b and

$$\int_0^b \frac{dx}{e^x+2} \leq \int_0^b \frac{dx}{e^x},$$

$S(b)$ also converges as $b \to \infty$.
The convergence of (B) is shown in Example 19, page 241.

55.   A.

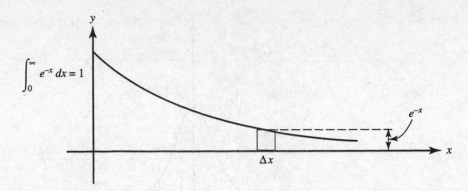

$$\int_0^\infty e^{-x}\, dx = 1$$

56.   C.

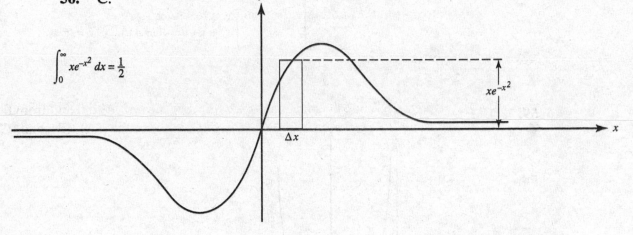

$$\int_0^\infty xe^{-x^2}\, dx = \tfrac{1}{2}$$

57.   E.

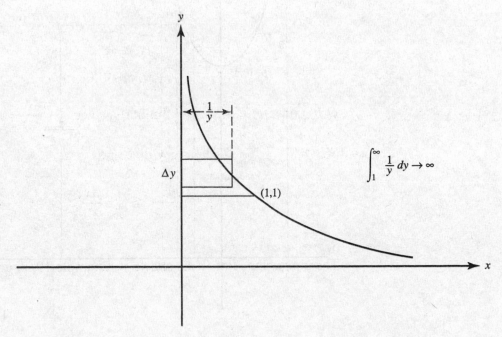

$$\int_1^\infty \frac{1}{y}\, dy \to \infty$$

58. B.

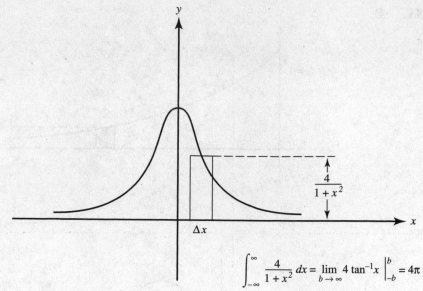

$$\int_{-\infty}^{\infty} \frac{4}{1+x^2}\,dx = \lim_{b \to \infty} 4\tan^{-1}x \,\Big|_{-b}^{b} = 4\pi$$

59. D.

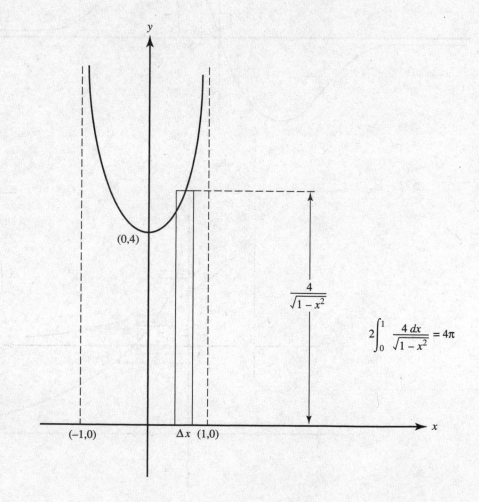

$$2\int_{0}^{1} \frac{4\,dx}{\sqrt{1-x^2}} = 4\pi$$

60.   B.

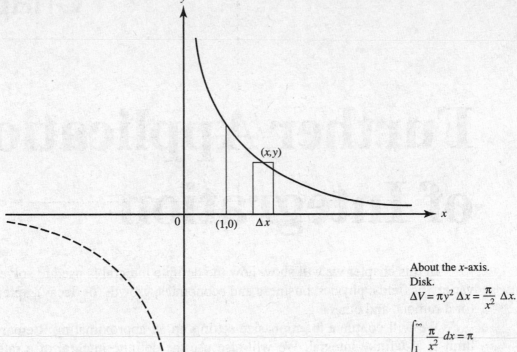

About the x-axis.
Disk.
$\Delta V = \pi y^2\, \Delta x = \dfrac{\pi}{x^2}\, \Delta x.$

$\displaystyle\int_1^\infty \dfrac{\pi}{x^2}\, dx = \pi$

61.   A.

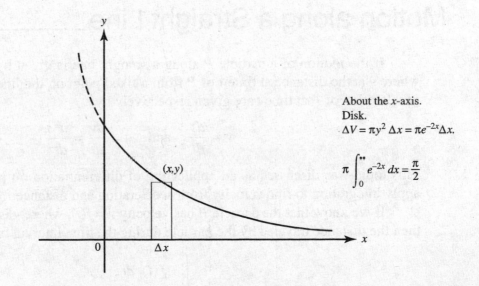

About the x-axis.
Disk.
$\Delta V = \pi y^2\, \Delta x = \pi e^{-2x}\Delta x.$

$\displaystyle \pi\int_0^\infty e^{-2x}\, dx = \dfrac{\pi}{2}$

Chapter 8

Further Applications of Integration

In this chapter we will show how the definite integral is used to solve problems in a variety of fields: physics, business and economics, growth (or decay), spread of a disease (or a rumor), and others.

We will continue to emphasize setting up an approximating Riemann sum, whose limit is a definite integral. We will also use the definite integral of a rate of change to obtain net change.

A. Motion along a Straight Line

If the motion of a particle P along a straight line is given by the equation $s = F(t)$, where s is the distance at time t of P from a fixed point on the line, then the velocity and acceleration of P at time t are given respectively by

$$v = \frac{ds}{dt} \quad \text{and} \quad a = \frac{dv}{dt} = \frac{d^2 s}{dt^2}.$$

This topic was discussed as an application of differentiation on page 113. Here we will apply integration to find velocity from acceleration and distance from velocity.

If we know that the particle P has velocity $v = f(t)$, where v is a continuous function, then the distance traveled by the particle during the time interval from $t = a$ to $t = b$ is

$$\int_a^b |f(t)| \, dt. \tag{1}$$

If $f(t) \geqq 0$ for all t on $[a, b]$ (i.e., P moves only in the positive direction), then (1) is equivalent to $\int_a^b f(t) \, dt$; similarly, if $f(t) \leqq 0$ on $[a, b]$ (P moves only in the negative direction), then (1) yields $- \int_a^b f(t) \, dt$. If $f(t)$ changes sign on $[a, b]$ (i.e., the direction of motion changes), then (1) gives the total distance traveled. Suppose, for example, that the situation is as follows:

$$a \leq t \leq c \qquad f(t) \geq 0;$$
$$c \leq t \leq d \qquad f(t) \leq 0;$$
$$d \leq t \leq b \qquad f(t) \geq 0.$$

Then the total distance traveled during the time interval from $t = a$ to $t = b$ is exactly

$$\int_a^c f(t)\, dt - \int_c^d f(t)\, dt + \int_d^b f(t)\, dt.$$

The displacement or *net change* in the particle's position from $t = a$ to $t = b$ is equal, by the Fundamental Theorem of Calculus (FTC), to

$$\int_a^b v(t)\, dt.$$

Example 1. If a body moves along a straight line with velocity $v = t^3 + 3t^2$, then the distance traveled between $t = 1$ and $t = 4$ is given by

$$\int_1^4 (t^3 + 3t^2)\, dt = \left(\frac{t^4}{4} + t^3 \right) \Bigg|_1^4 = \frac{507}{4}.$$

Note that $v > 0$ for all t on $[1, 4]$.

Example 2. A particle moves along the x-axis so that its velocity at time t is given by $v(t) = 6t^2 - 18t + 12$. (a) Find the total distance covered between $t = 0$ and $t = 4$. (b) Find the displacement of the particle from $t = 0$ to $t = 4$.

(a) Since $v(t) = 6t^2 - 18t + 12 = 6(t - 1)(t - 2)$, we see that:

$$\text{if} \qquad t < 1, \qquad \text{then } v > 0;$$
$$\text{if } 1 < t < 2, \qquad \text{then } v < 0;$$
$$\text{if } 2 < t, \qquad \text{then } v > 0.$$

Thus, the total distance covered between $t = 0$ and $t = 4$ is

$$\int_0^1 v(t)\, dt - \int_1^2 v(t)\, dt + \int_2^4 v(t)\, dt. \tag{2}$$

When we replace $v(t)$ by $6t^2 - 18t + 12$ in (2) and evaluate, we obtain 34 units for the total distance covered between $t = 0$ and $t = 4$. This can be verified by calculating

$$\int_0^4 |v(t)|\, dt$$

keying in

```
fnInt(abs(6X²-18X+12),X,0,4).
```

This example is the same as Example 26 on page 113, in which the required distance is computed by another method.

(b) To find the *displacement* of the particle from $t = 0$ to $t = 4$, we use the FTC, evaluating

$$\int_0^4 v(t)\, dt \qquad \text{or} \qquad \int_0^4 (6t^2 - 18t + 12)\, dt.$$

This yields $(2t^3 - 9t^2 + 12t)\Big|_0^4$ or $128 - 144 + 48 = 32$. This is the net change in position from $t = 0$ to $t = 4$, sometimes referred to as "position shift."

Example 3. The acceleration of an object moving on a line is given at time t by $a = \sin t$; when $t = 0$ the object is at rest. Find the distance s it travels from $t = 0$ to $t = \dfrac{5\pi}{6}$.

Since $a = \dfrac{d^2s}{dt^2} = \dfrac{dv}{dt} = \sin t$, it follows that

$$v(t) = \frac{ds}{dt} = \int \sin t\, dt; \qquad v(t) = -\cos t + C.$$

Also, $v(0) = 0$ yields $C = 1$. Thus $v(t) = 1 - \cos t$; and since $\cos t \le 1$ for all t we see that $v(t) \ge 0$ for all t. Thus, the distance traveled is

$$\int_0^{5\pi/6} (1 - \cos t)\, dt = (t - \sin t)\Big|_0^{5\pi/6} = \frac{5\pi}{6} - \frac{1}{2}.$$

*B.   Motion along a Plane Curve

In Chapter 4, §K2, it was pointed out that, if the motion of a particle P along a curve is given parametrically by the equations $x = x(t)$ and $y = y(t)$, then at time t the position vector $\mathbf{R}$, the velocity vector $\mathbf{v}$, and the acceleration vector $\mathbf{a}$ are:

$$\mathbf{R} = x\mathbf{i} + y\mathbf{j};$$
$$\mathbf{v} = \frac{d\mathbf{R}}{dt} = \frac{dx}{dt}\mathbf{i} + \frac{dy}{dt}\mathbf{j} = \dot{x}\mathbf{i} + \dot{y}\mathbf{j} = v_x\mathbf{i} + v_y\mathbf{j};$$
$$\mathbf{a} = \frac{d^2\mathbf{R}}{dt^2} = \frac{d\mathbf{v}}{dt} = \frac{d^2x}{dt^2}\mathbf{i} + \frac{d^2y}{dt^2}\mathbf{j} = \ddot{x}\mathbf{i} + \ddot{y}\mathbf{j} = a_x\mathbf{i} + a_y\mathbf{j}.$$

The components in the horizontal and vertical directions of $\mathbf{R}$, $\mathbf{v}$, and $\mathbf{a}$ are given respectively by the coefficients of $\mathbf{i}$ and $\mathbf{j}$ in the corresponding vector. The slope of $\mathbf{v}$ is $\dfrac{dy}{dx}$; its magnitude,

$$|\mathbf{v}| = \sqrt{\left(\frac{dx}{dt}\right)^2 + \left(\frac{dy}{dt}\right)^2} = \frac{ds}{dt},$$

is the speed of the particle, and the velocity vector is tangent to the path. The slope of $\mathbf{a}$ is $\dfrac{d^2y}{dt^2} \Big/ \dfrac{d^2x}{dt^2}$; its magnitude is $|\mathbf{a}| = \sqrt{a_x^2 + a_y^2}$.

*An asterisk denotes a topic covered only in Calculus BC.

How integration may be used to solve problems of curvilinear motion is illustrated in the following examples.

Example 4.   The motion of a particle satisfies the equations

$$\frac{d^2x}{dt^2} = 0, \qquad \frac{d^2y}{dt^2} = -g \ .$$

Find parametric equations for the motion if the initial conditions are $x = 0$, $y = 0$, $\frac{dx}{dt} = v_0 \cos \alpha$, and $\frac{dy}{dt} = v_0 \sin \alpha$, where v_0 and α are constants.

We integrate each of the given equations twice and determine the constants as indicated:

$$\frac{dx}{dt} = C_1 = v_0 \cos \alpha \ ; \qquad\qquad \frac{dy}{dt} = -gt + C_2;$$

$$v_0 \sin \alpha = C_2;$$

$$x = (v_0 \cos \alpha)t + C_3; \qquad\qquad y = -\frac{1}{2}gt^2 + (v_0 \sin \alpha)t + C_4;$$

$$x(0) = 0 \text{ yields } C_3 = 0. \qquad\qquad y(0) = 0 \text{ yields } C_4 = 0.$$

Finally, then, we have

$$x = (v_0 \cos \alpha)t; \qquad\qquad y = -\frac{1}{2}gt^2 + (v_0 \sin \alpha)t.$$

These are the equations for the path of a projectile that starts at the origin with initial velocity v_0 and at an angle of elevation α. If desired, t can be eliminated from this pair of equations to yield a parabola in rectangular coordinates.

Example 5.   A particle $P(x, y)$ moves along a curve so that

$$\frac{dx}{dt} = 2\sqrt{x} \qquad \text{and} \qquad \frac{dy}{dt} = \frac{1}{x} \text{ at any time } t \geq 0.$$

At $t = 0$, $x = 1$ and $y = 0$. Find the parametric equations of motion.

Since $\frac{dx}{\sqrt{x}} = 2 \, dt$, we integrate to get $2\sqrt{x} = 2t + C$, and use $x(0) = 1$ to find that $C = 2$. Therefore, $\sqrt{x} = t + 1$ and

$$x = (t + 1)^2. \tag{1}$$

Then $\frac{dy}{dt} = \frac{1}{x} = \frac{1}{(t+1)^2}$ by (1), so $dy = \frac{dt}{(t+1)^2}$ and

$$y = -\frac{1}{t+1} + C' . \tag{2}$$

Since $y(0) = 0$, this yields $C' = 1$, and so (2) becomes

$$y = 1 - \frac{1}{t+1} = \frac{t}{t+1} .$$

Thus the parametric equations are

$$x = (t+1)^2 \qquad \text{and} \qquad y = \frac{t}{t+1}.$$

Example 6. If a particle moves on a curve so that the acceleration vector is always perpendicular to the position vector, prove that the speed of the particle is constant.

The hypothesis is equivalent to the statement

$$\frac{\dfrac{d^2y}{dt^2}}{\dfrac{d^2x}{dt^2}} \cdot \frac{\dfrac{dy}{dt}}{\dfrac{dx}{dt}} = -1$$

or

$$\frac{\ddot{y}}{\ddot{x}} \cdot \frac{\dot{y}}{\dot{x}} = -1. \tag{3}$$

Equation (3) is equivalent to

$$\dot{y}\ddot{y} = -\dot{x}\ddot{x}, \tag{4}$$

and (4) is simply

$$\frac{d}{dt}\left(\frac{1}{2}\dot{y}^2\right) = \frac{d}{dt}\left(-\frac{1}{2}\dot{x}^2\right).$$

Then $\dfrac{1}{2}\,\dot{y}^2 = -\dfrac{1}{2}\,\dot{x}^2 + C'$, or

$$\dot{y}^2 + \dot{x}^2 = C. \tag{5}$$

But since the left-hand member of (5) is precisely $\left(\dfrac{ds}{dt}\right)^2$ or $|\mathbf{v}|^2$, (5) says that the square of the speed (and consequently the speed) is constant.

Example 7. A particle $P(x, y)$ moves along a curve so that its acceleration is given by

$$\mathbf{a} = -4\cos 2t\mathbf{i} - 2\sin t\mathbf{j} \quad \left(-\frac{\pi}{2} \leqq t \leqq \frac{\pi}{2}\right);$$

when $t = 0$, then $x = 1$, $y = 0$, $\dot{x} = 0$, and $\dot{y} = 2$. (a) Find the position vector $\mathbf{R}$ at any time t. (b) Find a Cartesian equation for the path of the particle and identify the conic on which P moves.

(a) $\mathbf{v} = (-2\sin 2t + c_1)\mathbf{i} + (2\cos t + c_2)\mathbf{j}$, and since $\mathbf{v} = 2\mathbf{j}$ when $t = 0$, it follows that $c_1 = c_2 = 0$. So $\mathbf{v} = -2\sin 2t\mathbf{i} + 2\cos t\mathbf{j}$. Also $\mathbf{R} = (\cos 2t + c_3)\mathbf{i} + (2\sin t + c_4)\mathbf{j}$; and since $\mathbf{R} = \mathbf{i}$ when $t = 0$, we see that $c_3 = c_4 = 0$. Finally, then,

$$\mathbf{R} = \cos 2t\mathbf{i} + 2\sin t\mathbf{j}.$$

(b) From (a) the parametric equations of motion are

$$x = \cos 2t, \qquad y = 2 \sin t.$$

By a trigonometric identity,

$$x = 1 - 2 \sin^2 t = 1 - \frac{y^2}{2}.$$

P travels in a counterclockwise direction along *part* of a parabola that has its vertex at (1, 0) and opens to the left. The path of the particle is sketched in Figure N8–1; note that $-1 \leqslant x \leqslant 1, -2 \leqslant y \leqslant 2$.

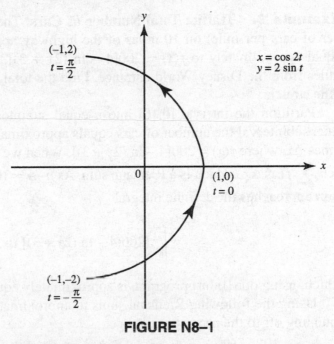

FIGURE N8–1

C. Other Applications of Riemann Sums _____

We will continue to set up Riemann sums to calculate a variety of quantities using definite integrals. In many of these examples we will partition into *n* equal subintervals a given interval (or region or ring or solid or the like), approximate the quantity over each small subinterval (and assume it is constant there), then add up all these small quantities. Finally, as $n \to \infty$, we will replace the sum by its equivalent definite integral to calculate the desired quantity.

Example 8. **Amount of Leaking Water.** Water is draining from a cylindrical pipe of radius 2 inches. At *t* seconds the water is flowing out with velocity $v(t)$ inches per second. Express the amount of water that has drained from the pipe in the first 3 minutes as a definite integral in terms of $v(t)$.

We first express 3 min as 180 sec. We then partition [0,180] into *n* subintervals each of length Δt. In Δt sec, approximately $v(t) \, \Delta t$ in. of water have drained from the pipe. Since a typical cross section has area 4π in.^{2} (Figure N8-2), in Δt sec the amount that has drained is

$$(4\pi \text{ in.}^2)\,(v(t) \text{ in./sec})(\Delta t \text{ sec}) = 4\pi v(t)\,\Delta t \text{ in.}^3.$$

The sum of the n amounts of water that drain from the pipe, as $n \to \infty$, is $\displaystyle\int_0^{180} 4\pi v(t)\,dt$; the units are cubic inches (in.^{3}).

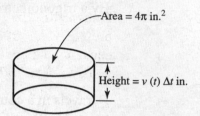

FIGURE N8–2

Example 9.　Traffic: Total Number of Cars. The density of cars (the number of cars per mile) on 10 miles of the highway approaching Disney World is equal approximately to $f(x) = 200[4 - \ln(2x + 3)]$, where x is the distance in miles from the Disney World entrance. Find the total number of cars on this 10-mile stretch.

Partition the interval [0,10] into n equal subintervals each of width Δx. In each subinterval the number of cars equals approximately the density of cars $f(x)$ times Δx, where $f(x) = 200[4 - \ln(2x + 3)]$. When we add n of these products we get $\sum f(x)\Delta x$, which is a Riemann sum. As $n \to \infty$ (or as $\Delta x \to 0$), the Riemann sum approaches the definite integral

$$\int_0^{10} [200(4 - \ln(2x + 3)]\,dx,$$

which, using our [fnInt] program, is approximately equal to 3118 cars.

Using the following Riemann sums to approximate the total number of cars, rounding off to the nearest car, we get

$$L(50) = 3159, \qquad R(50) = 3078, \qquad M(50) = 3118.$$

As usual, the midpoint rule does an excellent job.

Example 10.　Present and Future Value of a Single Payment. If a deposit of $\$P$ is compounded continuously at an annual rate r for t years, then the *future value* of $\$P$ is defined as the amount A after t years, given by

$$A = Pe^{rt}.$$

The *present value* (PV) P of future value A is obtained by solving the equation above for P:

$$P = Ae^{-rt}.$$

The present value, $\$P$, is the amount that must be invested now to produce $\$A$ in t years.

(a) If $\$1000$ is invested at 6% compounded continuously, then after 5 years it has grown to

$$1000e^{(0.06)\cdot 5} = \$1349.86.$$

(b) To determine the amount that must be invested today to accumulate $10,000 in 4 years at 5% compounded continuously, we evaluate P:

$$P = 10{,}000e^{(-0.05) \cdot 4} = \$8187.31.$$

Example 11. Present and Future Value of Several Payments. Find the present value of a lottery jackpot prize of $1,000,000 if it is to be paid in 10 equal annual installments of $100,000 each and 7% interest, compounded continuously, is available.

Note that
- the PV of the first payment is $100{,}000e^{-(0.07) \cdot 0}$,
- the PV of the second payment is $100{,}000e^{-(0.07) \cdot 1}$,
- the PV of the third payment is $100{,}000e^{-(0.07) \cdot 2}$,
....
- the PV of the tenth payment is $100{,}000e^{-(0.07) \cdot 9}$.

The PV of all 10 payments equals the sum of these:

$$100{,}000(e^{-(0.07) \cdot 0} + e^{-(0.07) \cdot 1} + \cdots + e^{-(0.07) \cdot 9}).$$

Keying in `100,000 sum seq (e`$^{-(0.07)N}$`,N,0,9,1)` on our calculator, we get $744,628—quite a bit less than the inviting million-dollar prize advertised.

Example 12. Present and Future Value of a Continuous Income Stream. Economists often visualize future income as continuous, which is realistic for large businesses, since payments arrive virtually all the time. To find the PV of an income stream (that is, the amount needed for investment now to deliver the amounts in the [future] income stream), we let

$P(t)$ be the income that arrives at time t (with $t = 0$ denoting the time now);
r be the annual interest rate, compounded continuously;
a and b be the times when the stream starts and finishes, respectively.

Then, as in our already well-established pattern, we partition the interval from a to b into n equal subintervals, approximate the PV of income to arrive in an interval of duration Δt by $P(t)\,\Delta t \cdot e^{-rt}$, sum all n small present values, getting the Riemann sum $\sum P(t)e^{-rt}\Delta t$. Then, as $n \to \infty$ (or as $\Delta t \to 0$), we obtain

$$\begin{matrix} \text{Present Value} \\ \text{of} \\ \text{Income Stream} \end{matrix} = \int_a^b P(t)e^{-rt}\,dt.$$

Example 13. Continuous Income Stream; Another Example. Suppose a company offers a 65-year-old a choice of retirement plans: (1) $20,000 paid continuously for 10 years; (2) $15,000 paid continuously for 15 years; (3) $150,000 paid now in a lump sum. If money is worth 6% compounded continuously, which option is best for her?

(1) $\text{PV} = \displaystyle\int_0^{10} 20{,}000e^{-0.06t}\,dt \approx \$150{,}396$;

(2) $\text{PV} = \displaystyle\int_0^{15} 15{,}000e^{-0.06t}\,dt \approx \$148{,}358.$

(3) $\text{PV} = \$150{,}000.$

Thus, plan (1) is best.

Example 14.   Continuous Income Stream: Third Example. A manufacturer estimates that his latest product will earn $15,000 a year. He further estimates that the income from it will be received continuously, at a constant rate, over the next 8 years. If the interest rate is 7%, compounded continuously, what is the present value of the anticipated future earnings?

We have

$$PV = \int_0^8 15{,}000 e^{-0.07t} \, dt \approx \$91{,}884.$$

Example 15.   Resource Depletion. In 1990 the yearly world petroleum consumption was about 22 billion barrels and the yearly exponential rate of increase in use was 5%. How many years after 1990 are the world's total estimated oil reserves of 900 billion barrels likely to last?

Given the yearly consumption in 1990 and the projected exponential rate of increase in consumption, the anticipated consumption during the Δtth part of a year (after 1990) is $22 e^{0.05t} \Delta t$ billion barrels. The total to be used during the following N years is therefore $\int_0^N 22 e^{0.05t} \, dt = 900$ billion barrels.

We must now solve this equation for N. We get

$$440 e^{0.05t} \Big|_0^N = 900,$$

$$440(e^{0.05N} - 1) = 900,$$

$$e^{0.05N} - 1 = \frac{900}{440},$$

$$e^{0.05N} = 1 + \frac{90}{44},$$

$$0.05N = \ln\left(1 + \frac{99}{44}\right),$$

$$N = \frac{1}{0.05} \ln\left(1 + \frac{90}{44}\right) \approx 22.3 \, \text{yr}.$$

Either more oil (or alternative sources of energy) must be found, or the world consumption must be sharply reduced.

D. FTC: Definite Integral of a Rate Is Net Change

If f is continuous and $f(t) = \dfrac{dF}{dt}$, then we know from the FTC that

$$\int_a^b f(t)\, dt = F(b) - F(a).$$

The definite integral of the rate of change of a quantity over an interval is the *net change* or *net accumulation* of the quantity over that interval. Thus, $F(b) - F(a)$ is the net change in $F(t)$ as t varies from a to b.

We've already illustrated this principle many times. Here are more examples.

Example 16. Let $G(t)$ be the rate of growth of a population at time t. Then the increase in population between times $t = a$ and $t = b$ is given by $\int_a^b G(t)\, dt$. The population may consist of people, deer, fruit flies, bacteria, and so on.

Example 17. Suppose a rumor is spreading at the rate of $f(t) = 100e^{-0.2t}$ new people per day. Then the number of people who hear the rumor during the 5th and 6th days is

$$\int_4^6 100e^{-0.2t}\, dt$$

or 74 people. If we let $F'(t) = f(t)$, then the integral above is the net change in $F(t)$ from $t = 4$ to $t = 6$, or the number of people who hear the rumor from the beginning of the 5th day to the end of the 6th.

Example 18. Economists define the *marginal cost of production* as the additional cost of producing one additional unit at a specified production level. It can be shown that if $C(x)$ is the cost at production level x then $C'(x)$ is the marginal cost at that production level.

If the marginal cost, in dollars, is $\dfrac{1}{x}$ per unit when x units are being produced, then the change in cost when production increases from 50 to 75 units is

$$\int_{50}^{75} \frac{1}{x}\, dx \approx \$0.41.$$

We replace "cost" above by "revenue" or "profit" to find total change in these quantities.

Example 19. If after t minutes a chemical is decomposing at the rate of $10e^{-t}$ grams per minute, then during the first 3 minutes the amount that has decomposed is

$$\int_0^3 10e^{-t}\, dt \approx 9.5 \text{ g}.$$

Example 20.   An official of the Environmental Protection Agency estimates that t years from now the level of a particular pollutant in the air will be increasing at the rate of $(0.3 + 0.4t)$ parts per million per year (ppm/yr). According to this estimate, the change in the pollutant level during the second year will be

$$\int_1^2 (0.3 + 0.4t)\, dt \simeq 0.9 \text{ ppm.}$$

Example 21.   Suppose an epidemic is spreading through a city at the rate of $f(t)$ new people per week. Then

$$\int_0^4 f(t)\, dt$$

is the number of people who will become infected during the next 4 weeks (or the total change in the number of infected people).

Work[†]

Work is defined as force times distance: $W = F \times d$. When a variable force $F(x)$ moves an object along the x-axis from a to b, we approximate an element of work done by the force over a short distance Δx by

$$\Delta W = F(x_k)\, \Delta x,$$

where $F(x_k)$ is the force acting at some point in the kth subinterval. We then use the FTC to get

$$W = \lim_{n \to \infty} \sum_{k=0}^n F(x_k)\, \Delta x = \int_a^b F(x)\, dx.$$

If the force is given in pounds and the distance in feet, then the work is given in foot-pounds (ft-lb). Problems typical of those involving computation of work are given in the following examples.

Example 22.   Find the work done by a force F, in pounds, which moves a particle along the x-axis from $x = 4$ feet to $x = 9$ feet, if $F(x) = \dfrac{1}{2\sqrt{x}}$.

For the work, W, we have

$$W = \int_4^9 \frac{dx}{2\sqrt{x}} = \sqrt{x}\ \Big|_4^9 = 3 - 2 = 1 \text{ ft-lb.}$$

[†]The topic work is not included in the Topical Outline but is an important application of integration.

Example 23.   A cylindrical reservoir of diameter 4 feet and height 6 feet is half-full of water weighing w pounds per cubic foot (Figure N8–3). Find the work done in emptying the water over the top.

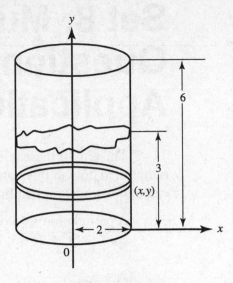

The volume of a slice of water is $\Delta V = \pi x^2 \, \Delta y$, where $x = 2$. A slice at height y is lifted $(6 - y)$ ft.

$$\Delta W = w \cdot \pi \cdot 4 \Delta y (6 - y);$$

$$W = 4\pi w \int_0^3 (6 - y) \, dy = 54\pi w \text{ ft-lb.}$$

We used 3 as the upper limit since the reservoir is only half full.

FIGURE N8–3

Example 24.   A hemispherical tank with flat side up has radius 4 feet and is filled with a liquid weighing w pounds per cubic foot. Find the work done in pumping all the liquid just to the top of the tank.

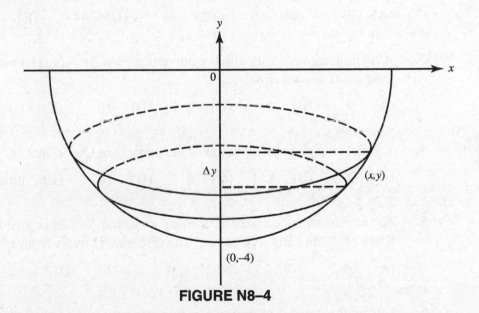

FIGURE N8–4

In Figure N8–4, the generating circle has equation $x^2 + y^2 = 16$. Note that over the interval of integration y is negative, and that a slice must be lifted a distance of $(-y)$ feet. Then for the work, W, we have

$$W = \pi w \int_{-4}^0 (-y)x^2 \, dy = -\pi w \int_{-4}^0 y(16 - y^2) \, dy = 64\pi w \text{ ft-lb.}$$

Set 8: Multiple-Choice Questions on Further Applications of Integration

The aim of these questions is mainly to reinforce how to set up definite integrals, rather than how to integrate or evaluate them. Therefore we encourage using a graphing calculator wherever helpful.

1. A particle moves along a line in such a way that its position at time t is given by $s = t^3 - 6t^2 + 9t + 3$. Its direction of motion changes when

    (A) $t = 1$ only   **(B)** $t = 2$ only   **(C)** $t = 3$ only
    (D) $t = 1$ and $t = 3$ **(E)** $t = 1, 2,$ and 3

2. A body moves along a straight line so that its velocity v at time t is given by $v = 4t^3 + 3t^2 + 5$. The distance the body covers from $t = 0$ to $t = 2$ equals

    (A)   34   **(B)**   55   **(C)**   24   **(D)**   44   **(E)**   none of these

3. A particle moves along a line with velocity $v = 3t^2 - 6t$. The total distance traveled from $t = 0$ to $t = 3$ equals

    (A)   9   **(B)**   4   **(C)**   2   **(D)**   16   **(E)**   none of these

4. The net change in the position of the particle in Question 3 is

    (A)   2   **(B)**   4   **(C)**   9   **(D)**   16   **(E)**   none of these

5. The acceleration of a particle moving on a straight line is given by $a = \cos t$, and when $t = 0$ the particle is at rest. The distance it covers from $t = 0$ to $t = 2$ is

    (A) $\sin 2$ **(B)** $1 - \cos 2$ **(C)** $\cos 2$ **(D)** $\sin 2 - 1$ **(E)** $-\cos 2$

6. During the worst 4-hr period of a hurricane the wind velocity, in miles per hour, is given by $v(t) = 5t - t^2 + 100$, $0 \leqslant t \leqslant 4$. The average wind velocity during this period (in mph) is

    (A)   10   **(B)**   100   **(C)**   102   **(D)** $104\frac{2}{3}$ **(E)** $108\frac{2}{3}$

7. A car accelerates from 0 to 60 mph in 10 sec, with constant acceleration. (Note that 60 mph = 88 ft/sec). The acceleration (in ft/sec^2) is

    (A)   5.3   **(B)**   6   **(C)**   8   **(D)**   8.8   **(E)**   none of these

For Questions 8–10 use the following information: The velocity **v** of a particle moving on a curve is given, at time t, by $\mathbf{v} = t\mathbf{i} - (1 - t)\mathbf{j}$. When $t = 0$, the particle is at point $(0,1)$.

*8. At time t the position vector **R** is

(A) $\dfrac{t^2}{2}\mathbf{i} - \dfrac{(1 - t^2)}{2}\mathbf{j}$ (B) $\dfrac{t^2}{2}\mathbf{i} + \dfrac{(1 - t)^2}{2}\mathbf{j}$

(C) $\dfrac{t^2}{2}\mathbf{i} + \dfrac{t^2 - 2t}{2}\mathbf{j}$ (D) $\dfrac{t^2}{2}\mathbf{i} + \dfrac{t^2 - 2t + 2}{2}\mathbf{j}$

(E) $\dfrac{t^2}{2}\mathbf{i} + (1 - t)^2\mathbf{j}$

*9. The acceleration vector at time $t = 2$ is

(A) $\mathbf{i} + \mathbf{j}$ (B) $\mathbf{i} - \mathbf{j}$ (C) $\mathbf{i} + 2\mathbf{j}$ (D) $2\mathbf{i} - \mathbf{j}$ (E) none of these

*10. The speed of the particle is a minimum when t equals

(A) 0   (B) $\dfrac{1}{2}$ (C) 1   (D) 1.5   (E) 2

*11. A particle moves along a curve in such a way that its position vector and velocity vector are perpendicular at all times. If the particle passes through the point $(4,3)$, then the equation of the curve is

(A) $x^2 + y^2 = 5$ (B) $x^2 + y^2 = 25$ (C) $x^2 + 2y^2 = 34$

(D) $x^2 - y^2 = 7$ (E) $2x^2 - y^2 = 23$

*12. The acceleration of a particle is given by the vector $\mathbf{a} = e^{-t}\mathbf{i} + e^t\mathbf{j}$. When $t = 0$, the particle has velocity $\mathbf{v} = 2\mathbf{i}$. Then the velocity vector at any time t is

(A) $(3 - e^{-t})\mathbf{i}$ (B) $(1 - e^{-t})\mathbf{i} + (e^t - 1)\mathbf{j}$ (C) $(2 - e^{-t})\mathbf{i} + e^t\mathbf{j}$

(D) $(2 - e^{-t})\mathbf{i} + (e^t - 1)\mathbf{j}$ (E) $(3 - e^{-t})\mathbf{i} + (e^t - 1)\mathbf{j}$

*13. The velocity of a particle is given by $(3 - e^{-t})\mathbf{i} + (e^t - 1)\mathbf{j}$. When $t = 0$, the particle is at $(1,2)$. The position vector **R** at time t is

(A) $(3t + e^{-t})\mathbf{i} + (e^t - t)\mathbf{j}$ (B) $(3t + e^{-t} + 1)\mathbf{i} + (e - t + 1)\mathbf{j}$

(C) $(3t + e^{-t})\mathbf{i} + (e^t - t + 1)\mathbf{j}$ (D) $(3t + e^{-t} + 1)\mathbf{i} + (e^t - t + 2)\mathbf{j}$

(E) none of these

14. Suppose the current world population is 6 billion and the population t yr from now is estimated to be $P(t) = 6e^{0.024t}$. Based on this supposition, the average population of the world, in billions, over the next 25 yr will be approximately

(A) 6.75   (B) 7.2   (C) 7.8   (D) 8.2   (E) 9.0

*An asterisk denotes a topic covered only in Calculus BC.

15. If a quantity $Q(t)$ is growing at the rate of 5% per year and Q now equals Q_0, then in t yr Q will equal

(A) $Q_0(1.05t)$ (B) $Q_0(1.05)^t$ (C) $Q_0e^{0.05t}$
(D) $Q_0e^{0.045t}$ (E) none of these

16. A growth rate of 3% per year is equivalent to a continuous growth rate of

(A) 3%   (B) 2.99%   (C) 2.98%   (D) 2.97%   (E) 2.96%

17. A stone is thrown upward from the ground with an initial velocity of 96 ft/sec. Its average velocity (given that $a(t) = -32$ ft/sec^2) during the first 2 sec is

(A) 16 ft/sec   (B) 32 ft/sec   (C) 64 ft/sec
(D) 80 ft/sec   (E) 96 ft/sec

18. Suppose the amount of a drug in a patient's bloodstream t hr after intravenous administration is $30/(t + 1)^2$ mg. The average amount in the bloodstream during the first 4 hr is

(A) 6.0 mg   (B) 11.0 mg   (C) 16.6 mg
(D) 24.0 mg   (E) none of these

19. A rumor spreads through a town at the rate of $(t^2 + 10t)$ new people per day. Approximately how many people hear the rumor during the second week after it was first heard?

(A) 1535   (B) 1894   (C) 2000
(D) 2219   (E) none of these

20. The present value of a continuous income stream of $1000 per year over a period of 15 yr, at an interest rate of 8% compounded continuously, is approximately equal to

(A) $3320.99   (B) $6740.29   (C) $8735.07
(D) $6740.29   (E) none of these

21. An author is setting up a fund that will pay her son $1000 per year. How much must she put into the fund if it is to last for the next 10 yr? Assume that money is to be paid out continuously, at a constant rate, and that the interest rate is 5% compounded continuously.

(A) $1649   (B) $2719   (C) $6065   (D) $7538   (E) $7869

22. The future value of a continuous income stream of $2000 per year for a period of 10 yr, at an annual interest rate of 5% compounded continuously, is approximately

(A) $32,974   (B) $25,949   (C) $21,025
(D) $15,739   (E) none of these

23.   A violin that sold for $300 in 1930 appreciated to $7000 in 1990. If the appreciation was at a continuous and constant rate, then that rate equals approximately

(A)   5.25%   (B)   5.73%   (C)   6.08%

(D)   6.30%   (E)   none of these

24.   Oil is leaking from a tanker at the rate of $1000\,e^{-0.3t}$ gal/hr, where t is given in hours. A general Riemann sum for the amount of oil that leaks out in the next 8 hr, where the interval [0,8] has been partitioned into n subintervals, is

(A) $\displaystyle\sum_{k=1}^{n} e^{-0.3t_k}\,\Delta t$ (B) $\displaystyle\lim_{n\to\infty}\sum_{k=1}^{n} e^{-0.3t_k}\,\Delta t$

(C) $\displaystyle\lim_{n\to\infty}\sum_{k=1}^{n} 1000e^{-0.3t_k}\,\Delta t$ (D) $\displaystyle\sum_{k=1}^{n} 1000e^{-0.3t_k}\,\Delta t$

(E) $\displaystyle 1000\sum_{k=1}^{n} e^{0.3t_k}\,\Delta t$

25.   In Question 24, the total number of gallons of oil that will leak out during the next 8 hr is approximately

(A)   1271   (B)   3031   (C)   3161   (D)   4323   (E)   11,023

26.   Assume that the density of vehicles (number per mile) during morning rush hour, for the 20-mi stretch along the New York State Thruway southbound from the Tappan Zee Bridge, is given by $f(x)$, where x is the distance, in miles, south of the bridge. Which of the following gives the number of vehicles (on this 20-mi stretch) from the bridge to a point x mi south of the bridge?

(A) $\displaystyle\int_0^x f(t)\,dt$ (B) $\displaystyle\int_x^{20} f(t)\,dt$ (C) $\displaystyle\int_0^{20} f(x)\,dx$

(D) $\displaystyle\sum_{k=1}^{n} f(x_k)\Delta x$ (where the 20-mi stretch has been partitioned into n equal subintervals)

(E)   none of these

27.   The center of a city that we will assume is circular is on a straight highway. The radius of the city is 3 mi. The density of the population, in thousands of people per square mile, is given approximately by $f(r) = 12 - 2r$ at a distance r mi from the highway. The population of the city is given by the integral

(A) $\displaystyle\int_0^3 (12 - 2r)\,dr$ (B) $\displaystyle 2\int_0^3 (12 - 2r)\sqrt{9 - r^2}\,dr$

(C) $\displaystyle 4\int_0^3 (12 - 2r)\sqrt{9 - r^2}\,dr$ (D) $\displaystyle\int_0^3 2\pi r(12 - 2r)\,dr$

(E) $\displaystyle 2\int_0^3 2\pi r(12 - 2r)\,dr$

28. The population density of Winnipeg, which is located in the middle of the Canadian prairie, drops dramatically as distance from the center of town increases. This is shown in the following table:

x = distance (in mi) from the center	0	2	4	6	8	10
$f(x)$ = density (hundreds of people/mi²)	50	45	40	30	15	5

The population living within a 10-mi radius of the center, using a Riemann sum, is approximately

(A) 608,500   (B) 650,000   (C) 691,200
(D) 702,000   (E) 850,000

29. If a factory continuously dumps pollutants into a river at the rate of $\dfrac{\sqrt{t}}{180}$ tons per day, then the amount dumped after 7 weeks is approximately

(A) 0.07 ton   (B) 0.90 ton   (C) 1.55 tons
(D) 1.9 tons   (E) 1.27 tons

30. A roast at 160°F is put into a refrigerator whose temperature is 45°F. The temperature of the roast is cooling at time t at the rate of $(-9e^{-0.08t})$°F per minute. The temperature, to the nearest degree F, of the roast 20 min after it is put in the refrigerator is

(A) 45°   (B) 70°   (C) 81°   (D) 90°   (E) 115°

31. How long will it take to release 9 tons of pollutant if the rate at which pollutant is being released is $te^{-0.3t}$ tons per week?

(A) 10.2 weeks   (B) 11.0 weeks   (C) 12.1 weeks
(D) 12.9 weeks   (E) none of these

32. If you deposit $1000 today, it will grow at the rate of $80e^{0.08t}$ dollars per year. In 6 years it will be worth (in dollars)

(A) 616.07   (B) 1129.29   (C) 1292.86
(D) 1616.07   (E) 6160.74

33. Which of the following is a Riemann sum for payment of a continuous income stream over the next year of $100 per year at 9% compounded continuously? (The 1-yr interval is partitioned into n equal subintervals.)

(A) $\displaystyle\sum_{k=1}^{n} 100e^{-0.09t_k}\,\Delta t$ (B) $\displaystyle\sum_{k=1}^{n} 100e^{-0.09}t_k\Delta t$

(C) $\displaystyle\sum_{k=1}^{n} \frac{100}{12}e^{-0.09t_k}\,\Delta t$ (D) $\displaystyle\lim_{n\to\infty}\sum_{k=1}^{n} 100e^{-0.09t_k}\,\Delta t$

(E) none of these

34. The average area of all circles with radii between 2 and 5 in. is

    (A) 7π **(B)** 11π **(C)** 13π **(D)** $\dfrac{29\pi}{2}$ **(E)** 17π

35. What is the exact total area bounded by the curve $f(x) = x^3 - 4x^2 + 3x$ and the x-axis?

    (A) -2.25 **(B)** 2.25 **(C)** 3 **(D)** $3\dfrac{1}{12}$ **(E)** none of these

36. Water is leaking from a tank at the rate of $(-0.1t^2 - 0.3t + 2)$ gal/hr. The total amount, in gallons, that will leak out in the next 3 hr is approximately

    (A) 1.00 **(B)** 2.08 **(C)** 3.13 **(D)** 3.48 **(E)** 3.75

37. A bacterial culture is growing at the rate of $1000e^{0.03t}$ bacteria in t hr. The total increase in bacterial population during the second hour is approximately

    (A) 46 **(B)** 956 **(C)** 1046 **(D)** 1061 **(E)** 2046

38. An 18-wheeler traveling at speed v mph gets about $(4 + 0.01v)$ mpg (miles per gallon) of diesel fuel. If its speed is $80\dfrac{t+1}{t+2}$ mph at time t, then the amount, in gallons, of diesel fuel used during the first 2 hr is approximately

    (A) 20 **(B)** 21.5 **(C)** 23.1 **(D)** 24 **(E)** 25

Answers for Set 8: Further Applications of Integration

1.	D	9.	A	17.	C	25.	B	33.	A
2.	A	10.	B	18.	A	26.	A	34.	C
3.	E	11.	B	19.	A	27.	C	35.	D
4.	E	12.	E	20.	C	28.	C	36.	E
5.	B	13.	C	21.	E	29.	E	37.	C
6.	D	14.	D	22.	B	30.	B	38.	C
7.	D	15.	B	23.	A	31.	A		
8.	D	16.	E	24.	D	32.	D		

1. D. Velocity $v(t) = \dfrac{ds}{dt} = 3(t-1)(t-3)$, and changes sign both when $t = 1$ and when $t = 3$.

2. A. Since $v > 0$ for $0 \leq t \leq 2$, the distance is equal to $\displaystyle\int_0^2 (4t^3 + 3t^2 + 5)\,dt$.

3. E. The answer is 8. Since the particle reverses direction when $t = 2$, and $v > 0$ for $t > 2$ but $v < 0$ for $t < 2$, therefore, the total distance is

$$-\int_0^2 (3t^2 - 6t)\,dt + \int_2^3 (3t^2 - 6t)\,dt.$$

4. E. $\displaystyle\int_0^3 (3t^2 - 6t)\,dt = 0$, so there is no change in position.

5. B. Since $v = \sin t$ is positive on $0 < t \leq 2$, the distance covered is

$$\int_0^2 \sin t\,dt = 1 - \cos 2.$$

6. D. Average velocity $= \dfrac{1}{4-0}\displaystyle\int_0^4 (5t - t^2 + 100)\,dt = 104\,\dfrac{2}{3}$ mph.

7. D. The velocity v of the car is linear since its acceleration is constant:

$$a = \frac{dv}{dt} = \frac{(60 - 0)\,\text{mph}}{10\,\text{sec}} = \frac{88\,\text{ft/sec}}{10\,\text{sec}} = 8.8\,\text{ft/sec}^2$$

8. D. $\mathbf{R} = \dfrac{t^2}{2}\mathbf{i} + \dfrac{t^2 - 2t + 2}{2}\mathbf{j}$.

9. A. $\mathbf{a} = \mathbf{i} + \mathbf{j}$ for all t.

10. B. $\mathbf{v} = t\mathbf{i} + (t-1)\mathbf{j}$. $|\mathbf{v}| = \sqrt{t^2 + (t-1)^2}$. $\dfrac{d|\mathbf{v}|}{dt} = \dfrac{2t-1}{|\mathbf{v}|}$.

11. B. Since $\mathbf{R} = x\mathbf{i} + y\mathbf{j}$, its slope is $\frac{y}{x}$; since $\mathbf{v} = \frac{dx}{dt}\mathbf{i} + \frac{dy}{dt}\mathbf{j}$, its slope is $\frac{dy}{dx}$.

If $\mathbf{R}$ is perpendicular to $\mathbf{v}$, then $\frac{y}{x} \cdot \frac{dy}{dx} = -1$, so

$$\frac{y^2}{2} = -\frac{x^2}{2} + C \quad \text{and} \quad x^2 + y^2 = k \ (k > 0).$$

Since (4,3) is on the curve, its equation must be

$$x^2 + y^2 = 25.$$

12. E. When $t = 0$, $\mathbf{v} = 2\mathbf{i} + 0\mathbf{j}$. Since $\mathbf{a} = e^{-t}\mathbf{i} + e^t\mathbf{j}$, therefore $\mathbf{v} = (-e^{-t} + c_1)\mathbf{i} + (e^t + c_2)\mathbf{j}$ at any time t. Now, $-e^{-t} + c_1$ must equal 2 when $t = 0$, so $c_1 = 3$; and $e^t + c_2$ must equal 0 when $t = 0$, so $c_2 = -1$.

13. C. When $t = 0$, $\mathbf{R} = \mathbf{i} + 2\mathbf{j}$. Since $\mathbf{v} = (3 - e^{-t})\mathbf{i} + (e^t - 1)\mathbf{j}$, $\mathbf{R} = (3t + e^{-t} + c_1)\mathbf{i} + (e^t - t + c_2)\mathbf{j}$ at any time t. Since $3t + e^{-t} + c_1$ must equal 1 when $t = 0$, it follows that $c_1 = 0$; since $e^t - t + c_2$ must equal 2 when $t = 0$, it follows that $c_2 = 1$.

14. D. $\frac{1}{25 - 0} \int_0^{25} P(t)\ dt = \frac{1}{25} \int_0^{25} 6e^{0.024t} \simeq 8.2$ (billion people).

15. B. "A quantity growing at the rate of 5% per year" means that after 1 yr the quantity Q_0 will have grown to $1.05Q_0$, after 2 yr to $1.05(1.05Q_0)$, ..., after t yr to $(1.05)^t \cdot Q_0$.

16. E. We need to solve the equation $(1.03)^t = e^{kt}$ for k. We have $1.03 = e^k$, or $k = \ln 1.03 \simeq 0.0296 = 2.96\%$.

17. C. Average velocity $= \frac{1}{2 - 0} \int_0^2 (-32t + 96)\ dt = 64$ ft/sec.

18. A. Average volume $= \frac{1}{4 - 0} \int_0^4 \frac{30}{(t + 1)^2}\ dt \simeq 6$ (mg).

19. A. The number of new people who hear the rumor during the second week is

$$\int_7^{14} (t^2 + 10t)\ dt \simeq 1535.$$

Be careful with the units! The answer is the total change, of course, in $F(t)$ from $t = 7$ to $t = 14$ days, where $F'(t) = t^2 + 10t$.

20. C. $\text{PV} = \int_0^{15} 1000e^{-0.08t}\ dt \simeq \$8735.07$.

21. E. $\text{PV} = 1000 \int_0^{10} e^{-0.05t}\ dt \simeq \$7869$.

22. B. $\text{PV} = \int_0^{10} 2000e^{-0.05t}\ dt \simeq \$15,738.77$. The future value of this amount is $\text{PV} \cdot e^{(0.05 \cdot 10)}$, since this is the amount that the PV earns in 10 yr at an annual rate of 5% compounded continuously. The answer is therefore approximately \$25,949. Verify that the correct answer is also obtained by evaluating $\int_0^{10} 2000e^{0.05t}\ dt$.

23.   A.   Let the rate of appreciation be $r\%$. Then

$$7000 = 300e^{0.01r \cdot 60},$$

$$\frac{70}{3} = e^{0.01r \cdot 60},$$

$$100 \cdot \frac{\ln(70/3)}{60} = r \simeq 5.25\%.$$

Note that integration is not called for here!

24.   D.   After partitioning $[0,8]$ into n subintervals, we approximate the amount of oil leaking in a typical subinterval of duration Δt: it is

$$1000e^{-0.3t_k}\Delta t \text{ gal/hr} \times \text{ hr.}$$

For the general Riemann sum we can choose any point in a subinterval, not necessarily the leftmost or rightmost; we've denoted this arbitrary choice by the subscript k. This sum now adds all n of the small quantities of oil (in gallons).

25.   B.   Total gallons $= \displaystyle\lim_{n\to\infty}\sum_{k=1}^{n}1000e^{-0.3t_k}\Delta t = \int_0^8 1000e^{-0.3t}\,dt.$

26.   A.   Be careful! The number of cars is to be measured over a distance of x (not 20) mi. The answer to the question is a function, not a number. Note that choice (C) gives the total number of cars on the entire 20-mi stretch.

27.   C.   Since the strip of the city shown in the figure is at a distance r mi from the highway, it is $2\sqrt{9-r^2}$ mi long and its area is $2\sqrt{9-r^2}\,\Delta r$. The strip's population is approximately $2(12-2r)\sqrt{9-r^2}\,\Delta r$. The total population of the entire city is *twice* the integral $2\displaystyle\int_0^3 (12-2r)\sqrt{9-r^2}\,dr$ as it includes both halves of the city.

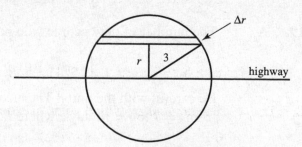

28.   C.

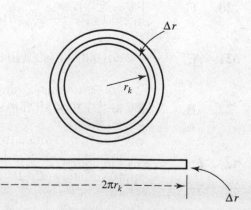

The population equals Σ (area $\cdot$ density). We partition the interval [0,10] along a radius from the center of town into 5 equal subintervals each of width $\Delta r = 2$ mi. We will divide Winnipeg into 5 rings. Each has area equal to (circumference $\times$ width), so the area is $2\pi r_k \, \Delta r$ or $4\pi r_k$. The population in the ring is

$$(4\pi r_k) \cdot (\text{density at } r_k) = 4\pi r_k \cdot f(r_k).$$

A Riemann sum, using left-hand values, is $4\pi \cdot 0 \cdot 50 + 4\pi \cdot 2 \cdot 45 + 4\pi \cdot 4 \cdot 40 + 4\pi \cdot 6 \cdot 30 + 4\pi \cdot 8 \cdot 15 = 4\pi(90 + 160 + 180 + 120) \approx 6912$ hundred people—or about 691,200 people.

29. E. The total amount dumped after 7 weeks is

$$\int_0^{49} \frac{\sqrt{t}}{180} \, dt \approx 1.27 \text{ tons}.$$

30. B. The total change in temperature of the roast 20 min after it is put in the refrigerator is

$$\int_0^{20} -9e^{-0.08t} \, dt \quad \text{or} \quad -89.7°\text{F}.$$

Since its temperature was 160°F when placed in the refrigerator, then 20 min later it is (160 − 89.7)°F or about 70°F. Note that the temperature of the refrigerator (45°F) is not used in answering the question because it is already "built into" the cooling rate.

31. A. Let T be the number of weeks required to release 9 tons. We can use parts to integrate $\int_0^T te^{0.3t} \, dt$, then substitute the limits. We must then set the resulting expression equal to 9 and solve for T. A faster, less painful alternative is to use a graphing calculator. We solved

```
0 = fnInt (Xe^(-0.3X), X, 0, T) - 9
```

with initial guess $T = 15$. The answer is about 10.2 weeks.

32. D. The *amount* by which your investment of $1000 today will *increase* in 6 yr is $\int_0^6 80e^{0.08t} \, dt$ or $616.07, so it will be worth $(1000 + 616.07)$.

Note: $1000e^{0.08(6)} = \$1616.07$.

33. A. In a typical subinterval of time Δt you receive $100\left(\frac{1}{n}\right)$ or $100 \, \Delta t$ dollars.

The present value of this payment is approximately $(100 \, \Delta t)e^{-0.09t_k}$ dollars. The total PV of all these PV's is approximated by the Riemann sum given as choice (A).

34. C. Average area $= \dfrac{1}{5-2} \displaystyle\int_2^5 \pi x^2 \, dx = \dfrac{\pi}{3}\left(\dfrac{125 - 8}{3}\right) = 13\pi.$

35.   D.   Using our calculator, we keyed in `fnInt(absY₁,X,0,3)` and got 3.08333 . . . , which equals $3\frac{1}{12}$. Note that the curve is above the x-axis on [0,1], but below on [1,3], and that the areas for $x < 0$ and $x > 3$ are unbounded.

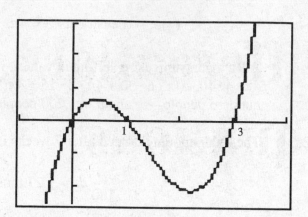

36.   E.   The FTC yields total change:

$$\int_0^3 (-0.1t^2 - 0.3t + 2)\, dt \simeq 3.75 \text{ (gal).}$$

37.   C.   The total change (increase) in population during the second hour is given by $\int_1^2 1000e^{0.03t}\, dt$.

38.   C.   We partition [0,2] into n equal subintervals each of time Δt hr. Since the 18-wheeler gets $(4 + 0.01v)$ mi/gal of diesel, it uses $\dfrac{1}{4 + 0.01v}$ gal/mi.

Since it covers $v \cdot \Delta t$ mi during Δt hr, it uses $\dfrac{1}{4 + 0.01v} \cdot v \cdot \Delta t$ gal in Δt hr.

Since $v = 80\,\dfrac{t+1}{t+2}$, we see that the diesel fuel used in the first 2 hr is

$$\int_0^2 \frac{80 \cdot \dfrac{t+1}{t+2}}{4 + (0.01) \cdot \left(80 \cdot \dfrac{t+1}{t+2}\right)}\, dt = \int_0^2 \frac{80(t+1)}{4.8t + 8.8}\, dt \simeq 23.1 \text{ gal.}$$

Chapter 9

Differential Equations

Review of Definitions and Methods

A.   Basic Definitions _____

A *differential equation* (d.e.) is any equation involving a derivative. In §F of Chapter 5 we solved some simple differential equations. In Example 54, page 164, we were given the velocity at time t of a particle moving along the x-axis:

$$v(t) = \frac{dx}{dt} = 4t^3 - 3t^2 . \tag{1}$$

From this we found the antiderivative:

$$x(t) = t^4 - t^3 + C. \tag{2}$$

If the initial position (at time $t = 0$) of the particle is $x = 3$, then

$$x(0) = 0 - 0 + C = 3,$$

and $C = 3$. So the solution to the initial-value problem is

$$x(t) = t^4 - t^3 + 3. \tag{3}$$

A *solution* of a d.e. is any function that satisfies it. We see from (2) above that the d.e. (1) has an infinite number of solutions—one for each real value of C. We call the family of functions (2) the *general solution* of the d.e. (1). With the given initial condition $x(0) = 3$, we determined C, thus finding the unique solution (3). This is called the *particular solution.*

In §A of Chapter 8 we solved more differential equations involving motion along a straight line. In §B we found parametric equations for the motion of a particle along a plane curve, given d.e.'s for the components of its acceleration and velocity.

A1. Rate of Change

A differential equation contains derivatives. A derivative gives information about the rate of change of a function. For example:

(1) If P is the size of a population at time t, then we can interpret the d.e.

$$\frac{dP}{dt} = 0.0325P$$

as saying that at any time t the rate at which the population is growing is proportional (3.25%) to its size at that time.

(2) The d.e. $\frac{dQ}{dt} = -(0.000275)Q$ tells us that at any time t the rate at which the quantity Q is decreasing is proportional (0.0275%) to the quantity existing at that time.

(3) In psychology, one typical stimulus-response situation, known as *logarithmic response,* is that in which the response y changes at a rate inversely proportional to the strength of the stimulus x. This is expressed neatly by the differential equation

$$\frac{dy}{dx} = \frac{k}{x} \qquad (k \text{ a constant}).$$

If we suppose, further, that there is no response when $x = x_0$, then we have the condition $y = 0$ when $x = x_0$.

(4) We are familiar with the d.e.

$$a = \frac{d^2s}{dt^2} = -32 \text{ ft/sec}^2$$

for the acceleration due to gravity acting on an object at a height s above ground level at time t. The acceleration is the rate of change of the object's velocity.

A2. Order of a Differential Equation

The *order* of a d.e. is the order of the highest derivative that occurs in it: $y' = 4y$ and $\frac{dy}{dx} = x^3 + 2$ are *first-order* d.e.'s; $y'' = -y$ and $\frac{d^2s}{dt^2} = -32$ are *second-order* d.e.'s.

†B.   Slope Fields _____

In this section we solve differential equations by obtaining a *slope field* or calculator picture that approximates the general solution. We call the graph of a solution of a d.e. a *solution curve*.

The slope field of a first-order d.e. is based on the fact that the d.e. can be interpreted as a statement about the slopes of its solution curves.

Example 1.   The d.e. $\dfrac{dy}{dx} = y$ tells us that at any point (x, y) on a solution curve the slope of the curve is equal to its y-coordinate. Since the d.e. says that y is a function whose derivative is also y, we know that

$$y = e^x$$

is a solution. In fact, $y = Ce^x$ is a solution of the d.e. for every constant C, since $y' = Ce^x = y$.

The d.e. $y' = y$ says that, at any point where $y = 1$, say $(0, 1)$ or $(1, 1)$ or $(5, 1)$, the slope of the solution curve is 1; at any point where $y = 3$, say $(0, 3)$, $(\ln 3, 3)$, or $(\pi, 3)$, the slope equals 3; and so on.

In Figure N9–1a we see some small line segments of slope 1 at several points where $y = 1$, and some segments of slope 3 at several points where $y = 3$. In Figure N9–1b we see the curve of $y = e^x$ with slope segments drawn in as follows:

POINT	SLOPE
$\left(-1, \dfrac{1}{e}\right)$	$\dfrac{1}{e} \simeq 0.4$
$\left(-\dfrac{1}{2}, \dfrac{1}{\sqrt{e}}\right)$	$\dfrac{1}{\sqrt{e}} \simeq 0.6$
$(0, 1)$	1
$\left(\dfrac{1}{2}, \sqrt{e}\right)$	$\sqrt{e} \simeq 1.6$
$(1, e)$	$e \simeq 2.7$

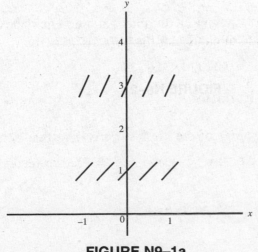

FIGURE N9–1a

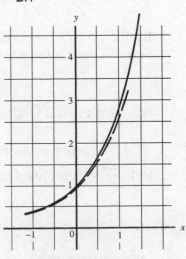

FIGURE N9–1b

†The topic slope fields will not be tested on the AB examination until 2004. See page x.

Figure N9–1c is the slope field for the d.e. $\dfrac{dy}{dx} = y$ obtained from our calculator program. The program computes slopes at many points (x, y) and draws small segments of the tangents at those points. The small segments approximate the solution curves. If we start at any point in the slope field and move so that the slope segments are always tangent to your motion, we will trace a solution curve. The slope field, as mentioned above, closely approximates the family of solutions.

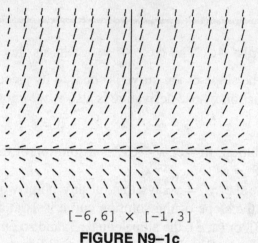

$[-6,6] \times [-1,3]$

FIGURE N9–1c

Example 2. The slope field for the d.e. $\dfrac{dy}{dx} = \dfrac{1}{x}$ is shown in Figure N9–2.
(a) Carefully draw the solution curve that passes through the point $(1, 0.5)$. (b) Find the general solution for the equation.

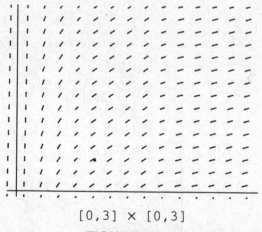

$[0,3] \times [0,3]$

FIGURE N9–2

(a) In Figure N9–2a we started at the point (1, 0.5), then moved from segment to segment drawing the curve to which these segments were tangent. The particular solution curve shown is the member of the family of solution curves

$$y = \ln x + C$$

that goes through the point (1, 0.5).

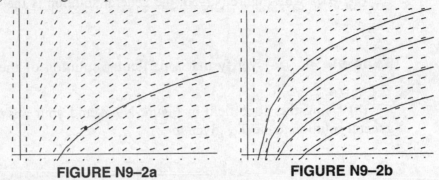

FIGURE N9–2a   FIGURE N9–2b

(b) Since we already know that, if $\dfrac{dy}{dx} = \dfrac{1}{x}$, then $y = \displaystyle\int \dfrac{1}{x}\,dx = \ln x + C$, we are assured of having found the correct general solution in (a).

In Figure N9–2b we have drawn several particular solution curves of the given d.e. Note that the vertical distance between any pair of curves is constant.

Example 3.   Match each slope field in Figure N9–3 with the proper d.e. from the following set. Find the general solution for each d.e. The particular solution that goes through (0,0) has been sketched in.

(A) $y' = \cos x$ (B) $\dfrac{dy}{dx} = 2x$

(C) $\dfrac{dy}{dx} = 3x^2 - 3$ (D) $y' = -\dfrac{\pi}{2}$

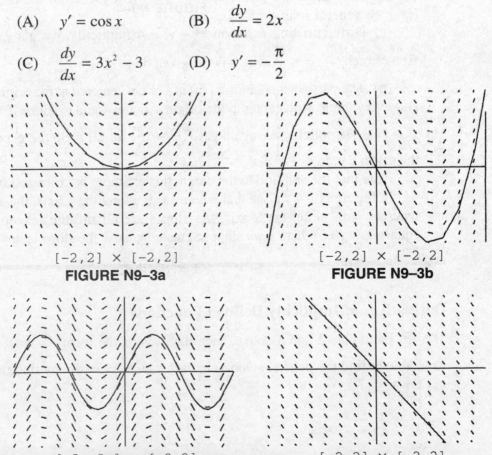

[−2,2] × [−2,2]   [−2,2] × [−2,2]
FIGURE N9–3a   FIGURE N9–3b

[−2π, 2π] × [−2,2]   [−2,2] × [−2,2]
FIGURE N9–3c   FIGURE N9–3d

(A) goes with Figure N9–3c. The solution curves in the family $y = \sin x + C$ are quite obvious.

(B) goes with Figure N9–3a. The general solution is the family of parabolas $y = x^2 + C$.

For (C) the slope field is shown in Figure N9–3b. The general solution is the family of cubics $y = x^3 - 3x + C$.

(D) goes with Figure N9–3d; the general solution is the family of lines $y = -\dfrac{\pi}{2}x + C$.

Example 4.   (a) Verify that a particular solution of the d.e. $\dfrac{dy}{dx} = -\dfrac{x}{y}$ is $x^2 + y^2 = 4$.

(b) Using the slope field in Figure N9–4 and your answer to (a), find the general solution to the d.e. given in (a).

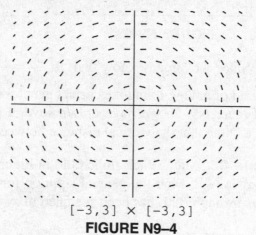

$$[-3,3] \times [-3,3]$$

FIGURE N9–4

(a) Differentiating equation $x^2 + y^2 = 4$ implicitly, we get $2x + 2y\dfrac{dy}{dx} = 0$, from which $\dfrac{dy}{dx} = -\dfrac{x}{y}$, which is the given d.e.

(b) The particular solution in (a) is a circle centered at the origin with radius 2. But every circle with center at the origin satisfies the d.e., since if we differentiate $x^2 + y^2 = r^2$ implicitly we again get $\dfrac{dy}{dx} = -\dfrac{x}{y}$ for all r. So the general solution of the given d.e. is $x^2 + y^2 = r^2$.

Note that the solution to the d.e. in this example is expressed implicitly.

In Figure N9–4 we see that where $x = 0$, along the y-axis, the slopes are all 0; where $y = 0$, along the x-axis, the slopes are all infinite (except where x also equals 0: at (0,0) there is *no* slope segment because the slope is not defined there).

Derivatives of Implicitly Defined Functions

In Examples 2 and 3 above, each d.e. was of the form $\dfrac{dy}{dx} = f(x)$ or $y' = f(x)$. We were able to find the general solution in each case very easily by finding the antiderivative $y = \displaystyle\int f(x)\,dx$.

We now consider d.e.'s of the form $\dfrac{dy}{dx} = f(x,y)$, where $f(x,y)$ is an expression in x and y; that is, $\dfrac{dy}{dx}$ is an implicitly defined function. Example 4 illustrates such a differential equation. Here is another example.

Example 5.   Figure N9–5 shows the slope field for

$$y' = x + y. \tag{1}$$

At each point (x,y) the slope is the sum of its coordinates. Three particular solutions have been added, through the points

(a) (0,0)   (b) (0, –1)   (c) (0, –2)

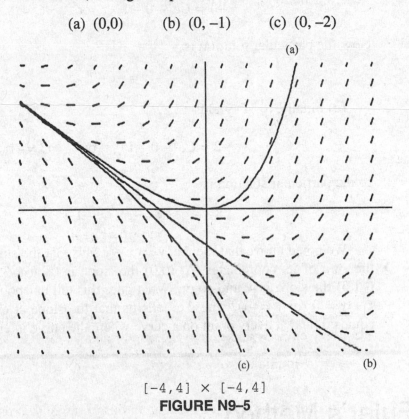

$$[-4,4] \times [-4,4]$$
FIGURE N9–5

Although determination of the general solution of the given d.e. is beyond the advanced placement course, it is easy to verify that the solution is

$$y = Ce^x - x - 1. \tag{2}$$

Differentiating (2) yields

$$y' = Ce^x - 1. \tag{3}$$

Since the right side of (3) is equal to $x + y$ by (2), it follows that

$$y' = x + y,$$

which is the given d.e. (1).

We will now identify the three particular solutions shown in Figure N9–5. In each case we will find C from the general solution $y = Ce^x - x - 1$ in (2) by replacing (x,y) by the given point:

(a) Using point (0,0), we get

$$0 = Ce^0 - 0 - 1 \qquad \text{or} \qquad C = 1,$$

so the particular solution is

$$y = e^x - x - 1.$$

(b) Point (0,–1) yields

$$-1 = Ce^0 - 0 - 1 \qquad \text{or} \qquad C = 0.$$

Now the particular solution is

$$y = -x - 1.$$

(c) Finally, (0,–2) yields

$$-2 = Ce^0 - 0 - 1 \qquad \text{or} \qquad C = -1,$$

so the particular solution is

$$y = -e^x - x - 1.$$

We noted above that the d.e. $y' = x + y$ says that the slope at any point equals the sum of its coordinates. At (0,0) the slope is 0; at (–3,2) the slope is –1; at (–1,5) the slope is 4; and so on. What does this tell us about points along the line $x + y = 0$ (or $y = -x$)? The d.e. tells us that the slope at every point on the line equals 0 (and at every point on a curve where the curve intersects the line $y = -x$).

*C.   Euler's Method

In §B we found solution curves to first-order differential equations *graphically,* using slope fields. Here we will find solutions *numerically,* using *Euler's method* to find points on solution curves.

When we use a slope field we start at an initial point, then move step by step so the slope segments are always tangent to the solution curve. With Euler's method we again select a starting point; but now we *calculate* the slope at that point (from the given d.e.), use the initial point and that slope to locate a new point, use the new point and calculate the slope at it (again from the d.e.) to locate still another point, and so on. The method is illustrated in Example 6.

*An asterisk denotes a topic covered only in Calculus BC.

Example 6.   Let $\dfrac{dy}{dx} = \dfrac{1}{x}$. Use Euler's method to approximate the y-values with ten steps, starting at point $P_0(1, 0)$ and letting $\Delta x = 0.1$.

The slope at $P_0 = (x_0, y_0) = (1, 0)$ is $\dfrac{dy}{dx} = \dfrac{1}{x_0} = \dfrac{1}{1} = 1$. To find the y-coordinate of $P_1(x_1, y_1)$, we add Δy to y_0, where

$$\Delta y = (\text{slope at } P_0) \cdot \Delta x = 1 \cdot (0.1) = 0.1.$$

Then
$$y_1 = y_0 + \Delta y = 0 + 0.1 = 0.1$$

and
$$P_1 = (1.1, 0.1).$$

To find the y-coordinate of $P_2(x_2, y_2)$ we add Δy to y_1, where

$$\Delta y = (\text{slope at } P_1) \cdot \Delta x = \dfrac{1}{x_1} \cdot \Delta x = \dfrac{1}{1.1} \cdot (0.1) = 0.091.$$

Then
$$y_2 = y_1 + \Delta y = 0.1 + 0.091 = 0.191$$

and
$$P_2 = (1.2, 0.191).$$

To find the y-coordinate of $P_3(x_3, y_3)$ we add Δy to y_2, where

$$\Delta y = (\text{slope at } P_2) \cdot \Delta x = \dfrac{1}{x_2} \cdot \Delta x = \dfrac{1}{1.2} \cdot (0.1) = 0.083.$$

Then
$$y_3 = y_2 + \Delta y = 0.191 + 0.083 = 0.274,$$

$$P_3 = (1.3, 0.274),$$

and so on.

The table summarizes all the data, for the 10 steps specified, from $x = 1$ to $x = 2$:

TABLE FOR $\dfrac{dy}{dx} = \dfrac{1}{x}$

	x	y	$(\text{SLOPE}) \cdot (0.1)$	$= \Delta y$	TRUE y^*
P_0	1	0	$(1/1) \cdot (0.1)$	0.1	0
P_1	1.1	0.1	$(1/1.1) \cdot (0.1)$	0.091	0.095
P_2	1.2	0.191	$(1/1.2) \cdot (0.1)$	0.083	0.183
P_3	1.3	0.274	$\cdot$	0.077	0.262
P_4	1.4	0.351	$\cdot$	0.071	0.336
P_5	1.5	0.422	$\cdot$	0.067	0.405
P_6	1.6	0.489	$\cdot$	0.062	0.470
P_7	1.7	0.551	$\cdot$	0.059	0.531
P_8	1.8	0.610	$\cdot$	0.056	0.588
P_9	1.9	0.666	$(1/1.9) \cdot (0.1)$	0.053	0.642
P_{10}	2.	0.719			0.693

*To three decimal places.

The table gives us the numerical solution of $\frac{dy}{dx} = \frac{1}{x}$ using Euler's method. Figure N9–6a shows the graphical solution, which agrees with the data from the table, for x increasing from 1 to 1.5 by five steps with Δx equal to 0.1 Figure N9–6b shows the Euler graph with ten steps, starting at (1,0) and the same Δx. Note that the solution of $\frac{dy}{dx} = \frac{1}{x}$ passing through the point (1,0) is $y = \ln x + C$, with $C = 0$. In Figure N9–6b we have added the particular solution $y = \ln x$ to the Euler graph.

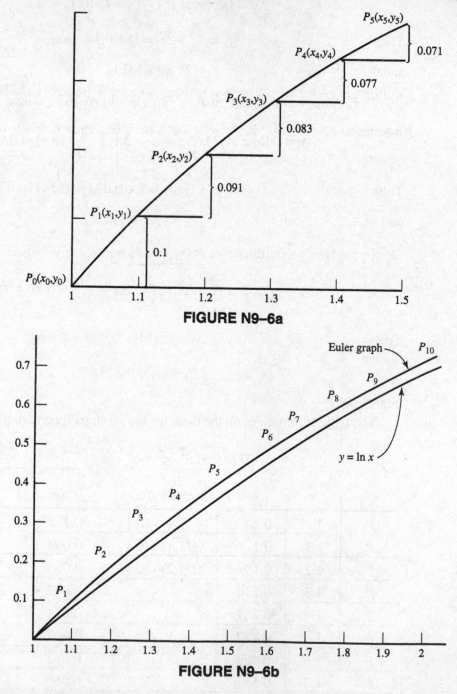

FIGURE N9–6a

FIGURE N9–6b

We observe that, since y'' for $\ln x$ equals $-\frac{1}{x^2}$, the true curve is concave down and below the Euler graph.

The last column in the table on page 305 shows the true value (to three decimal places) of y. The Euler approximation for ln 2 is 0.719; the true value is 0.693. The Euler approximation with ten steps is not very good! However, see what happens as we increase the number n of steps:

n	EULER APPROXIMATION	ERROR
10	0.719	0.026
20	0.706	0.013
40	0.699	0.006
80	0.696	0.003
160	0.695	0.002

Doubling the number of steps cuts the error approximately in half.

Example 7.   Show that Euler's method in Example 6 yields the same answer as the left Riemann sum used to approximate $\int_1^2 \frac{1}{x}\, dx$.

In Example 6 we started at $P_0\,(1,0)$ and used subintervals of length $\Delta x = 0.1$. The following table shows how we found the y-values:

	x	y	(SLOPE) $\cdot \Delta x =$ $\frac{1}{x} \cdot \Delta x = \Delta y$
P_0	$x_0 = 1$	$y_0 = 0$	$\frac{1}{x_0}\Delta x$
P_1	x_1	$\frac{1}{x_0}\Delta x$	$\frac{1}{x_1}\Delta x$
P_2	x_2	$\frac{1}{x_0}\Delta x + \frac{1}{x_1}\Delta x$	$\frac{1}{x_2}\Delta x$
P_3	x_3	$\frac{1}{x_0}\Delta x + \frac{1}{x_1}\Delta x + \frac{1}{x_2}\Delta x$	$\frac{1}{x_3}\Delta x$
...	...	...	...
P_9	x_9	$\frac{1}{x_0}\Delta x + \frac{1}{x_1}\Delta x + \cdots + \frac{1}{x_8}\Delta x$	$\frac{1}{x_9}\Delta x$
P_{10}	x_{10}	$\frac{1}{x_0}\Delta x + \frac{1}{x_1}\Delta x + \cdots + \frac{1}{x_9}\Delta x$	

We see that $y_{10} = \sum_{k=0}^{9} \frac{1}{x_k} \Delta x$, the left Riemann sum that is an approximation for $\int_1^2 \frac{1}{x}\, dx$. The Riemann sum is, of course, equal to y_{10} obtained by the Euler method, 0.719. At the end of Example 6, we used left Riemann sums to find the Euler approximations for $n = 10, 20, 40, 80,$ and 160.

Since a calculator program supplies Riemann sums quickly and effortlessly, why use Euler's method? The answer is that to evaluate a Riemann sum we need the d.e. to be expressible as $\frac{dy}{dx} = f(x)$. Euler's method can be applied also to equations of the type $\frac{dy}{dx} = F(x,y)$, where F is given implicitly (see Example 9). In addition, the latter method not only gives the sum but also approximates a solution to a d.e. at any point between x_0 and x_n, as the next example shows.

Example 8. Approximate (a) $y(1.15)$ and (b) $y(1.64)$ for the d.e. $\frac{dy}{dx} = \frac{1}{x}$ of Example 6, using the table obtained there from Euler's method. (See page 305.)

(a) To approximate $y(1.15)$ we use segment P_1P_2, finding the y-coordinate of its midpoint:

$$y(1.15) \simeq 0.5[y(1.1) + y(1.2)]$$
$$= 0.5(0.1 + 0.191) = 0.146.$$

(b) To approximate $y(1.64)$ we use points P_6 and P_7 as follows:

$$y(1.64) \simeq y(1.6) + 0.4[y(1.7) - y(1.60)]$$
$$\simeq 0.489 + 0.4(0.551 - 0.489)$$
$$= 0.489 + 0.025 = 0.514.$$

Example 9. Given the d.e. $\frac{dy}{dx} = x + y$ with initial condition $y(0) = 0$, use Euler's method with $\Delta x = 0.1$ to estimate y when $x = 0.5$.

Here are the relevant computations:

	x	y	(SLOPE) $\cdot \Delta x =$ $(x + y) \cdot (0.1) = \Delta y$
P_0	0	0	$0(0.1) = 0$
P_1	0.1	0	$(0.1)(0.1) = 0.01$
P_2	0.2	0.01	$(0.21)(0.1) = 0.021$
P_3	0.3	0.031	$(0.331)(0.1) = 0.033$
P_4	0.4	0.064	$(0.464)(0.1) = 0.046$
P_5	0.5	0.110	

In Figure N9–5, page 303, we showed the slope field for $y' = x + y$ with the solution curve through $(0,0)$ superimposed. We stated that the solution for this curve is

$$y = e^x - x - 1.$$

Let's see how our Euler approximation from the table compares with the exact value (to three decimal places). Euler value: $y(0.5) = 0.110$; exact value: $y(0.5) = e^{0.5} - 0.5 - 1 = 0.149$. To get a better approximation, we must increase the number of steps considerably!

A Caution: Euler's method approximates the solution by substituting short line segments in place of the actual curve. It can be quite accurate whtn the step sizes are small, but only if the curve does not have discontinuities, cusps, or asymptotes.

For example, the reader may verify that the curve $y = \dfrac{1}{2x - 5}$ solves the differential equation $\dfrac{dy}{dx} = -2y^2$ with initial condition $y = -1$ when $x = 2$. If, however, we attempt to approximate this solution using Euler's method with step size $\Delta x = 1$, the first step carries us to point $(3, -3)$. The accompanying graph (Figure N9–7) shows that this is nowhere near the solution curve, on which $y = 1$ when $x = 3$. Here, Euler's method fails because it leaps blindly across the vertical asymptote at $x = \frac{5}{2}$; thus this method cannot be expected to provide good approximations for any larger values of x.

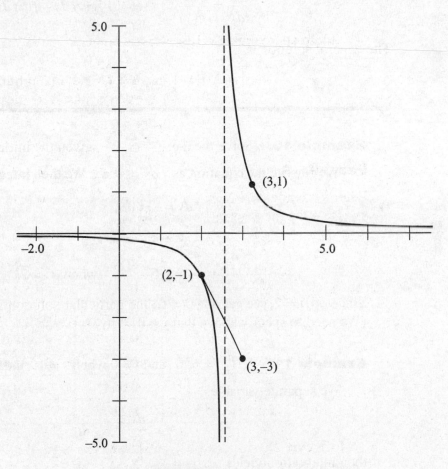

FIGURE N9–7

D. Solving First-Order Differential Equations Analytically

In the preceding sections we solved differential equations graphically, using slope fields, and numerically, using Euler's method. Both methods yield approximations. In this section we review how to solve some differential equations *exactly*.

Separating Variables

A first-order d.e. in x and y is *separable* if it can be written so that all the terms involving y are on one side and all the terms involving x are on the other.

A differential equation has variables separable if it is of the form

$$\frac{dy}{dx} = \frac{f(x)}{g(y)} \qquad \text{or} \qquad g(y)\,dy - f(x)\,dx = 0.$$

The general solution is

$$\int g(y)\,dy - \int f(x)\,dx = C \qquad (C \text{ arbitrary}).$$

Example 10.   Solve the d.e. $\dfrac{dy}{dx} = -\dfrac{x}{y}$, given the initial condition $y(0) = 2$.

We rewrite the equation as $y\,dy = -x\,dx$. We then integrate, getting

$$\int y\,dy = -\int x\,dx,$$

$$\frac{1}{2}y^2 = -\frac{1}{2}x^2 + k,$$

$$y^2 + x^2 = C \qquad (\text{where } C = 2k).$$

Since $y(0) = 2$, we get $4 + 0 = C$; the particular solution is therefore $x^2 + y^2 = 4$. (We need to specify above that $y \neq 0$. Why?)

Example 11.   If $\dfrac{ds}{dt} = \sqrt{st}$ and $t = 0$ when $s = 1$, find s when $t = 9$.

We separate variables:

$$\frac{ds}{\sqrt{s}} = \sqrt{t}\,dt;$$

then integration yields

$$2s^{1/2} = \frac{2}{3}t^{3/2} + C.$$

Using $s = 1$ and $t = 0$, we get $2 = \frac{2}{3}\cdot 0 + C$, so $C = +2$. Then

$$2s^{1/2} = \frac{2}{3}t^{3/2} + 2 \qquad \text{or} \qquad s^{1/2} = \frac{1}{3}t^{3/2} + 1.$$

When $t = 9$, we find that $s^{1/2} = 9 + 1$. So $s = 100$.

Example 12. If $(\ln y) \dfrac{dy}{dx} = \dfrac{y}{x}$, and $y = e$ when $x = 1$, find the value of y greater than 1 that corresponds to $x = e^4$.

Separating, we get $\dfrac{\ln y}{y}\, dy = \dfrac{dx}{x}$. We integrate:

$$\frac{1}{2}\ln^2 y = \ln|x| + C.$$

Using $y = e$ when $x = 1$ yields $C = \dfrac{1}{2}$. So

$$\frac{1}{2}\ln^2 y = \ln|x| + \frac{1}{2}.$$

When $x = e^4$, we have $\dfrac{1}{2}\ln^2 y = 4 + \dfrac{1}{2}$, $\ln^2 y = 9$, $\ln y = 3$ (where we chose $\ln y > 0$ because $y > 1$). So $y = e^3$.

Example 13. Find the general solution of the differential equation $\dfrac{du}{dv} = e^{v-u}$.

We rewrite:

$$\frac{du}{dv} = \frac{e^v}{e^u}; e^u\, du = e^v\, dv.$$

Taking antiderivatives yields $e^u = e^v + C$.

E. Exponential Growth and Decay

We now apply the method of separation of variables to three classes of functions associated with different rates of change. In each of the three cases, we describe the rate of change of a quantity, write the differential equation that follows from the description, then solve—or, in some cases, just give the solution of—the d.e. We list several applications of each case, and present relevant problems involving some of the applications.

Case I

An interesting special differential equation with wide applications is defined by the following statement: "A positive quantity y increases (or decreases) at a rate that at any time t is proportional to the amount present." It follows that the quantity y satisfies the d.e.

$$\frac{dy}{dt} = ky, \tag{1}$$

where $k > 0$ if y is increasing and $k < 0$ if y is decreasing.

From (1) it follows that

$$\frac{dy}{y} = k\,dt,$$

$$\int \frac{1}{y}\,dy = \int k\,dt,$$

$$\ln y = kt + C \qquad (C \text{ a constant}).$$

Then

$$y = e^{kt+C} = e^{kt} \cdot e^C$$
$$= ce^{kt} \qquad (\text{where } c = e^C).$$

If we are given an initial amount y, say y_0 at time $t = 0$, then

$$y_0 = c \cdot e^{k \cdot 0} = c \cdot 1 = c,$$

and our law of exponential change

$$y = ce^{kt} \qquad\qquad (2)$$

tells us that c is the initial amount of y (at time $t = 0$). If the quantity grows with time, then $k > 0$; if it decays (or diminishes, or decomposes), then $k < 0$. Equation (2) is often referred to as the *law of natural growth or decay*.

The length of time required for a quantity that is decaying exponentially to be reduced by half is called its *half-life*.

Example 14.   The population of a country is growing at a rate proportional to its population. If the growth rate per year is 4% of the current population, how long will it take for the population to double?

If the population at time t is P, then we are given that $\frac{dP}{dt} = 0.04P$. Substituting in equation (2), we see that the solution is

$$P = P_0 e^{0.04t},$$

where P_0 is the initial population. We seek t when $P = 2P_0$.

$$2P_0 = P_0 e^{0.04t},$$
$$2 = e^{0.04t},$$
$$\ln 2 = 0.04t,$$
$$t = \frac{\ln 2}{0.04} \simeq 17.33 \text{ yr.}$$

Example 15.   The bacteria in a certain culture increase continuously at a rate proportional to the number present. (a) If the number triples in 6 hours, how many will there be in 12 hours? (b) In how many hours will the original number quadruple?

We let N be the number at time t and N_0 the number initially. Then

$$\frac{dN}{dt} = kN, \qquad \frac{dN}{N} = k\,dt, \qquad \ln N = kt + C, \qquad \text{and} \qquad \ln N_0 = 0 + C,$$

so that $C = \ln N_0$. The general solution is then $N = N_0 e^{kt}$, with k still to be determined.

Since $N = 3N_0$ when $t = 6$, we see that $3N_0 = N_0 e^{6k}$ and that $k = \dfrac{1}{6}\ln 3$. Thus

$$N = N_0 e^{(t\ln 3)/6}.$$

(a) When $t = 12$, $N = N_0 e^{2\ln 3} = N_0 e^{\ln 3^2} = N_0 e^{\ln 9} = 9N_0$.

(b) We let $N = 4N_0$ in the centered equation above, and get

$$4 = e^{(t\ln 3)/6}, \qquad \ln 4 = \frac{t}{6}\ln 3, \qquad \text{and} \qquad t = \frac{6\ln 4}{\ln 3} \approx 7.6 \text{ hr.}$$

Example 16.   Radium-226 decays at a rate proportional to the quantity present. Its half-life is 1612 years. How long will it take for one quarter of a given quantity of radium-226 to decay?

If $Q(t)$ is the amount present at time t, then it satisfies the equation

$$Q(t) = Q_0 e^{kt}, \tag{1}$$

where Q_0 is the initial amount and k is the (negative) factor of proportionality. Since it is given that $Q = \dfrac{1}{2}Q_0$ when $t = 1612$, equation (1) yields

$$\frac{1}{2}Q_0 = Q_0 e^{k(1612)},$$

$$\frac{1}{2} = e^{1612k},$$

$$k = \frac{\ln\dfrac{1}{2}}{1612} = -0.00043$$

$$= -0.043\% \text{ to the nearest thousandth of a percent.}$$

We now have

$$Q = Q_0 e^{-0.00043t}. \tag{2}$$

When one quarter of Q_0 has decayed, three quarters of the initial amount remains. We use this fact in equation (2) to find t:

$$\frac{3}{4}Q_0 = Q_0 e^{-0.00043t},$$

$$\frac{3}{4} = e^{-0.00043t},$$

$$t = \frac{\ln\dfrac{3}{4}}{-0.00043} \approx 669 \text{ yr.}$$

Applications of Case I.

(1) A colony of bacteria may grow at a rate proportional to its size. (2) Other populations, such as those of humans, rodents, or fruit flies, whose supply of food is unlimited may also grow at a rate proportional to the size of the population. (3) Money invested at interest that is compounded continuously accumulates at a rate proportional to the amount present. The constant of proportionality is the interest rate. (4) The demand for certain precious commodities (gas, oil, electricity, valuable metals) has been growing in recent decades at a rate proportional to the existing demand.

Each of the above quantities (population, amount, demand) is a function of the form ce^{kt} (with $k > 0$). (See Figure N9–7a.)

(5) Radioactive isotopes, such as uranium-235, strontium-90, iodine-131, and carbon-14, decay at a rate proportional to the amount still present. (6) If P is the *present value* of a fixed sum of money A due t years from now, where the interest is compounded continuously, then P decreases at a rate proportional to the value of the investment. (7) It is common for the concentration of a drug in the bloodstream to drop at a rate proportional to the existing concentration. (8) As a beam of light passes through murky water or air, its intensity at any depth (or distance) decreases at a rate proportional to the intensity at that depth.

Each of the above four quantities (5 through 8) is a function of the form ce^{-kt} ($k > 0$). (See Figure N9–7b.)

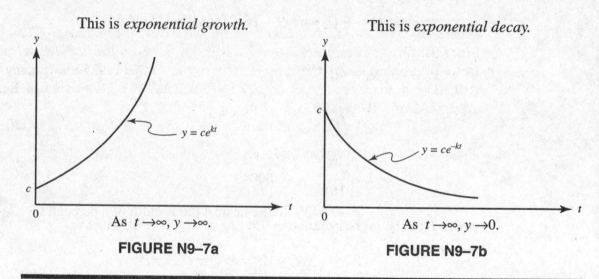

This is *exponential growth*.   This is *exponential decay*.

As $t \to \infty$, $y \to \infty$.   As $t \to \infty$, $y \to 0$.

FIGURE N9–7a   **FIGURE N9–7b**

Example 17. At a yearly rate of 5% compounded continuously, how long does it take for an investment to triple?

If P dollars are invested for t yr at 5%, the amount will grow to $A = Pe^{0.05t}$ in t yr. We seek t when $A = 3P$:

$$3 = e^{0.05t},$$

$$\frac{\ln 3}{0.05} = t \simeq 22\,\text{yr}.$$

Example 18. One important method of dating fossil remains is to determine what portion of the carbon content of a fossil is the radioactive isotope carbon-14. During life, any organism exchanges carbon with its environment. Upon death this circulation ceases, and the ^{14}C in the organism then decays at a rate proportional to the amount present. The proportionality factor is 0.012% per year.

When did an animal die if an archaeologist determines that only 25% of the original amount of ^{14}C is still present in its fossil remains?

The quantity Q of ^{14}C present at time t satisfies the equation

$$\frac{dQ}{dt} = -0.00012Q$$

with solution

$$Q(t) = Q_0 e^{-0.00012t}$$

(where Q_0 is the original amount). We are asked to find t when $Q(t) = 0.25Q_0$.

$$0.25Q_0 = Q_0 e^{-0.00012t},$$
$$0.25 = e^{-0.00012t},$$
$$\ln 0.25 = -0.00012t,$$
$$-1.386 = -0.00012t,$$
$$t \simeq 11{,}550.$$

Rounding to the nearest 500 yr, we see that the animal died approximately 11,500 yr ago.

Example 19. In 1970 the world population was approximately 3.5 billion. Since then it has been growing at a rate proportional to the population, and the factor of proportionality has been 1.9% per year. At that rate, in how many years will there be one person per square foot of land? (The land area of Earth is approximately 200,000,000 mi^2, or about 5.5×10^{15} ft^2.)

If $P(t)$ is the population at time t, the problem tells us that P satisfies the equation $\frac{dP}{dt} = 0.019P$. Its solution is the exponential growth equation

$$P(t) = P_0 e^{0.019t},$$

where P_0 is the initial population. Letting $t = 0$ for 1970, we have

$$3.5 \times 10^9 = P(0) = P_0 e^0 = P_0.$$

So

$$P(t) = (3.5 \times 10^9)e^{0.019t}.$$

The question is: for what t does $P(t)$ equal 5.5×10^{15}? We solve

$$(3.5)(10^9)e^{0.019t} = (5.5)10^{15},$$
$$e^{0.019t} \simeq (1.6)10^6.$$

Taking the logarithm of each side yields

$$0.019t \simeq \ln 1.6 + 6 \ln 10 \simeq 14.3,$$
$$t \simeq 750 \text{ yr},$$

where it seems reasonable to round off as we have. So, if the human population continues to grow at the present rate, there will be one person for every square foot of land in just 750 yr.

Two Approaches to the Same Problem.

In §D of Chapter 8, on the Fundamental Theorem of Calculus, we considered many examples illustrating the following principle:

The definite integral of a rate of change

gives net change: if $\dfrac{dF}{dt} = f(t)$, then

$$\int_a^b f(t)\,dt = F(b) - F(a).$$

The differential equations we have been considering in this chapter are reminiscent of the examples in Chapter 8. In Example 21 of Chapter 8 (page 284) we said: "Suppose an epidemic is spreading through a city at the rate of $f(t)$ new people per week. Then $\int_0^4 f(t)\,dt$ is the number of people who will become infected during the next 4 weeks."

If $F(t)$ is an antiderivative of $f(t)$—that is, $\dfrac{dF}{dt} = f(t)$—then

$$\int_0^4 f(t)\,dt = F(4) - F(0).$$

But now, start over and consider Example 21 as giving us the d.e. $\dfrac{dy}{dt} = f(t)$, where $y(t)$ is the number of people infected at time t weeks and $y(0) = 0$. If $\dfrac{dF}{dt} = f(t)$, we have, solving the d.e.,

$$y(t) = F(t) + C,$$
$$0 = F(0) + C,$$
$$C = -F(0),$$
$$y(t) = F(t) - F(0),$$

and

$$y(4) = F(4) - F(0),$$

as above.

Case II

The rate of change of a quantity $y = f(t)$ may be proportional, not to the amount present, but to a difference between that amount and a fixed constant. Two situations are to be distinguished: The rate of change is proportional to

(a) a fixed constant A minus the amount of the quantity present:

$$f'(t) = k[A - f(t)]$$

(b) the amount of the quantity present minus a fixed constant A:

$$f'(t) = -k[f(t) - A]$$

where (in both) $f(t)$ is the amount at time t and k and A are both positive. We may conclude that

(a) $f(t)$ is increasing (Fig. N9–8a):

$$f(t) = A - ce^{-kt}$$

(b) $f(t)$ is decreasing (Fig. N9–8b):

$$f(t) = A + ce^{-kt}$$

for some positive constant c.

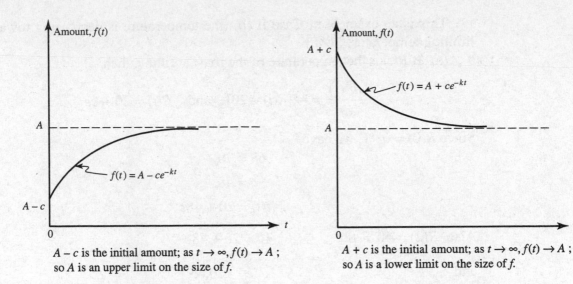

FIGURE N9–8a

$A - c$ is the initial amount; as $t \to \infty$, $f(t) \to A$; so A is an upper limit on the size of f.

FIGURE N9–8b

$A + c$ is the initial amount; as $t \to \infty$, $f(t) \to A$; so A is a lower limit on the size of f.

Here is how we solve the d.e. for Case II(a), where $A - y > 0$. If the quantity at time t is denoted by y and k is the positive constant of proportionality, then

$$y' = \frac{dy}{dt} = k(A - y),$$

$$\frac{dy}{A - y} = k \, dt,$$

$$-\ln (A - y) = kt + C,$$

$$\ln (A - y) = -kt - C,$$

$$A - y = e^{-kt} \cdot e^{-C}$$

$$= ce^{-kt}, \text{ where } c = e^{-C},$$

and

$$y = A - ce^{-kt}.$$

Case II (b) can be solved similarly.

Example 20.   According to Newton's law of cooling, a hot object cools at a rate proportional to the difference between its own temperature and that of its environment. If a roast at room temperature 68°F is put into a 20°F freezer, and if, after 2 hours, the temperature of the roast is 40°F: (a) What is its temperature after 5 hours? (b) How long will it take for the temperature of the roast to fall to 21°F?

This is an example of Case II (b) (the temperature is decreasing toward the limiting temperature 20°F).

(a) If $R(t)$ is the temperature of the roast at time t, then

$$\frac{dR(t)}{dt} = -k[R(t) - 20] \quad \text{and} \quad R(t) = 20 + ce^{-kt}.$$

Since $R(0) = 68°F$, we have

$$68 = 20 + c,$$
$$c = 48,$$
$$R(t) = 20 + 48e^{-kt}.$$

Also, $R(2) = 40°F$. So $\qquad 40 = 20 + 48e^{-k \cdot 2}$

and $\qquad\qquad\qquad\qquad e^{-k} \simeq 0.65,$

yielding $\qquad\qquad\qquad R(t) = 20 + 48(0.65)^t \qquad (*)$

and, finally, $\qquad\qquad R(5) = 20 + 48(0.65)^5 \simeq 26°F.$

(b) Equation (*) in part (a) gives the roast's temperature at time t. We must find t when $R = 21$:

$$21 = 20 + 48 (0.65)^t,$$
$$\frac{1}{48} = (0.65)^t,$$
$$-\ln 48 = t \ln(0.65),$$
$$t \simeq 9 \text{ hr.}$$

Example 21.   Advertisers generally assume that the rate at which people hear about a product is proportional to the number of people who have not yet heard about it. Suppose that the size of a community is 15,000, that to begin with no one has heard about a product, but that after 6 days 1500 people know about it. How long will it take for 2700 people to have heard of it?

Let $N(t)$ be the number of people aware of the product at time t. Then we are given that

$$N'(t) = k[15,000 - N(t)],$$

which is case IIa. The solution of this d.e. is

$$N(t) = 15,000 - ce^{-kt}.$$

Since $N(0) = 0$, $c = 15,000$ and

$$N(t) = 15,000(1 - e^{-kt}).$$

Since 1500 people know of the product after 6 days, we have

$$1500 = 15,000(1 - e^{-6k}),$$
$$e^{-6k} = 0.9,$$
$$k = \frac{\ln 0.9}{-6} = 0.018.$$

We now seek t when $N = 2700$:

$$2700 = 15{,}000(1 - e^{-0.018t}),$$
$$0.18 = 1 - e^{-0.018t},$$
$$e^{-0.018t} = 0.82,$$
$$t \simeq 11 \text{ days}.$$

Applications of Case II.

(1) Newton's law of heating says that a cold object warms up at a rate proportional to the difference between its temperature and that of its environment. If you put a roast at 68°F into an oven of 400°F, then the temperature at time t is $R(t) = 400 - 332e^{-kt}$.

(2) Because of air friction, the velocity of a falling object approaches a limiting value L (rather than increasing without bound). The acceleration (rate of change of velocity) is proportional to the difference between the limiting velocity and the object's velocity. If initial velocity is zero, then at time t the object's velocity $V(t) = L(1 - e^{-kt})$.

(3) If a tire has a small leak, then the air pressure inside drops at a rate proportional to the difference between the inside pressure and the fixed outside pressure O. At time t the inside pressure $P(t) = O + ce^{-kt}$.

*Case III: Logistic Growth

The rate of change of a quantity (for example, a population) may be proportional both to the amount (size) of the quantity and to the difference between a fixed constant A and its amount (size). If $y = f(t)$ is the amount, then

$$y' = ky(A - y), \tag{1}$$

where k and A are both positive. Equation (1) is called the *logistic differential equation*; it is used to model logistic growth.

The solution of the d.e. (1) is

$$y = \frac{A}{1 + ce^{-Akt}} \tag{2}$$

for some positive constant c.

In most applications, $c > 1$. In these cases, the initial amount $A/(1 + c)$ is *less* than $A/2$. In all applications, since the exponent of e in the expression for $f(t)$ is negative for all positive t, therefore, as $t \to \infty$,

(1) $ce^{-Akt} \to 0$;
(2) the denominator of $f(t) \to 1$;
(3) $f(t) \to A$.

Thus, A is an upper limit of f in this growth model. When applied to populations, A is called the *carrying capacity* or the *maximum sustainable population*.

Shortly we will solve specific examples of the logistic d.e. (1), instead of obtaining the general solution (2), since the latter is algebraically rather messy. (It is somewhat less complicated to verify that y' in (1) can be obtained by taking the derivative of (2).)

*An asterisk denotes a topic covered only in Calculus BC.

Unrestricted versus Restricted Growth

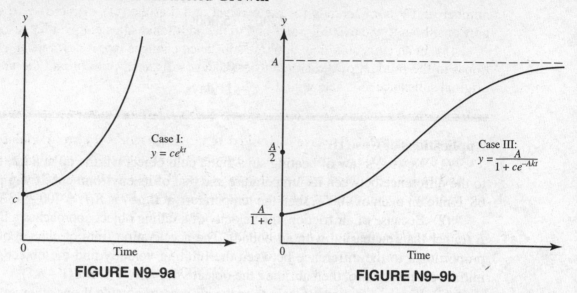

FIGURE N9–9a   FIGURE N9–9b

In Figures N9–9a and N9–9b we see the graphs of the growth functions of Cases I and III. The growth function of Case I is known as the *unrestricted* (or *uninhibited* or *unchecked*) model. It is not a very realistic one for most populations. It is clear, for example, that human populations cannot continue endlessly to grow exponentially. Not only is Earth's land area fixed, but also there are limited supplies of food, energy, and other natural resources. The growth function in Case III allows for such factors, which serve to check growth. It is therefore referred to as the *restricted* (or *inhibited*) model.

The two graphs are quite similar close to 0. This similarity implies that logistic growth is exponential at the start—a reasonable conclusion, since populations are small at the outset.

The S-shaped curve in Case III is often called a *logistic curve*. It shows that the rate of growth y':

    (1) increases slowly for a while; i.e., $y'' > 0$;

    (2) attains a maximum when $y = A/2$, at the point of inflection;

    (3) then decreases ($y'' < 0$), approaching 0 as y approaches its upper limit.

It is not difficult to verify these statements.

*Applications of Case III.

    (1) Some diseases spread through a (finite) population P at a rate proportional to the number of people, $N(t)$, infected by time t and the number, $P - N(t)$, not yet infected. Thus $N'(t) = kN(P - N)$ and, for some positive c and k,

$$N(t) = \frac{P}{1 + ce^{-Pkt}}.$$

    (2) A rumor (or fad or new religious cult) often spreads through a population P according to the formula in (1), where $N(t)$ is the number of people who have heard the rumor (acquired the fad, converted to the cult), and $P - N(t)$ is the number who have not.

    (3) Bacteria in a culture on a Petri dish grow at a rate proportional to the product of the existing population and the difference between the maximum sustainable population and the existing population. (Replace bacteria on a Petri dish by fish in a small lake, ants confined to a small receptacle, fruit flies supplied with only a limited amount of food, yeast cells, and so on.)

*An asterisk denotes a topic covered only in Calculus BC.

(4) Advertisers sometimes assume that sales of a particular product depend on the number of TV commercials for the product and that the rate of increase in sales is proportional both to the existing sales and to the additional sales conjectured as possible.

(5) In an autocatalytic reaction a substance changes into a new one at a rate proportional to the product of the amount of the new substance present and the amount of the original substance still unchanged.

Example 22. Because of limited food and space, a squirrel population cannot exceed 1000. It grows at a rate proportional both to the existing population and to the attainable additional population. If there were 100 squirrels 2 years ago, and 1 year ago the population was 400, about how many squirrels are there now?

Let P be the squirrel population at time t. It is given that

$$\frac{dP}{dt} = kP(1000 - P) \tag{3}$$

with $P(0) = 100$ and $P(1) = 400$. We seek $P(2)$.

We will find the general solution for the given d.e. (3) by separating the variables:

$$\frac{dP}{P(1000 - P)} = k\,dt.$$

It can easily be verified, using partial fractions, that

$$\frac{1}{P(1000 - P)} = \frac{1}{1000P} + \frac{1}{1000(1000 - P)}.$$

Now we integrate:

$$\int \frac{dP}{1000P} + \int \frac{dP}{1000(1000 - P)} = \int k\,dt,$$

getting

$$\ln P - \ln(1000 - P) = 1000kt + C$$

or

$$\ln \frac{(1000 - P)}{P} = -(1000kt + C),$$

$$\frac{1000 - P}{P} = ce^{-1000kt} \quad \text{(where } c = e^{-C}),$$

$$\frac{1000}{P} - 1 = ce^{-1000kt},$$

$$\frac{1000}{P} = 1 + ce^{-1000kt},$$

$$\frac{P}{1000} = \frac{1}{1 + ce^{-1000kt}},$$

and, finally (!),

$$P(t) = \frac{1000}{1 + ce^{-1000kt}}. \tag{4}$$

Please note that this is precisely the solution "advertised" on page 319 in equation (2), with A equal to 1000!

Now, using our initial condition $P(0) = 100$ in (4), we get

$$\frac{100}{1000} = \frac{1}{1 + c} \quad \text{and} \quad c = 9.$$

Using $P(1) = 400$, we get

$$400 = \frac{1000}{1 + 9e^{-1000k}},$$

$$1 + 9e^{-1000k} = 2.5,$$

$$e^{-1000k} = \frac{1.5}{9} = \frac{1}{6}. \tag{5}$$

So the particular solution is

$$P(t) = \frac{1000}{1 + 9(1/6)^t} \tag{6}$$

and $P(2) \simeq 800$ squirrels.

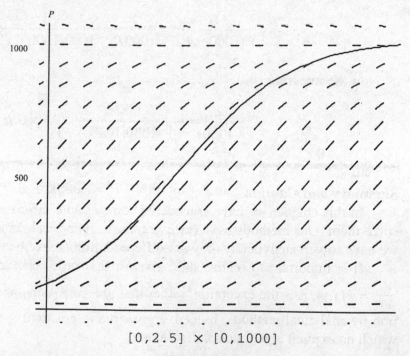

$$[0,2.5] \times [0,1000]$$

FIGURE N9–10

Figure N9–10 shows the slope field for equation (3), with $k = 0.00179$, which was obtained by solving equation (5) above. Note that the slopes are the same along any horizontal line, and that they are close to zero initially, reach a maximum at $P = 500$, then diminish again as P approaches its limiting value, 1000. We have superimposed the solution curve for $P(t)$ that we obtained in (6) above.

Example 23. Suppose a flu-like virus is spreading through a population of 50,000 at a rate proportional both to the number of people already infected and to the number still uninfected. If 100 people were infected yesterday and 130 are infected today:

(a) write an expression for the number of people $N(t)$ infected after t days;

(b) determine how many will be infected a week from today.

(a) We are told that $N'(t) = k \cdot N \cdot (50{,}000 - N)$, that $N(0) = 100$, and that $N(1) = 130$. The d.e. describing logistic growth leads to

$$N(t) = \frac{50{,}000}{1 + ce^{-50{,}000kt}}.$$

From $N(0) = 100$, we get

$$100 = \frac{50{,}000}{1 + c},$$

which yields $c = 499$. From $N(1) = 130$, we get

$$130 = \frac{50{,}000}{1 + 499e^{-50{,}000k}},$$

$$130(1 + 499e^{-50{,}000k}) = 50{,}000,$$

$$e^{-50{,}000k} = 0.77.$$

So

$$N(t) = \frac{50{,}000}{1 + 499(0.77)^t}.$$

(b) We must find $N(8)$. Since $t = 0$ represents yesterday:

$$N(8) = \frac{50{,}000}{1 + 499(0.77)^8} \approx 798 \text{ people}.$$

Summary and Caution

In this chapter we have considered some simple differential equations and ways to solve them. Our methods have been graphical, numerical, and analytical. Equations that we have solved analytically—by antidifferentiation—have been separable.

It is important to realize that, given a first-order differential equation of the type $\frac{dy}{dx} = F(x,y)$, it is the exception, rather than the rule, to be able to find the general solution by analytical methods. Indeed, a great many practical applications lead to d.e.'s for which no explicit algebraic solution exists.

Set 9: Multiple-Choice Questions on Differential Equations

In Questions 1 through 10, $a(t)$ denotes the acceleration function, $v(t)$ the velocity function, and $s(t)$ the position or height function at time t. (The acceleration due to gravity is -32 ft/sec^2.)

1. If $a(t) = 4t - 1$ and $v(1) = 3$, then $v(t)$ equals

   (A) $2t^2 - t$ (B) $2t^2 - t + 1$ (C) $2t^2 - t + 2$
   (D) $2t^2 + 1$ (E) $2t^2 + 2$

2. If $a(t) = 20t^3 - 6t$, $s(-1) = 2$, and $s(1) = 4$, then $v(t)$ equals

   (A) $t^5 - t^3$ (B) $5t^4 - 3t^2 + 1$ (C) $5t^4 - 3t^2 + 3$
   (D) $t^5 - t^3 + t + 3$ (E) $t^5 - t^3 + 1$

3. Given $a(t)$, $s(-1)$, and $s(1)$ as in Question 2, then $s(0)$ equals

   (A) 0   (B) 1   (C) 2   (D) 3   (E) 4

4. A stone is thrown straight up from the top of a building with initial velocity 40 ft/sec and hits the ground 4 sec later. The height of the building, in feet, is

   (A) 88   (B) 96   (C) 112   (D) 128   (E) 144

5. The maximum height is reached by the stone in Question 4 after

   (A) 4/5 sec   (B) 4 sec   (C) 5/4 sec   (D) 5/2 sec   (E) 2 sec

6. If a car accelerates from 0 to 60 mph in 10 sec, what distance does it travel in those 10 sec? (Assume the acceleration is constant and note that 60 mph = 88 ft/sec.)

   (A) 40 ft   (B) 44 ft   (C) 88 ft   (D) 400 ft   (E) 440 ft

7. A stone is thrown at a target so that its velocity after t sec is $(100 - 20t)$ ft/sec. If the stone hits the target in 1 sec, then the distance from the sling to the target is

   (A) 80 ft   (B) 90 ft   (C) 100 ft   (D) 110 ft   (E) 120 ft

8. What should the initial velocity be if you want a stone to reach a height of 100 ft when you throw it straight up?

   (A) 80 ft/sec   (B) 92 ft/sec   (C) 96 ft/sec
   (D) 112 ft/sec   (E) none of these

9. If the velocity of a car traveling in a straight line at time t is $v(t)$, then the difference in its odometer readings between times $t = a$ and $t = b$ is

(A) $\displaystyle\int_a^b |v(t)|\, dt$

(B) $\displaystyle\int_a^b v(t)\, dt$

(C) the net displacement of the car's position from $t = a$ to $t = b$
(D) the change in the car's position from $t = a$ to $t = b$
(E) none of these

10. If an object is moving up and down along the y-axis with velocity $v(t)$ and $s'(t) = v(t)$, then it is false that $\displaystyle\int_a^b v(t)\, dt$ gives

(A) $s(b) - s(a)$
(B) the net distance traveled by the object between $t = a$ and $t = b$
(C) the total change in $s(t)$ between $t = a$ and $t = b$
(D) the shift in the object's position from $t = a$ to $t = b$
(E) the total distance covered by the object from $t = a$ to $t = b$

11. A solution of the differential equation $y\, dy = x\, dx$ is

(A) $x^2 - y^2 = 4$ (B) $x^2 + y^2 = 4$ (C) $y^2 = 4x^2$
(D) $x^2 - 4y^2 = 0$ (E) $x^2 = 9 - y^2$

12. If $\dfrac{dy}{dx} = \dfrac{y}{2\sqrt{x}}$ and $y = 1$ when $x = 4$, then

(A) $y^2 = 4\sqrt{x} - 7$ (B) $\ln y = 4\sqrt{x} - 8$ (C) $\ln y = \sqrt{x} - 2$

(D) $y = e^{\sqrt{x}}$ (E) $y = e^{\sqrt{x}-2}$

13. If $\dfrac{dy}{dx} = e^y$ and $y = 0$ when $x = 1$, then

(A) $y = \ln |x|$ (B) $y = \ln |2 - x|$ (C) $e^{-y} = 2 - x$

(D) $y = -\ln |x|$ (E) $e^{-y} = x - 2$

14. If $\dfrac{dy}{dx} = \dfrac{x}{\sqrt{9 + x^2}}$ and $y = 5$ when $x = 4$, then y equals

(A) $\sqrt{9 + x^2} - 5$ (B) $\sqrt{9 + x^2}$ (C) $2\sqrt{9 + x^2} - 5$

(D) $\dfrac{\sqrt{9 + x^2} + 5}{2}$ (E) none of these

15.   If $\dfrac{ds}{dt} = \sin^2 \dfrac{\pi}{2} s$ and if $s = 1$ when $t = 0$, then, when $s = \dfrac{3}{2}$, t is equal to

(A) $\dfrac{1}{2}$ (B) $\dfrac{\pi}{2}$ (C)   1   (D) $\dfrac{2}{\pi}$ (E) $-\dfrac{2}{\pi}$

16.   The general solution of the differential equation $x\, dy = y\, dx$ is a family of

(A)   circles   (B)   hyperbolas   (C)   parallel lines
(D)   parabolas   (E)   lines passing through the origin

17.   The general solution of the differential equation $\dfrac{dy}{dx} = y$ is a family of

(A)   parabolas   (B)   straight lines   (C)   hyperbolas
(D)   ellipses   (E)   none of these

18.   A function $f(x)$ that satisfies the equations $f(x)f'(x) = x$ and $f(0) = 1$ is

(A) $f(x) = \sqrt{x^2 + 1}$ (B) $f(x) = \sqrt{1 - x^2}$ (C) $f(x) = x$
(D) $f(x) = e^x$ (E)   none of these

19.   The curve that passes through the point $(1, 1)$ and whose slope at any point (x, y) is equal to $\dfrac{3y}{x}$ has the equation

(A) $3x - 2 = y$ (B) $y^3 = x$ (C) $y = \left| x^3 \right|$
(D) $3y^2 = x^2 + 2$ (E) $3y^2 - 2x = 1$

20.   If radium decomposes at a rate proportional to the amount present, then the amount R left after t yr, if R_0 is present initially and c is the negative constant of proportionality, is given by

(A) $R = R_0 ct$ (B) $R = R_0 e^{ct}$ (C) $R = R_0 + \dfrac{1}{2} ct^2$

(D) $R = e^{R_0 ct}$ (E) $R = e^{R_0 + ct}$

21.   The population of a city increases continuously at a rate proportional, at any time, to the population at that time. The population doubles in 50 yr. After 75 yr the ratio of the population P to the initial population P_0 is

(A) $\dfrac{9}{4}$ (B) $\dfrac{5}{2}$ (C) $\dfrac{4}{1}$ (D) $\dfrac{2\sqrt{2}}{1}$ (E)   none of these

22.   If a substance decomposes at a rate proportional to the amount of the substance present, and if the amount decreases from 40 g to 10 g in 2 hr, then the constant of proportionality is

(A) $-\ln 2$ (B) $-\dfrac{1}{2}$ (C) $-\dfrac{1}{4}$ (D) $\ln \dfrac{1}{4}$ (E) $\ln \dfrac{1}{8}$

23. If $\frac{dy}{dx} = \frac{k}{x}$, k a constant, and if $y = 2$ when $x = 1$ and $y = 4$ when $x = e$, then, when $x = 2$, y equals

(A) 2   (B) 4   (C) ln 8   (D) ln 2 + 2   (E) ln 4 + 2

24. If $(g'(x))^2 = g(x)$ for all real x and $g(0) = 0$, $g(4) = 4$, then $g(1)$ equals

(A) $\frac{1}{4}$ (B) $\frac{1}{2}$ (C) 1   (D) 2   (E) 4

25. How much (to the nearest dollar) must a 25-year-old woman invest now at 8% compounded continuously to have \$40,000 when she is 60?

(A) \$243   (B) \$1055   (C) \$2432   (D) \$5413   (E) \$14,286

†26. The slope field shown at the right is for the differential equation

(A) $y' = x + 1$
(B) $y' = \sin x$
(C) $y' = -\sin x$
(D) $y' = \cos x$
(E) $y' = -\cos x$

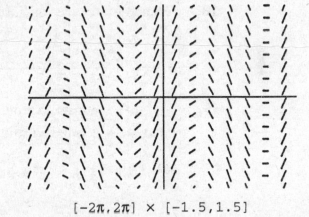

$[-2\pi, 2\pi] \times [-1.5, 1.5]$

†27. The slope field at the right is for the differential equation

(A) $y' = 2x$
(B) $y' = 2x - 4$
(C) $y' = 4 - 2x$
(D) $y' = y$
(E) $y' = x + y$

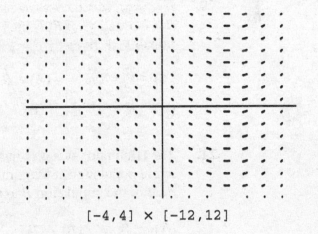

$[-4, 4] \times [-12, 12]$

†28. The solution curve of $y' = y$ that passes through point (2,3) is

(A) $y = e^x + 3$ (B) $y = \sqrt{2x + 5}$ (C) $y = 0.406e^x$

(D) $y = e^x - (e^2 + 3)$ (E) $y = e^x/(0.406)$

†The topic slope fields will not be tested on the AB examination before 2004.

†29. A solution curve has been superimposed
on the slope field shown at the right.
The solution is for the differential
equation and initial condition

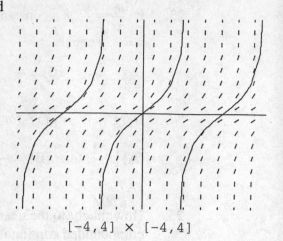

(A) $y' = \tan x;\ y(0) = 0$
(B) $y' = \cot x,\ y(\pi/4) = 1$
(C) $y' = 1 + x^2;\ y(0) = 0$
(D) $y' = \dfrac{1}{1 + x^2}\ ;\ y\left(\dfrac{\pi}{4}\right) = 1$
(E) $y' = 1 + y^2;\ y(0) = 0$

$[-4,4] \times [-4,4]$

The slope fields below are for Questions 30–35.

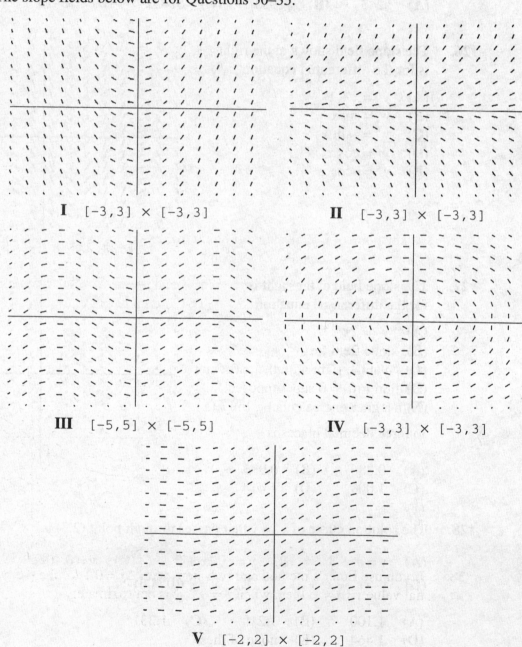

I $[-3,3] \times [-3,3]$

II $[-3,3] \times [-3,3]$

III $[-5,5] \times [-5,5]$

IV $[-3,3] \times [-3,3]$

V $[-2,2] \times [-2,2]$

†The topic slope fields will not be tested on the AB examination before 2004.

†**30.** Which slope field is for the differential equation $y' = y$?

   (A) I   **(B)** II   **(C)** III   **(D)** IV   **(E)** V

†**31.** Which slope field is for the differential equation $y' = -\dfrac{x}{y}$?

   (A) I   **(B)** II   **(C)** III   **(D)** IV   **(E)** V

†**32.** Which slope field is for the differential equation $y' = \sin x$?

   (A) I   **(B)** II   **(C)** III   **(D)** IV   **(E)** V

†**33.** Which slope field is for the differential equation $y' = 2x$?

   (A) I   **(B)** II   **(C)** III   **(D)** IV   **(E)** V

†**34.** Which slope field is for the differential equation $y' = e^{-x^2}$?

   (A) I   **(B)** II   **(C)** III   **(D)** IV   **(E)** V

†**35.** A particular solution curve of a differential equation whose slope field is shown above in II passes through the point $(0,-1)$. The equation is

   (A) $y = -e^x$ **(B)** $y = -e^{-x}$ **(C)** $y = x^2 - 1$ **(D)** $y = -\cos x$

   (E) $y = -\sqrt{1 - x^2}$

†**36.** At any point of intersection of a solution curve of the d.e. $y' = x + y$ and the line $x + y = 0$, the function y at that point

   (A) is equal to 0   **(B)** is a local maximum   **(C)** is a local minimum
   (D) has a point of inflection   **(E)** has a discontinuity

†**37.** The slope field for $F'(x) = e^{-x^2}$ is shown at the right with the particular solution $F(0) = 0$ superimposed. With a graphing calculator, $\lim\limits_{x \to \infty} F(x)$ to three decimal places is

   (A) 0.886   **(B)** 0.987
   (C) 1.000   **(D)** 1.414
   (E) ∞

$[-2,2] \times [-2,2]$

***38.** If you use Euler's method and two steps with $\Delta x = 0.1$ for the d.e. $y' = y$, with initial value $y(0) = 1$, then, when $x = 0.2$, y is approximately

   (A) 1.100   **(B)** 1.210   **(C)** 1.331
   (D) 1.464   **(E)** none of these

*An asterisk denotes a topic covered only in Calculus BC.
†The topic slope fields will not be tested on the AB examination before 2004.

*39.   The error in using Euler's method in Question 38 is approximately

    (A)   0.005   **(B)**   0.011   **(C)**   0.019   **(D)**   0.028   **(E)**   0.050

*40.   The table on page 308 lists the results of Euler's method for the initial-value problem

$$y' = x + y; \quad y(0) = 0$$

The approximate value of $y(0.47)$, according to the table, is

    (A)   0.045   **(B)**   0.055   **(C)**   0.068   **(D)**   0.096   **(E)**   none of these

41.   Which of these statements about Euler's method is(are) true?

    I.   It can be used to estimate solutions of differential equations numerically.

    II. It cannot be applied to an equation of the form $\dfrac{dy}{dx} = F(x, y)$, where F is defined implicitly.

    III. It should not be used on an interval on which the function becomes infinite.

    (A)   I only   **(B)**   II only   **(C)**   III only

    (D)   I and III only   **(E)**   I, II, and III

42.   Which statement about Euler's method is false?

    (A)   If you halve the step size, you approximately halve the error.
    (B)   Euler's method never gives exact solutions.
    (C)   Euler's method assumes that the slope of a solution curve is the same at all points in a short interval.
    (D)   Often, when applying Euler's method, the more steps you take the smaller the error.
    (E)   Euler's method is used to string together a set of linearizations that approximate a curve.

43.   A cup of coffee at temperature 180°F is placed on a table in a room at 68°F. The d.e. for its temperature at time t is $\dfrac{dy}{dt} = -0.11(y - 68)$; $y(0) = 180$. After 10 min the temperature (in °F) of the coffee is

    (A)   96   **(B)**   100   **(C)**   105   **(D)**   110   **(E)**   115

44.   Approximately how long does it take the temperature of the coffee in Question 43 to drop to 75°F?

    (A)   10 min   **(B)**   15 min   **(C)**   18 min   **(D)**   20 min   **(E)**   25 min

45.   The concentration of a medication injected into the bloodstream drops at a rate proportional to the existing concentration. If the factor of proportionality is 30% per hour, in how many hours will the concentration be one-tenth of the initial concentration?

    (A)   3   **(B)** $4\dfrac{1}{3}$ **(C)** $6\dfrac{2}{3}$

    (D) $7\dfrac{2}{3}$ **(E)**   none of these

*An asterisk denotes a topic covered only in Calculus BC.

*46. Which of the following statements characterize(s) the logistic growth of a population whose limiting value if L?

    I.   The rate of growth increases at first.

    II.   The growth rate attains a maximum when the population equals $\dfrac{L}{2}$.

    III.   The growth rate approaches 0 as the population approaches L.

    (A)   I only   **(B)**   II only   **(C)**   I and II only
    (D)   II and III only   **(E)**   I, II, and III

*47. Which of the following d.e.'s is not logistic?

    (A) $P' = P - P^2$ **(B)** $\dfrac{dy}{dt} = 0.01y(100 - y)$

    (C) $\dfrac{dx}{dt} = 0.8x - 0.004x^2$ **(D)** $\dfrac{dR}{dt} = 0.16(350 - R)$

    (E) $f'(t) = kf(t) \cdot [A - f(t)]$ (where k and A are constants)

*48. Suppose $P(t)$ denotes the size of an animal population at time t and its growth is described by the d.e. $\dfrac{dP}{dt} = 0.002P(1000 - P)$. The population is growing fastest

    (A)   initially   **(B)**   when $P = 500$ **(C)**   when $P = 1000$

    (D)   when $\dfrac{dP}{dt} = 0$ **(E)**   when $\dfrac{d^2P}{dt^2} > 0$

49. According to Newton's law of cooling, the temperature of an object decreases at a rate proportional to the difference between its temperature and that of the surrounding air. Suppose a corpse at a temperature of 32°C arrives at a mortuary where the temperature is kept at 10°C. Then the differential equation satisfied by the temperature T of the corpse t hr later is

    (A) $\dfrac{dT}{dt} = -k(T - 10)$ **(B)** $\dfrac{dT}{dt} = k(T - 32)$ **(C)** $\dfrac{dT}{dt} = 32e^{-kt}$

    (D) $\dfrac{dT}{dt} = -kT(T - 10)$ **(E)** $\dfrac{dT}{dt} = kT(T - 32)$

50. If the corpse in Question 49 cools to 27°C in 1 hr, then its temperature is given by the equation

    (A) $T = 22e^{0.205t}$ **(B)** $T = 10e^{1.163t}$ **(C)** $T = 10 + 22e^{-0.258t}$
    (D) $T = 32e^{-0.169t}$ **(E)** $T = 32 - 10e^{-0.093t}$

---

*An asterisk denotes a topic covered only in Calculus BC.

Answers for Set 9: Multiple-Choice Questions on Differential Equations

1. C	**11.** A	**21.** D	**31.** D	**41.** D
2. B	**12.** E	**22.** A	**32.** C	**42.** B
3. D	**13.** C	**23.** E	**33.** A	**43.** C
4. B	**14.** B	**24.** A	**34.** E	**44.** E
5. C	**15.** D	**25.** C	**35.** A	**45.** D
6. E	**16.** E	**26.** D	**36.** C	**46.** E
7. B	**17.** E	**27.** B	**37.** A	**47.** D
8. A	**18.** A	**28.** C	**38.** B	**48.** B
9. A	**19.** C	**29.** E	**39.** B	**49.** A
10. E	**20.** B	**30.** B	**40.** D	**50.** C

1. C. $v(t) = 2t^2 - t + C$; $v(1) = 3$; so $C = 2$.

2. B.   If $a(t) = 20t^3 - 6t$, then

$$v(t) = 5t^4 - 3t^2 + C_1,$$
$$s(t) = t^5 - t^3 + C_1 t + C_2,$$

Since

$$s(-1) = -1 + 1 - C_1 + C_2 = 2$$

and

$$s(1) = 1 - 1 + C_1 + C_2 = 4,$$

therefore

$$2C_2 = 6, \ C_2 = 3,$$
$$C_1 = 1.$$

So

$$v(t) = 5t^4 - 3t^2 + 1.$$

3. D.   From Answer 2, $s(t) = t^5 - t^3 + t + 3$, so $s(0) = C_2 = 3$.

4. B.   Since $a(t) = -32$, $v(t) = -32t + 40$, and the height of the stone $s(t) = -16t^2 + 40t + C$. When the stone hits the ground, 4 sec later, $s(t) = 0$, so

$$0 = -16(16) + 40(4) + C,$$
$$C = 96 \text{ ft.}$$

5. C.   From Answer 4

$$s(t) = -16t^2 + 40t + 96.$$

Then

$$s'(t) = -32t + 40,$$

which is zero if $t = 5/4$, and that yields maximum height, since $s''(t) = -32$.

6.   E.   The velocity $v(t)$ of the car is linear, since its acceleration is constant and

$$a(t) = \frac{dv}{dt} = \frac{(60-0) \text{ mph}}{10 \text{ sec}} = \frac{88 \text{ ft/sec}}{10 \text{ sec}} = 8.8 \text{ ft/sec}^2$$

$v(t) = 8.8t + C_1 \qquad \text{and} \qquad v(0) = 0, \quad \text{so } C_1 = 0;$

$s(t) = 4.4t^2 + C_2 \qquad \text{and} \qquad s(0) = 0, \quad \text{so } C_2 = 0;$

$$s(10) = 4.4(10^2) = 440 \text{ ft}.$$

7.   B.   Since $v = 100 - 20t$, $s = 100t - 10t^2 + C$ with $s(0) = 0$. So $s(1) = 100 - 10 = 90$ ft.

8.   A.   Since $v = -32t + v_0$ and $s = -16t^2 + v_0 t$, we solve simultaneously:

$$0 = -32t + v_0,$$
$$100 = -16t^2 + v_0 t.$$

These yield $t = 5/2$ and $v_0 = 80$ ft/sec.

9.   A.   The odometer measures the total trip distance from time $t = a$ to $t = b$ (whether the car moves forward or backward or reverses its direction one or more times from $t = a$ to $t = b$). This total distance is given exactly by

$$\int_a^b |v(t)| \, dt.$$

10.   E.   (A), (B), (C), and (D) are all true. See Answer 9.

11.   A.   Integrating yields $\frac{y^2}{2} = \frac{x^2}{2} + C$ or $y^2 = x^2 + 2C$ or $y^2 = x^2 + C'$, where we have replaced the arbitrary constant $2C$ by C'. Note that (A) is just $y^2 = x^2 + C'$, where $C' = -4$.

12.   E.   We separate variables. The particular solution is $\ln y = \sqrt{x} - 2$.

13.   C.   We separate variables. The particular solution is $-e^{-y} = x - 2$.

14.   B.   The general solution is $y = \sqrt{9 + x^2} + C$; $y = 5$ when $x = 4$ yields $C = 0$.

15.   D.   We separate variables to get $\csc^2 \left(\frac{\pi}{2} s \right) ds = dt$. We integrate:

$$-\frac{2}{\pi} \cot \left(\frac{\pi}{2} s \right) = t + C. \text{ With } t = 0 \text{ and } s = 1, C = 0. \text{ When } s = \frac{3}{2}, \text{ we get}$$

$$-\frac{2}{\pi} \cot \frac{3\pi}{4} = t.$$

16.   E.   Since $\int \frac{dy}{y} = \int \frac{dx}{x}$, it follows that

$$\ln y = \ln x + C \qquad \text{or} \qquad \ln y = \ln x + \ln k;$$

so $y = kx$.

17.   E.   The solution is $y = ke^x$, $k \neq 0$.

18.   A.   We rewrite and separate variables, getting $y \frac{dy}{dx} = x$. The general solution is

$$y^2 = x^2 + C \qquad \text{or} \qquad f(x) = \pm\sqrt{x^2 + C}.$$

19. C.   We are given that $\dfrac{dy}{dx} = \dfrac{3y}{x}$. The general solution is $\ln|y| = 3\ln|x| + C$.

Thus, $|y| = c\,|x^3|$; $y = \pm c\,x^3$. Since $y = 1$ when $x = 1$, we get $c = 1$.

20. B.   Since $\dfrac{dR}{dt} = cR$, $\dfrac{dR}{R} = c\,dt$, and $\ln R = ct + C$. When $t = 0$, $R = R_0$; so $\ln R_0 = C$ or $\ln R = ct + \ln R_0$. Thus

$$\ln R - \ln R_0 = ct; \quad \ln \frac{R}{R_0} = ct \qquad \text{or} \qquad \frac{R}{R_0} = e^{ct}.$$

21. D.   The question gives rise to the differential equation $\dfrac{dP}{dt} = kP$, where $P = 2P_0$ when $t = 50$. We seek $\dfrac{P}{P_0}$ for $t = 75$. We get $\ln \dfrac{P}{P_0} = kt$ with $\ln 2 = 50k$; then

$$\ln \frac{P}{P_0} = \frac{t}{50}\ln 2 \qquad \text{or} \qquad \frac{P}{P_0} = 2^{t/50}.$$

22. A.   We let S equal the amount present at time t; using $S = 40$ when $t = 0$ yields $\ln \dfrac{S}{40} = kt$. Since, when $t = 2$, $S = 10$, we get

$$k = \frac{1}{2}\ln\frac{1}{4} \qquad \text{or} \qquad \ln\frac{1}{2} \qquad \text{or} \qquad -\ln 2.$$

23. E.   The general solution is $y = k\ln|x| + C$, and the particular solution is $y = 2\ln|x| + 2$.

24. A.   We replace $g(x)$ by y and then solve the equation $\dfrac{dy}{dx} = \pm\sqrt{y}$. We use the constraints given to find the particular solution $2\sqrt{y} = x$ or $2\sqrt{g(x)} = x$.

25. C.   If P is the amount to be invested now, then

$$40{,}000 = Pe^{(0.08\cdot 35)}$$

and $P = \$2432.40$.

26. D.   We carefully(!) draw a curve for a solution to the d.e. represented by the slope field. It will be the graph of a member of the family $y = \sin x + C$. At the right we have superimposed the graph of the particular solution $y = \sin x - 0.5$.

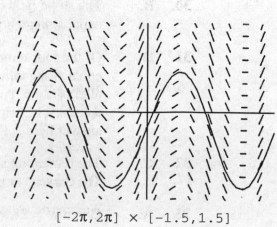

$$[-2\pi, 2\pi] \times [-1.5, 1.5]$$

27. **B.**

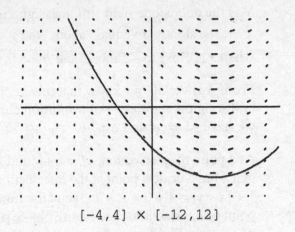

$$[-4,4] \times [-12,12]$$

It's easy to see that the answer must be choice (A), (B), or (C), because the slope field depends only on x: all the slope segments for a given x are parallel. Also, the solution curves in the slope field are all concave up, as they are only for choices (A) and (B). Finally, the solution curves all have a minimum at $x = 2$, which is true only for differential equation (B).

28. **C.** The general solution of $\dfrac{dy}{dx} = y$, or $\dfrac{dy}{y} = dx$ (with $y > 0$) is $\ln y = x + C$ or $y = ce^x$. For a solution to pass through (2,3), we have $3 = ce^2$ and $c = 3/e^2 \simeq 0.406$.

29. **E.** The *solution curve* is $y = \tan x$, which we can obtain from the differential equation $y' = 1 + y^2$ with the condition $y(0) = 0$ as follows:

$$\frac{dy}{1 + y^2} = dx, \qquad \tan^{-1} y = x, \qquad y = \tan x + C.$$

Since $y(0) = 0$, $C = 0$. Verify that (A) through (D) are incorrect.
NOTE: In matching slope fields and differential equations in Questions 30–34, keep in mind that if the slope segments along a vertical line are all parallel, signifying equal slopes for a fixed x, then the differential equation can be written as $y' = f(x)$. Replace "vertical" by "horizontal" and "x" by "y" in the preceding sentence to obtain a differential equation of the form $y' = g(y)$.

30. **B.** The slope field for $y' = y$ must by II; it is the only one whose slopes are equal along a horizontal line.

31. **D.** Of the four remaining slope fields, IV is the only one whose slopes are not equal along either a vertical or a horizontal line (the segments are *not* parallel). Its d.e. therefore cannot be either of type $y' = f(x)$ or $y' = g(y)$. The d.e. must be implicitly defined—that is, of the form $y' = F(x,y)$. So the answer here is IV.

32. **C.** The remaining slope fields, I, III, and V, all have d.e.'s of the type $y' = f(x)$. The curves "lurking" in III are trigonometric curves—not so in I and V.

33. **A.** Given $y' = 2x$, we immediately obtain the general solution, a family of parabolas, $y = x^2 + C$. (Trace the parabola in I through (0,0), for example.)

34. E. V is the only slope field still unassigned! Furthermore, the slopes "match" e^{-x^2}: the slopes are equal "about" the y-axis; slopes are very small when x is close to -2 and 2; and e^{-x^2} is a maximum at $x = 0$.

35. A. From Answer 30, we know that the d.e. for slope field II is $y' = y$. The general solution is $y = ce^x$. For a solution curve to pass through point $(0,-1)$, we have $-1 = ce^0$ and $c = -1$.

36. C. At a point of intersection, $y' = x + y$ and $x + y = 0$. So $y' = 0$, which implies that y has a critical point at the intersection. Since $y'' = 1 + y' = 1 + (x + y) = 1 + 0 = 1$, $y'' > 0$ and the function has a local minimum at the point of intersection. [See Figure N9–5, p. 303, showing the slope field for $y' = x + y$ and the curve $y = e^x - x - 1$ that has a local minimum at $(0,0)$.]

37. A. Although there is no elementary function (one made up of polynomial, trigonometric, or exponential functions or their inverses) that is an anti-derivative of $F'(x) = e^{-x^2}$, we know from the FTC, since $F(0) = 0$, that

$$F(x) = \int_0^x e^{-t^2}\, dt.$$

To approximate $\lim_{x \to \infty} F(x)$, we keyed in

`fnInt(e^{-x^2},X,0,50)`   and   `fnInt(e^{-x^2},X,0,60)`.

We got identical answers to 10 decimal places. Rounding to three decimal places yields 0.886.

38. B. Euler's method for $y' = y$, starting at $(0,1)$, with $\Delta x = 0.1$; yields

x	y	(SLOPE)* $\cdot \Delta x = \Delta y$	
0	1	$1 \cdot (0.1) = 0.1$	*The slope is y.
0.1	1.1	$(1.1) \cdot (0.1) = 0.11$	
0.2	1.21		

39. B. We want to compare $y(0.2)$ using Euler's method with the true value obtained by solving the d.e. $y' = y$. $\dfrac{dy}{y} = dx$, so $\ln y = x + C$. Then $y = ce^x$, and the initial value $y(0) = 1$ means $1 = ce^0 = c$, so $y = e^x$. Then $y = e^{0.2} \simeq 1.221$. The error is therefore

true value	–	Euler value	=	error
1.221	–	1.210	$\simeq$	0.011.

40. D. $y(0.47) = y(0.4) + 0.7(0.110 - 0.064) = 0.064 + 0.032 = 0.096.$

41. D. See Example 9 on page 308 — a counterexample to statement II.

42. B.   Statement (A) is justified on page 307. Statements (C) and (E) describe Euler's method. Statement (D) is illustrated by the examples presented in the text.

   (B) is false. Let $y' = c$, where c is a constant, and suppose $y(0) = 0$. Then Euler's method yields the solution curve for the linear function $y = cx$ (which is easy to verify). The exact solution of the given d.e., of course, is $y = cx$.

43. C.   We separate the variables in the given d.e., then solve:

$$\frac{dy}{y - 68} = -0.11\,dt,$$

$$\ln(y - 68) = -0.11t + c.$$

Since $y(0) = 180$, $\ln 112 = c$. Then

$$\ln\frac{y - 68}{112} = -0.11t,$$

$$y = 68 + 112e^{-0.11t}.$$

When $t = 10$, $y = 68 + 112e^{-1.1} \simeq 105°F$.

44. E.   The solution of the d.e. in Question 43, where y is the temperature of the coffee at time t, is

$$y = 68 + 112e^{-0.11t}.$$

We find t when $y = 75°F$:

$$75 = 68 + 112e^{-0.11t},$$

$$\frac{7}{112} = e^{-0.11t},$$

$$\frac{\ln 7 - \ln 112}{-0.11} = t \simeq 25 \text{ min.}$$

45. D.   If Q is the concentration at time t, then $\dfrac{dQ}{dt} = -0.30Q$. We separate variables and integrate:

$$\frac{dQ}{Q} = -0.30\,dt \quad \rightarrow \quad \ln Q = -0.30t + C.$$

We let $Q(0) = Q_0$. Then

$$\ln Q = -0.30t + \ln Q_0 \quad \rightarrow \quad \ln\frac{Q}{Q_0} = -0.30t \quad \rightarrow \quad \frac{Q}{Q_0} = e^{-0.30t}.$$

We now find t when $Q = 0.1Q_0$:

$$0.1 = e^{-0.30t},$$

$$t = \frac{\ln 0.1}{-0.3} \simeq 7\frac{2}{3} \text{ hr.}$$

46. E. See page 326 for the characteristics of the logistic model.

47. D. Verify that (A), (B), (C), and (E) are all logistic equations.

48. B. The rate of growth, $\dfrac{dP}{dt}$, is greatest when its derivative is 0 and the curve of $y' = \dfrac{dP}{dt}$ is concave down. Since

$$\frac{dP}{dt} = 2P - 0.002P^2,$$

therefore

$$\frac{d^2P}{dt^2} = 2 - 0.004P,$$

which is equal to 0 if $y'' = \dfrac{2}{0.004}$, or 500, animals. The curve of y' is concave down for all P, since

$$\frac{d}{dt}\left(\frac{d^2P}{dt^2}\right) = -0.004,$$

so $P = 500$ is the maximum population.

49. A. The description of temperature change here is an example of Case II (page 316): the rate of change is proportional to the amount or magnitude of the quantity present (i.e., the temperature of the corpse) minus a fixed constant (the temperature of the mortuary).

50. C. Since (A) is the correct answer to Question 49, we solve the d.e. in (A) given the initial condition $T(0) = 32$:

$$\frac{dT}{T-10} = -k\,dt,$$

$$\ln(T-10) = -kt + C.$$

Using $T(0) = 32$, we get $\ln(22) = C$, so

$$\ln(T-10) = -kt + \ln 22,$$

$$T - 10 = 22e^{-kt}.$$

To find k, we use the given information that $T(1) = 27$:

$$27 - 10 = 22e^{-k},$$
$$\frac{17}{22} = e^{-k},$$
$$k = 0.258.$$

Therefore $T = 10 + 22e^{-0.258t}$.

Chapter 10

Sequences and Series

Review of Principles

A. Sequences of Real Numbers‡

A1. DEFINITIONS.

An *infinite sequence* is a function whose domain is the set of positive integers. The sequence $\{n, f(n) | n = 1, 2, 3, \ldots\}$ is often denoted simply by $\{f(n)\}$. The sequence defined, for example, by $f(n) = \dfrac{1}{n}$ is the set of numbers $1, \dfrac{1}{2}, \dfrac{1}{3}, \ldots, \dfrac{1}{n}, \ldots$. The elements in this set are called the *terms* of the sequence, and the *nth* or *general* term of this sequence is $\dfrac{1}{n}$. Frequently the sequence is denoted by its *n*th term: $\{s_n\}$ is the sequence whose *n*th term is s_n. Thus,

$$\left\{\frac{n}{n^2 + 1}\right\} = \frac{1}{2}, \frac{2}{5}, \frac{3}{10}, \ldots;$$

$$\left\{\frac{2^n}{n!}\right\} = \frac{2}{1}, \frac{2^2}{2!}, \frac{2^3}{3!}, \ldots.$$

[Recall that for every nonnegative integer the definition of $n!$ (read "*n* factorial") is given by $0! = 1$, $(n + 1)! = (n + 1)n!$. So, for example, $1! = 1$, $2! = 2 \cdot 1$, $7! = 7 \cdot 6 \cdot 5 \cdot 4 \cdot 3 \cdot 2 \cdot 1$.]

$$\left\{1 + (-1)^n\right\} = 0, 2, 0, 2, \ldots;$$

$$\left\{1 + \frac{(-1)^n}{n}\right\} = 0, \frac{3}{2}, \frac{2}{3}, \frac{5}{4}, \frac{4}{5}, \frac{7}{6}, \frac{6}{7}, \ldots.$$

---

*An asterisk denotes a topic covered only in Calculus BC.
‡Topic will not be tested on the AP examination.

A *sequence* $\{s_n\}$ has the *limit L* if the difference between s_n and L can be made arbitrarily small by taking n sufficiently large. If $\{s_n\}$ has the limit L, we say it *converges* to L, and write

$$\lim_{n \to \infty} \{s_n\} = L.$$

Such a sequence is also said to be *convergent*. If $\{s_n\}$ does not have a (finite) limit, we say it is *divergent*.

A2. THEOREMS.

Several theorems follow from the definition of convergence:

THEOREM 2a. The limit of a convergent sequence is unique.
THEOREM 2b. $\lim_{n \to \infty} (c \cdot s_n) = c \lim_{n \to \infty} s_n$ (c a number).

THEOREM 2c. If two sequences converge, so do their sum, product, and quotient (with division by zero to be avoided).

Example 1. $\lim_{n \to \infty} \dfrac{1}{n} = 0.$

Example 2. $\lim_{n \to \infty} 1 + \dfrac{(-1)^n}{n} = 1.$

Example 3. $\lim_{n \to \infty} \dfrac{2n^2}{3n^3 - 1} = 0$ (by the Rational Function Theorem, page 30).

Example 4. $\lim_{n \to \infty} \dfrac{3n^4 + 5}{4n^4 - 7n^2 + 9} = \dfrac{3}{4}.$

Example 5. $\left\{ \dfrac{n^2 - 1}{n} \right\}$ diverges (to infinity), since $\lim_{n \to \infty} \dfrac{n^2 - 1}{n} = \infty.$

Example 6. $\{\sin n\}$ diverges since the sequence fails to have a limit, but note that it does not diverge to infinity.

Example 7. $\{(-1)^{n+1}\} = 1, -1, 1, -1, \ldots$ diverges because it oscillates.

Example 8. $\lim_{n \to \infty} \dfrac{\ln n}{n} = \lim_{n \to \infty} \dfrac{1/n}{1} = 0$ (by L'Hôpital's rule, page 63).

Example 9. $\lim_{n \to \infty} \dfrac{e^n}{n^2} = \lim_{n \to \infty} \dfrac{e^n}{2n} = \lim_{n \to \infty} \dfrac{e^n}{2} = \infty$ by repeated application of L'Hôpital's rule; the sequence diverges.

Example 10.   The sequence $\left\{ (-1)^{n+1} \cdot \dfrac{n}{n + 1} \right\}$ is graphed in Figure N10–1. On our calculator, we entered the SEQ and DOT modes. On the Y= screen, for Un, we entered

```
(-1)^(n+1)(n/(n+1))
```

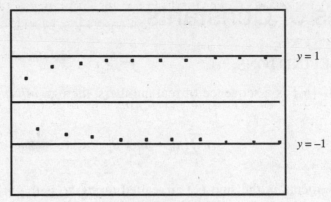

FIGURE N10–1

The sequence is $\frac{1}{2}, \; -\frac{2}{3}, \; +\frac{3}{4}, \; -\frac{4}{5}, \; +\frac{5}{6}, \ldots$. The positive (odd-numbered) terms of this sequence approach 1 as $n \to \infty$; the negative ones (even-numbered) terms approach -1. It is illuminating to TRACE the graph and see the cursor move up and down between positive and negative terms. It is clear that the sequence diverges.

DEFINITIONS. The sequence $\{s_n\}$ is said to be *increasing* if, for all n, $s_n \le s_{n+1}$, and to be *decreasing* if, for all n, $s_{n+1} \le s_n$; a *monotonic* sequence is either increasing or decreasing. If $|s_n| \le M$ for all n, then $\{s_n\}$ is said to be *bounded*.

Theorem 2d is the basic one on sequences:

THEOREM 2d. A monotonic sequence converges if and only if it is bounded.

In particular, if $\{s_n\}$ is increasing and $s_n \le M$ for all n, then $\{s_n\}$ converges to a limit that is less than or equal to M.

THEOREM 2e. Every unbounded sequence diverges. (Of course, a bounded sequence may also diverge; for example $1, -1, 1, -1, \ldots$, or $\{\cos n\}$.)

Example 11. Use Theorem 2d to prove that $\left\{1 - \frac{2}{n}\right\}$ converges.

If $s_n = 1 - \frac{2}{n}$, then $-1 \le s_n < 1$ for all n; also $s_{n+1} > s_n$, since

$$s_{n+1} - s_n = \left(1 - \frac{2}{n+1}\right) - \left(1 - \frac{2}{n}\right) = \frac{2}{n} - \frac{2}{n+1} = \frac{2}{n(n+1)},$$

which is positive for all n. Note that $\{s_n\}$ converges to 1.

Example 12. For what r's does $\{r^n\}$ converge?

If $|r| > 1$, then $\{r^n\}$ is unbounded, so diverges by Theorem 2e.

If $|r| < 1$, then $\lim_{n \to \infty} r^n = 0$.

If $r = 1$, then $\{r^n\}$ converges to 1.

If $r = -1$, then $\{r^n\}$ diverges by oscillation.

Finally, then, $\{r^n\}$ converges if $-1 < r \le 1$.

B.   Series of Constants

B1. DEFINITIONS.

If $\{u_n\}$ is a sequence of real numbers, then an *infinite series* is an expression of the form

$$\sum_{k=1}^{\infty} u_k = u_1 + u_2 + u_3 + \cdots + u_n + \cdots . \qquad (1)$$

The elements in the sum (1) are called *terms;* u_n is the *nth* or *general* term of series (1). Associated with series (1) is the sequence $\{s_n\}$, where

$$s_n = u_1 + u_2 + u_3 + \cdots + u_n; \qquad (2)$$

that is,

$$s_n = \sum_{k=1}^{n} u_k .$$

The sequence $\{s_n\}$ is called the *sequence of partial sums* of series (1). Note that

$$\{s_n\} = s_1, s_2, s_3, \ldots, s_n, \ldots \qquad (3)$$
$$= u_1, u_1 + u_2,\ u_1 + u_2 + u_3,\ \ldots,\ u_1 + u_2 + \cdots + u_n, \ldots .$$

If there is a finite number S such that

$$\lim_{n \to \infty} s_n = S ,$$

then we say that series (1) is *convergent,* or *converges to S,* or *has the sum S.* We write, in this case,

$$\sum_{k=1}^{\infty} u_k = S, \text{ where } \sum_{k=1}^{\infty} u_k = \lim_{n \to \infty} \sum_{k=1}^{n} u_k .$$

When there is no source of confusion, the infinite series (1) may be indicated simply by

$$\sum u_k \quad \text{or} \quad \sum u_n .$$

Note that, if the sequence $\{s_n\}$ of partial sums given by (3) converges, then so does series (1); if the sequence $\{s_n\}$ diverges, then so does the series.

Example 13.   Show that the geometric series

$$\frac{1}{2} + \frac{1}{4} + \cdots + \frac{1}{2^n} + \cdots \qquad (4)$$

converges to 1.

The series can be rewritten as

$$\left(1 - \frac{1}{2}\right) + \left(\frac{1}{2} - \frac{1}{4}\right) + \cdots + \left(\frac{1}{2^{n-1}} - \frac{1}{2^n}\right) + \cdots ; \qquad (5)$$

hence

$$s_n = \left(1 - \frac{1}{2}\right) + \left(\frac{1}{2} - \frac{1}{4}\right) + \cdots + \left(\frac{1}{2^{n-1}} - \frac{1}{2^n}\right)$$

$$= 1 - \frac{1}{2^n}.$$

Since $\lim_{n \to \infty} s_n = 1$,

$$\sum_{k=1}^{\infty} \frac{1}{2^k} = 1.$$

The trick above is equivalent to the following:

$$s_n = \frac{1}{2} + \frac{1}{4} + \frac{1}{8} + \cdots + \frac{1}{2^n}; \tag{6}$$

$$\frac{1}{2} s_n = \frac{1}{4} + \frac{1}{8} + \cdots + \frac{1}{2^n} + \frac{1}{2^{n+1}}. \tag{7}$$

Subtracting (7) from (6) yields

$$\frac{1}{2} s_n = \frac{1}{2} - \frac{1}{2^{n+1}} \quad \text{or} \quad s_n = 1 - \frac{1}{2^n}.$$

So, as before, $\lim_{n \to \infty} s_n = 1$.

We can see the convergence of $\displaystyle\sum_{1}^{\infty} \frac{1}{2^n}$ very nicely on a graphing calculator by graphing the first 10 partial sums. After selecting the [Par] and [Dot] modes, we entered $X_{1T} = T$, $Y_{1T} = $ sum seq $((1/2)\,\wedge\,N, N, 1, T, 1)$. The partial sums are shown in Figure N10–2.

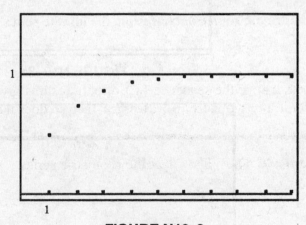

FIGURE N10–2

We've added the horizontal line at $y = 1$ to show how the partial sums approach it. Here, too, it is instructive to trace the graph "hopping" from dot to dot and observing the values of Y on the screen as T (or N) varies from 1 to 10.

Example 14.   Show that the *harmonic series*

$$1 + \frac{1}{2} + \frac{1}{3} + \frac{1}{4} + \cdots + \frac{1}{n} + \cdots \tag{8}$$

diverges.

The terms in (8) can be grouped as follows:

$$1 + \frac{1}{2} + \left(\frac{1}{3} + \frac{1}{4}\right) + \left(\frac{1}{5} + \frac{1}{6} + \frac{1}{7} + \frac{1}{8}\right) + \left(\frac{1}{9} + \frac{1}{10} + \cdots + \frac{1}{16}\right)$$
$$+ \left(\frac{1}{17} + \cdots + \frac{1}{32}\right) + \cdots . \tag{9}$$

The sum in (9) clearly exceeds

$$1 + \frac{1}{2} + 2\left(\frac{1}{4}\right) + 4\left(\frac{1}{8}\right) + 8\left(\frac{1}{16}\right) + 16\left(\frac{1}{32}\right) + \cdots ,$$

which equals

$$1 + \frac{1}{2} + \frac{1}{2} + \frac{1}{2} + \frac{1}{2} + \frac{1}{2} + \cdots . \tag{10}$$

Since the sum in (10) can be made arbitrarily large, it follows that $\sum \frac{1}{n}$ diverges.

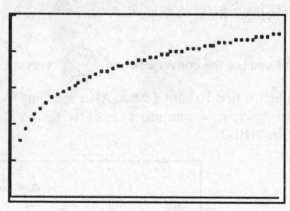

FIGURE N10–3a

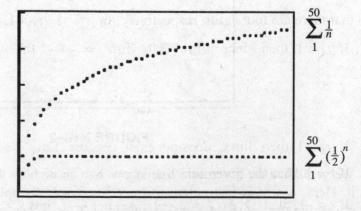

$$\sum_{1}^{50} \frac{1}{n}$$

$$\sum_{1}^{50} \left(\frac{1}{2}\right)^{n}$$

FIGURE N10–3b

In Figure N10–3a we see $\displaystyle\sum_{1}^{50}\frac{1}{n}$; when we TRACE this graph we find that

$$\sum_{1}^{10}\frac{1}{n}\simeq 2.9,\qquad \sum_{1}^{20}\frac{1}{n}\simeq 3.6,\qquad \sum_{1}^{50}\frac{1}{n}\simeq 4.5\,.$$

We showed above that $\displaystyle\sum\frac{1}{n}$ can be made to exceed any given number by making n sufficiently large.

Figure N10–3b compares the graphs of $\displaystyle\sum_{1}^{50}\left(\frac{1}{2}\right)^{n}$ and $\displaystyle\sum_{1}^{50}\frac{1}{n}$. The difference is remarkable! Tracing the graph of the geometric series, we see that the sum of 10 terms, to three decimal places, is 0.999, while the sum of 26 terms, to eight decimal places, is 0.99999999.

The geometric series

$$\sum_{k=1}^{\infty}ar^{k-1}=a+ar+ar^{2}+\cdots+ar^{k-1}+\cdots \qquad (a\neq 0) \tag{11}$$

converges if $|r|<1$ but diverges if $|r|\geqslant 1$.

Series (4) in Example 13 is a special case of (11) with $a=r=\dfrac{1}{2}$; the technique applied there is used here. For (11),

$$s_{n}=a+ar+ar^{2}+\cdots+ar^{n-1};$$
$$rs_{n}=\quad ar+ar^{2}+\cdots+ar^{n-1}+ar^{n};$$
$$s_{n}(1-r)=a-ar^{n}=a(1-r^{n});$$

and

$$s_{n}=\begin{cases}\dfrac{a(1-r^{n})}{1-r} & \text{if } r\neq 1,\\[2mm] a+a+\cdots+a=na & \text{if } r=1.\end{cases} \tag{12}$$

We now investigate $\displaystyle\lim_{n\to\infty}s_{n}$ to determine the behavior of series (11). We see from (12) that there are four cases, respectively for $|r|<1$, $|r|>1$, $r=1$, and $r=-1$.

If $|r|<1$, then, since $\displaystyle\lim_{n\to\infty}r^{n}=0$, $\displaystyle\lim_{n\to\infty}s_{n}=\frac{a}{1-r}$; thus

$$\sum ar^{k-1}=\frac{a}{1-r}. \tag{13}$$

If $|r|>1$, then $\displaystyle\lim_{n\to\infty}s_{n}$ does not exist, because $\displaystyle\lim_{n\to\infty}r^{n}=\infty$.

If $r=1$, then the given series diverges.

If $r=-1$, then $\displaystyle\sum ar^{k-1}=a-a+a-a+\cdots$; thus $s_{1}=a,\ s_{2}=0,\ s_{3}=a,\ s_{4}=0$. In this case $\{s_{n}\}$ oscillates and thus diverges; and so does the series.

The geometric series therefore converges if $|r|<1$ but diverges otherwise.

Example 15.   The series $0.3 + 0.03 + 0.003 + \cdots$ is geometric with $a = 0.3$ and $r = 0.1$. Its sum is therefore

$$\frac{a}{1-r} \quad \text{or} \quad \frac{0.3}{1-0.1} = \frac{0.3}{0.9} = \frac{1}{3}.$$

Of course $\frac{1}{3} = 0.333\ldots$, which is the given series.

B2.   THEOREMS ABOUT CONVERGENCE OR DIVERGENCE OF SERIES OF CONSTANTS.

The following theorems are important.

THEOREM 2a.   If $\sum u_k$ converges, then $\lim\limits_{n \to \infty} u_n = 0$.

This provides a convenient and useful test for divergence, since it is equivalent to the statement: If u_n does not approach zero, then the series $\sum u_n$ diverges. Note, however, particularly that the converse of Theorem 2a is *not* true. The condition that u_n approach zero is *necessary but not sufficient* for the convergence of the series. The harmonic series $\sum \frac{1}{n}$ is an excellent example of a series whose nth term goes to zero but that diverges (see Example 14 above). The series $\sum \frac{n}{n+1}$ diverges because $\lim\limits_{n \to \infty} u_n = 1$, not zero; the series $\sum \frac{n}{n^2+1}$ does not converge even though $\lim\limits_{n \to \infty} u_n = 0$.

THEOREM 2b.   A finite number of terms may be added to or deleted from a series without affecting its convergence or divergence; thus

$$\sum_{k=1}^{\infty} u_k \quad \text{and} \quad \sum_{k=m}^{\infty} u_k$$

(where m is any positive integer) both converge or both diverge. (Of course, the sums may differ.)

THEOREM 2c.   The terms of a series may be multiplied by a nonzero constant without affecting the convergence or divergence; thus

$$\sum_{k=1}^{\infty} a_k \quad \text{and} \quad \sum_{k=1}^{\infty} ca_k \quad (c \neq 0)$$

both converge or both diverge. (Again, the sums may differ.)

THEOREM 2d.   If $\sum a_n$ and $\sum b_n$ both converge, so does $\sum (a_n + b_n)$.

THEOREM 2e.   If the terms of a convergent series are regrouped, the new series converges.

B3.   TESTS FOR CONVERGENCE OF POSITIVE SERIES.

The series $\sum u_n$ is called a positive series if $u_n > 0$ for all n. Note for such a series that the associated sequence of partial sums $\{s_n\}$ is increasing. It then follows from Theorem 2d, page 341, that a positive series converges if and only if $\{s_n\}$ has an upper bound.

Example 16. $\sum \dfrac{1}{k!} = 1 + \dfrac{1}{2!} + \dfrac{1}{3!} + \cdots + \dfrac{1}{n!} + \cdots$ converges, since

$$s_n = 1 + \dfrac{1}{2!} + \dfrac{1}{3!} + \dfrac{1}{4!} + \cdots \dfrac{1}{n!}$$

$$< 1 + \dfrac{1}{2} + \dfrac{1}{2 \cdot 2} + \dfrac{1}{2 \cdot 2 \cdot 2} + \cdots + \dfrac{1}{2^n}$$

$$< 2$$

(using Equation (13), page 345, with $a = 1$ and $r = \dfrac{1}{2}$). Thus $\{s_n\}$ has an upper bound and the given series converges.

Example 17. $\sum \dfrac{1}{n}$ diverges since the sequence of sums $\{s_n\}$ is unbounded (see Example 14, page 344).

TESTS TO APPLY TO $\{u_n\}$ if $u_n > 0$.

TEST 3a.   If $\lim\limits_{n \to \infty} u_n \neq 0$, then $\sum u_n$ diverges. This is the contrapositive of Theorem 2a, page 346. A statement and its contrapositive are equivalent.

TEST 3b.   THE INTEGRAL TEST.   If $f(x)$ is a continuous, positive, decreasing function for which $f(n)$ is the nth term u_n of the series, then $\sum u_n$ converges if and only if the improper integral $\displaystyle\int_1^\infty f(x)\,dx$ converges.

PROOF.   First we show that if $\displaystyle\int_1^\infty f(x)\,dx$ diverges, then $\displaystyle\sum_1^\infty u_n$ diverges.

In Figure N10–4 we see that

$$\int_1^\infty f(x)\,dx \leq \sum_1^\infty u_n,$$

where u_n is the area of the nth rectangle, whose base is 1 and whose height is u_n. Thus, if $\displaystyle\int_1^\infty f(x)\,dx$ diverges, then so does $\displaystyle\sum_1^\infty u_n$.

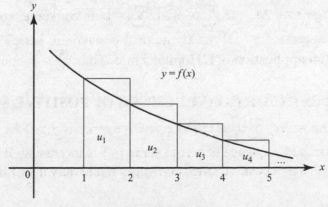

FIGURE N10–4

Next we show that if $\int_1^\infty f(x)\,dx$ converges, then $\sum_2^\infty u_n$ converges. Figure N10–5 shows that

$$\sum_2^\infty u_n \;\leq\; \int_1^\infty f(x)\,dx$$

so that

$$u_1 + \sum_2^\infty u_n \;\leq\; u_1 + \int_1^\infty f(x)\,dx$$

or

$$\sum_1^\infty u_n \;\leq\; u_1 + \int_1^\infty f(x)\,dx.$$

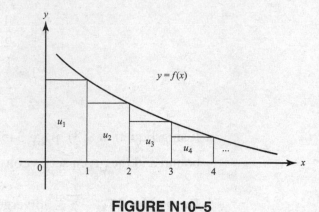

FIGURE N10–5

Since each term of the series is positive, the partial sums, which are increasing and bounded, must converge. Thus, if the integral converges, so does the series.

Example 18.   Does $\sum \dfrac{n}{n^2 + 1}$ converge?

The associated improper integral is

$$\int_1^\infty \frac{x\,dx}{x^2 + 1},$$

which equals

$$\lim_{b\to\infty} \frac{1}{2}\ln(x^2 + 1)\,\Big|_1^b = \infty.$$

The improper integral and the infinite series both diverge.

Example 19.   Test the series $\sum \dfrac{n}{e^n}$ for convergence.

$$\int_1^\infty \frac{x}{e^x}\,dx = \lim_{b\to\infty} \int_1^b xe^{-x}\,dx = \lim_{b\to\infty} -e^{-x}(1+x)\,\Big|_1^b$$

$$= -\lim_{b\to\infty}\left(\frac{1+b}{e^b} - \frac{2}{e}\right) = \frac{2}{e}$$

by an application of L'Hôpital's rule. Thus $\sum \dfrac{n}{e^n}$ converges.

Example 20.   Show that the p-series $\sum \dfrac{1}{n^p}$ converges if $p > 1$ but diverges if $p \leqq 1$.

Let $f(x) = \dfrac{1}{x^p}$ and investigate $\displaystyle\int_1^\infty \dfrac{dx}{x^p}$.

(a) If $p > 1$, then

$$\int_1^\infty x^{-p}dx = \lim_{b\to\infty}\frac{1}{1-p}\cdot x^{1-p}\ \bigg|_1^b = \lim_{b\to\infty}\frac{1}{1-p}(b^{1-p}-1)$$

$$= \frac{1}{p-1}\lim_{b\to\infty}\left(1-\frac{1}{b^{p-1}}\right) = \frac{1}{p-1}.$$

Consequently, $\sum \dfrac{1}{n^p}$ converges if $p > 1$.

(b) If $p = 1$, the improper integral is

$$\int_1^\infty \frac{dx}{x} = \lim_{b\to\infty}\ln b = \infty,$$

so $\sum \dfrac{1}{n}$ diverges, a result previously well established.

(c) If $p < 1$, then

$$\int_1^\infty x^{-p}dx = \lim_{b\to\infty}\frac{1}{1-p}(b^{1-p}-1) = \infty,$$

so $\sum \dfrac{1}{n^p}$ diverges.

TEST 3c. THE COMPARISON TEST.   We compare the general term of $\sum u_n$, the positive series we are investigating, with the general term of a series known to converge or diverge.

(1) If $\sum a_n$ converges and $u_n \leqq a_n$, then $\sum u_n$ converges.

(2) If $\sum b_n$ diverges and $u_n \geqq b_n$, then $\sum u_n$ diverges.

Any known series can be used for comparison. Particularly useful are the p-series, which converges if $p > 1$ but diverges if $p \leqq 1$ (see Example 20), and the geometric series, which converges if $|r| < 1$ but diverges if $|r| \geqq 1$.

Example 21. $1 + \dfrac{1}{2!} + \dfrac{1}{3!} + \cdots + \dfrac{1}{n!} + \cdots$ converges, since $\dfrac{1}{n!} < \dfrac{1}{2^{n-1}}$, and the

latter is the general term of the geometric series $\displaystyle\sum_1^\infty r^{n-1}$ with $r = \dfrac{1}{2}$.

Example 22. $\dfrac{1}{\sqrt{2}} + \dfrac{1}{\sqrt{5}} + \dfrac{1}{\sqrt{8}} + \cdots + \dfrac{1}{\sqrt{3n-1}} + \cdots$ diverges, since

$$\frac{1}{\sqrt{3n-1}} > \frac{1}{\sqrt{3n}} = \frac{1}{\sqrt{3}\cdot n^{1/2}} \; ;$$

the latter is the general term of the divergent p-series $\sum \dfrac{c}{n^p}$, where $c = \dfrac{1}{\sqrt{3}}$ and $p = \dfrac{1}{2}$.

Remember in using the Comparison Test that we may either discard a finite number of terms or multiply each term by a nonzero constant without affecting the convergence of the series we are testing.

Example 23. $\displaystyle\sum \frac{1}{n^n} = 1 + \frac{1}{2^2} + \frac{1}{3^3} + \cdots + \frac{1}{n^n} + \cdots$ converges, since $\dfrac{1}{n^n} < \dfrac{1}{2^{n-1}}$ and $\displaystyle\sum \frac{1}{2^{n-1}}$ converges.

TEST 3d. THE RATIO TEST. Let $\displaystyle\lim_{n\to\infty} \frac{u_{n+1}}{u_n} = L$, if it exists. Then $\displaystyle\sum u_n$ converges if $L < 1$ and diverges if $L > 1$. If $L = 1$, this test fails; apply one of the other tests.

Example 24. For $\displaystyle\sum \frac{1}{n!}$,

$$\lim_{n\to\infty} \frac{u_{n+1}}{u_n} = \lim_{n\to\infty} \frac{\frac{1}{(n+1)!}}{\frac{1}{n!}} = \lim_{n\to\infty} \frac{n!}{(n+1)!} = \lim_{n\to\infty} \frac{1}{n+1} = 0.$$

Therefore this series converges by the Ratio Test. (Compare Examples 16, page 347, and 21, page 349.)

Example 25. For $\displaystyle\sum \frac{n^n}{n!}$, we again use the Ratio Test:

$$\frac{u_n + 1}{u_n} = \frac{(n+1)^{n+1}}{(n+1)!} \cdot \frac{n!}{n^n} = \frac{(n+1)^n}{n^n}$$

and

$$\lim_{n\to\infty} \left(\frac{n+1}{n}\right)^n = \lim_{n\to\infty} \left(1 + \frac{1}{n}\right)^n = e.$$

(See §D2, page 32.) So the given series diverges since $e > 1$.

Example 26.   If the Ratio Test is applied to the p-series, $\sum \dfrac{1}{n^p}$, then

$$\frac{u_{n+1}}{u_n} = \frac{\dfrac{1}{(n+1)^p}}{\dfrac{1}{n^p}} = \left(\frac{n}{n+1}\right)^p \quad \text{and} \quad \lim_{n \to \infty}\left(\frac{n}{n+1}\right)^p = 1 \text{ for all } p.$$

But if $p > 1$ then $\sum \dfrac{1}{n^p}$ converges, while if $p \le 1$ then $\sum \dfrac{1}{n^p}$ diverges. This illustrates the failure of the Ratio Test to resolve the question of convergence when the limit of the ratio is 1.

B4. ALTERNATING SERIES AND ABSOLUTE CONVERGENCE.

Any test that can be applied to a positive series can be used for a series all of whose terms are negative. We consider here only one type of series with mixed signs, the so-called *alternating series*. This has the form:

$$\sum_{k=1}^{\infty} (-1)^{k+1} u_k = u_1 - u_2 + u_3 - u_4 + \cdots + (-1)^{k+1} u_k + \cdots, \tag{1}$$

where $u_k > 0$. The series

$$1 - \frac{1}{2} + \frac{1}{3} - \frac{1}{4} + \cdots + (-1)^{n+1} \cdot \frac{1}{n} + \cdots \tag{2}$$

is the *alternating harmonic* series.

THEOREM 4a.   The alternating series (1) converges if $u_{n+1} < u_n$ for all n and if $\lim_{n \to \infty} u_n = 0$.

Example 27.   The alternating harmonic series (2) converges, since $\dfrac{1}{n+1} < \dfrac{1}{n}$ for all n and since $\lim_{n \to \infty} \dfrac{1}{n} = 0$.

Example 28.   The series $\dfrac{1}{2} - \dfrac{2}{3} + \dfrac{3}{4} - \cdots$ diverges, since we see that $\lim_{n \to \infty} u_n = \lim_{n \to \infty} \dfrac{n}{n+1}$ is 1, not 0. (By Theorem 2a, page 346, if u_n does not approach 0, then $\sum u_n$ does not converge.)

THEOREM 4b.   If $u_{n+1} < u_n$ and $\lim_{n\to\infty} u_n = 0$ for the alternating series, then the numerical difference between the sum of the first n terms and the sum of the series is less than u_{n+1}; that is,

$$\left| \sum_{k=1}^{\infty} u_k - \sum_{k=1}^{n} u_k \right| < u_{n+1}.$$

Theorem 4b is exceedingly useful in computing the maximum possible error if a finite number of terms of a convergent series is used to approximate the sum of the series. If we stop with the nth term, the error is less than u_{n+1}, the first term omitted. The maximum possible or worst error, u_{n+1}, in the centered inequality above, is called the *error bound* for the alternating series. The error bound for an alternating series is therefore the first term omitted or dropped.

Example 29.   The sum $\displaystyle\sum_{k=1}^{\infty} \frac{(-1)^{k+1}}{k}$ differs from the sum

$$\left(1 - \frac{1}{2} + \frac{1}{3} - \frac{1}{4} + \frac{1}{5} - \frac{1}{6} \right)$$

by less than $\dfrac{1}{7}$, which is the error bound.

A series with mixed signs is said to *converge absolutely* (or to be *absolutely convergent*) if the series obtained by taking the absolute values of its terms converges; that is, $\displaystyle\sum u_n$ converges absolutely if $\displaystyle\sum |u_n| = |u_1| + |u_2| + \cdots + |u_n| + \cdots$ converges. A series that converges but not absolutely is said to converge *conditionally* (or to be *conditionally convergent*). The alternating harmonic series (2) converges conditionally since it converges, but does not converge absolutely. (The harmonic series diverges.)

THEOREM 4c.   If a series converges absolutely, then it converges.

Example 30.   Determine whether $\displaystyle\sum \frac{\sin \dfrac{n\pi}{3}}{n^2}$ converges absolutely, converges conditionally, or diverges.

Note that, since $\left| \sin n \dfrac{\pi}{3} \right| \leq 1$,

$$\left| \frac{\sin n \frac{\pi}{3}}{n^2} \right| \leq \frac{1}{n^2} \quad \text{for all } n.$$

But $\dfrac{1}{n^2}$ is the general term of a convergent p-series, so by the Comparison Test the given series converges absolutely (therefore, by Theorem 4c above, it converges).

Example 31.   The series

$$1 + \frac{1}{2} - \frac{1}{4} - \frac{1}{8} + \frac{1}{16} + \frac{1}{32} - \cdots (\pm) \frac{1}{2^n} \pm \cdots, \qquad (3)$$

whose signs are $+ + - - + + - - \ldots$, converges absolutely, since $\sum \frac{1}{2^n}$ is a convergent geometric series. Consequently, (3) converges.

B5.   COMPUTATIONS WITH SERIES OF CONSTANTS.

The nth partial sum of a convergent series, given by $s_n = \sum\limits_{k=1}^{n} u_k$, is an approximation to the sum of the whole series, $S = \sum\limits_{k=1}^{\infty} u_k$. The error introduced by using s_n for S, referred to as the remainder *after n* terms, is denoted by R_n. Note that

$$R_n = u_{n+1} + u_{n+2} + u_{n+3} + \cdots.$$

We summarize three useful theorems about the remainder for *convergent* series:

THEOREM 5a.   For the convergent geometric series $\sum\limits_{n=1}^{\infty} ar^{n-1}$, with $|r| < 1$,

$$|R_n| = \left| \frac{ar^n}{1-r} \right|.$$

This follows immediately from the results obtained on page 345.

$$s_n = \frac{a(1-r^n)}{1-r} \quad \text{and} \quad S = \frac{a}{1-r}.$$

THEOREM 5b.   For the convergent alternating series

$$u_1 - u_2 + u_3 - u_4 + \cdots + (-1)^{n+1} u_n + \cdots,$$

where $u_n > 0$ for all n, $R_n < u_{n+1}$. Furthermore, $R_n > 0$ if n is even but $R_n < 0$ if n is odd. Thus the error bound is the magnitude of the first term dropped. Note that Theorem 5b is a restatement of Theorem 4b, page 352.

THEOREM 5c.   Suppose the Integral Test (page 347) shows that the positive series $\sum\limits_{n=1}^{\infty} u_n$ converges. Then

$$R_n < \int_n^{\infty} f(x)\, dx \quad \text{(where } \int_1^{\infty} f(x)\, dx \text{ converges)}.$$

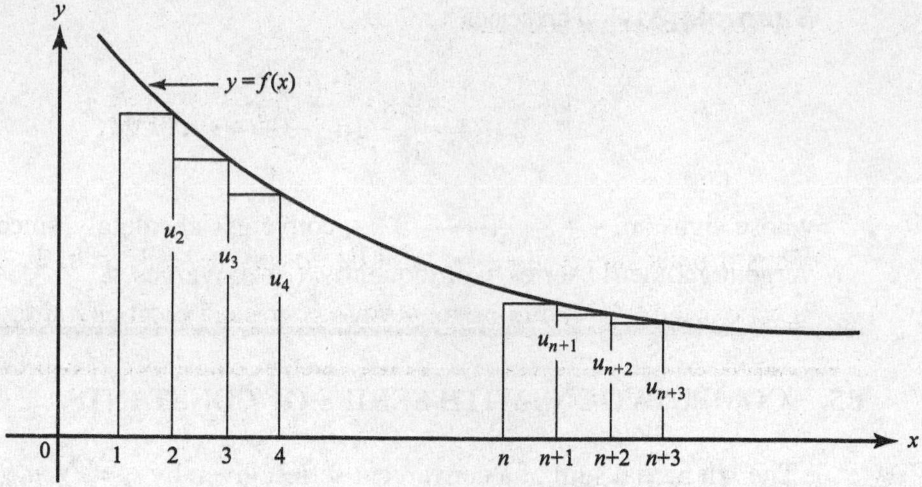

FIGURE N10–6

In Figure N10–6, $f(n) = u_n$, where $f(x)$ is a continuous, positive, decreasing function. Since the base of each rectangle is one unit, the rectangles, starting at the left, have areas $u_2, u_3, u_4, \ldots, u_{n+1}, u_{n+2}, \ldots$. We see that

$$R_n = u_{n+1} + u_{n+2} + u_{n+3} + \cdots < \int_n^\infty f(x)\, dx.$$

Example 32.   Estimate the error made in approximating $\displaystyle\sum_{n=1}^\infty \frac{n}{e^n}$ by the sum of its first 10 terms.

The series was shown to converge, using the Integral Test, in Example 19, page 348. Thus,

$$R_{10} < \int_{10}^\infty xe^{-x}\, dx = \lim_{b\to\infty} -e^{-x}(1+x)\ \Big|_{10}^{b}$$

$$= -\lim_{b\to\infty}\left(\frac{1+b}{e^b} - \frac{11}{e^{10}}\right) < 0.0005.$$

Example 33.   (a) Estimate the error introduced in approximating $\displaystyle\sum_{n=1}^\infty \frac{5}{3^n}$ by the sum of its first four terms. (b) How many terms are needed for the error to be less than 0.0001?

(a) The series is a geometric one with $a = \dfrac{5}{3}$ and $r = \dfrac{1}{3}$. Theorem 5a, page 353, yields

$$R_4 = \frac{ar^4}{1-r} = \frac{\dfrac{5}{3}\cdot\left(\dfrac{1}{3}\right)^4}{\dfrac{2}{3}} = \frac{5}{2\cdot 3^4} = 0.03 \text{ to two decimal places.}$$

(b) We seek n such that

$$R_n = \frac{ar^n}{1-r} = \frac{\frac{5}{3} \cdot \left(\frac{1}{3}\right)^n}{\frac{2}{3}} < 0.0001.$$

Thus n must satisfy $\dfrac{5}{2 \cdot 3^n} < 0.0001$. So $\dfrac{5}{0.0002} < 3^n$, $n \ln 3 > \ln 25{,}000$, and n must exceed 9.2. Therefore ten terms are needed for the required accuracy.

C.   Power Series

C1.   DEFINITIONS; CONVERGENCE.

An expression of the form

$$\sum_{k=0}^{\infty} a_k x^k = a_0 + a_1 x + a_2 x^2 + \cdots + a_n x^n + \cdots, \tag{1}$$

where the a's are constants, is called a *power series in x*; and

$$\sum_{k=0}^{\infty} a_k (x-a)^k = a_0 + a_1(x-a) + a_2(x-a)^2 + \cdots + a_n(x-a)^n + \cdots \tag{2}$$

is called a *power series in* $(x-a)$.

If in (1) or (2) x is replaced by a specific real number, then the power series becomes a series of constants that either converges or diverges. Note that series (1) converges if $x = 0$ and series (2) converges if $x = a$.

The set of values of x for which a power series converges is called its *interval of convergence*. If the interval of convergence of the power series (1) contains values of x other than 0, then 0 will be the midpoint of this interval; similarly, a is the midpoint of the interval of convergence of series (2).

The interval of convergence of a power series may be found by applying the Ratio Test to the series of absolute values.

See page 356 for the radius of convergence of a power series.

Example 34.   Find all x for which the following series converges:

$$1 + x + x^2 + \cdots + x^n + \cdots. \tag{3}$$

We find

$$\lim_{n \to \infty} \left| \frac{u_{n+1}}{u_n} \right| = \lim_{n \to \infty} \left| \frac{x^{n+1}}{x^n} \right| = \lim_{n \to \infty} |x| = |x|.$$

The series clearly converges absolutely (and therefore converges) if $|x| < 1$; it diverges if $|x| > 1$. The endpoints must be tested separately since the Ratio Test fails when the limit equals 1. When $x = 1$, (3) becomes $1 + 1 + 1 + \cdots$ and diverges; when $x = -1$, (3) becomes $1 - 1 + 1 - 1 + \cdots$ and diverges. Thus the series converges if $-1 < x < 1$.

Example 35. For what x's does $\sum_{n=1}^{\infty} \frac{(-1)^{n-1} x^{n-1}}{n+1}$ converge?

$$\lim_{n \to \infty} \left| \frac{u_{n+1}}{u_n} \right| = \lim_{n \to \infty} \left| \frac{x^n}{n+2} \cdot \frac{n+1}{x^{n-1}} \right| = \lim_{n \to \infty} |x| = |x| .$$

The series converges if $|x| < 1$ and diverges if $|x| > 1$. When $x = 1$ we have $\frac{1}{2} - \frac{1}{3} + \frac{1}{4} - \frac{1}{5} + \cdots$, an alternating convergent series; when $x = -1$ the series is $\frac{1}{2} + \frac{1}{3} + \frac{1}{4} + \cdots$, which diverges. So the series converges if $-1 < x \leqq 1$.

Example 36. $\sum_{n=1}^{\infty} \frac{x^n}{n!}$ converges for all x, since

$$\lim_{n \to \infty} \left| \frac{u_{n+1}}{u_n} \right| = \lim_{n \to \infty} \left| \frac{x^{n+1}}{(n+1)!} \cdot \frac{n!}{x^n} \right| = \lim_{n \to \infty} \frac{|x|}{n+1} = 0,$$

for all x. Thus the series converges if $-\infty < x < \infty$.

Example 37. Find all x for which the following series converges:

$$1 + \frac{x-2}{2^1} + \frac{(x-2)^2}{2^2} + \cdots + \frac{(x-2)^{n-1}}{2^{n-1}} + \cdots . \tag{4}$$

$$\lim_{n \to \infty} \left| \frac{u_{n+1}}{u_n} \right| = \lim_{n \to \infty} \left| \frac{(x-2)^n}{2^n} \cdot \frac{2^{n-1}}{(x-2)^{n-1}} \right| = \lim_{n \to \infty} \frac{|x-2|}{2} = \frac{|x-2|}{2},$$

which is less than 1 if $|x - 2| < 2$, that is, if $0 < x < 4$. Series (4) converges on this interval and diverges if $|x - 2| > 2$, that is, if $x < 0$ or $x > 4$.

When $x = 0$, (4) is $1 - 1 + 1 - 1 + \cdots$ and diverges. When $x = 4$, (4) is $1 + 1 + 1 + \cdots$ and diverges. So (4) converges if $0 < x < 4$.

Example 38. $\sum_{n=1}^{\infty} n! \, x^n$ converges only at $x = 0$, since

$$\lim_{n \to \infty} \frac{u_{n+1}}{u_n} = \lim_{n \to \infty} (n+1) x = \infty$$

unless $x = 0$.

RADIUS OF CONVERGENCE.   If the power series (1) converges when $|x| < r$ and diverges when $|x| > r$, then r is called the *radius of convergence*. Similarly, r is the radius of convergence of series (2) when (2) converges if $|x - a| < r$ but diverges if $|x - a| > r$. Note the radii of convergence for the series investigated in Examples 34–38:

EXAMPLE	INTERVAL OF CONVERGENCE	RADIUS OF CONVERGENCE
34	$-1 < x < 1$	1
35	$-1 < x \leqq 1$	1
36	$-\infty < x < \infty$	∞
37	$0 < x < 4$	2
38	$x = 0$ only	0

THEOREMS ABOUT CONVERGENCE.   The following theorems about convergence of power series are useful:

THEOREM 1a.   If $\sum a_n x^n$ converges for a nonzero number c, then it converges absolutely for all x such that $|x| < |c|$.

THEOREM 1b.   If $\sum a_n x^n$ diverges for $x = d$, then it diverges for all x such that $|x| > |d|$.

C2.   FUNCTIONS DEFINED BY POWER SERIES.

If $x = x_0$ is a number within the interval of convergence of the power series $\sum_{k=0}^{\infty} a_k(x-a)^k$, then the sum at $x = x_0$ is unique. We let the function f be defined by

$$f(x) = \sum_{k=0}^{\infty} a_k(x-a)^k$$
$$= a_0 + a_1(x-a) + \cdots + a_n(x-a)^n + \cdots; \tag{1}$$

its domain is the interval of convergence of the series; and it follows that

$$f(x_0) = \sum_{k=0}^{\infty} a_k(x_0-a)^k.$$

Functions defined by power series behave very much like polynomials, as indicated by the following *properties*:

PROPERTY 2a.   The function defined by (1) is continuous for each x in the interval of convergence of the series.

PROPERTY 2b.   The series formed by differentiating the terms of series (1) converges to $f'(x)$ for each x within the interval of convergence of (1); that is,

$$f'(x) = \sum_{1}^{\infty} ka_k(x-a)^{k-1}$$
$$= a_1 + 2a_2(x-a) + \cdots + na_n(x-a)^{n-1} + \cdots. \tag{2}$$

Note that we can conclude from property 2b that the power series (1) and its derived series (2) have the same radius of convergence but not necessarily the same interval of convergence.

Example 39. Let

$$f(x) = \sum_{k=1}^{\infty} \frac{x^k}{k(k+1)} = \frac{x}{1 \cdot 2} + \frac{x^2}{2 \cdot 3} + \cdots + \frac{x^n}{n(n+1)} + \cdots ; \tag{3}$$

then

$$f'(x) = \sum_{k=1}^{\infty} \frac{x^{k-1}}{k+1} = \frac{1}{2} + \frac{x}{3} + \frac{x^2}{4} + \cdots + \frac{x^{n-1}}{n+1} + \cdots . \tag{4}$$

From (3) we see that

$$\lim_{n\to\infty} \left| \frac{x^{n+1}}{(n+1)(n+2)} \cdot \frac{n(n+1)}{x^n} \right| = |x| ;$$

that

$$f(1) = \frac{1}{1 \cdot 2} + \frac{1}{2 \cdot 3} + \cdots + \frac{1}{n(n+1)} + \cdots ;$$

and that

$$f(-1) = -\frac{1}{1 \cdot 2} + \frac{1}{2 \cdot 3} - \cdots + \frac{(-1)^n}{n(n+1)} + \cdots ;$$

and conclude that (3) converges if $-1 \leqq x \leqq 1$.
     From (4) we see that

$$\lim_{n\to\infty} \left| \frac{x^n}{n+2} \cdot \frac{n+1}{x^{n-1}} \right| = |x| ;$$

that

$$f'(1) = \frac{1}{2} + \frac{1}{3} + \frac{1}{4} + \cdots ;$$

and that

$$f'(-1) = \frac{1}{2} - \frac{1}{3} + \frac{1}{4} - \cdots ;$$

and conclude that series (4) converges if $-1 \leqq x < 1$.
     Thus, the series given for $f(x)$ and $f'(x)$ have the same radius of convergence, but their intervals of convergence differ.

PROPERTY 2c.   The series obtained by integrating the terms of the given series (1) converges to $\int_a^x f(t)\,dt$ for each x within the interval of convergence of (1); that is,

$$\int_a^x f(t)\,dt = a_0(x - a) + \frac{a_1(x - a)^2}{2} + \frac{a_2(x - a)^3}{3}$$

$$+ \cdots + \frac{a_n(x - a)^{n+1}}{n+1} + \cdots \tag{5}$$

$$= \sum_{k=0}^{\infty} \frac{a_k(x - a)^{k+1}}{k+1} .$$

Example 40.   Obtain a series for $\dfrac{1}{(1-x)^2}$ by long division.

$$
\begin{array}{r}
1 + 2x + 3x^2 + 4x^3 + \cdots \\
1 - 2x + x^2 \overline{)\,1} \\
\underline{1 - 2x + x^2} \\
2x - x^2 \\
\underline{2x - 4x^2 + 2x^3} \\
+3x^2 - 2x^3 \\
\underline{3x^2 - 6x^3 + 3x^4} \\
4x^3 - 3x^4
\end{array}
$$

Then,

$$
\frac{1}{(1-x)^2} = 1 + 2x + 3x^2 + \cdots + (n+1)x^n + \cdots . \tag{6}
$$

Let us assume that the power series on the right in (6) converges to

$$
f(x) = \frac{1}{(1-x)^2} \quad \text{if } -1 < x < 1.
$$

Then by property 2c

$$
\int_0^x \frac{1}{(1-t)^2}\,dt = \int_0^x [1 + 2t + 3t^2 + \cdots + (n+1)t^n + \cdots]\,dt
$$

$$
= \left. \frac{1}{1-t} \right|_0^x = \frac{1}{1-x} - 1
$$

$$
= x + x^2 + x^3 + \cdots + x^{n+1} + \cdots ,
$$

so

$$
\frac{1}{1-x} = 1 + x + x^2 + x^3 + \cdots + x^n + \cdots . \tag{7}
$$

Note that the series on the right in (7) is exactly the one obtained for the function $\dfrac{1}{1-x}$ by long division:

$$
\begin{array}{r}
1 + x + x^2 + x^3 + \cdots \\
1 - x \overline{)\,1} \\
\underline{1 - x} \\
+ x \\
\underline{x - x^2} \\
x^2 \\
\underline{x^2 - x^3} \\
+ x^3
\end{array}
$$

Furthermore, the right-hand member of (7) is a geometric series with ratio $r = x$ and with $a = 1$; if $|x| < 1$, its sum is $\dfrac{a}{1-r} = \dfrac{1}{1-x}$.

PROPERTY 2d.   Two series may be added, substracted, multiplied, or divided (with division by zero to be avoided) for x's that lie within the intervals of convergence of both.

Example 41.   If

$$\frac{1}{1-x} = 1 + x + x^2 + \cdots + x^n + \cdots \quad (|x| < 1), \tag{8}$$

then $\dfrac{1}{(1-x)^2}$ can be obtained by multiplying (8) by itself.

$$
\begin{array}{l}
1 + x + x^2 + x^3 + \cdots \\
1 + x + x^2 + x^3 + \cdots \\
\hline
1 + x + x^2 + x^3 + \cdots \\
 + x + x^2 + x^3 + \cdots \\
 + + x^2 + x^3 + \cdots \\
 + + + x^3 + \cdots \\
\end{array}
$$

$$\frac{1}{(1-x)^2} = 1 + 2x + 3x^2 + 4x^3 + \cdots \quad (|x| < 1) \tag{9}$$

The series on the right in (9) is precisely the one obtained for $\dfrac{1}{(1-x)^2}$ by long division in Example 40.

Also, since the derivative of $\dfrac{1}{1-x}$ is exactly $\dfrac{1}{(1-x)^2}$, we can differentiate both sides of (8) above to get (9) once again.

C3.   TAYLOR POLYNOMIALS.

The function $f(x)$ at the point $x = a$ is approximated by the following *Taylor polynomial $P_n(x)$* of order n:

$$f(x) \simeq P_n(x) = f(a) + f'(a)(x - a) + \frac{f''(a)}{2!}(x - a)^2 + \cdots + \frac{f^{(n)}(a)}{n!}(x - a)^n. \tag{1}$$

The Taylor polynomial $P_n(x)$ and its first n derivatives all agree at a with f and its first n derivatives. The *order* of a Taylor polynomial is the order of the highest derivative, which is also the polynomial's last term.

In the special case where $a = 0$, the Taylor polynomial of order n that approximates $f(x)$ is

$$P_n(x) = f(0) + f'(0)x + \frac{f''(0)}{2!}x^2 + \cdots + \frac{f^{(n)}(0)}{n!}x^n. \tag{2}$$

Expression (2) is called the *Maclaurin polynomial* or the *polynomial at or about x = zero*.

The Taylor polynomial $P_1(x)$ at $x = 0$ is the tangent-line approximation to $f(x)$ near zero given by

$$f(x) \simeq P_1(x) = f(0) + f'(0)x.$$

It is the "best" linear approximation to f at 0, discussed at length in Chapter 4 §L.

Example 42.   Find the Taylor polynomial of order 4 at 0 for $f(x) = e^{-x}$. Use this to approximate $f(0.25)$.

The first four derivatives are $-e^{-x}$, e^{-x}, $-e^{-x}$, and e^{-x}; at $a = 0$, these equal -1, 1, -1, and 1 respectively. The approximating Taylor polynomial of order 4 is therefore

$$e^{-x} \approx 1 - x + \frac{1}{2!}x^2 - \frac{1}{3!}x^3 + \frac{1}{4!}x^4.$$

With $x = 0.25$ we have

$$e^{-0.25} \approx 1 - 0.25 + \frac{1}{2!}(0.25)^2 - \frac{1}{3!}(0.25)^3 + \frac{1}{4!}(0.25)^4$$

$$\approx 0.7788.$$

This is the precise value of $e^{-0.25}$ correct to four places.

In Figure N10–7 we see the graphs of $f(x)$ and of the Taylor polynomials:

$$P_0(x) = 1;$$

$$P_1(x) = 1 - x;$$

$$P_2(x) = 1 - x + \frac{x^2}{2!};$$

$$P_3(x) = 1 - x + \frac{x^2}{2!} - \frac{x^3}{3!};$$

$$P_4(x) = 1 - x + \frac{x^2}{2!} - \frac{x^3}{3!} + \frac{x^4}{4!}.$$

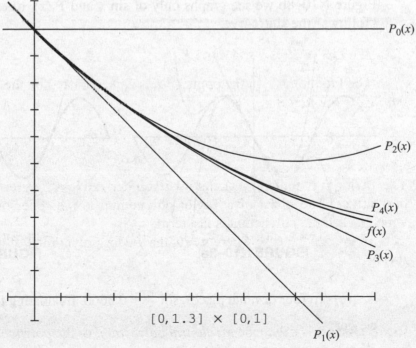

$$[0, 1.3] \times [0, 1]$$

FIGURE N10–7

Notice how closely $P_4(x)$ hugs $f(x)$ even as x approaches 1. In fact, $e^{-1} = 0.368$ to three decimal places; $P_4(1) = 0.375$.

Example 43.   (a) Find the Taylor polynomials P_1, P_3, P_5, and P_7 at $x = 0$ for $f(x) = \sin x$.

(b) Graph f and all four polynomials in $[-2\pi, 2\pi] \times [-2,2]$.

(c) Approximate $\sin \dfrac{\pi}{3}$ using each of the four polynomials.

(a) The derivatives of the sine function at 0 are given by the following table:

order of deriv	0	1	2	3	4	5	6	7
deriv of $\sin x$	$\sin x$	$\cos x$	$-\sin x$	$-\cos x$	$\sin x$	$\cos x$	$-\sin x$	$-\cos x$
deriv of $\sin x$ at 0	0	1	0	-1	0	1	0	-1

From the table we know that

$$P_1(x) = x;$$

$$P_3(x) = x - \frac{x^3}{3!};$$

$$P_5(x) = x - \frac{x^3}{3!} + \frac{x^5}{5!};$$

$$P_7(x) = x - \frac{x^3}{3!} + \frac{x^5}{5!} - \frac{x^7}{7!}.$$

(b) Figure N10–8a shows the graphs of $\sin x$ and the four polynomials. In Figure N10–8b we see graphs only of $\sin x$ and $P_7(x)$, to exhibit how closely P_7 "follows" the sine curve.

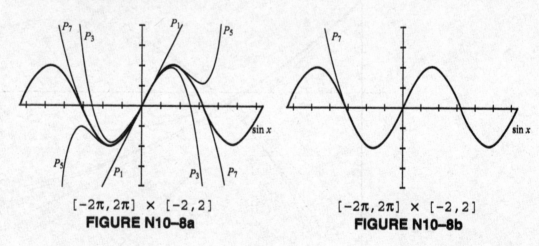

$[-2\pi, 2\pi] \times [-2, 2]$
FIGURE N10–8a

$[-2\pi, 2\pi] \times [-2, 2]$
FIGURE N10–8b

(c) To four decimal places, $\sin \dfrac{\pi}{3} = 0.8660$. Evaluating the polynomials at $\dfrac{\pi}{3}$, we get

$$P_1\left(\frac{\pi}{3}\right) = 1.0471, \quad P_3\left(\frac{\pi}{3}\right) = 0.8558, \quad P_5\left(\frac{\pi}{3}\right) = 0.8663, \quad P_7\left(\frac{\pi}{3}\right) = 0.8660.$$

We see that P_7 is correct to four decimal places.

Example 44.   (a) Find the Taylor polynomials of degrees 0, 1, 2, and 3 generated by $f(x) = \ln x$ at $x = 1$.

(b) Graph f and the four polynomials on the same set of axes.

(c) Using your calculator and graph from part (b), show that $|P_3 - f(x)| < 0.02$ when $0.53 < x < 1.58$.

(a) The derivatives of $\ln x$ at $x = 1$ are given by the table:

order of deriv	0	1	2	3
deriv of $\ln x$	$\ln x$	$\dfrac{1}{x}$	$-\dfrac{1}{x^2}$	$\dfrac{2}{x^3}$
deriv at $x = 1$	0	1	−1	2

From the table we have

$$P_0(x) = 0;$$

$$P_1(x) = (x-1);$$

$$P_2(x) = (x-1) - \frac{(x-1)^2}{2};$$

$$P_3(x) = (x-1) - \frac{(x-1)^2}{2} + \frac{(x-1)^3}{3}.$$

(b) Figure N10–9 shows the graphs of $\ln x$ and the four Taylor polynomials above, in $[0,2.5] \times [-1,1]$.

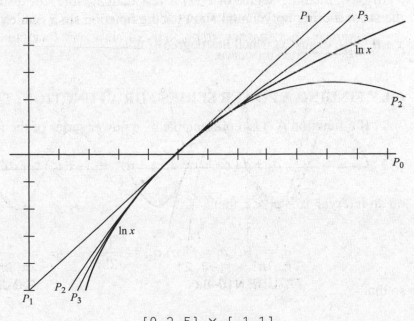

$[0,2.5] \times [-1,1]$

FIGURE N10–9

(c) Figure N10–10 is a graph of $Y_1 = P_3 - f(x) - 0.02$ in the viewing window $[0,2.5] \times [-0.02,0.02]$. If you zoom in on the graph in Figure N10–9, you will see that $P_3(x) > \ln x$, so $|P_3 - \ln x| = P_3 - \ln x$. For $P_3 - \ln x$ to be less than 0.02, $P_3 - \ln x - 0.02$ must therefore be less than 0 (that is, negative).

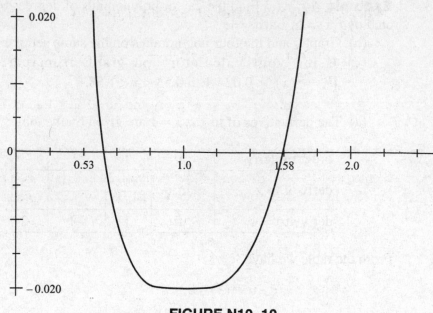

FIGURE N10–10

Using the root option twice on Y_1, we found, to two decimal places, that $|P_3 - \ln x| < 0.02$ when $0.53 < x < 1.58$.

A Note on Order and Degree

A Taylor polynomial has *degree n* if it has powers of $(x - a)$ up through the *n*th. If $f^{(n)}(a) = 0$, then the degree of $P_n(x)$ is less than *n*. Note, for instance, in Example 43, that the second-order polynomial $P_2(x)$ for the function $\sin x$ (which is identical with $P_1(x)$) is $x + 0 \cdot \dfrac{x^2}{2!}$, or just x, which has degree 1, not 2.

C4.   FINDING A POWER SERIES FOR A FUNCTION; TAYLOR SERIES.

If a function $f(x)$ is representable by a power series of the form

$$a_0 + a_1(x - a) + a_2(x - a)^2 + \cdots + a_n(x - a)^n + \cdots$$

on an interval $|x - a| < r$, then

$$a_0 = f(a), a_1 = f'(a), a_2 = \frac{f''(a)}{2!}, \ldots, a_n = \frac{f^{(n)}(a)}{n!}, \ldots, \tag{1}$$

so that

$$f(x) = f(a) + f'(a)(x - a) + \frac{f''(a)}{2!}(x - a)^2 + \cdots + \frac{f^{(n)}(a)}{n!}(x - a)^n + \cdots. \tag{2}$$

Series (2) is called the *Taylor series* of the function *f* about the number *a*. There is never more than one power series in $(x - a)$ for $f(x)$. Implicit in (2) is the requirement that the function and all its derivatives exist at $x = a$ if the function $f(x)$ is to generate a Taylor series expansion.

When $a = 0$ we have a special case of (2) given by

$$f(x) = f(0) + f'(0)x + \frac{f''(0)}{2!}x^2 + \cdots + \frac{f^{(n)}(0)}{n!}x^n + \cdots \tag{3}$$

Series (3) is called the *Maclaurin series* of the function f; this is the expansion of f about $x = 0$.

Example 45. If $f(x)$ and all its derivatives exist at $x = 0$ and if $f(x)$ is representable by a power series in x, show that $f(x)$ will have precisely the form given by (3).

Let $f(x) = a_0 + a_1 x + a_2 x^2 + a_3 x^3 + \cdots + a_n x^n + \cdots$ for all x in the interval of convergence. Then

$$f'(x) = a_1 + 2a_2 x + 3a_3 x^2 + \cdots \qquad + na_n x^{n-1} + \cdots,$$

$$f''(x) = 2a_2 \qquad + 3\cdot 2a_3 x + \cdots \qquad + n(n-1)a_n x^{n-2} + \cdots,$$

$$f'''(x) = \qquad\qquad 3\cdot 2a_3 + 4\cdot 3\cdot 2a_4 x + \cdots + n(n-1)(n-2)a_n x^{n-3} + \cdots,$$

$$f^{iv}(x) = \qquad\qquad\qquad 4\cdot 3\cdot 2a_4 + \cdots + n(n-1)(n-2)(n-3)a_n x^{n-4} + \cdots,$$

$$\cdot$$
$$\cdot$$
$$\cdot$$

$$f^{(n)}(x) = \qquad\qquad\qquad n!\, a_n + \cdots.$$

If we let $x = 0$ in each of the preceding equations, we get

$$f(0) = a_0 \qquad\qquad \text{so that} \qquad a_0 = f(0);$$

$$f'(0) = a_1 \qquad\qquad\qquad\qquad a_1 = f'(0);$$

$$f''(0) = 2a_2 \qquad\qquad\qquad\qquad a_2 = \frac{f''(0)}{2};$$

$$f'''(0) = 3!\, a_3 \qquad\qquad\qquad\qquad a_3 = \frac{f'''(0)}{3!};$$

$$f^{iv}(0) = 4!\, a_4 \qquad\qquad\qquad\qquad a_4 = \frac{f^{iv}(0)}{4!};$$

$$\cdot \qquad\qquad\qquad\qquad\qquad \cdot$$
$$\cdot \qquad\qquad\qquad\qquad\qquad \cdot$$
$$\cdot \qquad\qquad\qquad\qquad\qquad \cdot$$

$$f^{(n)}(0) = n!\, a_n \qquad\qquad\qquad\qquad a_n = \frac{f^{(n)}(0)}{n!}.$$

Example 46. Find the Maclaurin series for $f(x) = e^x$.

Here $f'(x) = e^x, \ldots, f^{(n)}(x) = e^x, \ldots$, for all n. So $f'(0) = 1, \ldots, f^{(n)}(0) = 1, \ldots$, for all n. By (3), then,

$$e^x = 1 + x + \frac{x^2}{2!} + \frac{x^3}{3!} + \cdots + \frac{x^n}{n!} + \cdots.$$

Example 47. Find the Maclaurin expansion for $f(x) = \sin x$.

$$f(x) = \sin x; \qquad\qquad\qquad f(0) = 0;$$

$$f'(x) = \cos x; \qquad\qquad\qquad f'(0) = 1;$$

$$f''(x) = -\sin x; \qquad\qquad\qquad f''(0) = 0;$$

$$f'''(x) = -\cos x; \qquad\qquad\qquad f'''(0) = -1;$$

$$f^{iv}(x) = \sin x; \qquad\qquad\qquad f^{iv}(0) = 0.$$

Thus,

$$\sin x = x - \frac{x^3}{3!} + \frac{x^5}{5!} - \cdots + (-1)^{n-1} \frac{x^{2n-1}}{(2n-1)!} + \cdots.$$

Example 48. Find the Maclaurin series for $f(x) = \dfrac{1}{1-x}$.

$$f(x) = (1-x)^{-1}; \qquad\qquad\qquad f(0) = 1;$$

$$f'(x) = (1-x)^{-2}; \qquad\qquad\qquad f'(0) = 1;$$

$$f''(x) = 2(1-x)^{-3}; \qquad\qquad\qquad f''(0) = 2;$$

$$f'''(x) = 3!(1-x)^{-4}; \qquad\qquad\qquad f'''(0) = 3!;$$

$$\vdots \qquad\qquad\qquad\qquad\qquad \vdots$$

$$f^{(n)}(x) = n!(1-x)^{-(n+1)}; \qquad\qquad f^{(n)}(0) = n!.$$

Then

$$\frac{1}{1-x} = 1 + x + x^2 + x^3 + \cdots + x^n + \cdots.$$

Note that this agrees exactly with the power series in x obtained by three different methods in Examples 40 and 41.

Example 49.   Find the Taylor series for the function $f(x) = \ln x$ about $x = 1$.

$$f(x) = \ln x; \qquad\qquad f(1) = \ln 1 = 0;$$

$$f'(x) = \frac{1}{x}; \qquad\qquad f'(1) = 1;$$

$$f''(x) = -\frac{1}{x^2}; \qquad\qquad f''(1) = -1;$$

$$f'''(x) = \frac{2}{x^3}; \qquad\qquad f'''(1) = 2;$$

$$f^{iv}(x) = \frac{-3!}{x^4}; \qquad\qquad f^{iv}(1) = -3!;$$

$$\vdots \qquad\qquad\qquad \vdots$$

$$f^{(n)}(x) = \frac{(-1)^{n-1}(n-1)!}{x^n}; \qquad f^{(n)}(1) = (-1)^{n-1}(n-1)!.$$

Then

$$\ln x = (x-1) - \frac{(x-1)^2}{2} + \frac{(x-1)^3}{3} - \frac{(x-1)^4}{4}$$

$$+ \cdots + \frac{(-1)^{n-1}(x-1)^n}{n} + \cdots.$$

FUNCTIONS THAT GENERATE NO SERIES.   Note that the following functions are among those that fail to generate a specific series in $(x - a)$ because the function and/or one or more derivatives do not exist at $x = a$:

FUNCTION	SERIES IT FAILS TO GENERATE
$\ln x$	about 0
$\ln (x - 1)$	about 1
$\sqrt{x - 2}$	about 2
$\sqrt{x - 2}$	about 0
$\tan x$	about $\frac{\pi}{2}$
$\sqrt{1 + x}$	about -1

C5. TAYLOR'S FORMULA WITH REMAINDER; LAGRANGE ERROR BOUND.

The following theorems resolve the problem of determining *when* a function is representable by a power series.

THEOREM 5a (TAYLOR'S THEOREM). If a function f and its first $(n + 1)$ derivatives are continuous on the interval $|x - a| < r$, then for each x in this interval

$$f(x) = f(a) + f'(a)(x - a) + \frac{f''(a)}{2!}(x - a)^2 + \cdots + \frac{f^{(n)}(a)}{n!}(x - a)^n + R_n(x),$$

where

$$R_n(x) = \frac{f^{(n+1)}(c)(x - a)^{n+1}}{(n + 1)!}$$

and c is some number between a and x. $R_n(x)$ is called the *Lagrange form of the remainder*, or the *Lagrange error bound*.

Note that the equation above expresses $f(x)$ as the sum of the Taylor polynomial $P_n(x)$ and the error that results when that polynomial is used as an approximation for $f(x)$.

When we truncate a series after the $(n + 1)$st term, we can compute the error bound R_n, according to Lagrange, if we know what to substitute for c. In practice we find, not R_n exactly, but only an upper bound for it by assigning to c the value between a and x that makes R_n as large as possible.

THEOREM 5b. If the function f has derivatives of all orders in some interval about $x = a$, then the power series

$$f(a) + f'(a)(x - a) + \frac{f''(a)}{2!}(x - a)^2 + \cdots + \frac{f^{(n)}(a)(x - a)^n}{n!} + \cdots$$

represents the function f for those x's, and only those, for which $\lim_{n \to \infty} R_x(x) = 0$.

Example 50. Prove that the Maclaurin series generated by e^x represents the function e^x for all real numbers.

From Example 46 we know that $f(x) = e^x$ generates the Maclaurin series

$$e^x = 1 + x + \frac{x^2}{2!} + \cdots + \frac{x^n}{n!} + \cdots. \tag{1}$$

The Lagrange error bound is

$$R_n(x) = \frac{e^c(x)^{n+1}}{(n + 1)!} \quad (0 < c < x).$$

If x is a positive fixed number, then $0 < e^c < e^x = C$, where C is a constant, and

$$|R_n(x)| < \frac{Cx^{n+1}}{(n + 1)!}.$$

We showed in Example 36, page 356, that $\displaystyle\sum_1^\infty \frac{x^k}{k!}$ converges for all x; consequently, by Theorem 2a of §B2 it follows that the general term must approach zero. So, if $x > 0$ then $R_n(x) \to 0$. By Theorem 5b just above, then, (1) represents e^x if $x > 0$.

If $x = 0$, then (1) reduces to $e^x = 1$ and $R_n(x) = 0$.

If $x < 0$, then $x < c < 0$ and $e^c < 1$, so

$$|R_n(x)| < \frac{|x|^{n+1}}{(n+1)!},$$

which approaches zero.

Consequently, (1) represents e^x for all real x.

Example 51. Find the Maclaurin series for $\ln(1 + x)$ and the associated Lagrange error bound.

$$f(x) = \ln(1+x); \qquad\qquad f(0) = 0;$$

$$f'(x) = \frac{1}{1+x}; \qquad\qquad f'(0) = 1;$$

$$f''(x) = -\frac{1}{(1+x)^2}; \qquad\qquad f''(0) = -1;$$

$$f'''(x) = \frac{2}{(1+x)^3}; \qquad\qquad f'''(0) = 2!;$$

$$\vdots \qquad\qquad\qquad\qquad \vdots$$

$$f^{(n)}(x) = \frac{(-1)^{n-1}(n-1)!}{(1+x)^n}; \qquad\qquad f^{(n)}(0) = (-1)^{n-1}(n-1)!$$

$$f^{(n+1)}(x) = \frac{(-1)^n \cdot n!}{(1+x)^{n+1}}.$$

Then

$$\ln(1+x) = x - \frac{x^2}{2} + \frac{x^3}{3} - \frac{x^4}{4} + \cdots + (-1)^{n-1} \cdot \frac{x^n}{n} + R_n(x),$$

where the Lagrange error bound is

$$R_n(x) = \frac{(-1)^n}{(1+c)^{n+1}} \cdot \frac{x^{n+1}}{n+1}.$$

Example 52. For what x may $\cos x$ be represented by its power series in $\left(x - \frac{\pi}{3}\right)$?

$$f(x) = \cos x; \qquad\qquad f\left(\frac{\pi}{3}\right) = \frac{1}{2};$$

$$f'(x) = -\sin x; \qquad\qquad f'\left(\frac{\pi}{3}\right) = -\frac{\sqrt{3}}{2};$$

$$f''(x) = -\cos x; \qquad\qquad f''\left(\frac{\pi}{3}\right) = -\frac{1}{2};$$

$$f'''(x) = \sin x; \qquad\qquad f'''\left(\frac{\pi}{3}\right) = \frac{\sqrt{3}}{2};$$

$$f^{iv}(x) = \cos x. \qquad\qquad f^{iv}\left(\frac{\pi}{3}\right) = \frac{1}{2}.$$

Note that $f^{(n)}(x) = \pm \sin x$ or $\pm \cos x$.

$$\cos x = \frac{1}{2} - \frac{\sqrt{3}}{2}\left(x - \frac{\pi}{3}\right) - \frac{1}{2}\cdot\frac{\left(x - \frac{\pi}{3}\right)^2}{2} + \frac{\sqrt{3}}{2}\cdot\frac{\left(x - \frac{\pi}{3}\right)^3}{3!} + \cdots.$$

The Lagrange error bound is

$$R_n(x) = \frac{f^{(n+1)}(c)\left(x - \frac{\pi}{3}\right)^{n+1}}{(n+1)!} \qquad \left(x < c < \frac{\pi}{3}\right).$$

Although we do not know exactly what c equals, we do know that

$$\left|f^{(n+1)}(c)\right| \leqq 1,$$

so that

$$|R_n(x)| \leqq \frac{\left|x - \frac{\pi}{3}\right|^{n+1}}{(n+1)!} \quad \text{and} \quad \lim_{n\to\infty} R_n(x) = 0$$

for all x.

C6. COMPUTATIONS WITH POWER SERIES.

We list here for reference some frequently used series expansions together with their intervals of convergence:

FUNCTION	SERIES EXPANSION	INTERVAL OF CONVERGENCE	
$\sin x$	$x - \dfrac{x^3}{3!} + \dfrac{x^5}{5!} - \cdots + \dfrac{(-1)^{n-1}x^{2n-1}}{(2n-1)!} + \cdots$	$-\infty < x < \infty$	(1)
$\sin x$	$\sin a + (x-a)\cos a - \dfrac{(x-a)^2}{2!}\sin a$ $-\dfrac{(x-a)^3}{3!}\cos a + \cdots$	$-\infty < x < \infty$	(2)
$\cos x$	$1 - \dfrac{x^2}{2!} + \dfrac{x^4}{4!} - \cdots + \dfrac{(-1)^{n-1}x^{2n-2}}{(2n-2)!} + \cdots$	$-\infty < x < \infty$	(3)
$\cos x$	$\cos a - (x-a)\sin a - \dfrac{(x-a)^2}{2!}\cos a$ $+\dfrac{(x-a)^3}{3!}\sin a + \cdots$	$-\infty < x < \infty$	(4)
e^x	$1 + x + \dfrac{x^2}{2!} + \dfrac{x^3}{3!} + \cdots + \dfrac{x^n}{n!} + \cdots$	$-\infty < x < \infty$	(5)
$\ln(1+x)$	$x - \dfrac{x^2}{2} + \dfrac{x^3}{3} - \dfrac{x^4}{4} + \cdots + \dfrac{(-1)^{n-1}x^n}{n} + \cdots$	$-1 \le x \le 1$	(6)
$\ln x$	$(x-1) - \dfrac{(x-1)^2}{2} + \dfrac{(x-1)^3}{3} - \cdots$ $+\dfrac{(-1)^{n-1}(x-1)^n}{n} + \cdots$	$0 < x \le 2$	(7)
$\tan^{-1} x$	$x - \dfrac{x^3}{3} + \dfrac{x^5}{5} - \dfrac{x^7}{7} + \cdots + \dfrac{(-1)^{n-1}x^{2n-1}}{2n-1} + \cdots$	$-1 \le x \le 1$	(8)

If we use the nth partial sum of a convergent power series for $f(x_0)$ as an approximation to $f(x_0)$, we need information about the magnitude of the error involved. We know that, if the first $(n+1)$ terms of the Taylor series are used for $f(x_0)$, the error is $R_n(x_0)$ as given in §C5. We have also indicated that it is necessary, not that we know exactly how large the error is, but only that we find an upper bound for it. If a convergent geometric series is involved, then

$$\sum_{n=1}^{\infty} ar^n = a + ar + \cdots + ar^{n-1} + R_n,$$

where

$$|R_n| = \left|\frac{ar^n}{1-r}\right|.$$

Recall, also, that for a convergent alternating series the error is less absolutely than the first term dropped; that is, if

$$S = a_1 - a_2 + a_3 - \cdots + (-1)^{n-1} a_n + R_n,$$

then

$$|R_n| < a_{n+1};$$

in fact, R_n has the same sign as the $(n + 1)$st term.

Example 53. Compute $\dfrac{1}{\sqrt{e}}$ to four decimal places.

We can use the Maclaurin series (5),

$$e^x = 1 + x + \frac{x^2}{2!} + \frac{x^3}{3!} + \frac{x^4}{4!} + \cdots,$$

and let $x = -\dfrac{1}{2}$ to get

$$e^{-1/2} = 1 - \frac{1}{2} + \frac{1}{4\cdot2} - \frac{1}{8\cdot3!} + \frac{1}{16\cdot4} - \frac{1}{32\cdot5!} + R_5$$

$$= 1 - 0.50000 + 0.12500 - 0.02083 + 0.00260 - 0.00026 + R_5$$

$$= 0.60651 + R_5 .$$

Note that, since we have a convergent alternating series, R_5 is less than the first term dropped:

$$R_5 < \frac{1}{64\cdot6!} < 0.00003,$$

so $\dfrac{1}{\sqrt{e}} = 0.6065$, correct to four decimal places.

Example 54. Compute $\cos 32°$ correct to four decimal places.

The Taylor series in powers of $\left(x - \dfrac{\pi}{6}\right)$ is obtained immediately from series (4), page 371, with $a = \dfrac{\pi}{6}$:

$$\cos x = \cos\frac{\pi}{6} - \left(x - \frac{\pi}{6}\right)\sin\frac{\pi}{6} - \frac{\left(x - \frac{\pi}{6}\right)^2}{2!}\cos\frac{\pi}{6}$$

$$+ \frac{\left(x - \frac{\pi}{6}\right)^3}{3!}\sin\frac{\pi}{6} + \cdots .$$

Since $x = 32°$, $\left(x - \dfrac{\pi}{6}\right) = 2° = 2(0.01745) = 0.03490$, and

$$\cos 32° = \frac{\sqrt{3}}{2} - (0.03490)\frac{1}{2} - \frac{(0.03490)^2}{2} \cdot \frac{\sqrt{3}}{2} + R_2$$
$$= 0.86602 - 0.01745 - 0.00053 + R_2$$
$$= 0.84804 + R_2.$$

If we use the Lagrange error bound here, then

$$R_2 = \frac{f'''(c)\left(x - \dfrac{\pi}{3}\right)^3}{3!} = \frac{(\sin c)(0.03490)^3}{6},$$

where $30° < c < 32°$. Although we do not know exactly what c is, we do know that $\sin c < 1$; so $R_2 < 0.00001$ and $\cos 32° = 0.8480$ to four decimal places.

Example 55.   For what values of x is the approximate formula

$$\ln(1 + x) = x - \frac{x^2}{2} \tag{9}$$

correct to three decimal places?
   We can use series (6), page 371:

$$\ln(1 + x) = x - \frac{x^2}{2} + \frac{x^3}{3} - \cdots .$$

Since this is a convergent alternating series, the error committed by using the first two terms is less than $\dfrac{|x|^3}{3}$. If $\dfrac{|x|^3}{3} < 0.0005$, then the approximate formula (9) will yield accuracy to three decimal places. We therefore require that $|x|^3 < 0.0015$ or that $|x| < 0.115$.

Example 56.   Estimate the error if the approximate formula

$$\sqrt{1 + x} = 1 + \frac{x}{2} \tag{10}$$

is used and $|x| < 0.02$.
   We obtain the first few terms of the Maclaurin series generated by $f(x) = \sqrt{1 + x}$:

$$f(x) = \sqrt{1 + x}; \qquad\qquad f(0) = 1;$$

$$f'(x) = \frac{1}{2}(1 + x)^{-1/2}; \qquad\qquad f'(0) = \frac{1}{2};$$

$$f''(x) = -\frac{1}{4}(1 + x)^{-3/2}; \qquad\qquad f''(0) = -\frac{1}{4};$$

$$f'''(x) = \frac{3}{8}(1 + x)^{-5/2}. \qquad\qquad f'''(0) = \frac{3}{8}.$$

Then

$$\sqrt{1+x} = 1 + \frac{x}{2} - \frac{1}{4} \cdot \frac{x^2}{2} + \frac{3}{8} \cdot \frac{x^3}{6} - \cdots.$$

If the first term is omitted, the series is strictly alternating; so if formula (10) is used, then $|R_1| < \frac{1}{4} \cdot \frac{x^2}{2}$. With $|x| < 0.02$, $|R_1| < 0.00005$.

Notice that the Lagrange error bound $|R_1|$ here is $\left| \frac{f'(c)x^2}{2!} \right|$, where $0 < c < 0.02$. Since $|c| < 0.02$, we see that

$$|R_1| < \frac{(0.02)^2}{8(1+0.02)^{3/2}} < 0.00005.$$

INDETERMINATE FORMS.   Series may also be used to evaluate indeterminate forms, as indicated in Examples 57–59.

Example 57.   Use series to evaluate $\lim\limits_{x \to 0} \frac{\sin x}{x}$.

From series (1), page 371,

$$\sin x = x - \frac{x^3}{3!} + \frac{x^5}{5!} - \cdots$$

$$\lim_{x \to 0} \frac{\sin x}{x} = \lim_{x \to 0} \left(1 - \frac{x^2}{3!} + \frac{x^4}{5!} - \cdots \right) = 1,$$

a well-established result obtained previously.

Example 58.   Use series to evaluate $\lim\limits_{x \to 0} \frac{\ln(x+1)}{3x}$.

We can use series (6), page 371, and write

$$\lim_{x \to 0} \frac{x - \frac{x^2}{2} + \frac{x^3}{3} - \frac{x^4}{4} + \cdots}{3x} = \lim_{x \to 0} \frac{1}{3} - \frac{x}{6} + \frac{x^2}{9} - \cdots$$

$$= \frac{1}{3}.$$

Example 59. $\lim\limits_{x \to 0} \frac{e^{-x^2} - 1}{x^2} = \lim\limits_{x \to 0} \frac{\left(1 - x^2 + \frac{x^4}{2!} - \frac{x^6}{3!} + \cdots \right) - 1}{x^2}$

$$= \lim_{x \to 0} \frac{-x^2 + \frac{x^4}{2!} - \cdots}{x^2} = \lim_{x \to 0} -1 + \frac{x^2}{2!} - \frac{x^4}{4!} + \cdots$$

$$= -1.$$

Example 60. Show how series may be used to evaluate π.

Since $\frac{\pi}{4} = \tan^{-1} 1$, a series for $\tan^{-1} x$ may prove helpful. Note that

$$\tan^{-1} x = \int_0^x \frac{dt}{1+t^2}$$

and that a series for $\frac{1}{1+t^2}$ is obtainable easily by long division to yield

$$\frac{1}{1+t^2} = 1 - t^2 + t^4 - t^6 + \cdots \tag{11}$$

If we integrate (11) term by term and then evaluate the definite integral we get

$$\tan^{-1} x = x - \frac{x^3}{3} + \frac{x^5}{5} - \frac{x^7}{7} + \cdots + \frac{(-1)^{n-1} x^{2n-1}}{2n-1} + \cdots \tag{12}$$

Compare with series (8) on page 371 and note especially that (12) converges on $-1 \le x \le 1$. We replace x by 1 in (12) to get

$$\tan^{-1} 1 = 1 - \frac{1}{3} + \frac{1}{5} - \frac{1}{7} + \cdots$$

So

$$\frac{\pi}{4} = 1 - \frac{1}{3} + \frac{1}{5} - \frac{1}{7} + \cdots \tag{13}$$

and

$$\pi = 4\left(1 - \frac{1}{3} + \frac{1}{5} - \frac{1}{7} + \cdots\right).$$

In Figure N10–11 we show the calculator graph of the series given just above for π. We keyed in

$$X_{1T} = T \qquad Y_{1T} = 4 \text{ sum seq} \left((-1)^{(N+1)}/(2N-1), N, 1, T, 1\right).$$

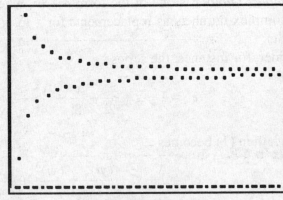

FIGURE N10–11

Here are some approximations for π when we TRACE the graph:

N	1	2	5	10	25	50	59	60
S_N	4	2.67	3.34	3.04	3.18	3.12	3.16	3.12

Since the series is alternating, the odd sums are greater, the even ones less, than the value of π. It is clear that several hundred terms of (13) may be required to get even two-place accuracy. There are series expressions for π that converge much more rapidly. (See Miscellaneous Free-Response Practice, Problem 12, page 432.)

Example 61.   Use series to evaluate $\displaystyle\int_0^{0.1} e^{-x^2}\, dx$ to four decimal places.

Although $\displaystyle\int e^{-x^2}\, dx$ cannot be expressed in terms of elementary functions, we can write a series for e^u, replace u by $(-x^2)$, and integrate term by term. Thus,

$$e^{-x^2} = 1 - x^2 + \frac{x^4}{2!} - \frac{x^6}{3!} + \cdots,$$

so

$$\int_0^{0.1} e^{-x^2}\, dx = x - \frac{x^3}{3} + \frac{x^5}{5\cdot 2!} - \frac{x^7}{7\cdot 3!} + \cdots \Bigg|_0^{0.1}$$

$$= 0.1 - \frac{0.001}{3} + \frac{0.00001}{10} - \frac{0.0000001}{42} + \cdots \qquad (14)$$

$$= 0.1 - 0.00033 + 0.000001 + R_6$$

$$= 0.09967 + R_6.$$

Since (14) is a convergent alternating series, $|R_6| < \dfrac{10^{-7}}{42}$, which will not affect the fourth decimal place. Then, correct to four decimal places,

$$\int_0^{0.1} e^{-x^2}\, dx = 0.0997 .$$

†C7.   POWER SERIES OVER COMPLEX NUMBERS.

A *complex number* is one of the form $a + bi$, where a and b are real and $i^2 = -1$. If we allow complex numbers as replacements for x in power series, we obtain some interesting results.

Consider, for instance, the series

$$e^x = 1 + x + \frac{x^2}{2!} + \frac{x^3}{3!} + \frac{x^4}{4!} + \cdots + \frac{x^n}{n!} \cdots . \qquad (1)$$

When $x = yi$, then (1) becomes

$$e^{yi} = 1 + yi + \frac{(yi)^2}{2!} + \frac{(yi)^3}{3!} + \frac{(yi)^4}{4!} + \cdots$$

$$= 1 + yi - \frac{y^2}{2!} - \frac{y^3 i}{3!} + \frac{y^4}{4!} + \cdots$$

$$= \left(1 - \frac{y^2}{2!} + \frac{y^4}{4!} + \cdots\right) + i\left(y - \frac{y^3}{3!} + \frac{y^5}{5!} - \cdots\right). \qquad (2)$$

---

†This is an optional topic not included in the BC Course Description.

So

$$e^{yi} = \cos y + i \sin y, \tag{3}$$

since the series within the parentheses of equation (2) converge respectively to cos *y* and sin *y*. Equation (3) is called *Euler's formula*. It follows from (3) that

$$e^{\pi i} = -1 \quad \text{and} \quad e^{2\pi i} = 1;$$

the latter is sometimes referred to as *Euler's magic formula*.

*Set 10: Multiple-Choice Questions on Sequences and Series

The asterisk applies to all the questions in this set: this topic is covered only in Calculus BC.

Note: No questions on sequences will appear on the BC examination. We have nevertheless chosen to include the topic in questions 1–7 because a series and its convergence are defined in terms of sequences. Review of sequences will enhance understanding of series.

1. Which sequence converges?

   (A) $\left\{n + \dfrac{3}{n}\right\}$ **(B)** $\left\{-1 + \dfrac{(-1)^n}{n}\right\}$ **(C)** $\left\{\sin\dfrac{n\pi}{2}\right\}$

   (D) $\left\{\dfrac{n!}{3^n}\right\}$ **(E)** $\left\{\dfrac{n}{\ln n}\right\}$

2. If $\{s_n\} = \left\{1 + \dfrac{(-1)^n}{n}\right\}$, then

   (A) $\{s_n\}$ diverges by oscillation   **(B)** $\{s_n\}$ converges to zero
   (C) $\lim\limits_{n\to\infty} s_n = 1$ **(D)** $\{s_n\}$ diverges to infinity
   (E) None of the above is true.

3. The sequence $\left\{\sin\dfrac{n\pi}{6}\right\}$

   (A) is unbounded   **(B)** is monotonic
   (C) converges to a number less than 1   **(D)** is bounded
   (E) diverges to infinity

4. Which of the following sequences diverges?

   (A) $\left\{\dfrac{1}{n}\right\}$ **(B)** $\left\{\dfrac{(-1)^{n+1}}{n}\right\}$ **(C)** $\left\{\dfrac{2^n}{e^n}\right\}$

   (D) $\left\{\dfrac{n^2}{e^n}\right\}$ **(E)** $\left\{\dfrac{n}{\ln n}\right\}$

5. Which of the following statements about sequences is false?

    (A) If $\{s_n\}$ is bounded, then it is convergent.

    (B) If $\lim\limits_{n\to\infty} s_n = L$, then $|s_n - L| < 0.001$ except for at most a finite number of n's.

    (C) If $\{s_n\}$ converges, then $\{s_n\}$ is bounded.

    (D) If $\{s_n\}$ is unbounded, then it diverges.

    (E) none of these

6. The sequence $\{r^n\}$ converges if and only if

    (A) $|r| < 1$ **(B)** $|r| \leq 1$ **(C)** $-1 < r \leq 1$

    (D) $0 < r < 1$ **(E)** $|r| > 1$

7. Which of the following statements about a convergent sequence is false?

    (A) The limit of a convergent sequence is unique.

    (B) The product of two convergent sequences converges.

    (C) The quotient of two convergent sequences converges.

    (D) If c is any number, then $\lim\limits_{n\to\infty} (c \cdot s_n) = c \lim\limits_{n\to\infty} s_n$.

    (E) none of these

8. $\sum u_n$ is a series of constants for which $\lim\limits_{n\to\infty} u_n = 0$. Which of the following statements is always true?

    (A) $\sum u_n$ converges to a finite sum.   **(B)** $\sum u_n$ equals zero.

    (C) $\sum u_n$ does not diverge to infinity.   **(D)** $\sum u_n$ is a positive series.

    (E) none of these

9. Note that $\dfrac{1}{n(n+1)} = \dfrac{1}{n} - \dfrac{1}{n+1}$ $(n \geq 1)$. $\sum\limits_{n=1}^{\infty} \dfrac{1}{n(n+1)}$ equals

    (A) 0   **(B)** 1   **(C)** $\dfrac{3}{2}$ **(D)** $\dfrac{3}{4}$ **(E)** ∞

10. The sum of the geometric series $\left(2 - 1 + \dfrac{1}{2} - \dfrac{1}{4} + \dfrac{1}{8} - \cdots\right)$ is

    (A) $\dfrac{4}{3}$ **(B)** $\dfrac{5}{4}$ **(C)** 1   **(D)** $\dfrac{3}{2}$ **(E)** $\dfrac{3}{4}$

11. Which of the following statements about series is true?

(A) If $\lim_{n\to\infty} u_n = 0$, then $\sum u_n$ converges.

(B) If $\lim_{n\to\infty} u_n \neq 0$, then $\sum u_n$ diverges.

(C) If $\sum u_n$ diverges, then $\lim_{n\to\infty} u_n \neq 0$.

(D) $\sum u_n$ converges if and only if $\lim_{n\to\infty} u_n = 0$.

(E) none of these

12. Which of the following statements about series is false?

(A) $\sum_{k=1}^{\infty} u_k = \sum_{k=m}^{\infty} u_k$, where m is any positive integer.

(B) If $\sum u_n$ converges, so does $\sum c u_n$ if $c \neq 0$.

(C) If $\sum a_n$ and $\sum b_n$ converge, so does $\sum (ca_n + b_n)$, where $c \neq 0$.

(D) If 1000 terms are added to a convergent series, the new series also converges.

(E) Rearranging the terms of a positive convergent series will not affect its convergence or its sum.

13. Which of the following series converges?

(A) $\sum \dfrac{1}{\sqrt[3]{n}}$ (B) $\sum \dfrac{1}{\sqrt{n}}$ (C) $\sum \dfrac{1}{n}$

(D) $\sum \dfrac{1}{10n-1}$ (E) $\sum \dfrac{2}{n^2-5}$

14. Which of the following series diverges?

(A) $\sum_{n=1}^{\infty} \dfrac{1}{n(n+1)}$ (B) $\sum_{n=1}^{\infty} \dfrac{n+1}{n!}$ (C) $\sum_{n=2}^{\infty} \dfrac{1}{n \ln n}$

(D) $\sum_{n=1}^{\infty} \dfrac{\ln n}{2^n}$ (E) $\sum_{n=1}^{\infty} \dfrac{n}{2^n}$

15. Which of the following series diverges?

(A) $\sum \dfrac{1}{n^2}$ (B) $\sum \dfrac{1}{n^2+n}$ (C) $\sum \dfrac{n}{n^3+1}$

(D) $\sum \dfrac{n}{\sqrt{4n^2-1}}$ (E) none of these

16. For which of the following series does the Ratio Test fail?

(A) $\sum \dfrac{1}{n!}$ (B) $\sum \dfrac{n}{2^n}$ (C) $1 + \dfrac{1}{2^{3/2}} + \dfrac{1}{3^{3/2}} + \dfrac{1}{4^{3/2}} + \cdots$

(D) $\dfrac{\ln 2}{2^2} + \dfrac{\ln 3}{2^3} + \dfrac{\ln 4}{2^4} + \cdots$ (E) $\sum \dfrac{n^n}{n!}$

17. Which of the following alternating series diverges?

(A) $\sum \dfrac{(-1)^{n-1}}{n}$ (B) $\sum \dfrac{(-1)^{n+1}(n-1)}{n+1}$ (C) $\sum \dfrac{(-1)^{n+1}}{\ln(n+1)}$

(D) $\sum \dfrac{(-1)^{n-1}}{\sqrt{n}}$ (E) $\sum \dfrac{(-1)^{n-1}(n)}{n^2+1}$

18. Which of the following series diverges?

(A) $3 - 1 + \dfrac{1}{9} - \dfrac{1}{27} + \cdots$ (B) $\dfrac{1}{\sqrt{2}} - \dfrac{1}{\sqrt{3}} + \dfrac{1}{\sqrt{4}} - \cdots$

(C) $\dfrac{1}{2^2} - \dfrac{1}{3^2} + \dfrac{1}{4^2} - \cdots$ (D) $1 - 1.1 + 1.21 - 1.331 + \cdots$

(E) $\dfrac{1}{1 \cdot 2} - \dfrac{1}{2 \cdot 3} + \dfrac{1}{3 \cdot 4} - \dfrac{1}{4 \cdot 5} + \cdots$

19. Let $S = \displaystyle\sum_{n=1}^{\infty} \left(\dfrac{2}{3}\right)^n$; then S equals

(A) 1   (B) $\dfrac{3}{2}$ (C) $\dfrac{4}{3}$ (D) 2   (E) 3

20. Which of the following statements is true?

(A) If $\sum u_n$ converges, then so does the series $\sum |u_n|$.
(B) If a series is truncated after the nth term, then the error is less than the first term omitted.
(C) If the terms of an alternating series decrease, then the series converges.
(D) If $r < 1$, then the series $\sum r^n$ converges.
(E) none of these

21. Which of the following expansions is impossible?

(A) $\sqrt{x-1}$ in powers of x (B) $\sqrt{x+1}$ in powers of x

(C) $\ln x$ in powers of $(x-1)$ (D) $\tan x$ in powers of $\left(x - \dfrac{\pi}{4}\right)$

(E) $\ln(1-x)$ in powers of x

22. The power series $x + \dfrac{x^2}{2} + \dfrac{x^3}{3} + \cdots + \dfrac{x^n}{n} + \cdots$ converges if and only if

(A) $-1 < x < 1$ (B) $-1 \leqq x \leqq 1$ (C) $-1 \leqq x < 1$
(D) $-1 < x \leqq 1$ (E) $x = 0$

23. The power series

$$(x+1) - \frac{(x+1)^2}{2!} + \frac{(x+1)^3}{3!} - \frac{(x+1)^4}{4!} + \cdots$$

diverges

(A) for no real x (B) if $-2 < x \leqq 0$ (C) if $x < -2$ or $x > 0$
(D) if $-2 \leqq x < 0$ (E) if $x \neq -1$

24. The series $\displaystyle\sum_{n=0}^{\infty} n!(x-3)^n$ converges if and only if

(A) $x = 0$ (B) $2 < x < 4$ (C) $x = 3$ (D) $2 \leqq x \leqq 4$
(E) $x < 2$ or $x > 4$

25. The series obtained by differentiating term by term the series

$$(x-2) + \frac{(x-2)^2}{4} + \frac{(x-2)^3}{9} + \frac{(x-2)^4}{16} + \cdots$$

converges for

(A) $1 \leqq x \leqq 3$ (B) $1 \leqq x < 3$ (C) $1 < x \leqq 3$
(D) $0 \leqq x \leqq 4$ (E) none of these

26. Let $f(x) = \displaystyle\sum_{n=0}^{\infty} x^n$. The radius of convergence of $\displaystyle\int_0^x f(t)\, dt$ is

(A) 0   (B) 1   (C) 2   (D) ∞ (E) none of these

27. The coefficient of x^4 in the Maclaurin series for $f(x) = e^{-x/2}$ is

(A) $-\dfrac{1}{24}$ (B) $\dfrac{1}{24}$ (C) $\dfrac{1}{96}$ (D) $-\dfrac{1}{384}$ (E) $\dfrac{1}{384}$

28. The Taylor polynomial of order 3 at $x = 0$ for $f(x) = \sqrt{1+x}$ is

(A) $1 + \dfrac{x}{2} - \dfrac{x^2}{4} + \dfrac{3x^3}{8}$ (B) $1 + \dfrac{x}{2} - \dfrac{x^2}{8} + \dfrac{x^3}{16}$

(C) $1 - \dfrac{x}{2} + \dfrac{x^2}{8} - \dfrac{x^3}{16}$ (D) $1 + \dfrac{x}{2} - \dfrac{x^2}{8} + \dfrac{x^3}{8}$

(E) $1 - \dfrac{x}{2} + \dfrac{x^2}{4} - \dfrac{3x^3}{8}$

29. The Taylor polynomial of order 3 at $x = 1$ for e^x is

(A) $1 + (x - 1) + \dfrac{(x - 1)^2}{2} + \dfrac{(x - 1)^3}{3}$

(B) $e\left[1 + (x - 1) + \dfrac{(x - 1)^2}{2} + \dfrac{(x - 1)^3}{3}\right]$

(C) $e\left[1 + (x + 1) + \dfrac{(x + 1)^2}{2!} + \dfrac{(x + 1)^3}{3!}\right]$

(D) $e\left[1 + (x - 1) + \dfrac{(x - 1)^2}{2!} + \dfrac{(x - 1)^3}{3!}\right]$

(E) $e\left[1 - (x - 1) + \dfrac{(x - 1)^2}{2!} + \dfrac{(x - 1)^3}{3!}\right]$

30. The coefficient of $\left(x - \dfrac{\pi}{4}\right)^3$ in the Taylor series about $\dfrac{\pi}{4}$ of $f(x) = \cos x$ is

(A) $\dfrac{\sqrt{3}}{12}$ (B) $-\dfrac{1}{12}$ (C) $\dfrac{1}{12}$

(D) $\dfrac{1}{6\sqrt{2}}$ (E) $-\dfrac{1}{3\sqrt{2}}$

31. Which of the following series can be used to compute ln 0.8?

(A) $\ln (x - 1)$ expanded about $x = 0$
(B) $\ln x$ about $x = 0$
(C) $\ln x$ in powers of $(x - 1)$
(D) $\ln (x - 1)$ in powers of $(x - 1)$
(E)   none of these

32. If $e^{-0.1}$ is computed using series, then, correct to three decimal places, it equals

(A)   0.905   (B)   0.950   (C)   0.904
(D)   0.900   (E)   0.949

33. The coefficient of x^2 in the Maclaurin series for $e^{\sin x}$ is

(A)   0   (B)   1   (C) $\dfrac{1}{2!}$

(D) -1 (E) $\dfrac{1}{4}$

34. Let $f(x) = \sum_{n=0}^{\infty} a_n x^n$, $g(x) = \sum_{n=0}^{\infty} b_n x^n$. Suppose both series converge for $|x| < R$. Let x_0 be a number such that $|x_0| < R$. Which of the following statements is false?

(A) $\sum_{n=0}^{\infty} (a_n + b_n)(x_0)^n$ converges to $f(x_0) + g(x_0)$.

(B) $\left[\sum_{n=0}^{\infty} a_n (x_0)^n\right]\left[\sum_{n=0}^{\infty} b_n (x_0)^n\right]$ converges to $f(x_0)g(x_0)$.

(C) $f(x) = \sum_{n=0}^{\infty} a_n x^n$ is continuous at $x = x_0$.

(D) $\sum_{n=1}^{\infty} n a_n x^{n-1}$ converges to $f'(x_0)$.

(E) none of these

35. The coefficient of $(x - 1)^5$ in the Taylor series for $x \ln x$ about $x = 1$ is

(A) $-\dfrac{1}{20}$ (B) $\dfrac{1}{5!}$ (C) $-\dfrac{1}{5!}$ (D) $\dfrac{1}{4!}$ (E) $-\dfrac{1}{4!}$

36. The radius of convergence of the series $\sum_{n=1}^{\infty} \dfrac{x^n}{2^n} \cdot \dfrac{n^n}{n!}$ is

(A) 0   (B) 2   (C) $\dfrac{2}{e}$ (D) $\dfrac{e}{2}$ (E) ∞

37. If the approximate formula $\sin x = x - \dfrac{x^3}{3!}$ is used and $|x| < 1$ (radian), then the error is numerically less than

(A) 0.001   (B) 0.003   (C) 0.005   (D) 0.008   (E) 0.009

38. If an appropriate series is used to evaluate $\int_0^{0.3} x^2 e^{-x^2}\, dx$, then, correct to three decimal places, the definite integral equals

(A) 0.009   (B) 0.082   (C) 0.098   (D) 0.008   (E) 0.090

39. If a suitable series is used, then $\int_0^{0.2} \dfrac{e^{-x} - 1}{x}\, dx$, correct to three decimal places, is

(A) −0.200   (B) 0.180   (C) 0.190   (D) −0.190   (E) −0.990

40. The function $f(x) = \sum_{n=0}^{\infty} a_n x^n$ and $f'(x) = -f(x)$ for all x. If $f(0) = 1$, then $f(0.2)$, correct to three decimal places, is

(A) 0.905   (B) 1.221   (C) 0.819   (D) 0.820   (E) 1.220

41. The sum of the series $\sum_{n=1}^{\infty} \left(\dfrac{\pi^3}{3^{\pi}} \right)^n$ is equal to

(A) 0   (B) 1   (C) $\dfrac{3^{\pi}}{\pi^3 - 3^{\pi}}$ (D) $\dfrac{\pi^3}{3^{\pi} - \pi^3}$ (E) none of these

42. When $\sum_{1}^{\infty} \dfrac{(-1)^{n-1}}{3n-1}$ is approximated by the sum of its first 300 terms, the error is closest to

(A) 0.001   (B) 0.002   (C) 0.005   (D) 0.01   (E) 0.02

43. If the series $\tan^{-1} 1 = 1 - \dfrac{1}{3} + \dfrac{1}{5} - \dfrac{1}{7} + \cdots$ is used to approximate $\dfrac{\pi}{4}$ with an error less than 0.001, then the smallest number of terms needed is

(A) 100   (B) 200   (C) 300   (D) 400   (E) 500

44. The Taylor polynomial of order 3 at $x = 0$ for $(1 + x)^p$, where p is a constant, is

(A) $1 + px + p(p-1)x^2 + p(p-1)(p-2)x^3$

(B) $1 + px + \dfrac{p(p-1)}{2}x^2 + \dfrac{p(p-1)(p-2)}{3}x^3$

(C) $1 + px + \dfrac{p(p-1)}{2!}x^2 + \dfrac{p(p-1)(p-2)}{3!}x^3$

(D) $px + \dfrac{p(p-1)}{2!}x^2 + \dfrac{p(p-1)(p-2)}{3!}x^3$

(E)   none of these

45. The Taylor series for $\ln(1 + 2x)$ about $x = 0$ is

(A) $2x - \dfrac{(2x)^2}{2} + \dfrac{(2x)^3}{3} - \dfrac{(2x)^4}{4} + \cdots$

(B) $2x - 2x^2 + 8x^3 - 16x^4 + \cdots$

(C) $2x - 4x^2 + 16x^3 + \cdots$

(D) $2x - x^2 + \dfrac{8}{3}x^3 - 4x^4 + \cdots$

(E) $2x - \dfrac{(2x)^2}{2!} + \dfrac{(2x)^3}{3!} - \dfrac{(2x)^4}{4!} + \cdots$

46. The set of all values of x for which $\sum_{n=1}^{\infty} \dfrac{n \cdot 2^n}{x^n}$ converges is

(A) only $x = 0$ (B) $|x| = 2$ (C) $-2 < x < 2$

(D) $|x| > 2$ (E) none of these

47. The third-order Taylor polynomial $P_3(x)$ for $\sin x$ about $\frac{\pi}{4}$ is

   (A) $\frac{1}{\sqrt{2}}\left((x - \frac{\pi}{4}) - \frac{1}{3!}(x - \frac{\pi}{4})^3\right)$

   (B) $\frac{1}{\sqrt{2}}\left(1 + (x - \frac{\pi}{4}) - \frac{1}{2}(x - \frac{\pi}{4})^2 + \frac{1}{3!}(x - \frac{\pi}{4})^3\right)$

   (C) $\frac{1}{\sqrt{2}}\left(1 + (x - \frac{\pi}{4}) - \frac{1}{2!}(x - \frac{\pi}{4})^2 - \frac{1}{3!}(x - \frac{\pi}{4})^3\right)$

   (D) $1 + (x - \frac{\pi}{4}) - \frac{1}{2}(x - \frac{\pi}{4})^2 - \frac{1}{6}(x - \frac{\pi}{4})^3$

   (E) $\frac{1}{\sqrt{2}}(1 + x - \frac{x^2}{2!} - \frac{x^3}{3!})$

48. Let f be the Taylor polynomial $P_7(x)$ of order 7 for $\tan^{-1} x$ about $x = 0$. Then it follows that, if $-1.5 < x < 1.5$,

   (A) $f(x) = \tan^{-1} x$
   (B) $f(x) \leq \tan^{-1} x$
   (C) $f(x) \geq \tan^{-1} x$
   (D) $f(x) > \tan^{-1} x$ if $x < 0$ but $< \tan^{-1} x$ if $x > 0$
   (E) $f(x) < \tan^{-1} x$ if $x < 0$ but $> \tan^{-1} x$ if $x > 0$

49. Replace the first sentence in Question 48 by "Let f be the Taylor polynomial $P_9(x)$ of order 9 for $\tan^{-1} x$ about $x = 0$." Which choice given in Question 48 is now the correct one?

Answers for Set 10: Sequences and Series

1.	B	**11.**	B	**21.**	A	**31.**	C	**41.**	D
2.	C	**12.**	A	**22.**	C	**32.**	A	**42.**	A
3.	D	**13.**	E	**23.**	A	**33.**	C	**43.**	E
4.	E	**14.**	C	**24.**	C	**34.**	E	**44.**	C
5.	A	**15.**	D	**25.**	B	**35.**	A	**45.**	A
6.	C	**16.**	C	**26.**	B	**36.**	C	**46.**	D
7.	C	**17.**	B	**27.**	E	**37.**	E	**47.**	C
8.	E	**18.**	D	**28.**	B	**38.**	A	**48.**	D
9.	B	**19.**	D	**29.**	D	**39.**	D	**49.**	E
10.	A	**20.**	E	**30.**	D	**40.**	C		

1. **B.** $\left\{-1+\dfrac{(-1)^n}{n}\right\}$ converges to -1.

2. **C.** Note that $\lim\limits_{n\to\infty}\dfrac{(-1)^n}{n}=0$.

3. **D.** The sine function varies continuously between -1 and 1 inclusive.

4. **E.** Note that $\left\{\dfrac{2}{e}\right\}^n$ is a sequence of the type $\{r^n\}$ with $|r|<1$; also that

$\lim \dfrac{n^2}{e^n}=0$ by repeated application of L'Hôpital's rule.

5. **A.** The sequence $1,-1,1,-1,\ldots$ is a good counterexample for statement (A).

6. **C.** See Example 12, page 341.

7. **C.** Division by zero must be avoided.

8. **E.** The harmonic series $\displaystyle\sum_{1}^{\infty}\dfrac{1}{k}$ is a counterexample for (A), (B), and (C).

$\displaystyle\sum_{1}^{\infty}\dfrac{(-1)^{k+1}}{k}$ shows that (D) does not follow.

9. **B.** $\displaystyle\sum_{1}^{\infty}\dfrac{1}{n(n+1)}=\dfrac{1}{1\cdot2}+\dfrac{1}{2\cdot3}+\dfrac{1}{3\cdot4}+\cdots+\dfrac{1}{n(n+1)}+\cdots;$ so

$s_n=1-\dfrac{1}{2}+\dfrac{1}{2}-\dfrac{1}{3}+\dfrac{1}{3}-\cdots+\dfrac{1}{n}-\dfrac{1}{n+1}=1-\dfrac{1}{n+1},$

and $\lim\limits_{n\to\infty}s_n=1.$

10. **A.** $S=\dfrac{a}{1-r}=\dfrac{2}{1-(-\frac{1}{2})}=\dfrac{4}{3}.$

11. B.   Find counterexamples for statements (A), (C), and (D).

12. A.   If $\displaystyle\sum_{k=1}^{\infty} u_k$ converges, so does $\displaystyle\sum_{k=1}^{\infty} u_k$, where m is any positive integer; but their *sums* are probably different.

13. E.   See Example 20, page 349. Each series given is essentially a *p*-series. Only in (E) is $p > 1$.

14. C.   Use the Integral Test on page 347.

15. D. $\dfrac{n}{\sqrt{4n^2-1}} > \dfrac{n}{\sqrt{4n^2}} = \dfrac{1}{2}$, the general term of a divergent series.

16. C.   The limit of the ratio for the series $\displaystyle\sum \dfrac{1}{n^{3/2}}$ is 1, so this test fails; note for (E) that

$$\lim_{n\to\infty}\frac{u_{n+1}}{u_n} = \lim_{n\to\infty}\left(\frac{n+1}{n}\right)^n = \lim_{n\to\infty}\left(1+\frac{1}{n}\right)^n = e.$$

17. B. $\displaystyle\lim_{n\to\infty}\dfrac{(-1)^{n+1}(n-1)}{n+1}$ does not equal 0.

18. D.   (A), (B), (C), and (E) all converge; (D) is the divergent geometric series with $r = -1.1$.

19. D. $S = \dfrac{a}{1-r} = \dfrac{\frac{2}{3}}{1-\frac{2}{3}} = \dfrac{\frac{2}{3}}{\frac{1}{3}} = 2$.

20. E.   Note the following counterexamples:

(A) $\displaystyle\sum \frac{(-1)^{n-1}}{n}$ (B) $\displaystyle\sum \frac{1}{n}$ (C) $\displaystyle\sum \frac{(-1)^{n-1}\cdot n}{2n-1}$

(D) $\displaystyle\sum \left(-\frac{3}{2}\right)^{n-1}$

21. A.   If $f(x) = \sqrt{x-1}$, then $f(0)$ is not defined.

22. C.   Since $\displaystyle\lim_{n\to\infty}\left|\dfrac{u_{n+1}}{u_n}\right| = |x|$, the series converges if $|x| < 1$. We must test the endpoints: when $x = 1$, we get the divergent harmonic series; $x = -1$ yields the convergent alternating harmonic series.

23. A. $\displaystyle\lim_{n\to\infty}\left|\dfrac{x+1}{n+1}\right| = 0$ for all $x \neq -1$; since the given series converges to 0 if $x = -1$, it therefore converges for *all* x.

24. C. $\displaystyle\lim_{n\to\infty}(n+1)(x-3) = \infty$ unless $x = 3$.

25. B.   The differentiated series is $\displaystyle\sum_{n=1}^{\infty}\frac{(x-2)^{n-1}}{n}$; so

$$\lim_{n\to\infty}\left|\frac{u_{n+1}}{u_n}\right|=|x-2| \ .$$

26. B.   The integrated series is $\displaystyle\sum_{n=0}^{\infty}\frac{x^{n+1}}{n+1}$ or $\displaystyle\sum_{n=1}^{\infty}\frac{x^{n}}{n}$. See Question 22.

27. E. $e^{-x/2}=1+\left(-\dfrac{x}{2}\right)+\left(-\dfrac{x}{2}\right)^{2}\cdot\dfrac{1}{2!}+\left(-\dfrac{x}{2}\right)^{3}\cdot\dfrac{1}{3!}+\left(-\dfrac{x}{2}\right)^{4}\cdot\dfrac{1}{4!}+\cdots .$

28. B.   Use (2) on page 366.

29. D.   Note that every derivative of e^x is e^x, which equals e at $x=1$. Use (1) on page 360 with $a=1$.

30. D.   Use (2) on page 364 with $a=\dfrac{\pi}{4}$ and series expansion (4) on page 371.

31. C.   Note that $\ln q$ is defined only if $q>0$, and that the derivatives must exist at $x=a$ in formula (2) for the Taylor series on page 364.

32. A.   Use

$$e^{-x}=1-x+\frac{x^2}{2!}-\cdots ; \qquad e^{-0.1}=1-(+0.1)+\frac{0.01}{2!}+R_2 \ .$$

$|R_2|<\dfrac{0.001}{3!}<0.0005$. Or use the series for e^x and let $x=-0.1$.

33. C. $e^{u}=1+u+\dfrac{u^2}{2!}+\cdots$; and $\sin x=x-\dfrac{x^3}{3!}+\cdots$, so

$$e^{\sin x}=1+\left(x-\frac{x^3}{3!}+\cdots\right)+\frac{1}{2!}\left(x-\frac{x^3}{3!}+\cdots\right)^{2}+\cdots .$$

Or generate the Maclaurin series for $e^{\sin x}$.

34. E.   (A), (B), (C), and (D) are all true statements.

35. A. $f(x)=x\ln x,$

$f'(x)=1+\ln x,$

$f''(x)=\dfrac{1}{x},$

$f'''(x)=-\dfrac{1}{x^2},$

$f^{(4)}(x)=\dfrac{2}{x^3},$

$f^{(5)}(x)=-\dfrac{3\cdot2}{x^4};\quad f^{(5)}(1)=-3\cdot2.$

So the coefficient of $(x-1)^5$ is $-\dfrac{3\cdot2}{5!}=-\dfrac{1}{20}.$

36.   **C.** $\lim\limits_{n\to\infty}\left|\dfrac{u_{n+1}}{u_n}\right| = \lim\limits_{n\to\infty}\left|\dfrac{x^{n+1}(n+1)^{n+1}}{2^{n+1}(n+1)!}\cdot\dfrac{2^n\cdot n!}{x^n\cdot n^n}\right|$

$\qquad\qquad = \lim\limits_{n\to\infty}\left|\dfrac{x(n+1)^{n+1}}{2(n+1)n^n}\right| = \lim\limits_{n\to\infty}\left|\dfrac{x}{2}\left(\dfrac{n+1}{n}\right)^n\right|$

$\qquad\qquad = \lim\limits_{n\to\infty}\left|\dfrac{x}{2}\left(1+\dfrac{1}{n}\right)^n\right| = \left|\dfrac{x}{2}\cdot e\right|.$

Since the series converges when $\left|\dfrac{x}{2}\cdot e\right| < 1$, that is, when $|x| < \dfrac{2}{e}$, the radius of convergence is $\dfrac{x}{e}$.

37.   **E.**   The error, $|R_4|$, is less than $\dfrac{1}{5!} < 0.009.$

38.   **A.** $\displaystyle\int_0^{0.3} x^2 e^{-x^2}\,dx = \int_0^{0.3} x^2\left(1 - x^2 + \dfrac{x^4}{2!} - \cdots\right)dx$

$\qquad\qquad = \displaystyle\int_0^{0.3}\left(x^2 - x^4 + \dfrac{x^6}{2!} - \cdots\right)dx$

$\qquad\qquad = \dfrac{x^3}{3} - \dfrac{x^5}{5} + \cdots \Big|_0^{0.3}$

$\qquad\qquad = 0.009$ to three decimal places.

39.   **D.** $\displaystyle\int_0^{0.2}\dfrac{e^{-x}-1}{x}\,dx = \int \dfrac{\left(1 - x + \dfrac{x^2}{2!} - \dfrac{x^3}{3!} + \cdots\right) - 1}{x}\,dx$

$\qquad\qquad = \displaystyle\int_0^{0.2}\left(-1 + \dfrac{x}{2} - \dfrac{x^2}{3!} + \cdots\right)dx$

$\qquad\qquad = -x + \dfrac{x^2}{2\cdot2} - \dfrac{x^3}{3\cdot6} + \cdots \Big|_0^{0.2}$

$\qquad\qquad = -0.190$; the error $< \dfrac{(0.2)^4}{96} < 0.0005.$

40.   **C.** $f(x) = a_0 + a_1 x + a_2 x^2 + a_3 x^3 + \cdots$; if $f(0) = 1$, then $a_0 = 1.$

$f'(x) = a_1 + 2a_2 x + 3a_3 x^2 + 4a_4 x^3 + \cdots$; $f'(0) = -f(0) = -1,$

so $a_1 = -1$. Since $f'(x) = -f(x)$, $f(x) = -f'(x)$:

$1 - x + a_2 x^2 + a_3 x^3 + \cdots = -(-1 + 2a_2 x + 3a_3 x^2 + 4a_4 x^3 + \cdots)$

identically. Thus,

$$-2a_2 = -1, \qquad a_2 = \frac{1}{2},$$

$$-3a_3 = a_2, \qquad a_3 = -\frac{1}{3!},$$

$$-4a_4 = a_3, \qquad a_4 = \frac{1}{4!},$$

$$\vdots \qquad\qquad \vdots$$

It is clear, then, that

$$f(x) = 1 - x + \frac{x^2}{2!} - \frac{x^3}{3!} + \frac{x^4}{4!} - \cdots;$$

$$f(0.2) = 1 - 0.2 + \frac{(0.2)^2}{2!} - \frac{(0.2)^3}{3!} + R_3$$

$$= 0.819; \quad |R_3| < 0.0005.$$

41. D.   Use a calculator to verify that the ratio $\dfrac{\pi^3}{3^\pi}$ (of the given geometric series) equals approximately 0.98. Since the ratio $r < 1$, the sum of the series equals

$$\frac{a}{1-r} \qquad \text{or} \qquad \frac{\pi^3}{3^\pi} \cdot \frac{1}{1 - \dfrac{\pi^3}{3^\pi}}.$$

Simplify to get (D).

42. A.   The error (of the given alternating series) is less in absolute value than the first term dropped; that is, less than $\left| \dfrac{(-1)^{301}}{903 - 1} \right| \approx 0.0011$. Choice (A) is closest to this approximation.

43. E.   The error is less than the first term dropped, namely, $\dfrac{1}{2(n+1)-1}$, or $\dfrac{1}{2n+1}$ (see (8) on page 371), so $n \geqslant 500$.

44. C.   This polynomial is associated with the binomial series $(1 + x)^p$. Verify that $f(0) = 1, f'(0) = p, f''(0) = p(p-1), f'''(0) = p(p-1)(p-2)$.

45. A.   The fastest way to find the series for $\ln(1 + 2x)$ about $x = 0$ is to substitute $2x$ for x in the series

$$\ln(1 + x) = x - \frac{x^2}{2} + \frac{x^3}{3} - \frac{x^4}{4} + \cdots.$$

46.   D. $\lim\limits_{n\to\infty}\left|\dfrac{u_{n+1}}{u_n}\right|=\left|\dfrac{2}{x}\right|$. The series therefore converges if $\left|\dfrac{2}{x}\right|<1$. If $x>0$, $\left|\dfrac{2}{x}\right|=\dfrac{2}{x}$,

which is less than 1 if $2<x$. If $x<0$, $\left|\dfrac{2}{x}\right|=-\dfrac{2}{x}$, which is less than 1 if $-2>x$. Now for the endpoints:

$x=2$ yields $1+1+1+1+\ldots$, which diverges;
$x=-2$ yields $-1+1-1+1-\ldots$, which diverges.

The answer is $|x|>2$.

47.   C.   The function and its first three derivatives at $\dfrac{\pi}{4}$ are $\sin\dfrac{\pi}{4}=\dfrac{1}{\sqrt{2}}$;

$\cos\dfrac{\pi}{4}=\dfrac{1}{\sqrt{2}}$; $-\sin\dfrac{\pi}{4}=-\dfrac{1}{\sqrt{2}}$; and $-\cos\dfrac{\pi}{4}=-\dfrac{1}{\sqrt{2}}$. $P_3(x)$ is choice C.

48.   D.   Graph $f(x)$ and $\tan^{-1}x$, as shown below.

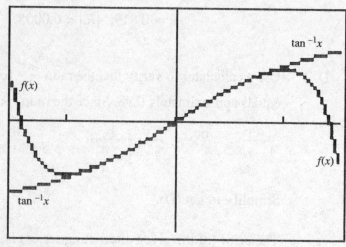

$$[-1.5,1.5]\times[-1.5,1.5]$$

Note that $f(x)=x-\dfrac{x^3}{3}+\dfrac{x^5}{5}-\dfrac{x^7}{7}=P_7(x)$. If $x<0$, then $P_7(x)$ exceeds $\tan^{-1}x$. The last statement can be verified without drawing the graphs: compare $P_7(x)$ above and $P_9(x)$ for $\tan^{-1}x$.

49.   E.   Use your calculator to graph $\tan^{-1}x$ and $f(x)$, the 9th-degree Taylor polynomial, and compare them. Or examine both analytically as suggested in Question 48.

Chapter 11

Miscellaneous Multiple-Choice Practice Questions

These questions provide further practice for Parts A and B of Section I of the examination. Answers begin on page 409.

Part A

You may *not* use a calculator for any question in this part.

1. Which of the following functions is continuous at $x = 0$?

(A) $f(x)\begin{cases} = \sin\dfrac{1}{x} & \text{for } x \neq 0 \\ = 0 & \text{for } x = 0 \end{cases}$
  (B) $f(x) = [x]$ (greatest-integer function)

(C) $f(x)\begin{cases} = \dfrac{x}{x} & \text{for } x \neq 0 \\ = 0 & \text{for } x = 0 \end{cases}$
  (D) $f(x)\begin{cases} = x\sin\dfrac{1}{x} & \text{for } x \neq 0 \\ = 0 & \text{for } x = 0 \end{cases}$

(E) $f(x) = \dfrac{x+1}{x}$

2. Which of the following statements about the graph of $y = \dfrac{x^2 + 1}{x^2 - 1}$ is *not* true?
(A) The graph is symmetric to the y-axis.
(B) The graph has two vertical asymptotes.
(C) There is no y-intercept.
(D) The graph has one horizontal asymptote.
(E) There is no x-intercept.

3. $\lim\limits_{x \to 1^-} ([x] - |x|) =$
(A) -1 (B) 0 (C) 1 (D) 2 (E) none of these

4. The x-coordinate of the point on the curve $y = x^2 - 2x + 3$ at which the tangent is perpendicular to the line $x + 3y + 3 = 0$ is
(A) $-\dfrac{5}{2}$ (B) $-\dfrac{1}{2}$ (C) $\dfrac{7}{6}$ (D) $\dfrac{5}{2}$ (E) none of these

5. $\lim\limits_{x \to 1} \dfrac{\dfrac{3}{x} - 3}{x - 1}$ is

(A) -3 (B) -1 (C) 1   (D) 3   (E) nonexistent

6. $\ln(3e^{x+2}) - \ln(3e^x) =$
    (A) 2   (B) e^2 (C) $2 \ln 3$ (D) $\ln 3 + 2$ (E) $2 \ln 3 + 2$

7. $\displaystyle\int_0^6 |x - 4| \, dx =$
    (A) 6   (B) 8   (C) 10   (D) 11   (E) 12

8. $\lim\limits_{x \to \infty} \dfrac{3 + x - 2x^2}{4x^2 + 9}$ is

(A) $-\dfrac{1}{2}$ (B) $\dfrac{1}{2}$ (C) 1   (D) 3   (E) nonexistent

9. The maximum value of the function $f(x) = x^4 - 4x^3 + 6$ on the closed interval $[1, 4]$ is
    (A) 1   (B) 0   (C) 3   (D) 6   (E) none of these

10. Let $\begin{cases} f(x) = \dfrac{\sqrt{x + 4} - 3}{x - 5} & \text{if } x \neq 5, \\ f(5) = c, \end{cases}$

and let f be continuous at $x = 5$. Then $c =$
    (A) $-\dfrac{1}{6}$ (B) 0   (C) $\dfrac{1}{6}$ (D) 1   (E) 6

11. $\displaystyle\int_0^{\pi/2} \cos^2 x \sin x \, dx =$

(A) -1 (B) $-\dfrac{1}{3}$ (C) 0   (D) $\dfrac{1}{3}$ (E) 1

12. If $\sin x = \ln y$ and $0 < x < \pi$, then, in terms of x, $\dfrac{dy}{dx}$ equals

(A) $e^{\sin x} \cos x$ (B) $e^{-\sin x} \cos x$ (C) $\dfrac{e^{\sin x}}{\cos x}$ (D) $e^{\cos x}$ (E) $e^{\sin x}$

13. If $f(x) = x \cos \dfrac{1}{x}$, then $f'\left(\dfrac{2}{\pi}\right)$ equals

(A) $\dfrac{\pi}{2}$ (B) $-\dfrac{2}{\pi}$ (C) -1 (D) $-\dfrac{\pi}{2}$ (E) 1

14. The equation of the tangent to the curve $y = e^x \ln x$, where $x = 1$, is
    (A) $y = ex$ (B) $y = e^x + 1$ (C) $y = e(x - 1)$ (D) $y = ex + 1$
    (E) $y = x - 1$

15. If the displacement from the origin of a particle on a line is given by $s = 3 + (t - 2)^4$, then the number of times the particle reverses direction is

(A) 0   (B) 1   (C) 2   (D) 3   (E) none of these

16. $\displaystyle\int_{-1}^{0} e^{-x}\, dx$ equals

(A) $1 - e$ (B) $\dfrac{1 - e}{e}$ (C) $e - 1$ (D) $1 - \dfrac{1}{e}$ (E) $e + 1$

17. If
$$f(x) = \begin{cases} x^2 & \text{for } x \leqq 2 \\ 4x - x^2 & \text{for } x > 2 \end{cases},$$

then $\displaystyle\int_{-1}^{4} f(x)\, dx$ equals

(A) 7   (B) $\dfrac{23}{3}$ (C) $\dfrac{25}{3}$ (D) 9   (E) $\dfrac{65}{3}$

18. If the position of a particle on a line at time t is given by $s = t^3 + 3t$, then the speed of the particle is decreasing when

(A) $-1 < t < 1$ (B) $-1 < t < 0$ (C) $t < 0$ (D) $t > 0$
(E) $|t| > 1$

19. A rectangle with one side on the x-axis is inscribed in the triangle formed by the lines $y = x$, $y = 0$, and $2x + y = 12$. The area of the largest such rectangle is

(A) 6   (B) 3   (C) $\dfrac{5}{2}$ (D) 5   (E) 7

20. The abscissa of the first-quadrant point that is on the curve of $x^2 - y^2 = 1$ and closest to the point $(3, 0)$ is

(A) 1   (B) $\dfrac{3}{2}$ (C) 2   (D) 3   (E) none of these

21. If $y = \sqrt{x^2 + 16}$, then $\dfrac{d^2 y}{dx^2}$ is

(A) $-\dfrac{1}{4(x^2 + 16)^{3/2}}$ (B) $4(3x^2 + 16)$ (C) $\dfrac{16}{\sqrt{x^2 + 16}}$
(D) $\dfrac{2x^2 + 16}{(x^2 + 16)^{3/2}}$ (E) $\dfrac{16}{(x^2 + 16)^{3/2}}$

22. The region bounded by the parabolas $y = x^2$ and $y = 6x - x^2$ is rotated about the x-axis so that a vertical line segment cut off by the curves generates a ring. The value of x for which the ring of largest area is obtained is

(A) 4   (B) 3   (C) $\dfrac{5}{2}$ (D) 2   (E) $\dfrac{3}{2}$

23. $\displaystyle\int \dfrac{dx}{x \ln x}$ equals

(A) $\ln(\ln x) + C$ (B) $-\dfrac{1}{\ln^2 x} + C$ (C) $\dfrac{(\ln x)^2}{2} + C$ (D) $\ln x + C$

(E) none of these

24. The volume obtained by rotating the region bounded by $x = y^2$ and $x = 2 - y^2$ about the y-axis is equal to

(A) $\dfrac{16\pi}{3}$ (B) $\dfrac{32\pi}{3}$ (C) $\dfrac{32\pi}{15}$ (D) $\dfrac{64\pi}{15}$ (E) $\dfrac{8\pi}{3}$

25. The general solution of the differential equation $\dfrac{dy}{dx} = \dfrac{1 - 2x}{y}$ is a family of

(A) straight lines   (B) circles   (C) hyperbolas   (D) parabolas
(E) ellipses

26. Estimate $\displaystyle\int_0^4 \sqrt{25 - x^2}\, dx$ using the left rectangular rule and two subintervals.

(A) $3 + \sqrt{21}$ (B) $5 + \sqrt{21}$ (C) $6 + 2\sqrt{21}$ (D) $8 + 2\sqrt{21}$
(E) $10 + 2\sqrt{21}$

27. $\displaystyle\int_0^7 \sin \pi x\, dx =$

(A) -2 (B) $-\dfrac{2}{\pi}$ (C) 0 (D) $\dfrac{1}{\pi}$ (E) $\dfrac{2}{\pi}$

***28.** $\displaystyle\lim_{x \to 0} \dfrac{\tan 3x}{2x} =$

(A) 0 (B) $\dfrac{1}{2}$ (C) $\dfrac{2}{3}$ (D) $\dfrac{3}{2}$ (E) ∞

29. $\displaystyle\lim_{h \to 0} \dfrac{\tan (\pi/4 + h) - 1}{h} =$

(A) 0 (B) $\dfrac{1}{2}$ (C) 1 (D) 2 (E) ∞

30. The number of values of k for which $f(x) = e^x$ and $g(x) = k \sin x$ have a common point of tangency is
(A) 0 (B) 1 (C) 2 (D) large but finite   (E) infinite

31. The curve $2x^2 y + y^2 = 2x + 13$ passes through $(3, 1)$. Use the local linearization of the curve to find the approximate value of y at $x = 2.8$.
(A) 0.5 (B) 0.9 (C) 0.95 (D) 1.1 (E) 1.4

32. $\displaystyle\int \cos^3 x\, dx =$

(A) $\dfrac{\cos^4 x}{4} + C$ (B) $\dfrac{\sin^4 x}{4} + C$ (C) $\sin x - \dfrac{\sin^3 x}{3} + C$

(D) $\sin x + \dfrac{\sin^3 x}{3} + C$ (E) $\cos x - \dfrac{\cos^3 x}{3} + C$

33. The region bounded by $y = \tan x$, $y = 0$, and $x = \dfrac{\pi}{4}$ is rotated about the x-axis. The volume generated equals

(A) $\pi - \dfrac{\pi^2}{4}$ (B) $\pi(\sqrt{2} - 1)$ (C) $\dfrac{3\pi}{4}$ (D) $\pi\left(1 + \dfrac{\pi}{4}\right)$
(E) none of these

34. $\displaystyle\lim_{h \to 0} \dfrac{a^h - 1}{h}$, if $a > 0$, equals

(A) 1 (B) a (C) $\ln a$ (D) $\log_{10} a$ (E) $a \ln a$

*An asterisk denotes a topic covered only in Calculus BC.

35. Solutions of the differential equation whose slope field is shown here are most likely to be

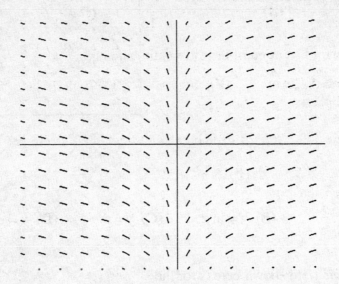

(A) quadratic   (B) cubic   (C) sinusoidal   (D) exponential
(E) logarithmic

36. $\lim\limits_{h \to 0} \dfrac{1}{h} \displaystyle\int_{\frac{\pi}{4}}^{\frac{\pi}{4}+h} \dfrac{\sin x}{x}\, dx =$

(A) 0   (B) 1   (C) $\dfrac{\sqrt{2}}{2}$ (D) $\dfrac{2\sqrt{2}}{\pi}$ (E) $\dfrac{2\sqrt{2}(\pi-4)}{\pi^2}$

37. The graph of g, shown below, consists of the arcs of two quarter-circles and two straight-line segments. The value of $\displaystyle\int_0^{12} g(x)\, dx$ is

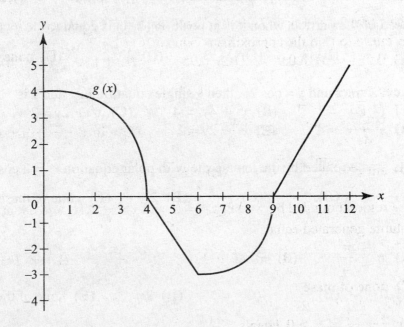

(A) $\pi + 2$ (B) $\dfrac{7\pi}{4} + \dfrac{9}{2}$ (C) $\dfrac{7\pi}{4} + 8$ (D) $7\pi + \dfrac{9}{2}$ (E) $\dfrac{25\pi}{4} + \dfrac{21}{2}$

38. Which of these could be a particular solution of the differential equation whose slope field is shown here?

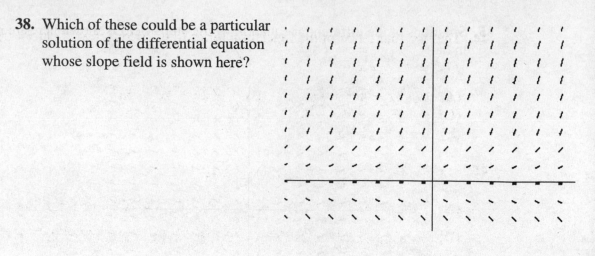

(A) $y = \dfrac{1}{x}$ (B) $y = \ln x$ (C) $y = e^x$ (D) $y = e^{-x}$ (E) $y = e^{x^2}$

39. The slope field shown here is for the differential equation

(A) $y' = \dfrac{1}{x}$ (B) $y' = \ln x$

(C) $y' = e^x$ (D) $y' = y$

(E) $y' = -y^2$

*40. A particle moves along the parabola $x = 3y - y^2$ so that $\dfrac{dy}{dt} = 3$ at all time t. The speed of the particle when it is at position (2, 1) is equal to

(A) 0   (B) 3   (C) $\sqrt{13}$ (D) $3\sqrt{2}$ (E) none of these

*41. If $x = 2 \sin u$ and $y = \cos 2u$, then a single equation in x and y is

(A) $x^2 + y^2 = 1$ (B) $x^2 + 4y^2 = 4$ (C) $x^2 + 2y = 2$

(D) $x^2 + y^2 = 4$ (E) $x^2 - 2y = 2$

*42. The area bounded by the lemniscate with polar equation $r^2 = 2 \cos 2\theta$ is equal to

(A) 4   (B) 1   (C) $\dfrac{1}{2}$ (D) 2   (E) none of these

*43. $\displaystyle\int_{-\infty}^{\infty} \dfrac{dx}{x^2 + 1} =$

(A) 0   (B) $\dfrac{\pi}{2}$ (C) π (D) 2π (E) none of these

---

* An asterisk denotes a topic covered only in Calculus BC.

*44. The first four terms of the Maclaurin series (the Taylor series about $x = 0$) for
$f(x) = \dfrac{1}{1 - 2x}$ are
(A) $1 + 2x + 4x^2 + 8x^3$ (B) $1 - 2x + 4x^2 - 8x^3$
(C) $-1 - 2x - 4x^2 - 8x^3$ (D) $1 - x + x^2 - x^3$
(E) $1 + x + x^2 + x^3$

*45. $\displaystyle\int x^2 e^{-x}\, dx =$

(A) $\dfrac{1}{3} x^3 e^{-x} + C$ (B) $-\dfrac{1}{3} x^3 e^{-x} + C$ (C) $-x^2 e^{-x} + 2x e^{-x} + C$
(D) $-x^2 e^{-x} - 2x e^{-x} - 2e^{-x} + C$ (E) $-x^2 e^{-x} + 2x e^{-x} - 2e^{-x} + C$

*46. $\displaystyle\int_0^{\pi/2} \sin^2 x\, dx$ is equal to

(A) $\dfrac{1}{3}$ (B) $\dfrac{\pi}{4} - \dfrac{1}{4}$ (C) $\dfrac{\pi}{2}$ (D) $\dfrac{\pi}{2} - \dfrac{1}{3}$ (E) $\dfrac{\pi}{4}$

*47. A curve is given parametrically by the equations $x = t$, $y = 1 - \cos t$. The area bounded by the curve and the x-axis on the interval $0 \le t \le 2\pi$ is equal to
(A) $2(\pi + 1)$ (B) π (C) 4π (D) $\pi + 1$ (E) 2π

*48. If $x = a \cot \theta$ and $y = a \sin^2 \theta$, then $\dfrac{dy}{dx}$, when $\theta = \dfrac{\pi}{4}$, is equal to

(A) $\dfrac{1}{2}$ (B) -1 (C) 2 (D) $-\dfrac{1}{2}$ (E) $-\dfrac{1}{4}$

*49. Which of the following improper integrals diverges?
(A) $\displaystyle\int_0^{\infty} e^{-x^2} dx$ (B) $\displaystyle\int_{-\infty}^{0} e^x dx$ (C) $\displaystyle\int_0^1 \dfrac{dx}{x}$ (D) $\displaystyle\int_0^{\infty} e^{-x} dx$ (E) $\displaystyle\int_0^1 \dfrac{dx}{\sqrt{x}}$

*50. $\displaystyle\int_2^4 \dfrac{du}{\sqrt{16 - u^2}}$ equals

(A) $\dfrac{\pi}{12}$ (B) $\dfrac{\pi}{6}$ (C) $\dfrac{\pi}{4}$ (D) $\dfrac{\pi}{3}$ (E) $\dfrac{2\pi}{3}$

*51. $\displaystyle\lim_{x \to 0^+} \left(\dfrac{1}{x}\right)^x$ is
(A) $-\infty$ (B) 0 (C) 1 (D) ∞ (E) nonexistent

*52. If we substitute $x = \tan \theta$, which of the following is equivalent to $\displaystyle\int_0^1 \sqrt{1 + x^2}\, dx$?

(A) $\displaystyle\int_0^1 \sec \theta\, d\theta$ (B) $\displaystyle\int_0^1 \sec^3 \theta\, d\theta$ (C) $\displaystyle\int_0^{\pi/4} \sec \theta\, d\theta$

(D) $\displaystyle\int_0^{\pi/4} \sec^3 \theta\, d\theta$ (E) $\displaystyle\int_0^{\tan 1} \sec^3 \theta\, d\theta$

* An asterisk denotes a topic covered only in Calculus BC.

*53. $\displaystyle\lim_{x \to 0^+} \frac{\cot x}{\ln x} =$

(A) $-\infty$ (B) -1 (C) 0 (D) 1 (E) ∞

*54. When rewritten as partial fractions, $\dfrac{3x + 2}{x^2 - x - 12}$ includes which of the following?

I. $\dfrac{1}{x + 3}$ II. $\dfrac{1}{x - 4}$ III. $\dfrac{2}{x - 4}$

(A) none   (B) I only   (C) II only   (D) III only   (E) I and III

*55. Using two terms of an appropriate Maclaurin series, estimate $\displaystyle\int_0^1 \frac{1 - \cos x}{x}\,dx$.

(A) $\dfrac{1}{96}$ (B) $\dfrac{23}{96}$ (C) $\dfrac{1}{4}$ (D) $\dfrac{25}{96}$

(E) undefined; the integral is improper

*56. The slope of the spiral $r = \theta$ at $\theta = \dfrac{\pi}{4}$ is

(A) $-\sqrt{2}$ (B) -1 (C) 1 (D) $\dfrac{4 + \pi}{4 - \pi}$ (E) undefined

Part B

A graphing calculator is required for some questions.

57. The graph of function h is shown here. Which of these statements is (are) true?

I. The first derivative is never negative.
II. The second derivative is constant.
III. The first and second derivatives equal 0 at the same point.

(A) I only   (B) III only   (C) I and II
(D) I and III   (E) I, II, and III

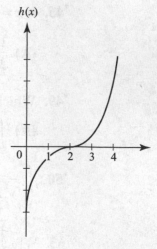

58. Graphs of functions $f(x)$, $g(x)$, and $h(x)$ are shown on page 401. Consider the following statements:

I. $g(x) = f'(x)$
II. $f(x) = g'(x)$
III. $h(x) = g''(x)$

Which of these statements is (are) true?

(A) I only   (B) II only   (C) II and III only   (D) all three
(E) none of these

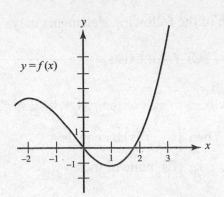

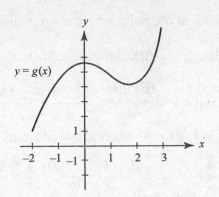

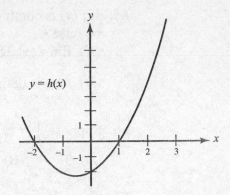

59. If $y = \int_3^x \dfrac{1}{\sqrt{3+2t}}\, dt$, then $\dfrac{d^2 y}{dx^2} =$

(A) $-\dfrac{1}{(3+2x)^{\frac{3}{2}}}$

(B) $\dfrac{3}{(3+2x)^{\frac{5}{2}}}$

(C) $\dfrac{\sqrt{3+2x}}{2}$

(D) $\dfrac{(3+2x)^{\frac{3}{2}}}{3}$

(E) 0

60. If $\int_{-3}^4 f(x)\, dx = 6$, then $\int_{-4}^3 f(x+1)\, dx =$

(A) -6 (B) -5 (C) 5 (D) 6 (E) 7

61. At what point in the interval $[1, 1.5]$ is the rate of change of $f(x) = \sin x$ equal to its average rate of change on the interval?

(A) 0.995 (B) 1.058 (C) 1.239 (D) 1.253 (E) 1.399

62. Suppose $f'(x) = x^2(x-1)$. Then $f''(x) = x(3x-2)$. Over which interval(s) is the graph of f both increasing and concave up?

I. $x < 0$ II. $0 < x < \dfrac{2}{3}$ III. $\dfrac{2}{3} < x < 1$ IV. $x > 1$

(A) I only   (B) II only   (C) II and IV   (D) I and IV
(E) IV only

63. Which of the following statements is true about the graph of $f(x)$ in Question 62?
(A) The graph has no relative extrema.
(B) The graph has one relative extremum and one inflection point.
(C) The graph has two relative extrema and one inflection point.
(D) The graph has two relative extrema and two inflection points.
(E) None of the preceding statements is true.

64. The nth derivative of $\ln(x+1)$ at $x = 2$ equals

(A) $\dfrac{(-1)^{n-1}}{3^n}$

(B) $\dfrac{(-1)^n \cdot n!}{3^{n+1}}$

(C) $\dfrac{(-1)^{n-1}(n-1)!}{3^n}$

(D) $\dfrac{(-1)^n \cdot n!}{3^{n+1}}$

(E) $\dfrac{(-1)^{n+1}}{3^{n+1}}$

65. If $f(x)$ is continuous at the point where $x = a$, which of the following statements may be false?

(A) $\lim\limits_{x \to a} f(x)$ exists. **(B)** $\lim\limits_{x \to a} f(x) = f(a)$. **(C)** $f'(a)$ exists.

(D) $f(a)$ is defined. **(E)** $\lim\limits_{x \to a^-} f(x) = \lim\limits_{x \to a^+} f(x)$.

66. Suppose $\int_0^3 f(x + k)\, dx = 4$, where k is a constant. Then $\int_k^{3+k} f(x)\, dx$ equals

(A) 3 **(B)** $4 - k$ **(C)** 4 **(D)** $4 + k$ **(E)** none of these

67. The volume, in cubic feet, of an "inner tube" with inner diameter 4 ft and outer diameter 8 ft is

(A) $4\pi^2$ **(B)** $12\pi^2$ **(C)** $8\pi^2$ **(D)** $24\pi^2$ **(E)** $6\pi^2$

68. If $f(u) = \tan^{-1} u^2$ and $g(u) = e^u$, then the derivative of $f(g(u))$ is

(A) $\dfrac{2ue^u}{1 + u^4}$ **(B)** $\dfrac{2ue^{u^2}}{1 + u^4}$ **(C)** $\dfrac{2e^u}{1 + 4e^{2u}}$ **(D)** $\dfrac{2e^{2u}}{1 + e^{4u}}$

(E) $\dfrac{2e^{2u}}{\sqrt{1 - e^{4u}}}$

69. If $\sin(xy) = y$, then $\dfrac{dy}{dx}$ equals

(A) $\sec(xy)$ **(B)** $y\cos(xy) - 1$ **(C)** $\dfrac{1 - y\cos(xy)}{x\cos(xy)}$

(D) $\dfrac{y\cos(xy)}{1 - x\cos(xy)}$ **(E)** $\cos(xy)$

70. Let $x > 0$. Suppose

$$\frac{d}{dx}f(x) = g(x) \quad \text{and} \quad \frac{d}{dx}g(x) = f(\sqrt{x});$$

then $\dfrac{d^2}{dx^2}f(x^2) =$

(A) $f(x^4)$ **(B)** $f(x^2)$ **(C)** $2xg(x^2)$ **(D)** $\dfrac{1}{2x}f(x)$
(E) $2g(x^2) + 4x^2 f(x)$

71. The region bounded by $y = e^x$, $y = 1$, and $x = 2$ is rotated about the x-axis. The volume of the solid generated is given by the integral

(A) $\pi \int_0^2 e^{2x}\, dx$ **(B)** $2\pi \int_1^{e^2} (2 - \ln y)(y - 1)\, dy$ **(C)** $\pi \int_0^2 (e^{2x} - 1)\, dx$

(D) $2\pi \int_0^{e^2} y\,(2 - \ln y)\, dy$ **(E)** $\pi \int_0^2 (e^x - 1)^2\, dx$

72. Suppose the function f is continuous on $1 \leqq x \leqq 2$, that $f'(x)$ exists on $1 < x < 2$, that $f(1) = 3$, and that $f(2) = 0$. Which of the following statements is *not* necessarily true?

(A) The Mean-Value Theorem applies to f on $1 \leqq x \leqq 2$.

(B) $\displaystyle\int_1^2 f(x)\,dx$ exists.

(C) There exists a number c in the closed interval $[1, 2]$ such that $f'(c) = 3$.

(D) If k is any number between 0 and 3, there is a number c between 1 and 2 such that $f(c) = k$.

(E) If c is any number such that $1 < c < 2$, then $\displaystyle\lim_{x \to c} f(x)$ exists.

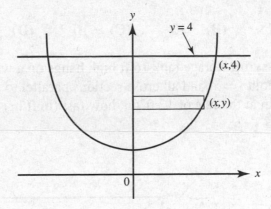

73. The region S in the figure is bounded by $y = \sec x$, the y–*axis,* and $y = 4$. What is the volume of the solid formed when S is rotated about the y-axis?

(A) 0.791   (B) 2.279   (C) 5.692   (D) 11.385   (E) 17.217

74. If 40 g of a radioactive substance decomposes to 20 g in 2 yr, then, to the nearest gram, the amount left after 3 yr is

(A) 10   (B) 12   (C) 14   (D) 16   (E) 17

75. An object in motion along a line has acceleration $a(t) = \pi t + \dfrac{2}{1 + t^2}$ and is at rest when $t = 1$. Its average velocity from $t = 0$ to $t = 2$ is

(A) 0.362   (B) 0.274   (C) 3.504   (D) 7.008   (E) 8.497

76. Find the area bounded by $y = \tan x$ and $x + y = 2$, and above the x-axis on the interval $[0, 2]$.

(A) 0.919   (B) 0.923   (C) 1.013   (D) 1.077   (E) 1.494

77. An ellipse has major axis 20 and minor axis 10. Rounded off to the nearest integer, the maximum area of an inscribed rectangle is

(A) 50   (B) 79   (C) 80   (D) 82   (E) 100

78. The average value of $y = x \ln x$ on the interval $1 \leqq x \leqq e$ is

(A) 0.772   (B) 1.221   (C) 1.359   (D) 1.790   (E) 2.097

79. Let $f(x) = \int_0^x (1 - 2\cos^3 t)\, dt$ for $0 \le x \le 2\pi$. On which interval is f increasing?

 (A) $0 < x < \pi$ (B) $0.654 < x < 5.629$ (C) $0.654 < x < 2\pi$

 (D) $\pi < x < 2\pi$ (E) none of these

80. The table shows the speed of an object (in ft/sec) during a 3-sec period. Estimate its acceleration (in ft/sec^2) at $t = 1.5$ sec.

time, sec	0	1	2	3
speed, ft/sec	30	22	12	0

 (A) -17 (B) -13 (C) -10 (D) -5 (E) 17

81. A maple-syrup storage tank 16 ft high hangs on a wall. The back is in the shape of the parabola $y = x^2$ and all cross sections parallel to the floor are squares. If syrup is pouring in at the rate of 12 ft^3/hr, how fast (in ft/hr) is the syrup level rising when it is 9 ft deep?

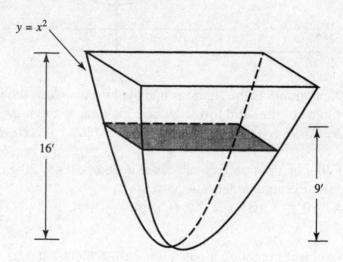

 (A) $\dfrac{2}{27}$ (B) $\dfrac{1}{3}$ (C) $\dfrac{4}{3}$ (D) 36   (E) 162

82. In a protected area (no predators, no hunters) the deer population increases at a rate of $\dfrac{dP}{dt} = k(1000 - P)$, where $P(t)$ represents the population of deer at t yr. If 300 deer were originally placed in the area, and a census showed the population had grown to 500 in 5 yr, how many deer will there be after 10 yr?

 (A) 608   (B) 643   (C) 700   (D) 833   (E) 892

83. Shown is the graph of $f(x) = \dfrac{4}{x^2 + 1}$.

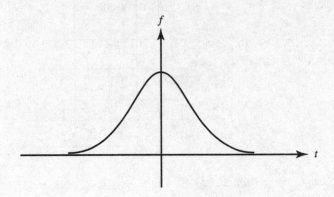

Let $H(x) = \displaystyle\int_0^x f(t)\, dt$. The local linearization of H at $x = 1$ is $H(x)$ equals

(A) $2x$ (B) $-2x - 4$ (C) $2x + \pi - 2$ (D) $-2x + \pi + 2$

(E) $2x + \ln 16 + 2$

84. A smokestack 100 ft tall is used to treat industrial emissions. The diameters, measured at 25-ft intervals, are shown in the table. Using the midpoint rule, estimate the volume of the smokestack to the nearest 100 ft^3.

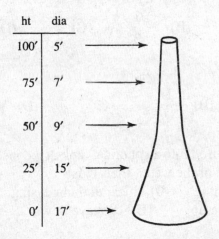

ht	dia
100'	5'
75'	7'
50'	9'
25'	15'
0'	17'

(A) 8100   (B) 9500   (C) 9800   (D) 12,500   (E) 39,300

For questions 85–89 the table shows the values of differentiable functions f and g.

x	f	f'	g	g'
1	2	$\frac{1}{2}$	-3	5
2	3	1	0	4
3	4	2	2	3
4	6	4	3	$\frac{1}{2}$

85. If $P(x) = \dfrac{f(x)}{g(x)}$, then $P'(3) =$

(A) -2 (B) $-\dfrac{8}{9}$ (C) $-\dfrac{1}{2}$ (D) $\dfrac{2}{3}$ (E) 2

86. If $H(x) = f(g(x))$, then $H'(3) =$
(A) 1   (B) 2   (C) 3   (D) 6   (E) 9

87. If $M(x) = f(x) \cdot g(x)$, then $M'(3) =$
(A) 2   (B) 6   (C) 8   (D) 14   (E) 16

88. If $K(x) = g^{-1}(x)$, then $K'(3) =$

(A) $-\dfrac{1}{2}$ (B) $-\dfrac{1}{3}$ (C) $\dfrac{1}{3}$ (D) $\dfrac{1}{2}$ (E) 2

89. If $R(x) = \sqrt{f(x)}$, then $R'(3) =$

(A) $\dfrac{1}{4}$ (B) $\dfrac{1}{2\sqrt{2}}$ (C) $\dfrac{1}{2}$ (D) $\sqrt{2}$ (E) 2

90. Water is poured into a spherical tank at a constant rate. If $W(t)$ is the rate of increase of the depth of the water, then W is
(A) constant   (B) linear and increasing   (C) linear and decreasing
(D) concave up   (E) concave down

91. The graph of f' is shown below. If $f(7) = 3$ then $f(1) =$

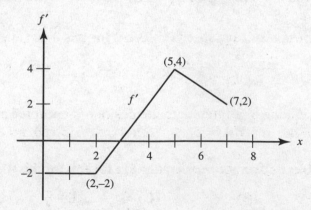

(A) -10 (B) -4 (C) -3 (D) 10   (E) 16

92. At an outdoor concert, the crowd stands in front of the stage filling a semicircular disk of radius 100 yd. The approximate density of the crowd x yd from the stage is given by

$$D(x) = \frac{20}{2\sqrt{x} + 1}$$

people per square yard. About how many people are at the concert?

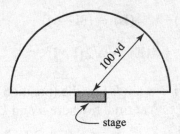

(A) 200   (B) 19,500   (C) 21,000   (D) 165,000   (E) 591,000

93. The Centers for Disease Control announced that, although more AIDS cases were reported this year, the rate of increase is slowing down. If we graph the number of AIDS cases as a function of time, the curve is currently
(A) increasing and linear   (B) increasing and concave down
(C) increasing and concave up   (D) decreasing and concave down
(E) decreasing and concave up

The graph is for Questions 94–96. It shows the velocity, in feet per second, for $0 < t < 8$, of an object moving along a straight line.

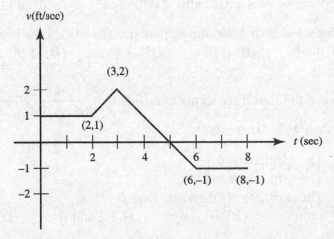

94. The object's average speed (in ft/sec) for this 8-sec interval was

(A) 0   (B) $\dfrac{3}{8}$ (C) 1   (D) $\dfrac{8}{3}$ (E) 8

95. When did the object return to the position it occupied at $t = 2$?
(A) $t = 4$ (B) $t = 5$ (C) $t = 6$ (D) $t = 8$ (E) never

96. The object's average acceleration (in ft/sec²) for this 8-sec interval was

(A) -2 (B) $-\dfrac{1}{4}$ (C) 0   (D) $\dfrac{1}{4}$ (E) 1

97. If a block of ice melts at the rate of $\dfrac{72}{2t + 3}$ cm^3/min, how much ice melts during the first 3 min?

(A) 8 cm^3 (B) 16 cm^3 (C) 21 cm^3 (D) 40 cm^3 (E) 79 cm^3

*98. A particle moves counterclockwise on the circle $x^2 + y^2 = 25$ with a constant speed of 2 ft/sec. Its velocity vector, **v**, when the particle is at (3, 4), equals

(A) $-\dfrac{1}{5}(8\mathbf{i} - 6\mathbf{j})$ (B) $\dfrac{1}{5}(8\mathbf{i} - 6\mathbf{j})$ (C) $-2\sqrt{3}\mathbf{i} + 2\mathbf{j}$

(D) $2\mathbf{i} - 2\sqrt{3}\mathbf{j}$ (E) $-2\sqrt{2}(\mathbf{i} - \mathbf{j})$

*99. Let $\mathbf{R} = a \cos kt\mathbf{i} + a \sin kt\mathbf{j}$ be the (position) vector $x\mathbf{i} + y\mathbf{j}$ from the origin to a moving point $P(x, y)$ at time t, where a and k are positive constants. The acceleration vector, **a**, equals

(A) $-k^2\mathbf{R}$ (B) $a^2 k^2 \mathbf{R}$ (C) $-a\mathbf{R}$ (D) $-ak^2(\cos t\mathbf{i} + \sin t\mathbf{j})$

(E) $-\mathbf{R}$

*100. The length of the curve $y = 2^x$ between (0, 1) and (2, 4) is

(A) 3.141   (B) 3.664   (C) 4.823   (D) 5.000   (E) 7.199

*101. The position of a moving object is given by $P(t) = (3t, e^t)$. Its acceleration is

(A) undefined   (B) constant in both magnitude and direction

(C) constant in magnitude only   (D) constant in direction only

(E) constant in neither magnitude nor direction

*102. Suppose we plot a particular solution of $\dfrac{dy}{dx} = 4y$ from initial point (0, 1) using Euler's method. After one step of size $\Delta x = 0.1$, how big is the error?

(A) 0.09   (B) 1.09   (C) 1.49   (D) 1.90   (E) 2.65

*103. We use the first three terms to estimate $\displaystyle\sum_{n=0}^{\infty} \frac{(-1)^n}{n^2 + 1}$. Which of the following statements is (are) true?

    I. The estimate is 0.7.

    II. The estimate is too low.

    III. The estimate is off by less than 0.1.

(A) I only   (B) III only   (C) I and II   (D) I and III   (E) all three

*104. Which of these diverges?

(A) $\displaystyle\sum_{n=1}^{\infty} \frac{2}{3^n}$ (B) $\displaystyle\sum_{n=1}^{\infty} \left(\frac{2}{3}\right)^n$ (C) $\displaystyle\sum_{n=1}^{\infty} \frac{2}{3n}$ (D) $\displaystyle\sum_{n=1}^{\infty} \frac{2}{n^3}$ (E) $\displaystyle\sum_{n=1}^{\infty} \frac{2n}{3^n}$

---

* An asterisk denotes a topic covered only in Calculus BC.

*105. Find the radius of convergence of $\sum_{n=1}^{\infty} \dfrac{n!}{n^n} x^n$.

(A) 0   (B) $\dfrac{1}{e}$ (C) 1   (D) e (E) ∞

*106. When we use $e^x \simeq 1 + x + \dfrac{x^2}{2}$ to estimate $\sqrt{e}$, the Lagrange remainder is no greater than

(A) 0.021   (B) 0.034   (C) 0.042   (D) 0.067   (E) 0.742

*107. An object in motion along a curve has position $P(t) = (\tan t, \cos 2t)$ for $0 \le t \le 1$. How far does it travel?

(A) 0.96   (B) 1.73   (C) 2.10   (D) 2.14   (E) 3.98

Answers to Miscellaneous Multiple-Choice Practice Questions _____

1. D	23. A	45. D	67. E	89. C
2. C	24. A	46. E	68. D	90. D
3. A	25. E	47. E	69. D	91. B
4. D	26. E	48. D	70. E	92. B
5. A	27. E	49. C	71. C	93. B
6. A	28. D	50. D	72. C	94. C
7. C	29. D	51. C	73. D	95. E
8. A	30. E	52. D	74. C	96. B
9. D	31. D	53. A	75. A	97. D
10. C	32. C	54. E	76. D	98. A
11. D	33. A	55. B	77. E	99. A
12. A	34. C	56. D	78. B	100. B
13. A	35. E	57. D	79. B	101. D
14. C	36. D	58. C	80. C	102. A
15. B	37. B	59. A	81. B	103. D
16. C	38. C	60. D	82. B	104. C
17. C	39. A	61. D	83. C	105. D
18. C	40. D	62. E	84. C	106. B
19. A	41. C	63. E	85. A	107. D
20. B	42. D	64. C	86. C	
21. E	43. C	65. C	87. E	
22. D	44. A	66. C	88. E	

* An asterisk denotes a topic covered only in Calculus BC.

Part A

1.   **D.**   If $f(x) = x \sin \dfrac{1}{x}$ for $x \neq 0$ and $f(0) = 0$, then

$$\lim_{x \to 0} f(x) = 0 = f(0);$$

thus this function is continuous at 0. Explain why each of the other functions is not continuous at $x = 0$.

2.   **C.**   To find the y-intercept, let $x = 0$; $y = -1$.

3.   **A.** $\displaystyle\lim_{x \to 1^-} [x] - \lim_{x \to 1^-} |x| = 0 - 1 = -1$.

4.   **D.**   The line $x + 3y + 3 = 0$ has slope $-\dfrac{1}{3}$; a line perpendicular to it has slope 3. The slope of the tangent to $y = x^2 - 2x + 3$ at any point is the derivative $2x - 2$. Set $2x - 2$ equal to 3.

5.   **A.** $\displaystyle\lim_{x \to 1} \dfrac{\dfrac{3}{x} - 3}{x - 1}$ is $f'(1)$, where $f(x) = \dfrac{3}{x}$, $f'(x) = -\dfrac{3}{x^2}$. Or, simplify the given fraction to

$$\dfrac{3 - 3x}{x(x-1)} = \dfrac{3(1-x)}{x(x-1)} = \dfrac{-3}{x} \quad (x \neq 1).$$

6.   **A.** $\ln(3e^{x+2}) - \ln(3e^x) = \ln \dfrac{3e^{x+2}}{3e^x} = \ln e^2 = 2 \ln e = 2.$

7.   **C.** $\displaystyle\int_0^6 |x - 4|\, dx = \int_0^4 (4 - x)\, dx + \int_4^6 (x - 4)\, dx = \left(4x - \dfrac{x^2}{2}\right)\Big|_0^4 + \left(\dfrac{x^2}{2} - 4x\right)\Big|_4^6$

$$= 8 + [(18 - 24) - (8 - 16)] = 8 + (-6 + 8) = 10.$$

Save time by finding the area under $y = |x - 4|$ from a sketch!

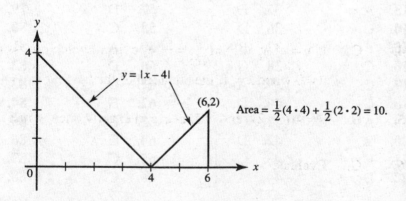

$y = |x - 4|$

$(6,2)$

Area $= \dfrac{1}{2}(4 \cdot 4) + \dfrac{1}{2}(2 \cdot 2) = 10.$

8.   **A.**   Since the degrees of numerator and denominator are the same, the limit as $x \to \infty$ is the ratio of the coefficients of the terms of highest degree: $\dfrac{-2}{4}$.

9.   **D.**   On the interval $[1, 4]$, $f'(x) = 0$ only for $x = 3$. Since $f(3)$ is a relative minimum, check the endpoints to find that $f(4) = 6$ is the absolute maximum of the function.

10.   **C.**   To find $\lim f$ as $x \to 5$ (if it exists), multiply f by $\dfrac{\sqrt{x + 4} + 3}{\sqrt{x + 4} + 3}$.

$$f(x) = \frac{x - 5}{(x - 5)(\sqrt{x + 4} + 3)}$$

and if $x \neq 5$ this equals $\dfrac{1}{\sqrt{x + 4} + 3}$. So $\lim f(x)$ as $x \to 5$ is $\dfrac{1}{6}$. For f to be continuous at $x = 5$, $f(5)$ or c must also equal $\dfrac{1}{6}$.

11.   **D.**   Evaluate $-\dfrac{1}{3}\cos^3 x \Big|_0^{\pi/2}$.

12.   **A.** $\cos x = \dfrac{1}{y}\dfrac{dy}{dx}$ and $\dfrac{dy}{dx} = y \cos x$. From the equation given, $y = e^{\sin x}$.

13.   **A.**   If $f(x) = x \cos \dfrac{1}{x}$, then

$$f'(x) = -x \sin \frac{1}{x}\left(-\frac{1}{x^2}\right) + \cos \frac{1}{x} = \frac{1}{x}\sin \frac{1}{x} + \cos \frac{1}{x},$$

and

$$f'\left(\frac{2}{\pi}\right) = \frac{\pi}{2} \cdot 1 + 0.$$

14.   **C.**   If $y = e^x \ln x$, then $\dfrac{dy}{dx} = \dfrac{e^x}{x} + e^x \ln x$, which equals e when $x = 1$. Since also $y = 0$ when $x = 1$, the equation of the tangent is $y = e(x - 1)$.

15.   **B.** $v = 4(t - 2)^3$ and changes sign exactly once, when $t = 2$.

16.   **C.**   Evaluate $-e^{-x}\Big|_{-1}^{0}$.

17.   **C.** $\displaystyle\int_{-1}^{4} f(x)\, dx = \int_{-1}^{2} x^2\, dx + \int_{2}^{4} (4x - x^2)\, dx.$

18.   **C.**   Since $v = 3t^2 + 3$, it is always positive, while $a = 6t$ and is positive for $t > 0$ but negative for $t < 0$. The speed therefore increases for $t > 0$ but decreases for $t < 0$.

19.   A.   Note from the figure that the area, A, of a typical rectangle is

$$A = (x_2 - x_1) \cdot y = \left(\frac{12 - y}{2} - y \right) \cdot y = 6y - \frac{3y^2}{2}.$$

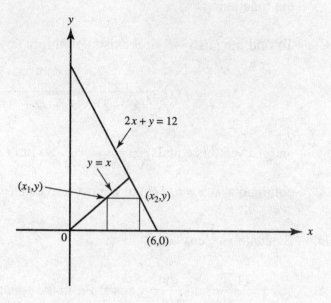

For $y = 2$, $\dfrac{dA}{dy} = 0$. Note that $\dfrac{d^2A}{dy^2}$ is always negative.

20.   B.   If S represents the square of the distance from $(3, 0)$ to a point (x, y) on the curve, then $S = (3 - x)^2 + y^2$. Setting $\dfrac{dS}{dx} = 0$ yields $\dfrac{dy}{dx} = \dfrac{3 - x}{y}$, while differentiating the relation $x^2 - y^2 = 1$ with respect to x yields $\dfrac{dy}{dx} = \dfrac{x}{y}$. These derivatives are equal for $x = \dfrac{3}{2}$.

21.   E.   Rewrite y as $(x^2 + 16)^{1/2}$, yielding

$$\frac{dy}{dx} = \frac{x}{(x^2 + 16)^{1/2}},$$

and apply the quotient rule, page 47.

22.   D.   See the figure. Since the area, A, of the ring equals $\pi\,(y_2{}^2 - y_1{}^2)$,

$$A = \pi\,[(6x - x^2)^2 - x^4] = \pi\,[36x^2 - 12x^3 + x^4 - x^4]$$

and

$$\frac{dA}{dx} = \pi\,(72x - 36x^2) = 36\pi x\,(2 - x),$$

where it can be verified that $x = 2$ produces the maximum area.

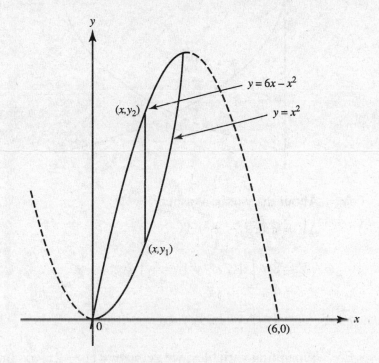

23.   A.   This is of type $\displaystyle\int \frac{du}{u}$ with $u = \ln x$.

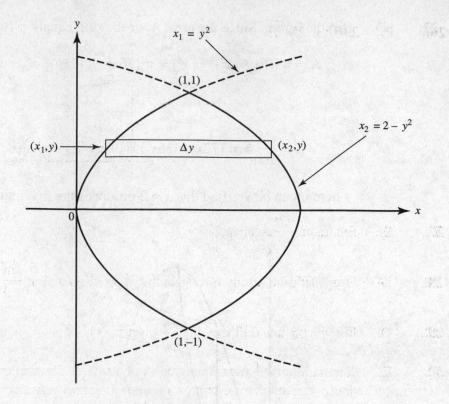

24. **A.** About the y-axis. Washer.

$$\Delta V = \pi \left(x_2^{\,2} - x_1^{\,2} \right) \Delta y,$$

$$V = 2\pi \int_0^1 \left[(2 - y^2)^2 - y^4 \right] dy$$

$$= 2\pi \int_0^1 (4 - 4y^2)\, dy.$$

25. **E.** Separating variables, we get $y\, dy = (1 - 2x)\, dx$. Integrating gives

$$\frac{1}{2} y^2 = x - x^2 + C$$

or

$$y^2 = 2x - 2x^2 + k$$

or

$$2x^2 + y^2 - 2x = k.$$

26.   E. $2(5) + 2\sqrt{21}.$

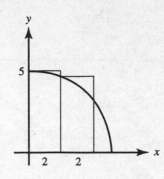

27.   E.   Evaluate $\dfrac{-1}{\pi}\cos(\pi x)\Big|_{0}^{7}.$

28.   D.   Use L'Hôpital's rule or rewrite the expression as $\displaystyle\lim_{x\to 0}\ \dfrac{\sin 3x}{3x}\cdot\dfrac{1}{\cos 3x}\cdot\dfrac{3}{2}.$

29.   D.   For $f(x) = \tan x$, this is $f'\left(\dfrac{\pi}{4}\right) = \sec^2\left(\dfrac{\pi}{4}\right).$

30.   E.   The parameter k determines the amplitude of the sine curve. For $f = k \sin x$ and $g = e^x$ to have a common point of tangency, say at $x = q$, the curves must both go through (q, y) and their slopes must be equal at q. Thus, we must have

$$k \sin q = e^q \quad \text{and} \quad k \cos q = e^q$$

and therefore

$$\sin q = \cos q.$$

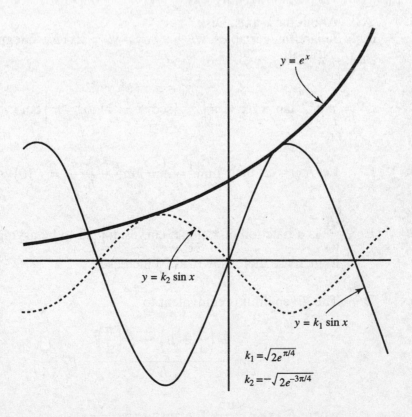

Thus,

$$q = \frac{\pi}{4} \pm n\pi$$

and

$$k = \frac{e^q}{\sin\left(\frac{\pi}{4} \pm n\pi\right)}.$$

The figure shows $k_1 = \sqrt{2}e^{\pi/4}$ and $k_2 = -\sqrt{2}e^{-3\pi/4}$.

31. **D.** We differentiate implicitly to find the slope $\frac{dy}{dx}$:

$$2\left(x^2 \frac{dy}{dx} + 2xy\right) + 2y\frac{dy}{dx} = 2,$$

$$\frac{dy}{dx} = \frac{1 - 2xy}{x^2 + y}.$$

At $(3, 1)$, $\frac{dy}{dx} = -\frac{1}{2}$. The linearization is $y \simeq -\frac{1}{2}(x - 3) + 1$.

32. **C.** $\displaystyle\int \cos^3 x\, dx = \int \cos^2 x \cdot \cos x\, dx = \int (1 - \sin^2 x)\cos x\, dx$

$$= \int \cos x\, dx - \int \sin^2 x \cdot \cos x\, dx = \sin x - \frac{\sin^3 x}{3} + C.$$

33. **A.** About the x-axis. Disk.

$$\Delta V = \pi y^2 \,\Delta x,$$

$$V = \pi \int_0^{\pi/4} \tan^2 x\, dx = \pi \int_0^{\pi/4} (\sec^2 x - 1)\, dx = \pi\left[\tan x - x\right]_0^{\pi/4} = \pi\left(1 - \frac{\pi}{4}\right).$$

34. **C.** Let $f(x) = a^x$; then $\displaystyle\lim_{h \to 0} \frac{a^h - 1}{h} = \lim_{h \to 0} \frac{a^{0+h} - a^0}{h} = f'(0) = a^0 \ln a = \ln a.$

35. **E.** $\dfrac{dy}{dx}$ is a function of x alone; curves appear to be asymptotic to the y-axis and to increase more slowly as $|x|$ increases.

36. **D.** The given limit is equivalent to

$$\lim_{h \to 0} \frac{F\left(\frac{\pi}{4} + h\right) - F\left(\frac{\pi}{4}\right)}{h} = F'\left(\frac{\pi}{4}\right),$$

where $F'(x) = \dfrac{\sin x}{x}$. The answer is $\dfrac{2\sqrt{2}}{\pi}$.

37.   B. $\int_0^{12} g(x)\, dx = \int_0^4 g(x)\, dx + \int_4^6 g(x)\, dx + \int_6^9 g(x)\, dx + \int_9^{12} g(x)\, dx$

$$= 4\pi - 3 - \frac{9\pi}{4} + \frac{15}{2}.$$

38.   C.   In the figure, the curve for $y = e^x$ has been superimposed on the slope field.

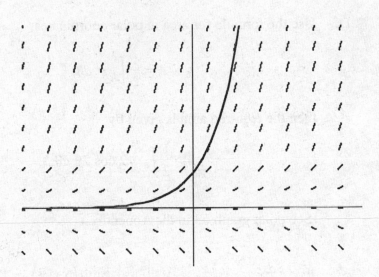

39.   A.   The solution curve shown is $y = \ln x$, so the differential equation is $y' = \dfrac{1}{x}$.

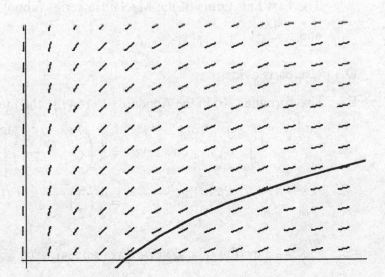

40.   D.   The speed, $|\mathbf{v}|$, equals $\sqrt{\left(\dfrac{dx}{dt}\right)^2 + \left(\dfrac{dy}{dt}\right)^2}$, and since $x = 3y - y^2$,

$$\frac{dx}{dt} = (3 - 2y)\frac{dy}{dt} = (3 - 2y)\cdot 3.$$

Then $|\mathbf{v}|$ is evaluated, using $y = 1$, and equals $\sqrt{(3)^2 + (3)^2}$.

41. **C.** The equations may be rewritten as

$$\frac{x}{2} = \sin u \quad \text{and} \quad y = 1 - 2\sin^2 u,$$

giving $y = 1 - 2 \cdot \frac{x^2}{4}$.

42. **D.** Use the formula for area in polar coordinates,

$$A = \frac{1}{2} \int_\alpha^\beta r^2 \, d\theta;$$

then the required area is given by

$$4 \cdot \frac{1}{2} \int_0^{\pi/4} 2\cos 2\theta \, d\theta.$$

(See polar graph 63 in the Appendix.)

43. **C.** $\displaystyle\int_{-\infty}^{\infty} \frac{dx}{x^2 + 1} = \lim_{b \to \infty} \tan^{-1} x \Big|_{-b}^{b} = \frac{\pi}{2} - \left(-\frac{\pi}{2}\right) = \pi.$

44. **A.** The first three derivatives of $\dfrac{1}{1 - 2x}$ are $\dfrac{2}{(1 - 2x)^2}, \dfrac{8}{(1 - 2x)^3},$ and $\dfrac{48}{(1 - 2x)^4}.$

The first four terms of the Maclaurin series (about $x = 0$) are $1, +2x, +\dfrac{8x^2}{2!},$ and $+\dfrac{48x^3}{3!}.$

45. **D.** Use parts twice.

46. **E.** Use formula (20) in the Appendix to rewrite the integral as

$$\frac{1}{2}\int_0^{\pi/2} (1 - \cos 2x)\, dx = \frac{1}{2}\left(x - \frac{\sin 2x}{2}\right)\Big|_0^{\pi/2}$$

$$= \frac{1}{2}\left(\frac{\pi}{2}\right).$$

47. **E.** The area, A, is represented by $\displaystyle\int_0^{2\pi} (1 - \cos t)\, dt = 2\pi.$

48.   D. $\dfrac{dy}{dx} = -2\sin^3\theta\cos\theta.$

49.   C.   Check to verify that each of the other improper integrals converges.

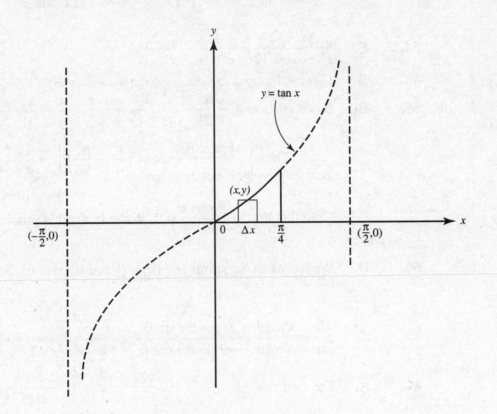

50.   D. $\displaystyle\int_2^4 \dfrac{du}{\sqrt{16-u^2}} = \sin^{-1}\dfrac{u}{4}\bigg|_2^4 = \dfrac{\pi}{3}.$ The given integral *is* improper. See

Example 21, page 242.

51.   C.   Let $y = \left(\dfrac{1}{x}\right)^x$. Then $\ln y = -x\ln x$ and

$$\lim_{x\to 0^+}\ln y = \lim_{x\to 0^+}\dfrac{-\ln x}{1/x}.$$

Now apply L'Hôpital's rule:

$$\lim_{x\to 0^+}\ln y = \lim_{x\to 0^+}\dfrac{-1/x}{-1/x^2} = 0.$$

So, if $\lim\limits_{x\to 0^+}\ln y = 0$, then $\lim\limits_{x\to 0^+} y = 1.$

52.   D. $\sqrt{1+\tan^2\theta} = \sec\theta;\ dx = \sec^2\theta;\ 0 \leqslant x \leqslant 1;\ \text{so } 0 \leqslant \theta \leqslant \dfrac{\pi}{4}.$

53. **A.** This is an indeterminate form of type $\frac{\infty}{\infty}$; use L'Hôpital's rule:

$$\lim_{x \to 0^+} \frac{\cot x}{\ln x} = \lim_{x \to 0^+} \frac{-\csc^2 x}{1/x} = \lim_{x \to 0^+} \frac{x}{\sin x} \cdot \frac{-1}{\sin x} = -\infty$$

54. **E.** $\dfrac{3x+2}{x^2-x-12} = \dfrac{1}{x+3} + \dfrac{2}{x-4}$.

55. **B.** Since $\cos x \simeq 1 - \dfrac{x^2}{2!} + \dfrac{x^4}{4!}$, $\dfrac{1-\cos x}{x} \simeq \dfrac{x}{2} - \dfrac{x^3}{4!}$.

Then $\displaystyle\int_0^1 \frac{1-\cos x}{x}\,dx \simeq \left(\frac{x^2}{4} - \frac{x^4}{96}\right)\bigg|_0^1 = \frac{1}{4} - \frac{1}{96}$.

Note that $\displaystyle\lim_{x \to 0^+} \frac{1-\cos x}{x} = 0$, so the integral is proper.

56. **D.** We represent the spiral as $P(\theta) = (\theta \cos \theta, \theta \sin \theta)$. So

$$\frac{dy}{dx} = \frac{dy/d\theta}{dx/d\theta} = \frac{\theta \cos \theta + \sin \theta}{-\theta \sin \theta + \cos \theta} = \frac{\frac{\pi}{4} \cdot \frac{\sqrt{2}}{2} + \frac{\sqrt{2}}{2}}{-\frac{\pi}{4} \cdot \frac{\sqrt{2}}{2} + \frac{\sqrt{2}}{2}} = \frac{\pi/4 + 1}{-\pi/4 + 1}.$$

Part B

57. **D.** Since h is increasing, $h' \geq 0$. The graph of h is concave downward for $x < 2$ and upward for $x > 2$, so h'' changes sign at $x = 2$, where it appears that $h' = 0$ also.

58. **C.** I is false since, for example, $f'(-2) = f'(1) = 0$ but neither $g(-2)$ nor $g(1)$ equals zero.

II is true. Verify this!

III is also true. Check the concavity of g: when the curve is concave down, $h < 0$; when up, $h > 0$.

59. **A.** If $y = \displaystyle\int_3^x \frac{1}{\sqrt{3+2t}}\,dt$, then $\dfrac{dy}{dx} = \dfrac{1}{\sqrt{3+2x}}$, so $\dfrac{d^2y}{dx^2} = -\dfrac{1}{2}(3+2x)^{-3/2}(2)$.

60. **D.** $\displaystyle\int_{-4}^3 f(x+1)\,dx$ represents the area of the same region as $\displaystyle\int_{-3}^4 f(x)\,dx$, translated one unit to the left.

61. **D.** According to the Mean Value Theorem, there exists a number c in the interval $[1, 1.5]$ such that $f'(c) = \dfrac{f(1.5) - f(1)}{1.5 - 1}$. Use the Solver (in radian mode) to find c such that $\cos c = \dfrac{\sin 1.5 - \sin 1}{0.5}$.

62. **E.** Here are the relevant sign lines:

$$\underset{0 \qquad 1}{\underline{-\qquad -\qquad +}} \quad \text{signs of } f'(x) \qquad\qquad \underset{0 \qquad \frac{2}{3}}{\underline{+\qquad -\qquad +}} \quad \text{signs of } f''(x)$$

We see that f' and f'' are both positive only if $x > 1$.

63. **E.** We see from the sign lines in Question 62 that f has a local minimum (at $x = 1$) and two points of inflection (at $x = 0$ and at $x = \frac{2}{3}$).

64. **C.** The derivatives of $\ln(x + 1)$ are $\dfrac{1}{x + 1}, \dfrac{-1}{(x + 1)^2}, \dfrac{+2!}{(x + 1)^3}, \dfrac{-(3!)}{(x + 1)^4}, \cdots$

The nth derivative at $x = 2$ is $\dfrac{(-1)^{n-1}(n - 1)!}{3^n}$.

65. **C.** The absolute-value function $f(x) = |x|$ is continuous at $x = 0$, but $f'(0)$ does not exist.

66. **C.** Let $F'(x) = f(x)$; then $F'(x + k) = f(x + k)$;

$$\int_0^3 f(x + k)\, dx = F(3 + k) - F(k);$$

$$\int_k^{3+k} f(x)\, dx = F(3 + k) - F(k).$$

Or let $u = x + k$. Then $dx = du$; when $x = 0$, $u = k$; when $x = 3$, $u = 3 + k$.

67. **E.** See the figure. The equation of the generating circle is $(x - 3)^2 + y^2 = 1$, which yields $x = 3 \pm \sqrt{1 - y^2}$.

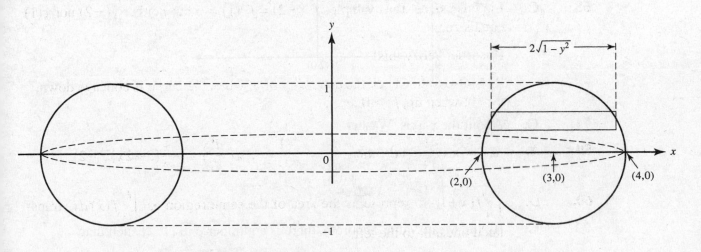

About the y-axis: $\Delta V = 2\pi \cdot 3 \cdot 2\sqrt{1 - y^2}\, \Delta y$.

Thus, $V = 2 \displaystyle\int_0^1 12\pi \sqrt{1 - y^2}\, dy$

$= 24\pi$ times the area of a quarter of a unit circle $= 6\pi^2$.

68.   D.   You want $\dfrac{d}{du}\tan^{-1}e^{2u}$, where

$$\frac{d}{dx}\tan^{-1}v=\frac{\dfrac{dv}{dx}}{1+v^{2}}.$$

69.   D.   Let $y'=\dfrac{dy}{dx}$. Then $\cos(xy)[xy'+y]=y'$. Solve for y'.

70.   E. $\dfrac{d^{2}}{dx^{2}}f(x^{2})=\dfrac{d}{dx}\left[\dfrac{d}{dx}f(x^{2})\right]=\dfrac{d}{dx}\left[\dfrac{d}{dx}f(x^{2})\cdot\dfrac{dx^{2}}{dx}\right]=\dfrac{d}{dx}\left[g(x^{2})\cdot2x\right]$

$$=g(x^{2})\frac{d}{dx}(2x)+2x\frac{d}{dx}g(x^{2})=g(x^{2})\cdot2+2x\frac{d}{dx^{2}}g(x^{2})\frac{dx^{2}}{dx}$$

$$=2g(x^{2})+2x\cdot f(\sqrt{x^{2}})\cdot2x=2g(x^{2})+4x^{2}\cdot f(x).$$

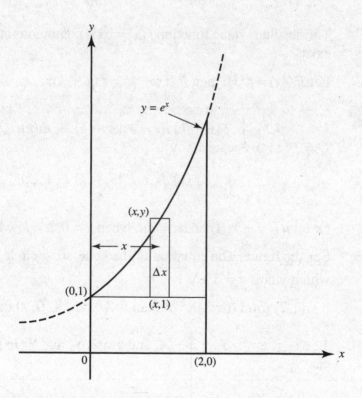

71.   C.   About the x-axis. Washer.

$$\Delta V=\pi(y^{2}-1^{2})\,\Delta x,$$

$$V=\pi\int_{0}^{2}(e^{2x}-1)\,dx.$$

72.   C.   By the Mean Value Theorem, there is a number c in $[1, 2]$ such that

$$f'(c)=\frac{f(2)-f(1)}{2-1}=-3.$$

73.   **D.**   The enclosed region, S, is bounded by $y = \sec x$, the y-axis, and $y = 4$. It is to be rotated about the y-axis.

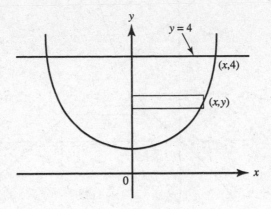

Use disks; then $\Delta V = \pi R^2 H = \pi$ (arc sec y)^2 Δy. With a change in variables, find V from

$$\pi \text{fnInt } ((\cos^{-1}(1/X))^2, X, 1, 4),$$

getting 11.385.

74.   **C.**   If Q is the amount at time t, then $Q = 40e^{-kt}$. Since $Q = 20$ when $t = 2$, $k = -0.3466$. Now find Q when $t = 3$, from $Q = 40e^{-(0.3466)3}$, getting $Q = 14$ to the nearest gram.

75.   **A.**   The velocity $v(t)$ is an antiderivative of $a(t)$, where $a(t) = \pi t + \dfrac{2}{1 + t^2}$. So $v(t) = \dfrac{\pi t^2}{2} + 2 \arctan t + C$. Since $v(1) = 0$, $C = -\pi$. The required average

velocity $= \dfrac{1}{2 - 0} \displaystyle\int_0^2 v(t)\, dt$, which is equal (with a change in variables) to

$$(1/2) \text{ fnInt } ((\pi/2)X^2 + 2\tan^{-1} X - \pi, X, 0, 2) \text{ or } 0.362.$$

76.   **D.**   Let $Y_1 = \tan X$, $Y_2 = 2 - X$, and graph Y_1 and Y_2 in $[-1, 3] \times [-1, 3]$. Note that

$$\Delta A = (x_{\text{line}} - x_{\text{curve}})\, \Delta y$$
$$= (2 - y - \text{Arctan } y)\, \Delta y.$$

The limits are $y = 0$ and $y = b$, where b is the ordinate of the intersection of the curve and the line. Find b by calculating

$$\text{solve } (2 - Y - \tan^{-1} Y, Y, 1).$$

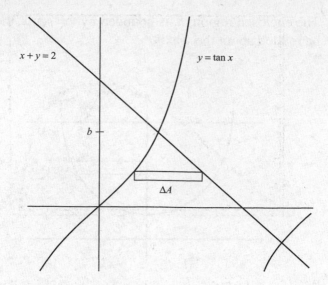

Store the answer to (1) in memory as B. The desired area is now found from

$$\text{fnInt}(2 - Y - \tan^{-1} Y, Y, 0, B).$$

The answer is 1.077.

77.   **E.**   Center the ellipse at the origin and let (x, y) be the coordinates of the vertex of the inscribed rectangle in the first quadrant, as shown in the figure.

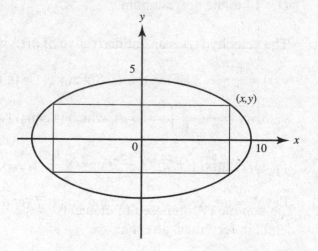

$$\frac{x^2}{100} + \frac{y^2}{25} = 1,$$

To maximize the rectangle's area $A = 4xy$, solve the equation of the ellipse, getting

$$x = \sqrt{100 - 4y^2} = 2\sqrt{25 - y^2}.$$

So $A = 8y\sqrt{25 - y^2}$. With a change of variables, let $Y_1 = 8X\sqrt{(25 - X^2)}$ and graph Y_1 in the window $[0,5] \times [0,150]$. Using the [maximum] option, read directly from the screen that the maximum area (the Y-coordinate) equals 100.

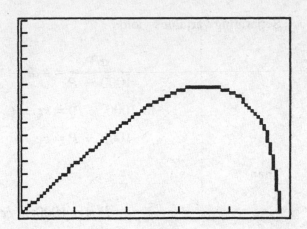

78. **B.** fnInt (Xln X,X,1,e)/(e − 1) yields 1.221.

79. **B.** When f' is positive, f increases. By the Fundamental Theorem of Calculus, $f'(x) = 1 - 2 (\cos x)^3$. Use Y_1 for $f'(X)$ and graph Y_1 in $[0, 2\pi] \times [-2, 4]$. It is clear that $f' > 0$ on the interval $a < x < b$. Using the ROOT option twice, quickly obtain $a = 0.654$ and $b = 5.629$.

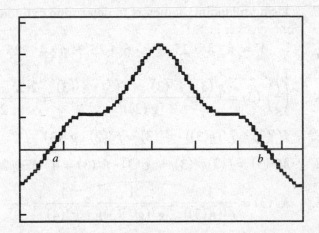

80. **C.** $a(1.5) \simeq \dfrac{v(2) - v(1)}{2 - 1} = \dfrac{12 - 22}{1}$.

81. **B.** The volume is composed of elements of the form $\Delta V = (2x)^2 \, \Delta y$. If h is the depth, in feet, then, after t hr,

$$V(h) = 4 \int_0^h y \, dy \text{ and } \frac{dV}{dt} = 4h \frac{dh}{dt}.$$

Thus, $12 = 4\,(9)\,\dfrac{dh}{dt}$

and $\dfrac{dh}{dt} = \dfrac{1}{3}$ ft/hr.

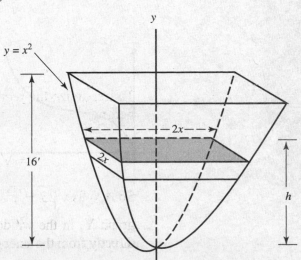

82.   B.   Separating variables yields

$$\frac{dP}{1000 - P} = k\, dt,$$

$$-\ln(1000 - P) = kt + C,$$

$$1000 - P = ce^{-kt}.$$

Then

$$P(t) = 1000 - ce^{-kt}.$$

$P(0) = 300$ gives $c = 700$. $P(5) = 500$ yields $500 = 1000 - 700e^{-5k}$, so $k \approx +0.0673$. Now $P(10) = 1000 - 700e^{-0.673} \approx 643$.

83.   C. $H(1) = \int_0^1 \frac{4}{x^2 + 1}\, dx = 4 \arctan 1 = \pi$. $H'(1) = f(1) = 2$.

The equation of the tangent line is $y - \pi = 2(x - 1)$.

84.   C.   Using midpoint diameters to determine cylinders, estimate the volume to be

$$V \approx \pi \cdot 8^2 \cdot 25 + \pi \cdot 6^2 \cdot 25 + \pi \cdot 4^2 \cdot 25 + \pi \cdot 3^2 \cdot 25.$$

85.   A. $\left(\dfrac{f}{g}\right)'(3) = \dfrac{g(3) \cdot f'(3) - f(3) \cdot g'(3)}{(g(3))^2} = \dfrac{2(2) - 4(3)}{2^2}$.

86.   C. $H'(3) = f'(g(3)) \cdot g'(3) = f'(2) \cdot g'(3)$.

87.   E. $M'(3) = f(3)g'(3) + g(3) \cdot f'(3) = 4 \cdot 3 + 2 \cdot 2$.

88.   E. $K'(3) = \dfrac{1}{g'(K(3))} = \dfrac{1}{g'(g^{-1}(3))} = \dfrac{1}{g'(4)} = \dfrac{1}{\frac{1}{2}}$.

89.   C. $R'(3) = \dfrac{1}{2}(f(3))^{-1/2} \cdot f'(3)$.

90.   D.   Here are the pertinent curves, with d denoting the depth of the water:

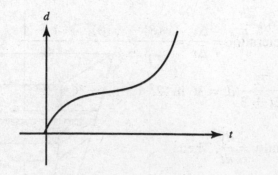
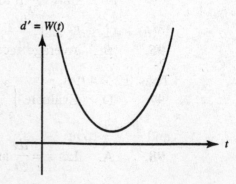

91. B. Use areas; then $\int_1^7 f' = -3 + 10 = 7$. Thus, $f(7) - f(1) = 7$.

92. B. The region x units from the stage can be approximated by the semicircular ring shown; its area is then the product of its circumference and its width.

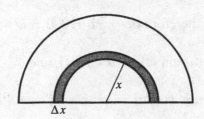

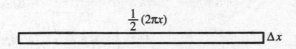

The number of people standing in the region is the product of the area and the density:

$$\Delta P = (\pi x \, \Delta x) \left(\frac{20}{2\sqrt{x} + 1} \right).$$

To find the total number of people, evaluate

$$20\pi \int_0^{100} \frac{x}{2\sqrt{x} + 1} \, dx$$

93. B. $\dfrac{dy}{dt}$ is positive, but decreasing; hence $\dfrac{dy^2}{dt^2} < 0$.

94. C. Average speed $= \dfrac{\text{total distance}}{\text{elapsed time}} = \dfrac{\text{total area}}{8} = \dfrac{8}{8}$.

95. E. On $2 \leqslant t \leqslant 5$, the object moved $3\frac{1}{2}$ ft to the right; then on $5 \leqslant t \leqslant 8$, it moved only $2\frac{1}{2}$ ft to the left.

96. B. Average acceleration $= \dfrac{\Delta v}{\Delta t} = \dfrac{v(8) - v(0)}{8 - 0} = \dfrac{-1 - 1}{8}$.

97. D. Evaluate $\int_0^3 \dfrac{72}{2t + 3} \, dt = 36 \ln(2t + 3) \Big|_0^3 = 36 \ln 3$.

98. A. Let $\dot{x} = \dfrac{dx}{dt}$ and $\dot{y} = \dfrac{dy}{dt}$. Then

$$2x\dot{x} + 2y\dot{y} = 0 \text{ and } \dot{y} = -\frac{3}{4}\dot{x} \text{ at the point } (3, 4).$$

Use, also, the facts that the speed is given by $|\mathbf{v}| = \sqrt{\dot{x}^2 + \dot{y}^2} = 2$ and that the point moves counterclockwise; then $\dot{x}^2 + \dot{y}^2 = 4$, yielding $\dot{x} = -\dfrac{8}{5}$ and $\dot{y} = +\dfrac{6}{5}$ at the given point. The velocity vector, $\mathbf{v}$, at $(3, 4)$

must therefore be $-\dfrac{8}{5}\mathbf{i} + \dfrac{6}{5}\mathbf{j}$.

99. **A.** $\mathbf{v} = -ak \sin kt\mathbf{i} + ak \cos kt\mathbf{j}$, and

$\mathbf{a} = -ak^2 \cos kt\mathbf{i} - ak^2 \sin kt\mathbf{j} = -k^2\mathbf{R}$.

100. **B.** The formula for length of arc is

$$L = \int_a^b \sqrt{1 + \left(\frac{dy}{dx}\right)^2}\, dx.$$

Since $y = 2^x$, here

$$L = \int_0^2 \sqrt{1 + (2^x \ln 2)^2}\, dx.$$

Thus L equals

fnInt $(\sqrt{(1 + (2^X \ln 2)^2)}, X, 0, 2)$ or 3.664.

101. **D.** $\mathbf{a}(t) = (0, e^t)$, the acceleration is always upward.

102. **A.** At $(0, 1)$, $\dfrac{dy}{dx} = 4$, so Euler's method yields $(0.1, 1+0.1(4)) = (0.1, 1.4)$. $\dfrac{dy}{dx} = 4y$ has particular solution $y = e^{4x}$; the error is $e^{4(0.1)} - 1.4$.

103. **D.** $1 - \dfrac{1}{2} + \dfrac{1}{5} = 0.7$. Since the first term dropped from this alternating convergent series is $-\dfrac{1}{10}$, the estimate is too high, but within 0.1 of the true sum.

104. **C.** $\displaystyle\sum_{n=1}^{\infty} \frac{2}{3n} = \frac{2}{3}\sum_{n=1}^{\infty} \frac{1}{n}$, which equals a constant times the harmonic series.

105. **D.** We seek x such that

$$\lim_{n\to\infty} \left| \frac{(n + 1)! \cdot x^{n+1}}{(n + 1)^{n+1}} \cdot \frac{n^n}{n! \cdot x^n} \right| < 1$$

or such that $|x| \cdot \displaystyle\lim_{n\to\infty} \left(\frac{n}{n + 1}\right)^n < 1$

or such that $|x| < \dfrac{1}{\displaystyle\lim_{n \to 0} \left(\dfrac{n}{n+1}\right)^n}$.

The fraction equals $\displaystyle\lim_{n \to \infty} \left(\dfrac{n+1}{n}\right)^n = \lim_{n \to \infty} \left(1 + \dfrac{1}{n}\right)^n = e$.

Then $|x| < e$ and the radius of convergence is e.

106.   B.   The error is less than the maximum value of $\dfrac{e^c}{3!}x^3$ for $0 \le x \le \dfrac{1}{2}$.

This maximum occurs at $c = x = \dfrac{1}{2}$.

107.   D.   Distance $= \displaystyle\int_0^1 \sqrt{\left(\dfrac{dx}{dt}\right)^2 + \left(\dfrac{dy}{dt}\right)^2}\, dt$

$= \displaystyle\int_0^1 \sqrt{(\sec^2 t)^2 + (-2\sin 2t)^2}\, dt.$

Note that the curve is traced exactly once by the parametric equations from $t = 0$ to $t = 1$.

Chapter 12

Miscellaneous Free-Response Practice Questions

These problems provide further practice for both parts of Section II of the examination. Solutions begin on page 436.

Part A

A graphing calculator is required for some of these problems. See instructions on page xi.

1. Use this table to answer the questions that follow.

x	2.5	3.0	3.5	4.0	4.5	5.0
$f(x)$	7.6	5.7	4.2	3.1	2.2	1.5

(a) Estimate $f'(4.0)$ and $f'(4.75)$.
(b) Estimate the concavity of the curve $y = f(x)$ at $x = 3.75$.

2. The equation of the tangent line to the curve $x^2y - x = y^3 - 8$ at the point (0,2) is $12y + x = 24$.
(a) Given that the point $(0.3, y_0)$ is on the curve, find y_0 approximately, using the tangent line.
(b) Find the true value of y_0.
(c) What can you conclude about the curve near $x = 0$ from your answers to parts (a) and (b)?

3. Draw a graph of $y = f(x)$, given that f satisfies all the following conditions:
(1) $f'(-1) = f'(1) = 0$.
(2) If $x < -1, f'(x) > 0$ but $f'' < 0$.
(3) If $-1 < x < 0, f'(x) > 0$ and $f'' > 0$.
(4) If $0 < x < 1, f'(x) > 0$ but $f'' < 0$.
(5) If $x > 1, f'(x) < 0$ and $f'' < 0$.

4. A differentiable function f defined on $-7 < x < 7$ has $f(0) = 0$ and
   $f'(x) = 2x \sin x - e^{-x^2} + 1$. (Note: The following questions refer to f, not to f'.)
   (a) Describe the symmetry of f.
   (b) On what intervals is f decreasing?
   (c) For what values of x does f have a relative maximum? Justify your answer.
   (d) How many points of inflection does f have? Justify your answer.

5. Let C represent the piece of the curve $y = \sqrt[3]{64 - 16x^2}$ that lies in the first quadrant.
   Let S be the region bounded by C and the coordinate axes.
   (a) Find the slope of the line tangent to C at $y = 1$.
   (b) Find the area of S.
   (c) Find the volume generated when S is rotated about the x-axis.

6. Find the x-coordinate of the point R on the curve of $y = x - x^2$ such that the line OR
   (where O is the origin) divides the area bounded by the curve and the x-axis into two
   regions of equal area.

7. Suppose $f'' = \sin(2^x)$ for $-1 < x < 3.2$.
   (a) On what intervals is f concave downward? Justify your answer.
   (b) Find the x-coordinates of all relative minima of f'.
   (c) How many points of inflection does f' have? Justify your answers.

8. Let $f(x) = \cos x$ and $g(x) = x^2 - 1$.
   (a) Find the coordinates of any points of intersection of f and g.
   (b) Find the area bounded by f and g.

9. (a) In order to investigate mail-handling efficiency, each hour one morning a local
   post office checked the rate (letters/min) at which an employee was sorting mail.
   Use the results shown in the table to estimate the total number of letters he may have
   sorted that morning.

Time	8	9	10	11	12
Letters/min	10	12	8	9	11

   (b) Hoping to speed things up a bit, the post office tested a sorting machine that can
   process mail at the constant rate of 20 letters per minute. The graph shows the rate at
   which letters arrived at the post office and were dumped into this sorter.

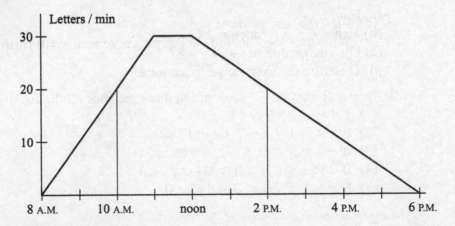

   (i)   When did letters start to pile up?
   (ii) When was the pile the biggest?
   (iii) How big was it then?
   (iv) At about what time did the pile vanish?

*10. A particle moves on the curve of $y^3 = 2x + 1$ so that its distance from the x-axis is increasing at the constant rate of 2 units/sec. When $t = 0$, the particle is at $(0,1)$.
   (a) Find a pair of parametric equations $x = x(t)$ and $y = y(t)$ that describe the motion of the particle for nonnegative t.
   (b) Find $|\mathbf{a}|$, the magnitude of the particle's acceleration, when $t = 1$.

*11. Find the area that the polar curves $r = 2 - \cos \theta$ and $r = 3 \cos \theta$ have in common.

*12. (a) Verify that $\pi/4$ and the expression

$$(4\tan^{-1}(1/5)) - (\tan^{-1}(1/239))$$

   agree on your calculator to 10 decimal places.
   (b) Use the Taylor polynomial of degree 7 about 0,

$$\tan^{-1} x = x - x^3/3 + x^5/5 - x^7/7,$$

   to approximate $\tan^{-1} 1/5$, and the polynomial of degree 1 to approximate $\tan^{-1} 1/239$.
   (c) Use part (b) to evaluate the expression in (a).
   (d) Explain how the approximation for $\pi/4$ given here compares with that obtained using $\pi/4 = \tan^{-1} 1$.

*13. (a) Show that the series $\displaystyle\sum_{n=1}^{\infty} (-1)^{n+1} \frac{1}{\ln(n + 1)}$ converges.
   (b) How many terms of the series are needed to get a partial sum within 0.1 of the sum of the whole series?
   (c) Tell whether the series $\displaystyle\sum_{n=2}^{\infty} (-1)^n \frac{1}{n \ln n}$ is absolutely convergent, conditionally convergent, or divergent. Justify your answer.

*14. Given $\dfrac{dy}{dt} = ky (10 - y)$ with $y = 2$ at $t = 0$ and $y = 5$ at $t = 2$:
   (a) Find k.
   (b) Express y as a function of t.
   (c) For what value of t will $y = 8$?
   (d) Describe the long-range behavior of y.

---

*An asterisk denotes a topic covered only in Calculus BC.

*15. An object P is in motion in the first quadrant along the parabola $y = 18 - 2x^2$ in such a way that at t sec the abscissa of its position is $x = \frac{1}{2}t$.
   (a) Where is P when $t = 4$?
   (b) What is the vertical component of its velocity there?
   (c) At what rate is its distance from the origin changing then?
   (d) When does it hit the x-axis?
   (e) How far did it travel altogether?

*16. A particle moves in the xy-plane in such a way that at any time $t \geq 0$ its position is given by $x(t) = 4 \arctan t$, $y(t) = \dfrac{12t}{t^2 + 1}$.

   (a) Sketch the path of the particle, indicating the direction of motion.
   (b) At what time t does the particle reach its highest point? Justify.
   (c) Find the coordinates of that highest point, and sketch the velocity vector there.
   (d) Describe the long-term behavior of the particle.

Part B

You may *not* use a calculator for any question in this part.

17. The figure below shows the graph of f', the derivative of f, with domain $-3 \leq x \leq 9$. The graph of f' has horizontal tangents at $x = 2$ and $x = 4$, and a corner at $x = 6$.

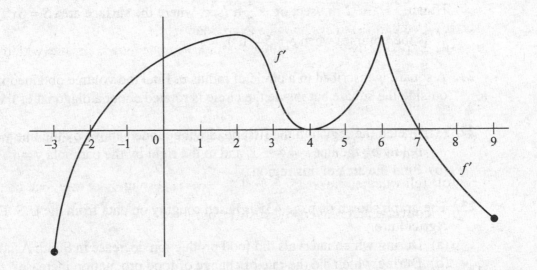

   (a) Is f continuous? Explain.
   (b) Find all values of x at which f attains a relative minimum. Justify.
   (c) Find all values of x at which f attains a relative maximum. Justify.
   (d) At what value of x does f attain its absolute maximum? Justify.
   (e) Find all values of x at which f has a point of inflection. Justify.

18. Find the area of the largest rectangle (with sides parallel to the coordinate axes) that can be inscribed in the region bounded by the graphs of $f(x) = 8 - 2x^2$ and $g(x) = x^2 - 4$.

---

* An asterisk denotes a topic covered only in Calculus BC.

19. Given the graph of $f(x)$, sketch the graph of $f'(x)$.

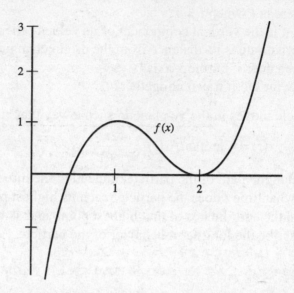

20. A cube is contracting so that its surface area decreases at the constant rate of 72 in.^{2}/sec. Determine how fast the volume is changing at the instant when the surface area is 54 ft^2.

Hint: $\dfrac{dS}{dt} = -72$ in.^{2}/sec, or $-\dfrac{1}{2}$ ft^2/sec, where the surface area $S = 6x^2$. Find $\dfrac{dV}{dt}$ (V is the volume) when $S = 54$ (ft^2).

21. A square is inscribed in a circle of radius a. Find the volume obtained if the region outside the square but inside the circle is rotated about a diagonal of the square.

22. (a) Sketch the region in the first quadrant bounded above by the line $y = x + 4$, below by the line $y = 4 - x$, and to the right by the parabola $y = x^2 + 2$.
   (b) Find the area of this region.

23. The graph shown on page 435 is based roughly on data from the U.S. Department of Agriculture.
   (a) During which intervals did food production decrease in South Asia?
   (b) During which did the rate of change of food production increase?
   (c) During which did the increase in food production accelerate?

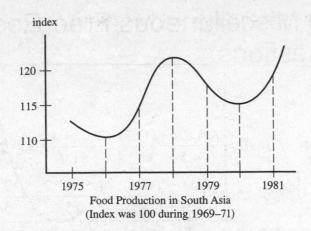

Food Production in South Asia
(Index was 100 during 1969–71)

24. A particle moves along a straight line so that its acceleration at any time t is given in terms of its velocity v by $a = -2v$.
(a) Find v in terms of t if $v = 20$ when $t = 0$.
(b) Find the distance the particle travels while v changes from $v = 20$ to $v = 5$.

*25. (a) Find the Taylor polynomial of order 4 about 0 for $\cos x$.

(b) Use part (a) to evaluate $\displaystyle\int_0^1 \cos x \, dx$ to five decimal places.

(c) Estimate the error in (b), justifying your answer.

*26. (a) Find the Maclaurin series for $f(x) = \ln(1 + x)$.
(b) What is the radius of convergence of the series in (a)?
(c) Use the first five terms in (a) to approximate $\ln(1.2)$.
(d) Estimate the error in (c), justifying your answer.

*27. A cycloid is given parametrically by $x = \theta - \sin \theta$, $y = 1 - \cos \theta$.

(a) Find the slope of the curve at the point where $\theta = \dfrac{2\pi}{3}$.

(b) Find the equation of the tangent to the cycloid at the point where $\theta = \dfrac{2\pi}{3}$.

*28. Find the area enclosed by both the polar curves $r = 4 \sin \theta$ and $r = 4 \cos \theta$.

*29. (a) By comparing $\displaystyle\sum_1^\infty \frac{1}{n!}$ with $\displaystyle\sum_1^\infty \frac{1}{2^{n-1}}$ show that, when $\displaystyle\sum_1^\infty \frac{1}{n!}$ is replaced by $\displaystyle\sum_1^{11} \frac{1}{n!}$, the error is less than 0.001.

(b) Does the series $\displaystyle\sum_{k=1}^\infty ke^{-k}$ converge? Justify your answer.

* An asterisk denotes a topic covered only in Calculus BC.

Solutions to Miscellaneous Free-Response Practice Questions

Part A

1. (a) $f'(4.0) \simeq \dfrac{f(4.5) - f(3.5)}{4.5 - 3.5} = \dfrac{2.2 - 4.2}{1.0} = -2.0.$

   $f'(4.75) \simeq \dfrac{f(5) - f(4.5)}{5 - 4.5} = \dfrac{1.5 - 2.2}{0.5} = -\dfrac{7}{5} = -1.4.$

   (b) You want an approximation for $f''(3.75)$. Since

   $$f''(3.75) \simeq \frac{f'(4) - f'(3.5)}{4 - 3.5},$$

   use $f'(4)$ from (a) and estimate $f'(3.5)$:

   $$f'(3.5) = \frac{f(4) - f(3)}{4 - 3} = \frac{3.1 - 5.7}{4 - 3} = -2.6.$$

   Then

   $$f''(3.75) \simeq \frac{-2.0 - (-2.6)}{0.5} = \frac{0.6}{0.5} = 1.2.$$

2. (a) $12y_0 + 0.3 = 24$ yields $y_0 \simeq 1.975$.

   (b) Replace x by 0.3 in the equation of the curve:

   $$(0.3)^2 y_0 - (0.3) = y_0^3 - 8 \text{ or}$$
   $$y_0^3 - 0.09y_0 - 7.7 = 0.$$

   Using the Solver, find the root of this equation to three decimal places: $y_0 = 1.990$.

   (c) Since the approximation exceeds the true value of y_0 for $x = 0.3$, conclude that the given curve is concave down near $x = 0$. (Therefore, it is below the tangent line at $x = 0$.)

3.   The graph shown satisfies all five conditions. So do many others!

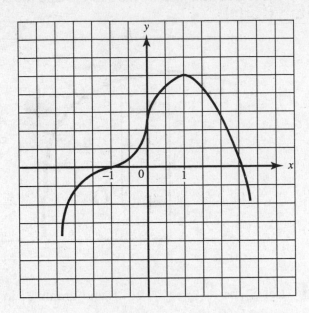

4.   Let $Y_1 = f'(X) = 2X\sin X - e^{(-x^2)} + 1$. Graph Y_1 in $[-7,7] \times [-10,10]$.

(a)   Since Y_1 (or f') is even and f contains $(0,0)$, f is odd and its graph is symmetric about the origin.

(b)   Since f is decreasing when $f' < 0$, f decreases on the intervals (a, c), and (j, l). Use the Solver or [root] option and symmetry to determine that f decreases on $-6.202 < x < -3.294$ and on $3.294 < x < 6.202$.

(c) f has a relative maximum at $x = q$ if $f'(q) = 0$ and if f changes from increasing ($f' > 0$) to decreasing ($f' < 0$) as x increases through q. There are two relative maxima here: at $x = a = -6.202$ and at $x = j = 3.294$.

(d) f has a point of inflection when f changes its concavity; that is, when f' changes from increasing to decreasing, as it does at points d and h, or when f changes from decreasing to increasing, as it does at points b, g, and k. So there are five points of inflection altogether.

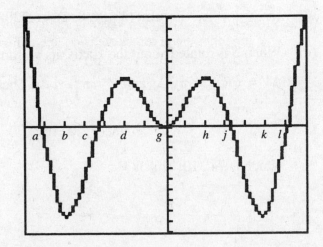

5.   In the graph below, C is the piece of the curve lying in the first quadrant. S is the region bounded by the curve C and the coordinate axes.

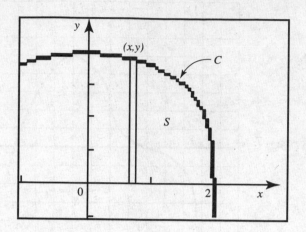

(a)   Let Y_1 be $\sqrt[3]{(64 - 16X^2)}$ and graph Y_1 in $[-1, 3] \times [-1, 5]$. Since you want dy/dx, the slope of the tangent, where $y = 1$, use the Solver on

$$Y_1 - 1 = 0, X = 2$$

(storing the answer at B). Then calculate

$$\text{nDeriv } (Y_1, X, B),$$

which equals -21.182.

Combine the calculations above, as follows:

$$\text{nDeriv } (Y_1, X, \text{solve } (Y_1 - 1, X, 1)) = -21.182.$$

The answer is the slope of the tangent to C at $y = 1$.

(b)   Since $\Delta A = y\Delta x$, $A = \displaystyle\int_0^2 y \, dx = $ the area of S. Evaluate

$$\text{fnInt } (Y_1, X, 0, 2) = 6.730.$$

(c)   When S is rotated about the x-axis, its volume can be obtained using disks:
$\Delta V = \pi R^2 H = \pi y^2 \Delta y$ and $V = \pi \displaystyle\int_0^2 y^2 \, dx$. Then calculate

$$\pi \text{fnInt } (Y_1{}^2, X, 0, 2),$$

which is 74.310. This is V.

6.   See the figure, where R is the point (a,b), and seek a such that

$$\int_0^a \left(x - x^2 - \frac{b}{a} \cdot x \right) dx = \frac{1}{2} \int_0^1 (x - x^2)\, dx.$$

This yields

$$\frac{a^2}{2} - \frac{a^3}{3} - \frac{ab}{2} = \frac{1}{4} - \frac{1}{6}.$$

Replace b by $a - a^2$; then

$$\frac{a^2}{2} - \frac{a^3}{3} - \frac{a}{2}(a - a^2) = \frac{1}{12},$$

$$a^3 = \frac{1}{2} \quad \text{and} \quad a = \frac{1}{\sqrt[3]{2}} = \frac{\sqrt[3]{4}}{2}.$$

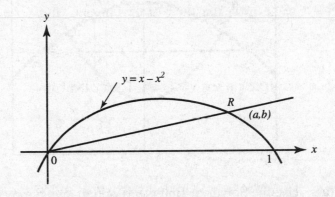

7.   Let $Y_1 = \sin 2^x$ and graph Y_1 in $[-1, 3.2] \times [-1, 1]$. Note that $Y_1 = f''$.

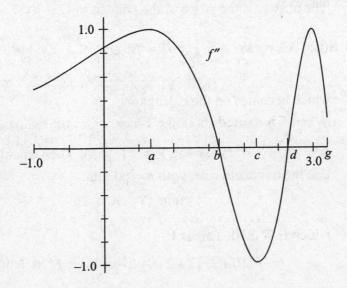

(a) f is concave downward where f'' is negative, namely, on (b, d). Use the Solver or [root] option to obtain $b = 1.651$ and $d = 2.651$. The answer to (a) is therefore $1.651 < x < 2.651$.

(b) f' has a relative minimum at $x = d$ because f'' equals 0 at d, is less than 0 on (b,d), and greater than 0 on (d,g). Thus f' has a relative minimum (from part a) at 2.651.

(c) f' has a point of inflection wherever its second derivative f''' changes from positive to negative or vice versa. This is equivalent to f'' changing from increasing to decreasing (as at a and g) or vice versa (as at c). Therefore, f' has three points of inflection on $[-1, 3.2]$.

8. Let $Y_1 = \cos X$ and $Y_2 = X^2 - 1$. Graph both in $[-2, 2] \times [-2, 2]$. Here, $Y_1 = f$ and $Y_2 = g$.

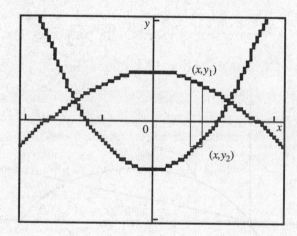

(a) Use the Solver or [intersect] option and symmetry to find the coordinates of the two points of intersection: $(1.177, 0.384)$ and $(-1.177, 0.384)$.

(b) Since $\Delta A = (y_1 - y_2)\, \Delta x = [f(x) - g(x)]\, \Delta x$, the area A bounded by the two curves is

$$A = 2\int_0^{1.177} (y_1 - y_2)\, dx,$$

which becomes on the calculator

$$2\,\texttt{fnInt}(Y_1 - Y_2, X, 0, 1.177) = 3.114.$$

9. (a) Use the trapezoid rule, with $h = 60$ min:

$$= \frac{h}{2}(y_0 + 2y_0 + 2y_0 + 2y_0 + y_5)$$

$$= \frac{60}{2}(10 + 2\cdot 12 + 2\cdot 8 + 2\cdot 9 + 11) = 2370 \text{ letters.}$$

(b) Draw a horizontal line at $y = 20$ (as shown on the graph on page 441) representing the rate at which letters are processed then.

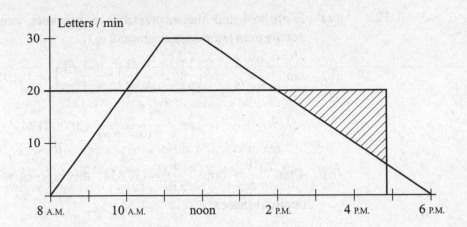

(i) Letters began to pile up when they arrived at a rate greater than that at which they were being processed, that is, at $t = 10$ A.M.

(ii) The pile was largest when the letters stopped piling up, at $t = 2$ P.M.

(iii) The number of letters in the pile is represented by the area of the small trapezoid above the horizontal line: $\frac{1}{2}(4 \cdot 60 + 1 \cdot 60)(10) = 1500$.

(iv) The pile began to diminish after 2 P.M., when letters were processed at a rate faster than they arrived, and vanished when the area of the shaded triangle represented 1500 letters. At 5 P.M. this area is

$$\frac{1}{2}(3 \cdot 60)(10) = 1800 \text{ letters, so the pile vanished shortly before 5 P.M.}$$

10.   (a)   Since $\dfrac{dy}{dt} = 2$, $y = 2t + 1$ and $x = 4t^3 + 6t^2 + 3t$.

(b)   Since $\dfrac{d^2y}{dt^2} = 0$ and $\dfrac{d^2x}{dt^2} = 24t + 12$, then, when $t = 1$, $|\mathbf{a}| = 36$.

11.   See the figure. The required area A is twice the sum of the following areas: that of the limaçon from 0 to $\dfrac{\pi}{3}$, and that of the circle from $\dfrac{\pi}{3}$ to $\dfrac{\pi}{2}$. Thus

$$A = 2\left[\frac{1}{2}\int_0^{\pi/3} (2 - \cos \theta)^2 \, d\theta + \frac{1}{2}\int_{\pi/3}^{\pi/2} (3\cos \theta)^2 \, d\theta\right]$$

$$= \frac{9\pi}{4} - 3\sqrt{3}.$$

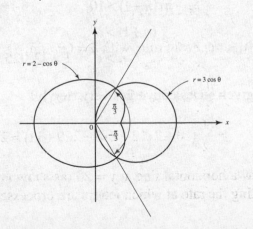

12.   (a)   Both $\pi/4$ and the expression in brackets yield 0.7853981634, which is accurate to ten decimal places.

(b) $\tan^{-1}\dfrac{1}{5} = \dfrac{1}{5} - \dfrac{1}{3}\left(\dfrac{1}{5}\right)^3 + \dfrac{1}{5}\left(\dfrac{1}{5}\right)^5 - \dfrac{1}{7}\left(\dfrac{1}{5}\right)^7 = 0.197396.$

$$\tan^{-1}\dfrac{1}{239} = \dfrac{1}{239} = 0.004184.$$

(c) $4\tan^{-1}\dfrac{1}{5} - \tan^{-1}\dfrac{1}{239} = 0.7854$; this agrees with the value of $\dfrac{\pi}{4}$ to four decimal places.

(d)   The series

$$\tan^{-1}1 = 1 - \dfrac{1}{3} + \dfrac{1}{5} - \dfrac{1}{7} + \dots$$

converges *very* slowly. Example 60, page 375, showed a calculator graph for 60 terms of the series for π (which equals $4\tan^{-1}1$) and evaluated the sum for 60 terms using the TRACE key. To four decimal places, we get $\pi = 3.1249$, which yields 0.7812 for $\pi/4$—not accurate even to two decimal places.

13.   (a)   The given series is alternating. Since $\lim\limits_{n\to\infty} \ln(n+1) = \infty$, $\lim\limits_{n\to\infty}\dfrac{1}{\ln(n+1)} = 0$. Since $\ln x$ is an increasing function,

$$\ln(n+1) > \ln n \quad \text{and} \quad \dfrac{1}{\ln(n+1)} < \dfrac{1}{\ln n}.$$

The series therefore converges.

(b)   For an alternating series, the error in using the first n terms for the sum of the whole series is less than the absolute value of the $(n + 1)$st term. Thus the error is less than $\dfrac{1}{\ln(n+1)}$. Solve for n using $\dfrac{1}{\ln(n+1)} < 0.1$:

$$\ln(n+1) > 10,$$
$$(n+1) > e^{10},$$
$$n > e^{10} - 1 > 22{,}025.$$

The given series converges very slowly!

(c)   The series $\sum_{2}^{\infty}(-1)^{n}\dfrac{1}{n\ln n}$ is conditionally convergent. The given alternating series converges since the nth term approaches 0 and $\dfrac{1}{(n+1)\ln(n+1)}<\dfrac{1}{n\ln n}$. However, the *positive* series diverges by the Integral Test, since

$$\int_{2}^{\infty}\frac{1}{x\ln x}\,dx=\lim_{b\to\infty}\ln(\ln x)\Big|_{2}^{b}=\infty.$$

14.   (a)   Solve by separation of variables:

$$\frac{dy}{y(10-y)}=k\,dt,$$

$$\frac{1}{10}\int\left(\frac{1}{y}+\frac{1}{10-y}\right)dy=\int k\,dt,$$

$$\frac{1}{10}\ln\left(\frac{y}{10-y}\right)=kt+C,$$

$$\ln\left(\frac{10-y}{y}\right)=-10(kt+C).$$

Let $c=e^{-10C}$; then

$$\frac{10-y}{y}=ce^{-10kt}.$$

Now use initial condition $y=2$ at $t=0$:

$$\frac{8}{2}=ce^{0}\text{ so }c=4;$$

and the other condition, $y=5$ at $t=2$, gives

$$\frac{5}{5}=4e^{-20k}\ \text{ or }\ k=\frac{1}{10}\ln 2.$$

(b)   Since $c=4$ and $k=\dfrac{1}{10}\ln 2$, then $\dfrac{10-y}{y}=4e^{-10\left(\frac{1}{10}\ln 2\right)t}$. Solving for y yields

$$y=\frac{10}{1+4\cdot 2^{-t}}.$$

(c) $8=\dfrac{10}{1+4\cdot 2^{-t}}$. means $1+4\cdot 2^{-t}=1.25$, so $t=4$.

(d) $\lim\limits_{t\to\infty}\dfrac{10}{1+4\cdot 2^{-t}}=10$, so the value of y approaches 10.

15. (a) Since $x = \dfrac{1}{2}t$, $x(4) = \dfrac{1}{2}(4) = 2$. Since $y = 18 - 2 \cdot 2^2 = 10$, P is at $(2,10)$.

(b) Since $y = 18 - 2x^2$, $\dfrac{dy}{dt} = -4x\dfrac{dx}{dt}$. Since $x = \dfrac{1}{2}t$, $\dfrac{dx}{dt} = \dfrac{1}{2}$. Therefore

$$\frac{dy}{dt} = -4x\frac{dx}{dt} = -4 \cdot 2 \cdot \frac{1}{2} = -4 \text{ units/sec.}$$

(c) Let D = the object's distance from the origin. Then

$$D^2 = x^2 + y^2, \text{ and at } (2,10)\ D = \sqrt{104}.$$

$$2D\frac{dD}{dt} = 2x\frac{dx}{dt} + 2y\frac{dy}{dt},$$

$$2\sqrt{104}\,\frac{dD}{dt} = 2 \cdot 2 \cdot \frac{1}{2} + 2 \cdot 10(-4),$$

$$\frac{dD}{dt} = \frac{-78}{2\sqrt{104}} = -3.824 \text{ unit/sec.}$$

(d) The object hits the x-axis when $y = 18 - 2x^2 = 0$, or $x = 3$. Since

$x = \dfrac{1}{2}t = 3, t = 6.$

(e) The length of the arc of $y = 18 - 2x^2$ for $0 \le x \le 3$ is given by

$$L = \int \sqrt{1 + \left(\frac{dy}{dx}\right)^2}\, dx = \int_0^3 \sqrt{1 + (-4x)^2}\, dx = 18.460 \text{ units.}$$

16. (a) See graph.

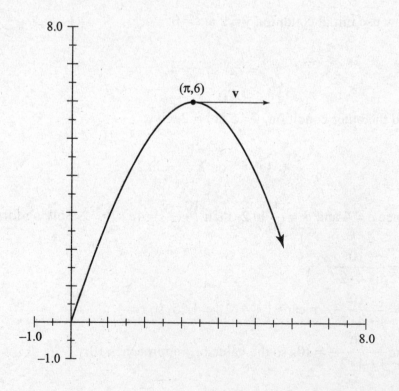

(b)   You want to maximize $y(t) = \dfrac{12t}{t^2 + 1}$.

$$y'(t) = \frac{(t^2 + 1)(12) - 12t(2t)}{(t^2 + 1)^2} = \frac{12(1 - t)(1 + t)}{(t^2 + 1)^2}.$$

See signs analysis.

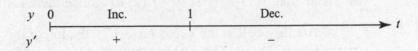

The maximum y occurs when $t = 1$.

(c)   Since $x(1) = 4\arctan 1 = \pi$ and $y(1) = \dfrac{12}{1 + 1} = 6$, the coordinates of the highest point are $(\pi, 6)$.

Since $x'(t) = \dfrac{4}{1 + t^2}$ and $y'(t) = \dfrac{12(1 - t^2)}{(t^2 + 1)^2}$, so $\mathbf{v}(1) = (2, 0)$. This vector is shown on the graph.

(d) $\displaystyle\lim_{t \to \infty} x(t) = \lim_{t \to \infty} 4\arctan t = 4\left(\dfrac{\pi}{2}\right) = 2\pi$, and $\displaystyle\lim_{t \to \infty} y(t) = \lim_{t \to \infty} \dfrac{12t}{t^2 + 1} = 0.$ Thus the particle approaches the point $(2, 0)$.

Part B

17.   (a) f' is defined for all x in the interval. Since f is therefore differentiable, it must also be continuous.

(b)   From the graph we see that $f' = 0$ at $x = -2$, 4, and 7. The sign of f' tells whether f is increasing or decreasing on the relevant intervals. f attains relative minima at $x = -2$ and $x = 9$.

(c)   Similarly, f attains relative maxima at $x = -3$ and $x = 7$.

(d)   Note that $f(7) - f(-3) = \int_{-3}^{7} f'(x)\, dx$. Since there is more area above the

x-axis than below the x-axis on $[-3,7]$, the integral is positive and $f(7) - f(-3) > 0$. This implies that $f(7) > f(-3)$, and that the absolute maximum occurs at $x = 7$.

(e)   Since f' changes from increasing to decreasing (or vice versa) at $x = 2$, 4, and 6, f'' changes sign at each of these points. Therefore, f has points of inflection there.

18.   Draw a sketch of the region bounded above by $y_1 = 8 - 2x^2$ and below by $y_2 = x^2 - 4$, and inscribe a rectangle in this region as described in the question. If (x, y_1) and (x, y_2) are the vertices of the rectangle in quadrants I and IV, respectively, then the area

$$A = 2x\,(y_1 - y_2) = 2x(12 - 3x^2), \quad \text{or} \quad A(x) = 24x - 6x^3.$$

Then $A'(x) = 24 - 18x^2 = 6(4 - 3x^2)$, which equals 0 when $x = \dfrac{2}{\sqrt{3}} = \dfrac{2\sqrt{3}}{3}$. Check to verify that $A''(x) < 0$ at this point. This assures that this value of x yields maximum area, which is given by $\dfrac{4\sqrt{3}}{3} \times 8$.

19. The graph of $f'(x)$ is shown here.

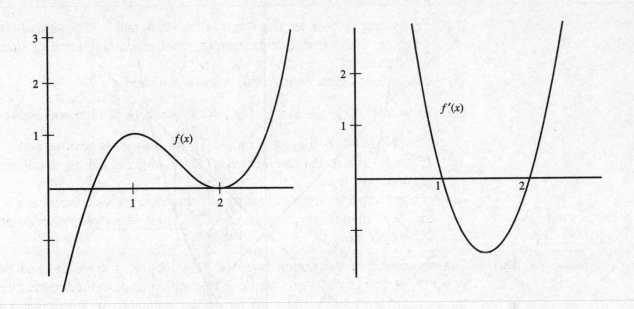

20.   The rate of change in volume when the surface area is 54 ft^3 is $-\dfrac{3}{8}$ ft^3/sec.

21.   See the figure. The equation of the circle is $x^2 + y^2 = a^2$; the equation of RS is $y = a - x$. If y_2 is an ordinate of the circle and y_1 of the line, then

$$\Delta V = \pi y_2{}^2 \Delta x - \pi y_1{}^2 \Delta x,$$

$$V = 2\pi \int_0^a [(a^2 - x^2) - (a - x)^2] dx = \frac{2}{3} \pi a^3.$$

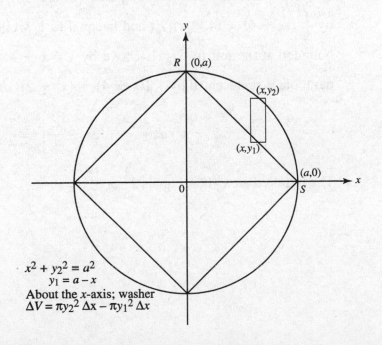

$x^2 + y_2{}^2 = a^2$
$y_1 = a - x$
About the x-axis; washer
$\Delta V = \pi y_2{}^2 \, \Delta x - \pi y_1{}^2 \, \Delta x$

22.   (a)   The region is sketched in the figure. The pertinent points of intersection are labeled.

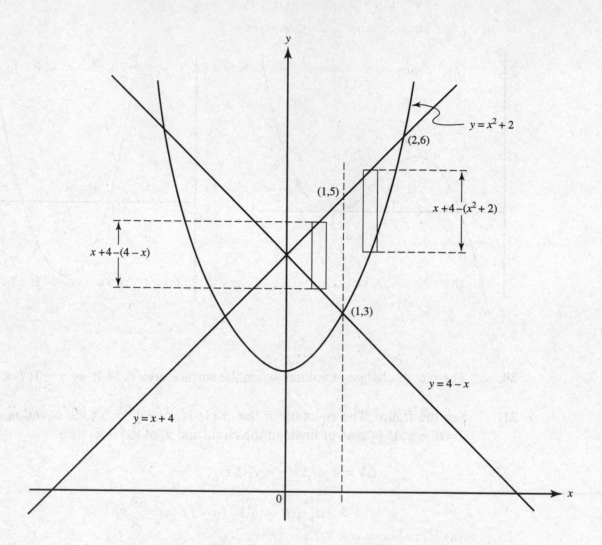

$y = x^2 + 2$

(2,6)

$x + 4 - (x^2 + 2)$

(1,5)

$x + 4 - (4 - x)$

(1,3)

$y = 4 - x$

$y = x + 4$

(b)   The required area consists of two parts. The area of the triangle is represented by $\int_0^1 [(x + 4) - (4 - x)]\, dx$ and is equal to 1, while the area of the region bounded at the left by $x = 1$, above by $y = x + 4$, and at the right by the parabola is represented by $\int_1^2 [(x + 4) - (x^2 + 2)]\, dx$. This equals

$$\int_1^2 (x + 2 - x^2)\, dx = \frac{x^2}{2} + 2x - \frac{x^3}{3}\Big|_1^2 = \frac{7}{6}.$$

The required area, thus, equals $2\frac{1}{6}$ or $\frac{13}{6}$.

23.   (a)   1975 to 1976 and 1978 to 1980.

       (b)   1975 to 1977 and 1979 to 1981.

       (c)   1976 to 1977 and 1980 to 1981.

24.   (a)   Since $a = \dfrac{dv}{dt} = -2v$, then, separating variables, $\dfrac{dv}{v} = -2dt$. Integrating gives

$$\ln v = -2t + C, \tag{1}$$

and, since $v = 20$ when $t = 0$, $C = \ln 20$. Then (1) becomes $\ln \dfrac{v}{20} = -2t$ or, solving for v,

$$v = 20e^{-2t}. \tag{2}$$

       (b)   Note that $v > 0$ for all t. Let s be the required distance traveled (as v decreases from 20 to 5); then

$$s = \int_{v=20}^{v=5} 20e^{-2t}\,dt = \int_{t=20}^{\ln 2} 20e^{-2t}\,dt, \tag{3}$$

where, when $v = 20$, $t = 0$. Also, when $v = 5$, use (2) to get $\dfrac{1}{4} = e^{-2t}$ or $-\ln 4 = -2t$. So $t = \ln 2$. Evaluating s in (3) gives

$$s = -10e^{-2t}\Big|_0^{\ln 2} = -10\left(\frac{1}{4} - 1\right) = \frac{15}{2}.$$

25.   (a)   For $f(x) = \cos x, f'(x) = -\sin x, f''(x) = -\cos x, f'''(x) = \sin x, f^{(4)}(x) = \cos x,$ $f^{(5)}(x) = -\sin x, f^{(6)}(x) = -\cos x$. The Taylor polynomial of order 4 about 0 is

$$\cos x = 1 - \frac{x^2}{2!} + \frac{x^4}{4!}.$$

Note that the next term of the alternating Maclaurin series for $\cos x$ is $-\dfrac{x^6}{6!}$.

       (b) $\displaystyle\int_0^1 \cos x\,dx = x - \frac{x^3}{3\cdot 2!} + \frac{x^5}{5\cdot 4!}\bigg|_0^1 = 1 - \frac{1}{6} + \frac{1}{120} \approx 0.84167.$

       (c)   The error in (b), a convergent alternating series, is less absolutely than the first term dropped:

$$\int_0^1 \frac{x^6}{6!}\,dx = \frac{x^7}{7!}\bigg|_0^1 < 0.00020.$$

26.   (a)   Let $f(x) = \ln(1 + x)$. Then $f'(x) = \dfrac{1}{1 + x}$, $f''(x) = -\dfrac{1}{(1 + x)^2}$,

$f'''(x) = \dfrac{2}{(1 + x)^3}$, $f^{(4)}(x) = -\dfrac{3!}{(1 + x)^4}$, $f^{(5)}(x) = \dfrac{4!}{(1 + x)^5}$. At $x = 0$,

$f(0) = 0, f'(0) = 1, f''(0) = -1, f'''(0) = 2, f^{(4)}(0) = -(3!)$, and $f^{(5)}(0) = 4!$. So

$$\ln(1 + x) = x - \frac{x^2}{2} + \frac{x^3}{3} - \frac{x^4}{4} + \frac{x^5}{5} - \cdots.$$

(b)   Using the ratio test, you know that the series converges when

$\lim\limits_{n \to \infty} \left| \dfrac{x^{n+1}}{n + 1} \cdot \dfrac{n}{x^n} \right| < 1$, that is, when $|x| < 1$, or $-1 < x < 1$. Thus, the

radius of convergence is 1.

(c) $\ln(1.2) = 0.2 - \dfrac{(0.2)^2}{2} + \dfrac{(0.2)^3}{3} - \dfrac{(0.2)^4}{4} + \dfrac{(0.2)^5}{5} = 0.18233.$

(d)   Since the series is convergent and alternating, the error in the answer for (c)

is less absolutely than $\dfrac{(0.2)^6}{6} \approx 0.00001.$

27.   From the equations for x and y,

$$dx = (1 - \cos \theta)\, d\theta \quad \text{and} \quad dy = \sin \theta\, d\theta.$$

(a)   The slope at any point is given by $\dfrac{dy}{dx}$, which here is $\dfrac{\sin \theta}{1 - \cos \theta}$. When

$\theta = \dfrac{2\pi}{3}$, the slope is $\dfrac{\sqrt{3}}{3}$.

(b)   When $\theta = \dfrac{2\pi}{3}$, $x = \dfrac{2\pi}{3} - \dfrac{\sqrt{3}}{2}$ and $y = 1 - \left(-\dfrac{1}{2}\right) = \dfrac{3}{2}$. The equation of the tangent is

$$9y - 3\sqrt{3} \cdot x = 18 - 2\pi\sqrt{3}.$$

28.   **(b)**   Both curves are circles with centers at, respectively, $(2, 0)$ and at $\left(2, \dfrac{\pi}{2}\right)$; the circles intersect at $\left(2\sqrt{2}, \dfrac{\pi}{4}\right)$. The common area is given by

$$2\int_0^{\pi/4} (4\sin\theta)^2 \, d\theta \quad \text{or} \quad 2\int_{\pi/4}^{\pi/2} (4\cos\theta)^2 \, d\theta.$$

The answer is $2(\pi - 2)$.

29.   **(a)**   If $n > 2$, $n! > 2^{n-1}$; $\dfrac{1}{n!} < \dfrac{1}{2^{n-1}}$; also $\displaystyle\sum_1^\infty \dfrac{1}{2^{n-1}}$ is a geometric series, with ratio r equal to $\dfrac{1}{2}$. When the first 11 terms of $\displaystyle\sum_1^\infty \dfrac{1}{n!}$ are used for $\displaystyle\sum_1^\infty \dfrac{1}{n!}$ the error is $\displaystyle\sum_{12}^\infty \dfrac{1}{n!}$, which is less than $\displaystyle\sum_{12}^\infty \dfrac{1}{2^{n-1}}$. Thus

$$T\sum_{12}^\infty \dfrac{1}{n!} < \sum_{12}^\infty \dfrac{1}{2^{n-1}} = \dfrac{1}{2^{11}} \cdot \dfrac{1}{1-\dfrac{1}{2}} = \dfrac{1}{2^{10}} < 0.001.$$

(See page 345 for the sum of a geometric series. Here $a = \dfrac{1}{2^{11}}$ and $r = \dfrac{1}{2}$.)

(b)   Use the Integral Test:

$$\int_1^\infty xe^{-x}\,dx = \lim_{b\to\infty} -e^{-x}(1+x)\Big|_1^b = \lim_{b\to\infty}\left(\dfrac{1+b}{e^b} - \dfrac{2}{e}\right).$$

Apply L'Hôpital's rule to the first fraction within the parentheses; then the integral equals $\dfrac{2}{e}$. Since the improper integral converges, so does the given series.

AB PRACTICE
EXAMINATIONS

AB Practice Examination 1

Section I Multiple-Choice Questions _____

Part A

The use of calculators is *not* permitted for this part of the examination.

There are 28 questions in Part A, for which 55 minutes are allowed. To compensate for possible guessing, the grade on this part is determined by subtracting one-fourth of the number of wrong answers from the number answered correctly.

Choose the best answer for each question.

(Answers are given on page 465).

1. $\lim\limits_{x \to \infty} \dfrac{3x^2 - 4}{2 - 7x - x^2}$ is

    (A) 3   **(B)** 1   **(C)** -3 **(D)** ∞ **(E)** 0

2. $\lim\limits_{h \to 0} \dfrac{\cos\left(\dfrac{\pi}{2} + h\right)}{h}$ is

    (A) 1   **(B)** nonexistent   **(C)** 0   **(D)** -1 **(E)** none of these

3. If, for all x, $f'(x) = (x - 2)^4(x - 1)^3$, it follows that the function f has
    (A) a relative minimum at $x = 1$
    (B) a relative maximum at $x = 1$
    (C) both a relative minimum at $x = 1$ and a relative maximum at $x = 2$
    (D) neither a relative maximum nor a relative minimum
    (E) relative minima at $x = 1$ and at $x = 2$

4. Let $(Fx) = \displaystyle\int_0^x \dfrac{10}{1 + e^t}\, dt$. Which of the following statements is (are) true?
    I. $F'(0) = 5$.   **II.** $F(2) < F(6)$.   **III.** F is concave upward.

    (A) I only   **(B)** II only   **(C)** III only
    (D) I and II   **(E)** I and III

5. If $f(x) = 10^x$ and $10^{1.04} \approx 10.96$, which is closest to $f'(1)$?

    (A) 0.24   **(B)** 0.92   **(C)** 0.96   **(D)** 10.5   **(E)** 24

6. If f is differentiable, we can use the line tangent to f at $x = a$ to approximate values of f near $x = a$. Suppose this method always underestimates the correct values. If so, then at $x = a$, f must be

    (A) positive   **(B)** increasing   **(C)** decreasing
    (D) concave upward   **(E)** concave downward

7. If $f(x) = \cos x \sin 3x$, then $f'\left(\dfrac{\pi}{6}\right)$ is equal to

    (A) $\dfrac{1}{2}$ **(B)** $-\dfrac{\sqrt{3}}{2}$ **(C)** 0   **(D)** 1   **(E)** $-\dfrac{1}{2}$

8. $\displaystyle\int_0^1 \dfrac{x\,dx}{x^2 + 1}$ is equal to

    (A) $\dfrac{\pi}{4}$ **(B)** $\ln\sqrt{2}$ **(C)** $\dfrac{1}{2}(\ln 2 - 1)$ **(D)** $\dfrac{3}{2}$ **(E)** $\ln 2$

9. The graph of f'' is shown below. If $f'(1) = 0$, then $f'(x) = 0$ at $x =$

    (A) 0   **(B)** 2   **(C)** 3   **(D)** 4   **(E)** 7

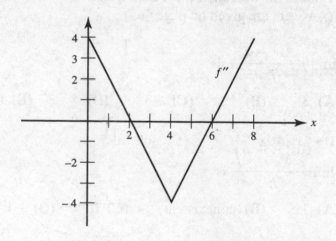

Use the table, which shows the values of differentiable functions f and g, for Questions 10 and 11.

x	f	f'	g	g'
1	2	$\dfrac{1}{2}$	-3	5
2	3	1	0	4
3	4	2	2	3
4	6	4	3	$\dfrac{1}{2}$

10. If $P(x) = g^2(x)$, then $P'(3)$ equals

    (A) 4   (B) 6   (C) 9   (D) 12   (E) 18

11. If $H(x) = f^{-1}(x)$, then $H'(3)$ equals

    (A) $-\dfrac{1}{16}$ (B) $-\dfrac{1}{8}$ (C) $-\dfrac{1}{2}$ (D) $\dfrac{1}{2}$ (E) 1

12. The total area of the region bounded by the graph of $y = x\sqrt{1 - x^2}$ and the x-axis is

    (A) $\dfrac{1}{3}$ (B) $\dfrac{1}{3}\sqrt{2}$ (C) $\dfrac{1}{2}$ (D) $\dfrac{2}{3}$ (E) 1

13. The curve of $y = \dfrac{1 - x}{x - 3}$ is concave upward when

    (A) $x > 3$ (B) $1 < x < 3$ (C) $x > 1$ (D) $x < 1$ (E) $x < 3$

14. The area of the largest isosceles triangle that can be drawn with one vertex at the origin and with the others on a line parallel to and above the x-axis and on the curve $y = 27 - x^2$ is

    (A) $12\sqrt{3}$ (B) 27   (C) $24\sqrt{3}$ (D) 54   (E) 108

15. The average (mean) value of $\tan x$ on the interval from $x = 0$ to $x = \dfrac{\pi}{3}$ is

    (A) $\ln \dfrac{1}{2}$ (B) $\dfrac{3}{\pi}\ln 2$ (C) $\ln 2$ (D) $\dfrac{\sqrt{3}}{2}$ (E) $\dfrac{9}{\pi}$

16. $\displaystyle\int \sin(x^2)\,dx =$

    (A) $-\cos(x^2) + C$ (B) $\cos(x^2) + C$ (C) $-\dfrac{\cos x^2}{2x} + C$

    (D) $2x \cos x^2 + C$ (E) none of these

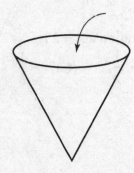

17. Water is poured at a constant rate into the conical reservoir shown in the figure. If the depth of the water is graphed as a function of time, the graph is
    (A) decreasing   (B) constant
    (C) linear   (D) concave upward
    (E) concave downward

18. If $f(x) = \begin{cases} x^2 & \text{for } x \le 1 \\ 2x - 1 & \text{for } x > 1 \end{cases}$, then

(A) $f(x)$ is not continuous at $x = 1$
(B) $f(x)$ is continuous at $x = 1$ but $f'(1)$ does not exist
(C) $f'(1)$ exists and equals 1
(D) $f'(1) = 2$
(E) $\lim\limits_{x \to 1} f(x)$ does not exist

19. $\lim\limits_{x \to 2^-} \dfrac{|x - 2|}{x - 2}$ is

(A) $-\infty$ (B) -1 (C) 1 (D) ∞ (E) nonexistent

The graph shown is for Questions 20 and 21. It consists of a quarter-circle and two line segments, and represents the velocity of an object during the 6-second interval.

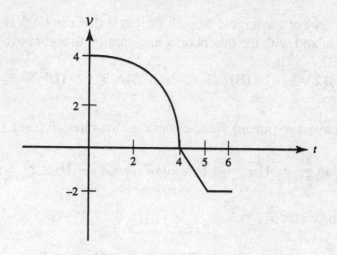

20. The object's average speed during the 6-second interval is

(A) $\dfrac{4\pi + 3}{6}$ (B) $\dfrac{4\pi - 3}{6}$ (C) -1 (D) $-\dfrac{1}{3}$ (E) 1

21. The object's acceleration at $t = 2$ is

(A) -1 (B) $-\dfrac{1}{2}$ (C) $-\dfrac{1}{3}$ (D) $-\dfrac{1}{\sqrt{3}}$ (E) $-\sqrt{3}$

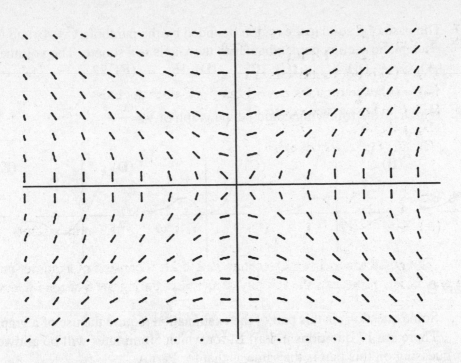

†22. Which of the following equations can be a solution of the differential equation whose slope field is shown above?

(A) $2xy = 1$ (B) $2x + y = 1$ (C) $2x^2 + y^2 = 1$ (D) $2x^2 - y^2 = 1$
(E) $y = 2x^2 + 1$

23. If y is a differentiable function of x, then the slope of the curve of $xy^2 - 2y + 4y^3 = 6$ at the point where $y = 1$ is

(A) $-\dfrac{1}{18}$ (B) $-\dfrac{1}{26}$ (C) $\dfrac{5}{18}$ (D) $-\dfrac{11}{18}$ (E) 2

24. In the following, $L(n)$, $R(n)$, $M(n)$, and $T(n)$ denote, respectively, left, right, midpoint, and trapezoidal sums with n subdivisions. Which of the following is equal exactly to

$$\int_{-1}^{1} |x|\, dx?$$

(A) $L(1)$ (B) $L(2)$ (C) $R(1)$ (D) $M(1)$ (E) none of these

25. $\displaystyle\int_{a}^{x} g(t)\, dt - \int_{b}^{x} g(t)\, dt$ is equal to the constant

(A) 0 (B) $b - a$ (C) $a - b$ (D) $\displaystyle\int_{a}^{b} g(t)\, dt$ (E) $g(b) - g(a)$

26. The solution of the differential equation $\dfrac{dy}{dx} = 2xy^2$ for which $y = -1$ when $x = 1$ is

(A) $y = -\dfrac{1}{x^2}$ (B) $\ln y^2 = x^2 - 1$ (C) $\dfrac{y^3}{3} = x^2 - \dfrac{4}{3}$

(D) $y = -\dfrac{1}{x}$ (E) none of these

†The topic slope fields will not be tested on the AB Examination before 2004. See p. x.

27. The base of a solid is the region bounded by the parabola $y^2 = 4x$ and the line $x = 2$. Each plane section perpendicular to the x-axis is a square. The volume of the solid is
(A) 6   (B) 8   (C) 10   (D) 16   (E) 32

28. Which of the following could be the graph of $y = \dfrac{x^2}{e^x}$?

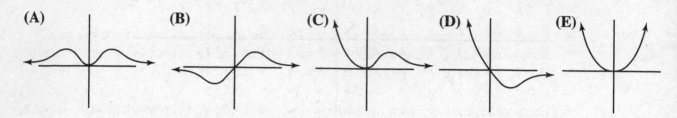

(A)   (B)   (C)   (D)   (E)

Part B

Some questions in this part of the examination require the use of a graphing calculator.
There are 17 questions in Part B, for which 50 minutes will be allowed. The penalty for guessing on this part is the same as that for Part A.
If the exact numerical value of the correct answer is not listed as a choice, select the choice that is closest to the exact numerical answer.
Choose the best answer for each question.
(Answers are given on page 469.)

29. If $f(3) = 8$ and $f'(3) = -4$ then $f(3.02)$ is approximately
(A) -8.08 (B) 7.92   (C) 7.98   (D) 8.02   (E) 8.08

30. An object moving along a line has velocity $v(t) = t \cos t - \ln(t + 2)$, where $0 \leqslant t \leqslant 10$. How many times does the object reverse direction?
(A) none   (B) one   (C) two   (D) three   (E) four

Use the graph of f' for Questions 31 and 32.

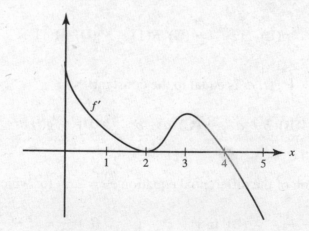

31. f has a local minimum at $x =$
(A) 0 only   (B) 4 only   (C) 0 and 4   (D) 0 and 5   (E) 0, 4, and 5

32. f has a point of inflection at $x =$
(A) 2 only   (B) 3 only   (C) 4 only   (D) 2 and 3 only
(E) 2, 3, and 4

33. For what value of c on $[0,1]$ is the tangent to the graph of $f(x) = e^x - x^2$ parallel to the secant line?

(A) -0.248 (B) 0.351 (C) 0.500 (D) 0.693 (E) 0.718

34. Find the volume of the solid generated when the region bounded by the y-axis, $y = e^x$, and $y = 2$ is rotated around the y-axis.

(A) 0.296 (B) 0.592 (C) 2.427 (D) 3.998 (E) 27.577

35. A bank account earns interest compounded continuously at a rate of 6% per year. How many years, approximately, will be required for the account to triple in value?

(A) 3   (B) 9   (C) 18   (D) 33   (E) 50

36. The acceleration of a particle moving along a straight line is given by $a = 6t$. If, when $t = 0$, its velocity, v, is 1 and its position, s, is 3, then at any time t

(A) $s = t^3 + 3$ (B) $s = t^3 + 3t + 1$ (C) $s = t^3 + t + 3$

(D) $s = \dfrac{t^3}{3} + t + 3$ (E) $s = \dfrac{t^3}{3} + \dfrac{t^2}{2} + 3$

37. If $y = f(x^2)$ and $f'(x) = \sqrt{5x - 1}$ then $\dfrac{dy}{dx}$ is equal to

(A) $2x\sqrt{5x^2 - 1}$ (B) $\sqrt{5x - 1}$ (C) $2x\sqrt{5x - 1}$ (D) $\dfrac{\sqrt{5x - 1}}{2x}$

(E) none of these

38. If the area under $y = \sin x$ is equal to the area under $y = x^2$ between $x = 0$ and $x = k$, then $k =$

(A) -1.105 (B) 0.877 (C) 1.105 (D) 1.300 (E) 1.571

39. If the substitution $x = 2t + 1$ is used, which of the following is equivalent to $\displaystyle\int_0^3 \sqrt[4]{2t + 1}\ dt$?

(A) $\displaystyle\int_0^3 \sqrt[4]{x}\,dx$ (B) $\dfrac{1}{2}\displaystyle\int_0^3 \sqrt[4]{x}\ dx$ (C) $\dfrac{1}{2}\displaystyle\int_{-\frac{1}{2}}^1 \sqrt[4]{x}\ dx$ (D) $\displaystyle\int_1^7 \dfrac{1}{2}\sqrt[4]{x}\ dx$

(E) $2\displaystyle\int_1^7 \sqrt[4]{x}\ dx$

40. At noon, an experimenter has 50 grams of a radioactive isotope. At noon 9 days later only 45 grams remain. To the nearest day, how many days after the experiment started will there be only 20 grams?

(A) 54   (B) 59   (C) 60   (D) 75   (E) 78

41. A 26-foot ladder leans against a building so that its foot moves away from the building at the rate of 3 feet per second. When the foot of the ladder is 10 feet from the building, the top is moving down at the rate of r feet per second, where r is

(A) $\dfrac{46}{3}$ (B) $\dfrac{3}{4}$ (C) $\dfrac{5}{4}$ (D) $\dfrac{5}{2}$ (E) $\dfrac{4}{5}$

42. If $F(x) = \displaystyle\int_{1}^{2x} \dfrac{1}{1 - t^3}\, dt$, then $F'(x) =$

(A) $\dfrac{1}{1 - x^3}$ (B) $\dfrac{1}{1 - 2x^3}$ (C) $\dfrac{2}{1 - 2x^3}$ (D) $\dfrac{1}{1 - 8x^3}$ (E) $\dfrac{2}{1 - 8x^3}$

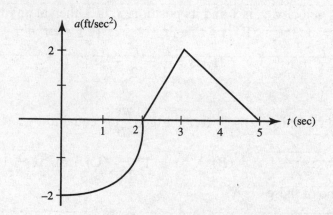

43. The graph shows an object's acceleration (in ft/sec^2). It consists of a quarter-circle and two line segments. If the object was at rest at $t = 5$ seconds, what was its initial velocity?

(A) -2 ft/sec   (B) $3 - \pi$ ft/sec   (C) 0 ft/sec   (D) $\pi - 3$ ft/sec
(E) $\pi + 3$ ft/sec

44. Water is leaking from a tank at the rate of $R(t) = 5 \arctan\left(\dfrac{t}{5}\right)$ gallons per hour, where t is the number of hours since the leak began. How many gallons will leak out during the first day?

(A) 7   (B) 82   (C) 124   (D) 141   (E) 164

45. Find the y-intercept of the line tangent to $y = (x^3 - 4x^2 + 8)e^{\cos x^2}$ at $x = 2$.

(A) -21.032 (B) -2.081 (C) 0 (D) 4.161 (E) 21.746

Section II:Free-Response Problems _____

Part A

A graphing calculator is required for some of these problems. See instructions on p. xi. Answers begin on p. 473.

1. Let h be a function that is even and continuous on the closed interval $[-4,4]$. The function h and its derivatives have the properties indicated in the table below. Use this information to sketch a possible graph of h on $[-4,4]$.

x	$h(x)$	$h'(x)$	$h''(x)$
0	$-$	0	$+$
$0 < x < 1$	$-$	$+$	$+$
1	0	$+$	0
$1 < x < 2$	$+$	$+$	$-$
2	$+$	0	0
$2 < x < 3$	$+$	$+$	$+$
3	$+$	undefined	undefined
$3 < x < 4$	$+$	$-$	$-$

2. An object in motion along the x-axis has velocity $v(t) = (t + e^t)\sin t^2$ for $1 \le t \le 3$.
   (a) Sketch the graph of velocity as a function of time in the window $[1,3] \times [-15,20]$.
   (b) When is the object moving to the left?
   (c) Give one value of t from the interval in part (b) at which the speed of the object is increasing.
   (d) At $t = 1$ this object's position was $x = 10$. Where is the object when $t = 3$?

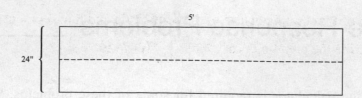

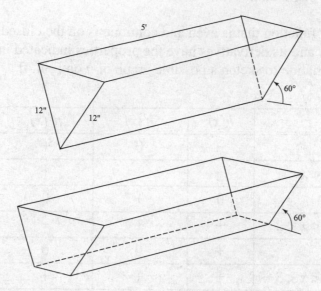

3. The sides of a watering trough are made by folding a sheet of metal 24 inches wide and 5 feet long at an angle of 60°, as shown in the figure. Ends are added, and then the trough is filled with water.

  (a) If water pours into the trough at the rate of 600 cubic inches per minute, how fast is the water level rising when the water is 4 inches deep?

  (b) Suppose, instead, the sheet of metal is folded twice, keeping the sides of equal height and inclined at an angle of 60°, as shown. Where should the folds be in order to maximize the volume of the trough? Justify your answer.

Part B

  No calculator is allowed for any of these problems.
  See instructions on p. xi.

4. Let C represent the curve determined by $f(x) = \dfrac{6}{\sqrt[3]{2x+5}}$ for $-2 \le x \le 11$.

  (a) Let R represent the region between C and the x-axis. Find the area of R.

  (b) Set up, but do not solve, an equation to find the value of k such that the line $x = k$ divides R into two regions of equal area.

  (c) Find the volume of the solid generated when C is rotated around the x-axis.

5. Let $y = f(x)$ be the function that has an x-intercept at $(2,0)$ and satisfies the differential equation $x^2 e^y \dfrac{dy}{dx} = 4$.

  (a) Solve the differential equation, expressing y as a function of x.

  (b) Find the domain of the function.

  (c) Find the equation of any horizontal asymptote.

6. The graph of function f consists of the semicircle and line segment shown in the figure. Define the area function $A(x) = \int_0^x f(t)\, dt$ for $0 \le x \le 18$.

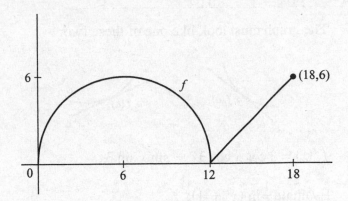

(a) Find $A(6)$ and $A(18)$.
(b) What is the average value of f on the given interval?
(c) Write the equation of the line tangent to the graph of A at $x = 6$.
(d) Use this line to estimate the area between f and the x-axis on $[0,7]$.
(e) Give the coordinates of any points of inflection on the graph of A. Justify your answer.

Answers to AB Practice Examination 1: Section I

1. C	10. D	19. B	28. C	37. A
2. D	11. E	20. A	29. B	38. D
3. A	12. D	21. D	30. C	39. D
4. D	13. E	22. D	31. D	40. E
5. E	14. D	23. A	32. D	41. C
6. D	15. B	24. B	33. B	42. E
7. E	16. E	25. D	34. B	43. D
8. B	17. E	26. A	35. C	44. C
9. C	18. D	27. E	36. C	45. D

Part A

1.   C.   Use the Rational Function Theorem (pages 30–31).

2.   D. $\lim\limits_{h \to 0} \dfrac{\cos\left(\dfrac{\pi}{2} + h\right)}{h} = \lim\limits_{h \to 0} \dfrac{-\sin h}{h} = -1.$

3.   A.   Although $f'(2) = f'(1) = 0$, $f'(x)$ changes sign only as x increases through 1, and in this case $f'(x)$ changes from negative to positive.

4.   D. $F'(x) = \dfrac{10}{1 + e^x} > 0$, and $F''(x) = \dfrac{-10\,e^x}{(1 + e^x)} < 0$.

5.   E. $\dfrac{f(1.04) - f(1)}{1.04 - 1} = \dfrac{0.96}{0.04}$.

6.   D.   The graph must look like one of these two:

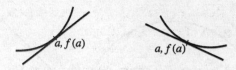

7.   E. $f'(x) = 3 \cos x \cos 3x - \sin x \sin 3x$.

8.   B.   Evaluate $\dfrac{1}{2} \ln (x^2 + 1)\Big|_0^1$.

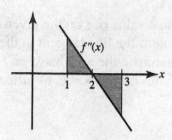

9.   C.   Let $f'(x) = \displaystyle\int_1^x f''(t)\,dt$. Then f' increases for $1 < x < 2$, then begins to decrease. In the figure the area below the x-axis, from 2 to 3, is equal in magnitude to that above the x-axis.

10.   D. $P'(x) = 2g(x) \cdot g'(x)$.

11.   E.   Note that $H(3) = f^{-1}(3) = 2$. Therefore

$$H'(3) = \frac{1}{f'(H(3))} = \frac{1}{f'(2)} = 1.$$

12.   D.   Note that the domain of y is all x such that $|x| \leqq 1$ and that the graph is symmetric to the origin. The area is given by

$$2\int_0^1 x\sqrt{1 - x^2}\,dx.$$

13.   E.   Since

$$y' = 2(x - 3)^{-2} \text{ and } y'' = -4(x - 3)^{-3} = \frac{-4}{(x - 3)^3},$$

y'' is positive when $x < 3$.

14.   D.   Draw a figure and let (x, y) be the point in the first quadrant where the line parallel to the x-axis meets the parabola. The area of the triangle is given by $A = xy = x(27-x^2)$. Show that A is a maximum for $x = 3$. The maximum area is therefore 54.

15.   B. $\dfrac{1}{\pi/3} \displaystyle\int_0^{\pi/3} \tan x \, dx = \dfrac{3}{\pi}\left[-\ln \cos x\right]_0^{\pi/3} = \dfrac{3}{\pi}\left(-\ln \dfrac{1}{2}\right).$

16.   E. $[\cos (x^2)]' = -\sin (x^2) \cdot 2x$. The missing factor $2x$ cannot be created through introduction of constants alone.

17.   E.   As the water gets deeper, the rate of change of depth decreases: $\dfrac{d^2h}{dt^2} < 0$.

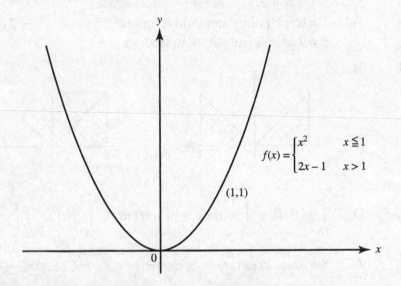

$$f(x) = \begin{cases} x^2 & x \le 1 \\ 2x - 1 & x > 1 \end{cases}$$

(1,1)

18.   D.   The graph of f is shown in the figure; f is defined and continuous at all x, including $x = 1$. Since

$$\lim_{x \to 1^-} f'(x) = 2 = \lim_{x \to 1^+} f'(x),$$

$f'(1)$ exists and is equal to 2.

19.   B.   Since $|x - 2| = 2 - x$ if $x < 2$, the limit as $x \to 2^-$ is $\dfrac{2-x}{x-2} = -1$.

20.   A.   Average speed $= \dfrac{\text{distance covered in 6 sec}}{\text{time elapsed}}$

$$= \dfrac{\frac{1}{4}\pi(4^2) + \frac{1}{2}(1 \cdot 2) + 1 \cdot 2}{6}.$$

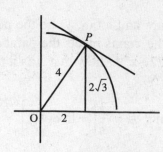

21.   **D.**   Acceleration is the slope of the velocity curve; since the slope of the radius $OP = \sqrt{3}$, the slope of the tangent is $-\dfrac{1}{\sqrt{3}}$.

22.   **D.**   Particular solutions appear to be branches of hyperbolas. See page 459.

23.   **A.**   Differentiating implicitly yields $2xyy' + y^2 - 2y' + 12y^2y' = 0$. When $y = 1$, $x = 4$. Substitute to find y'.

24.   **B.**

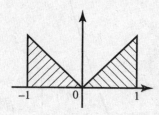   =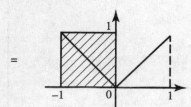

25.   **D.** $\displaystyle\int_a^x g(t)\,dt - \int_b^x g(t)\,dt = \int_a^x g(t)\,dt + \int_x^b g(t)\,dt = \int_a^b g(t)\,dt.$

26.   **A.**   Separate to get $\dfrac{dy}{y^2} = 2x\,dx$, $-\dfrac{1}{y} = x^2 + C$. Since $-(-1) = 1 + C$ implies that $C = 0$, the solution is

$$-\frac{1}{y} = x^2 \qquad \text{or} \qquad y = -\frac{1}{x^2}.$$

27.   **E.**

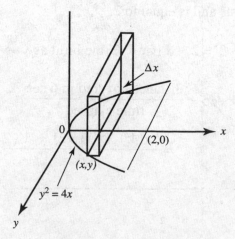

$\Delta V = (2y)^2\,\Delta x;$

$V = 4\displaystyle\int_0^2 y^2\,dx$

$\quad = 4\displaystyle\int_0^2 4x\,dx$

$\quad = 32.$

28.   **C.**   Note that $\displaystyle\lim_{x \to \infty} \frac{x^2}{e^x} = 0$, $\displaystyle\lim_{x \to -\infty} \frac{x^2}{e^x} = \infty$, and $\dfrac{x^2}{e^x} \geqslant 0$ for all x.

Part B

29.   B.   At $x = 3$, the equation of the tangent line is $y - 8 = -4(x - 3)$; so $f(x) \simeq -4(x - 3) + 8. \, f(3.02) \simeq -4(0.02) + 8$.

30.   C.

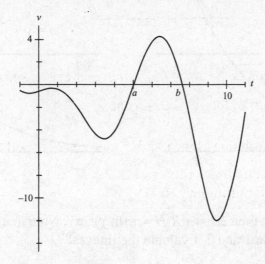

The velocity is graphed above in $[-1, 11] \times [-15, 5]$. The object reverses direction when the velocity changes sign, that is, when the graph crosses the x-axis. There are two such reversals—at $x = a$ and at $x = b$.

31.   D.   The sign diagram

$$
\begin{array}{ccccccc}
f & & \text{inc} & & \text{inc} & & \text{dec} \\
\vdash & & \dashv & & \vdash & & \dashv \\
& 0 & & 2 & & 4 & & 5 \\
f' & & + & & + & & - \\
\end{array}
$$

shows that f has a maximum at $x = 4$ and a minimum at the endpoints.

32.   D.   Since f' decreases, increases, then decreases, f'' changes from negative to positive, then back to negative. These sign changes occur at $x = 2$ and $x = 3$.

33.   B.

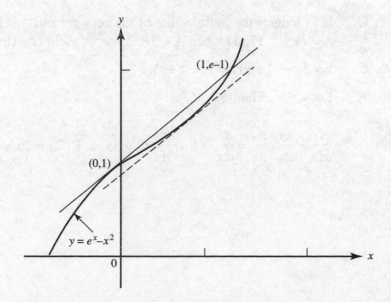

On the curve of $f(x) = e^x - x^2$, the two points labeled are $(0,1)$ and $(1, e - 1)$. The slope of the secant line is $m = \dfrac{\Delta y}{\Delta x} = \dfrac{e - 2}{1} - e - 2$. Find c in $[0,1]$ such that $f'(c) = e - 2$, or $f'(c) - (e - 2) = 0$. Since $f'(x) = e^x - 2x$, c can be calculated by solving $0 = e^x - 2x - (e - 2)$. The answer is 0.351.

34. **B.**

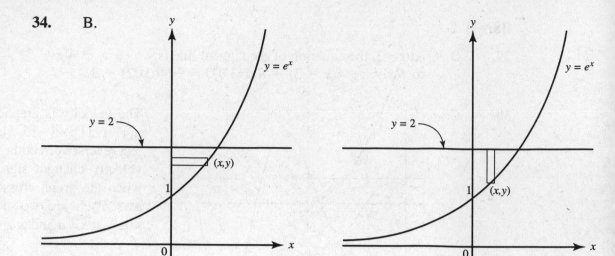

Use disks; then $\Delta V = \pi R^2 H = \pi(\ln y)^2 \, \Delta y$. Note that the limits of the definite integral are 1 and 2. Evaluate the integral

$$\pi \int_{1}^{2} (\ln x)^2 \, dx.$$

(Use shells; then $\Delta V = 2\pi RHT = 2\pi x(2 - e^x) \, \Delta x$. Here, the upper limit of integration is the value of x for which $e^x = 2$, namely, ln 2. Now evaluate

$$2\pi \int_{0}^{\ln 2} x(2 - e^x) \, dx.$$

Again, the answer is 0.592.

35. **C.** If P denotes the initial value of the account and A is the value at time t, then $A = Pe^{0.06t}$. Find t when $A = 3P$: $3 = e^{0.06t}$ yields $t = (\ln 3)/0.06 \approx 18$ yr.

36. **C.** $v = 3t^2 + 1$ and $s = t^3 + t + 3$.

37. **A.** Let $u = x^2$. Then

$$\frac{dy}{dx} = \frac{dy}{du} \cdot \frac{du}{dx} = \frac{df}{du} \cdot f'(u)\frac{du}{dx} = \sqrt{5u - 1} \cdot 2x = 2x\sqrt{5x^2 - 1}.$$

38.   **D.**

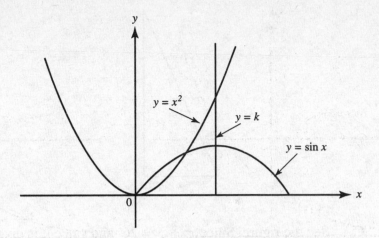

You want $\displaystyle\int_0^k \sin x\, dx$ to equal $\displaystyle\int_0^k x^2\, dx$. You have then

$$-\cos x \Big|_0^k = \frac{x^3}{3} \Big|_0^k,$$

$$-\cos k - (-\cos 0) = \frac{k^3}{3} - 0,$$

$$-\cos k + 1 = \frac{k^3}{3},$$

$$0 = \frac{k^3}{3} + \cos k - 1.$$

A calculator yields $k = 1.300$.

39.   **D.**   If $x = 2t + 1$, then $t = \dfrac{x-1}{2}$, so $dt = \dfrac{1}{2}\, dx$. When $t = 0$, $x = 1$; when $t = 3$, $x = 7$.

40.   **E.**   Use $A(t) = A_0 e^{kt}$, where the initial amount A_0 is 50. Then $A(t) = 50e^{kt}$. Since 45 g remain after 9 days, $45 = 50e^{k \cdot 9}$, which yields $k = \dfrac{\ln 0.9}{9}$.

To find t when 20 g remain, solve

$$20 = 50e^{\left(\frac{\ln 09}{9}\right) \cdot t}.$$

Thus,

$$t = \frac{9 \ln 0.4}{\ln 0.9} = 78.3.$$

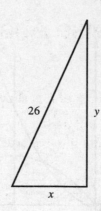

41. C. See the figure. Since $x^2 + y^2 = 26^2$ and since it is given that $\dfrac{dx}{dt} = 3$, it follows that

$$2x\frac{dx}{dt} + 2y\frac{dy}{dt} = 0 \quad \text{and} \quad \frac{dy}{dt} = -\frac{x}{y} \ (3)$$

at any time t. When $x = 10$, then $y = 24$ and $\dfrac{dy}{dt} = \dfrac{-5}{4}$.

42. E. Let $u = 2x$ and note that $F'(u) = \dfrac{1}{1 - u^3}$. Then

$$F'(x) = F'(u)u'(x) = 2F'(u) = 2 \cdot \frac{1}{1 - (2x)^3}.$$

43. D. $v(5) - v(0) = \displaystyle\int_0^5 a(t)\,dt = -\frac{1}{4}\pi \cdot 2^2 + \frac{1}{2}(3)(2) = -\pi + 3$. Since $v(5) = 0$, $-v(0) = -\pi + 3$; so $v(0) = \pi - 3$.

44. C. $\displaystyle\int_0^{24} 5\arctan\left(\frac{t}{5}\right)dt = 124.102$

45. D. Let $y = (x^3 - 4x^2 + 8)e^{\cos(x^2)}$. The equation of the tangent at point $(2, y(2))$ is $y - y(2) = y'(2)(x - 2)$. Note that $y(2) = 0$. To find the y-intercept, let $x = 0$ and solve for y: $y = -2y'(2)$. A calculator yields $y = 4.161$.

Section II:Free-Response Answers _____

Part A

AB 1.   One possible graph of *h* is shown; it has the following properties:

- continuity on [–4,4],
- symmetry about the *y*-axis,
- roots at $x = -1, 1$,
- horizontal tangents at $x = -2, 0, 2$,
- points of inflection at $x = -3, -2, -1, 1, 2, 3$,
- corners at $x = -3, 3$.

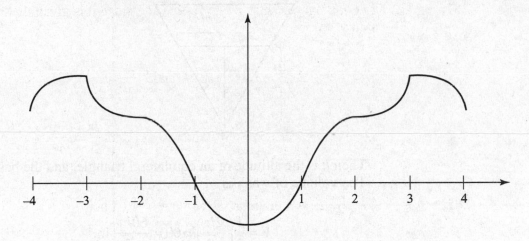

AB/BC 2. (a) Let $Y_1 = (X + e^x) \sin (X^2)$ and graph Y_1 in $[1, 3] \times [-15, 20]$. Note that Y_1 represents velocity *v* and X represents time *t*.

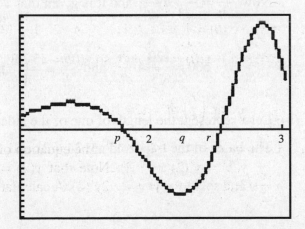

(b) The object moves to the left when the velocity is negative, namely, on the interval $p < t < r$. Use the calculator to solve; then $p = 1.772$ and $r = 2.507$. The answer is $1.772 < t < 2.507$.

(c) As the object moves to the left (with $v(t)$ negative), the speed of the object increases when its acceleration $v'(t)$ is also negative, that is, when $v(t)$ is decreasing. This is true when, for example, $t = 2$.

(d) The displacement of an object from time t_1 to time t_2 is equal to

$$\int_{t_1}^{t_2} v(t)\, dt.$$

Evaluate this integral on the calculator, noting that $t_1 = 1$ and $t_2 = 3$. The answer in 4.491. This means that at $t = 3$ the object is 4.491 units to the right of its position at $t = 1$, given to be $x = 10$. Hence, at $t = 3$ the object is at $x = 10 + 4.491 = 14.491$.

AB/BC3. (a) Let h represent the depth of the water, as shown.

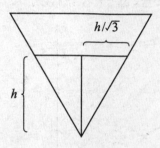

Then h is the altitude of an equilateral triangle, and the base $b = \dfrac{2h}{\sqrt{3}}$. The volume of water is

$$V = \frac{1}{2}\left(\frac{2h}{\sqrt{3}}\right) h \cdot 60 = \frac{60h^2}{\sqrt{3}} \text{ in.}^3$$

Now $\dfrac{dV}{dt} = \dfrac{120}{\sqrt{3}} h \dfrac{dh}{dt}$, and it is given that $\dfrac{dV}{dt} = 600$. Thus, when $h = 4$,

$$600 = \frac{120}{\sqrt{3}} 4 \frac{dh}{dt}, \text{ and } \frac{dh}{dt} = \frac{5\sqrt{3}}{4} \text{ in/min.}$$

(b) Let x represent the length of one of the sides, as shown.

The bases of the trapezoid are $24 - 2x$ and $24 - 2x + 2\dfrac{x}{2}$, and the height is $\dfrac{x}{2}\sqrt{3}$.

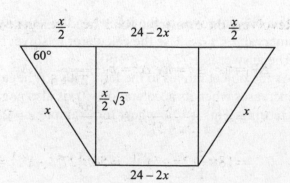

The volume of the trough (in in.^{3}) is given by

$$V = \frac{(24-2x)+(24-x)}{2} \cdot \frac{x}{2}\sqrt{3}\times 60 = 15\sqrt{3}\left(48x - 3x^2\right) \qquad (0 < x < 12),$$

$$V' = 15\sqrt{3}\left(48 - 6x\right) = 0 \text{ when } x = 8.$$

Since $V'' = 15\sqrt{3}(-6) < 0$, the maximum volume is attained by folding the metal 8 in. from the edges.

Part B

AB 4. (a) Draw elements as shown. Then

$$\Delta A = y\,\Delta x = \frac{6}{\sqrt[3]{2x+5}}\,\Delta x,$$

$$A = \int_{-2}^{11} \frac{6}{\sqrt[3]{2x+5}}\,dx = \frac{6}{2}\int_{-2}^{11} (2x+5)^{-1/3}(2dx)$$

$$= 3 \cdot \frac{3}{2}(2x+5)^{\frac{2}{3}}\Big|_{-2}^{11} = \frac{9}{2}\left(27^{2/3} - 1^{2/3}\right) = 36.$$

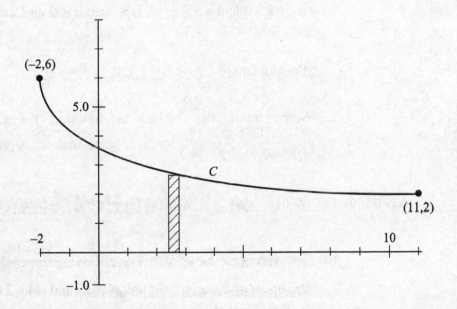

(b) $\displaystyle \int_{-2}^{k} \frac{6}{\sqrt[3]{2x+5}}\,dx = \int_{k}^{11} \frac{6}{\sqrt[3]{2x+5}}\,dx.$

(c) Revolving the element around the x-axis generates disks. Then

$$\Delta V = \pi r^2\,\Delta x = \pi y^2 \Delta x = \pi\left(\frac{6}{\sqrt[3]{2x+5}}\right)^2 \Delta x,$$

$$V = \pi \int_{-2}^{11} \frac{36}{(2x+5)^{2/3}}\,dx = \frac{36\pi}{2}\int_{-2}^{11} (2x+5)^{-2/3}(2dx)$$

$$= 18\pi\frac{3}{1}(2x+5)^{1/3}\Big|_{-2}^{11} = 54\pi\left[27^{1/3} - 1^{1/3}\right] = 108\pi.$$

AB 5.   (a) The differential equation $x^2 e^y \dfrac{dy}{dx} = 4$ is separable:

$$\int e^y \, dy = \int \frac{4}{x^2} \, dx,$$

$$e^y = -\frac{4}{x} + c.$$

If $y = 0$ when $x = 2$, then $e^0 = -\dfrac{4}{2} + c$; thus $c = 3$, and $e^y = -\dfrac{4}{2} + 3$.

Solving for y gives the solution: $y = \ln\left(3 - \dfrac{4}{x}\right)$.

(b) $y = \ln\left(3 - \dfrac{4}{x}\right)$ is defined only if $3 - \dfrac{4}{x} > 0$.

$\dfrac{3x - 4}{x} > 0$ only if the numerator and denominator have the same sign.

$(x > 0$ and $3x - 4 > 0)$ OR $(x < 0$ and $3x - 4 < 0)$,

$\left(x > 0 \text{ and } x > \dfrac{4}{3}\right)$ OR $\left(x < 0 \text{ and } x < \dfrac{4}{3}\right)$.

The domain is $\left\{x \mid x < 0 \quad \text{OR} \quad x > \dfrac{4}{3}\right\}$.

(c) Since $\displaystyle\lim_{x \to \pm\infty} \ln\left(3 - \dfrac{4}{x}\right) = \ln 3$, the function $y = \ln\left(3 - \dfrac{4}{x}\right)$ has a horizontal asymptote at $y = \ln 3$.

AB/BC 6. (a) $A(6) = \dfrac{1}{4}\left(\pi 6^2\right) = 9\pi$, $A(18) = \dfrac{1}{2}\left(\pi 6^2\right) + \dfrac{1}{2} 6 \times 6 = 18\pi + 18$.

(b) The average value of $f = \dfrac{\displaystyle\int_0^{18} f(x)\, dx}{18 - 0} = \dfrac{18\pi + 18}{18} = \pi + 1$.

(c) The line tangent to the graph of A at $x = 6$ passes through point $(6, A(6))$ or $(6, 9\pi)$. Since $A'(x) = f(x)$, the graph of f shows that $A'(6) = f(6) = 6$. Hence, an equation of the line is $y - 9\pi = 6(x - 6)$.

(d) Use the tangent line; then $A(x) = y \approx 6(x - 6) + 9\pi$, so
$A(7) \approx 6(7 - 6) + 9\pi = 6 + 9\pi$.

(e) Since f is increasing on $[0,6]$, f' is positive there. Because $f'(x) = A'(x)$, $f'(x) = A''(x)$; thus A is concave upward for $[0,6]$. Similarly, the graph of A is concave downward for $[6,12]$, and upward for $[12,18]$. There are points of inflection on the graph of A at $(6,9\pi)$ and $(12,18\pi)$.

AB Practice Examination 2

Section I Multiple-Choice Questions_____

Part A

Scc instructions, page 455. Answers begin on page 487.

1. $\displaystyle\lim_{x\to\infty} \frac{20x^2 - 13x + 5}{5 - 4x^3}$ is

   (A) -5 (B) ∞ (C) 0 (D) 5 (E) 1

2. $\displaystyle\lim_{h\to o} \frac{\ln(2 + h) - \ln 2}{h}$ is

   (A) 0 (B) $\ln 2$ (C) $\dfrac{1}{2}$ (D) $\dfrac{1}{\ln 2}$ (E) ∞

3. If $y = e^{-x^2}$, then $y''(0)$ equals

   (A) 2 (B) -2 (C) $\dfrac{2}{e}$ (D) 0 (E) -4

Use the table shown for Questions 4 and 5. The differentiable functions f and g have the values shown.

x	f	f'	g	g'
1	2	$\dfrac{1}{2}$	-3	5
2	3	1	0	4
3	4	2	2	3
4	6	4	3	$\dfrac{1}{2}$

4. The average rate of change of function f on $[1,4]$ is

   (A) 7/6   (B) 4/3   (C) 15/8   (D) 9/4   (E) 8/3

5. If $h(x) = g(f(x))$ then $h'(3) =$
   (A) 1/2   (B) 1   (C) 4   (D) 6   (E) 9

6. The derivative of a function f is given for all x by

$$f'(x) = x^2(x + 1)^3(x - 4)^2.$$

The set of x for which f is a relative maximum is
   (A) $\{0, -1, 4\}$ (B) $\{-1\}$ (C) $\{0, 4\}$ (D) $\{1\}$
   (E) none of these

7. If $y = \dfrac{x - 3}{2 - 5x}$, then $\dfrac{dy}{dx}$ equals
   (A) $\dfrac{17 - 10x}{(2 - 5x)^2}$ (B) $\dfrac{13}{(2 - 5x)^2}$ (C) $\dfrac{x - 3}{(2 - 5x)^2}$ (D) $\dfrac{17}{(2 - 5x)^2}$
   (E) $\dfrac{-13}{(2 - 5x)^2}$

8. A rectangle of perimeter 18 inches is rotated about one of its sides to generate a right circular cylinder. The rectangle that generates the cylinder of largest volume has an area, in square inches, of

   (A) 14   (B) 20   (C) $\dfrac{81}{4}$ (D) 18   (E) $\dfrac{77}{4}$

†9. Which equation has the slope field shown below?
   (A) $\dfrac{dy}{dx} = \dfrac{5}{y}$ (B) $\dfrac{dy}{dx} = \dfrac{5}{x}$ (C) $\dfrac{dy}{dx} = \dfrac{x}{y}$ (D) $\dfrac{dy}{dx} = 5y$

   (E) $\dfrac{dy}{dx} = x + y$

†The topic slope fields will not be tested on the AB examination before 2004. See page x.

For Questions 10–12, the graph shows the velocity of an object moving along a line, for $0 \leqslant t \leqslant 9$.

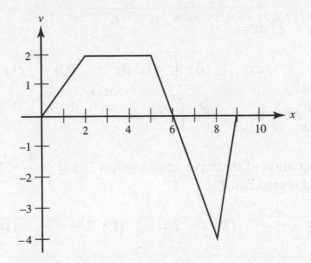

10. At what time does the object attain its maximum acceleration?

(A) $2 < t < 5$ (B) $5 < t < 8$ (C) $t=6$ (D) $t=8$ (E) $8 < t < 9$

11. The object is farthest from the starting point at $t =$

(A) 2   (B) 5   (C) 6   (D) 8   (E) 9

12. At $t = 8$, the object was at position $x = 10$. At $t = 5$, the object's position was $x =$

(A) -5 (B) 5   (C) 7   (D) 13   (E) 15

13. $\displaystyle\int_{\pi/4}^{\pi/2} \sin^3 \alpha \cos \alpha \, d\alpha$ is equal to

(A) $\dfrac{3}{16}$ (B) $\dfrac{1}{8}$ (C) $-\dfrac{1}{8}$ (D) $-\dfrac{3}{16}$ (E) $\dfrac{3}{4}$

14. $\displaystyle\int_{0}^{1} \dfrac{e^x}{(3 - e^x)^2} \, dx$ equals

(A) $3 \ln (e-3)$ (B) 1 (C) $\dfrac{1}{3 - e}$ (D) $\dfrac{e - 2}{3 - e}$ (E) none of these

15. A differentiable function has the values shown in this table:

x	2.0	2.2	2.4	2.6	2.8	3.0
$f(x)$	1.39	1.73	2.10	2.48	2.88	3.30

Estimate $f'(2.1)$.

(A) 0.34   (B) 0.59   (C) 1.56   (D) 1.70   (E) 1.91

16. If $\displaystyle\int_{0}^{1} (1 - e^{-x}) \, dx$ is approximated using Riemann sums and the same number of subdivisions, and if L, R, M, and T denote, respectively left, right, midpoint, and trapezoid sums, then it follows that

(A) $L \leqslant R \leqslant M \leqslant T$ (B) $L \leqslant T \leqslant M \leqslant R$ (C) $L \geqslant R \geqslant M \geqslant T$
(D) $L \leqslant M \leqslant T \leqslant R$ (E) None of these is true.

17. The number of vertical tangents to the graph of $y^2 = x - x^3$ is
(A) 4   (B) 3   (C) 2   (D) 1   (E) 0

18. $\displaystyle\int_0^6 f(x-1)\,dx =$

(A) $\displaystyle\int_{-1}^{7} f(x)\,dx$ (B) $\displaystyle\int_{-1}^{5} f(x)\,dx$ (C) $\displaystyle\int_{-1}^{5} f(x+1)\,dx$

(D) $\displaystyle\int_{1}^{5} f(x)\,dx$ (E) $\displaystyle\int_{1}^{7} f(x)\,dx$

19. The equation of the curve shown below is $y = \dfrac{4}{1+x^2}$. What does the area of the shaded region equal?

(A) $4 - \dfrac{\pi}{4}$ (B) $8 - 2\pi$ (C) $8 - \pi$ (D) $8 - \dfrac{\pi}{2}$ (E) $2\pi - 4$

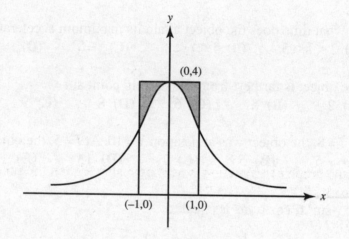

20. The average value of $\dfrac{1}{2}t^2 - \dfrac{1}{3}t^3$ over the interval $-2 \leq t \leq 1$ is

(A) $\dfrac{1}{36}$ (B) $\dfrac{1}{12}$ (C) $\dfrac{11}{12}$ (D) 2   (E) $\dfrac{33}{12}$

21. If $f'(x) = 2f(x)$ and $f(2) = 1$, then $f(x) =$
(A) e^{2x-4} (B) $e^{2x}+1-e^4$ (C) e^{4-2x} (D) e^{2x+1} (E) e^{x-2}

22. The table below shows values of $f''(x)$ for various values of x:

x	-1	0	1	2	3
$f''(x)$	-4	-1	2	5	8

The function f could be
(A) a linear function   (B) a quadratic function
(C) a cubic function   (D) a fourth-degree function
(E) an exponential function

23. The curve $x^3 + x \tan y = 27$ passes through $(3,0)$. Use local linearization to estimate the value of y at $x = 3.1$. The value is

(A) -2.7 (B) -0.9 (C) 0 (D) 0.1 (E) 3.0

24. At what value of h is the rate of increase of $\sqrt{h}$ twice the rate of increase of h?

(A) $\dfrac{1}{16}$ (B) $\dfrac{1}{4}$ (C) 1 (D) 2 (E) 4

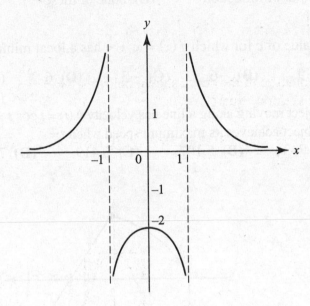

25. If the graph of a function is as shown above, then the function $f(x)$ could be given by which of the following?

(A) $f(x) = \dfrac{x+2}{x^2 - 1}$ (B) $f(x) = \dfrac{1}{1-x^2}$ (C) $f(x) = \dfrac{x^2 - 1}{x^2 + 1}$

(D) $f(x) = \dfrac{2}{x^2 - 1}$ (E) $f(x) = \dfrac{2}{1-x^2}$

26. A function $f(x)$ equals $\dfrac{x^2 - x}{x - 1}$ for all x except $x = 1$. For the function to be continuous at $x = 1$, the value of $f(1)$ must be

(A) 0 (B) 1 (C) 2 (D) ∞ (E) none of these

27. The number of inflection points of $f(x) = 3x^5 - 10x^3$ is

(A) 4 (B) 3 (C) 2 (D) 1 (E) 0

28. Suppose $f(x) = \displaystyle\int_0^x \dfrac{4+t}{t^2 + 4}\, dt$. It follows that

(A) f increases for all x (B) f increases only if $x < -4$
(C) f has a local min at $x = -4$ (D) f has a local max at $x = -4$
(E) f has no critical points

Part B

See instructions, page 460. Answers begin on page 487.

29. Let $G(x) = [f(x)]^2$. At $x = a$, f is increasing and concave downward, while G is decreasing. Which describes G at $x = a$?
  (A) concave downward   **(B)** concave upward   **(C)** linear
  (D) point of inflection   **(E)** none of these

30. The value of c for which $f(x) = x + \dfrac{c}{x}$ has a local minimum at $x = 3$ is

  (A) -9 **(B)** -6 **(C)** -3 **(D)** 6   **(E)** 9

31. An object moving along a line has velocity $v(t) = t \cos t - \ln(t + 2)$, where $0 \le t \le 10$. The object achieves its maximum speed when $t =$
  (A) 3.743   **(B)** 5.107   **(C)** 6.419   **(D)** 7.550   **(E)** 9.538

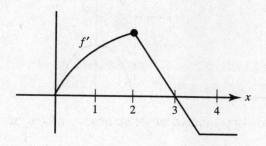

32. The graph of f', which consists of a quarter-circle and two line segments, is shown above. At $x = 2$ which of the following statements is true?
  (A) f is not continuous.
  (B) f is continuous but not differentiable.
  (C) f has a relative maximum.
  (D) f has a point of inflection.
  (E) none of these

33. Let $H(x) = \int_0^x f(t)\, dt$, where f is the function whose graph appears below.

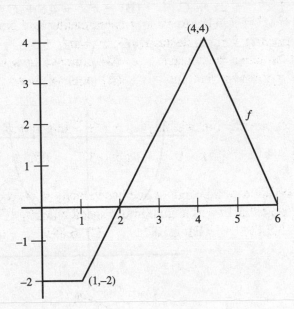

The local linearization of $H(x)$ near $x = 3$ is $H(x) \approx$

(A) $-2x + 8$ (B) $2x - 4$ (C) $-2x + 4$ (D) $2x - 8$ (E) $2x - 2$

34. The table shows the speed of an object, in feet per second, during a 3-second period.

time (sec)	0	1	2	3
speed (ft/sec)	30	22	12	0

Estimate the distance the object travels, using the trapezoid method.

(A) 34 ft   (B) 45 ft   (C) 48 ft   (D) 49 ft   (E) 64 ft

35. In a marathon, when the winner crosses the finish line many runners are still on the course, some quite far behind. If the density of runners x miles from the finish line is given by $R(x) = 20[1 - \cos(1 + .03x^2)]$ runners per mile, how many are within 8 miles of the finish line?

(A) 30   (B) 145   (C) 157   (D) 166   (E) 195

36. Which best describes the behavior of the function $y = \arctan\left(\dfrac{1}{\ln x}\right)$ at $x = 1$?

(A) It has a jump discontinuity.
(B) It has an infinite discontinuity.
(C) It has a removable discontinuity.
(D) It is both continuous and differentiable.
(E) It is continuous but not differentiable.

37. If $f(t) = \displaystyle\int_0^{t^2} \frac{1}{1 + x^2}\, dx$, then $f'(t)$ equals

(A) $\dfrac{1}{1 + t^2}$ (B) $\dfrac{2t}{1 + t^2}$ (C) $\dfrac{1}{1 + t^4}$ (D) $\dfrac{2t}{1 + t^4}$ (E) $\tan^{-1} t^2$

38. $\int (\sqrt{x} - 2) x^2 \, dx =$

(A) $\dfrac{2}{3} x^{3/2} - 2x + C$ (B) $\dfrac{5}{2} x^{3/2} - 4x + C$ (C) $\dfrac{2}{3} x^{3/2} - 2x + \dfrac{x^3}{3} + C$

(D) $\dfrac{2}{5} x^{5/2} - \dfrac{2}{3} x^3 + C$ (E) $\dfrac{2}{7} x^{7/2} - \dfrac{2}{3} x^3 + C$

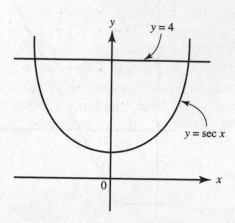

39. The region S in the figure shown above is bounded by $y = \sec x$ and $y = 4$. What is the volume of the solid formed when S is rotated about the x-axis?

(A) 0.304   (B) 39.867   (C) 53.126   (D) 54.088   (E) 108.177

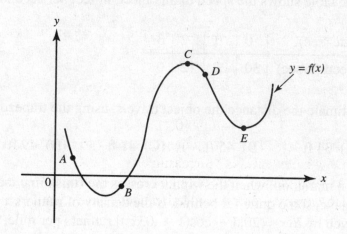

40. At which point on the graph of $y = f(x)$ shown above is $f'(x) < 0$ and $f''(x) > 0$?

(A) A (B) B (C) C (D) D (E) E

41. Let $f(x) = x^5 + 1$, and let g be the inverse function of f. What is the value of $g'(0)$?

(A) -1 (B) $\dfrac{1}{5}$ (C) 1 (D) $g'(0)$ does not exist.

(E) $g'(0)$ cannot be determined from the given information.

42. The hypotenuse AB of a right triangle ABC is 5 feet, and one leg, AC, is decreasing at the rate of 2 feet per second. The rate, in square feet per second, at which the area is changing when $AC = 3$ is

(A) $\dfrac{25}{4}$ (B) $\dfrac{7}{4}$ (C) $-\dfrac{3}{2}$ (D) $-\dfrac{7}{4}$ (E) $-\dfrac{7}{2}$

43. At how many points on the interval $[0,\pi]$ does $f(x) = 2 \sin x + \sin 4x$ satisfy the Mean Value Theorem?

(A) none   (B) 1   (C) 2   (D) 3   (E) 4

44. If the radius r of a sphere is increasing at a constant rate, then the rate of increase of the volume of the sphere is

(A) constant   (B) increasing   (C) decreasing

(D) increasing for $r < 1$ and decreasing for $r > 1$

(E) decreasing for $r < 1$ and increasing for $r > 1$

45. The rate at which a purification process can remove contaminants from a tank of water is proportional to the amount of contaminant remaining. If 20% of the contaminant can be removed during the first minute of the process and 98% must be removed to make the water safe, approximately how long will the decontamination process take?

(A) 2 min   (B) 5 min   (C) 18 min   (D) 20 min   (E) 40 min

Section II: Free-Response Questions _____

Part A

A graphing calculator is required for some of these problems.
See instructions on p. xi. Answers begin on p. 493.

1. A function f is defined on the interval $[0,4]$, and its derivative is
$f'(x) = e^{\sin x} - 2 \cos 3x$.

(a) Sketch f' in the window $[0,4] \times [-2,5]$
(Note that the following questions refer to f.)

(b) On what interval is f increasing?

(c) At what value(s) of x does f have local maxima? Justify your answer.

(d) How many points of inflection does f have? Justify your answer.

2. The rate of sales of a new software product is given by $S(t) = Ce^{kt}$, where S is measured in thousands of units sold per month and t is measured in months from the initial release of the product on January 1, 2000.

(a) This product initially sold at the rate of 2500 units per month, and the sales rate has doubled every 3 months. Find C and k.

(b) Find the average rate of sales for the first year.

(c) Using the midpoint rule with three equal subdivisions, write an expression that approximates $\int_4^7 S(t)\,dt$.

(d) Using correct units, explain the meaning of $\int_4^7 S(t)\,dt$ in terms of software sales.

3. (a) A spherical snowball melts so that its surface area shrinks at the rate of 10 square centimeters per minute. What is the rate of change of volume when the snowball is 12 centimeters in diameter?

   (b) The snowball is packed most densely nearest the center. Suppose that, when it is 12 centimeters in diameter, its density x centimeters from the center is given by $d(x) = \dfrac{1}{1 + \sqrt{x}}$ grams per cubic centimeter. How much does the snowball weigh then?

Part B

No calculator is allowed for any of these problems.
See instructions on p. xi.

4. The graph of function f passes through point $(2, 5)$ and satisfies the differential equation $\dfrac{dy}{dx} = \dfrac{6x^2 - 4}{y}$.

   (a) Write an equation of the linearization of f at $(2, 5)$.
   (b) Using this linearization, estimate $f(2.1)$.
   (c) Solve the differential equation, expressing f as a function of x.
   (d) Using your answer to part (c), find $f(2.1)$.

5. Let R represent the first-quadrant region bounded by the y-axis and the curves $y = 2^x$ and $y = 8 \cos \dfrac{\pi x}{6}$, as shown in the graph.

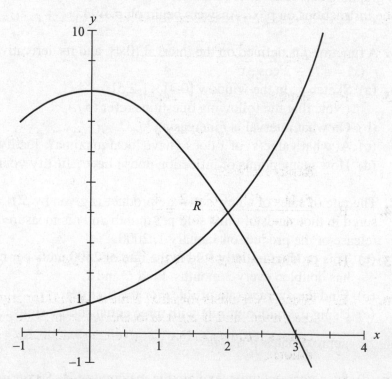

   (a) Find the area of region R.
   (b) Set up, but do not evaluate, integrals in terms of a single variable for:
       (i) the volume of the solid formed when R is rotated around the x-axis,
       (ii) the volume of the solid whose base is R, if all cross sections in planes perpendicular to the x-axis are squares.

6. Given the function $f(x) = e^{2x}(x^2 - 2)$:
   (a) For what values of x is f decreasing?
   (b) Does this decreasing arc reach a local or a global minimum? Justify your answer.
   (c) Does f have a global maximum? Justify your answer.

Answers to AB Practice Examination 2: Section I

1.	C	10.	E	19.	B	28.	C	37.	D
2.	C	11.	C	20.	C	29.	B	38.	E
3.	B	12.	D	21.	A	30.	E	39.	E
4.	B	13.	A	22.	C	31.	E	40.	A
5.	B	14.	E	23.	B	32.	D	41.	B
6.	E	15.	D	24.	A	33.	D	42.	D
7.	E	16.	B	25.	D	34.	D	43.	E
8.	D	17.	B	26.	B	35.	D	44.	B
9.	A	18.	B	27.	B	36.	A	45.	C

Part A

1. **C.** Use the Rational Function Theorem on pages 30 and 31.

2. **C.** Note that $\lim\limits_{h \to 0} \dfrac{\ln(2+h) - \ln 2}{h} = f'(2)$, where $f(x) = \ln x$.

3. **B.** Since $y' = -2xe^{-x^2}$, therefore
$$y'' = -2(x \cdot e^{-x^2} \cdot (-2x) + e^{-x^2}).$$

   Replace x by 0.

4. **B.** $\dfrac{f(4) - f(1)}{4 - 1} = \dfrac{6 - 2}{4 - 1} = \dfrac{4}{3}$.

5. **B.** $h'(3) = g'(f(3)) \cdot f'(3) = g'(4) \cdot f'(3) = \dfrac{1}{2} \cdot 2.$

6. **E.** Since $f'(x)$ exists for all x, it must equal 0 for any x_0 for which f is a relative maximum, and it must also change sign from positive to negative as x increases through x_0. For the given derivative, no x satisfies both of these conditions.

7. **E.** Use the quotient rule (formula (6) on page 47).

8. D. See the figure. The volume, V, of the cylinder equals $\pi x^2 y$, where $2x + 2y = 18$. Then, $V = \pi x^2 (9 - x)$ and $V' = \pi(18x - 3x^2)$. Since $x = 6$ yields maximum volume, the area of the rectangle, xy, equals 18.

9. A. Note that (1) on a horizontal line the slope segments are all parallel, so the slopes there are all the same and $\dfrac{dy}{dx}$ must depend only on y; (2) along the x-axis (where $y = 0$) the slopes are infinite; and (3) as y increases, the slope decreases.

10. E. Acceleration is the derivative (the slope) of velocity v; v is steepest on $8 < t < 9$.

11. C. Velocity v is the derivative of position; $v > 0$ until $t = 6$; $v < 0$ thereafter.

12. D. From $t = 5$ to $t = 8$ the displacement can be found by determining the areas of two triangles: $\dfrac{1}{2}(1)(2) + \dfrac{1}{2}(2)(-4) = -3$. Thus, if K is the object's position at $t = 5$, then $K - 3 = 10$ at $t = 8$.

13. A. Evaluate $\dfrac{1}{4}\sin^4 \alpha \Big|_{\pi/4}^{\pi/2}$.

14. E. Evaluating $\dfrac{1}{3 - e^x}\Big|_0^1$ yields $\dfrac{e - 1}{2(3 - e)}$.

15. D. $f'(2.1) \simeq \dfrac{f(2.2) - f(2.0)}{2.2 - 2.0}$.

16. B. $f(x) = (1 - e^{-x})$ is increasing and concave downward.

17. B. Implicit differentiation yields $2yy' = 1 - 3x^2$; so $\dfrac{dy}{dx} = \dfrac{1 - 3x^2}{2y}$. At a vertical tangent, $\dfrac{dy}{dx}$ is undefined; y must therefore equal 0. The original equation with $y = 0$ is $0 = x - x^3$, which has three solutions.

18. **B.** Let $t = x - 1$.

19. **B.** The required area, A, is given by the integral

$$2\int_0^1 \left(4 - \frac{4}{1 + x^2}\right) dx = 2(4x - 4\tan^{-1} x)\Big|_0^1 = 2\left(4 - 4 \cdot \frac{\pi}{4}\right).$$

20. **C.** The average value equals $\dfrac{1}{3}\left(\dfrac{t^3}{6} - \dfrac{t^4}{12}\right)\Big|_{-2}^{1}$.

21. **A.** Solve the differential equation $\dfrac{dy}{dx} = 2y$ by separation of variables: $\dfrac{dy}{y} = 2dx$

yields $y = ce^{2x}$. The initial condition yields $1 = ce^{2 \cdot 2}$; so $c = e^{-4}$ and $y = e^{2x-4}$.

22. **C.** Changes in values of f'' show that f''' is constant.

23. **B.** By implicit differentiation, $3x^2 + x\sec^2 y \dfrac{dy}{dx} + \tan y = 0$. At $(3,0)$, $\dfrac{dy}{dx} = -9$;

so the equation of the tangent line at $(3,0)$ is $y = -9(x-3)$.

24. **A.** $(h^{1/2})' = 2h'$ implies $\dfrac{1}{2}h^{-1/2} = 2$.

25. **D.** The graph shown has the following characteristics: it has no x-intercepts; the y-intercept is -2; it has vertical asymptotes $x = 1$ and $x = -1$; and it has the x-axis as horizontal asymptote.

26. **B.** Since $\lim\limits_{x \to 1} f(x) = 1$, to render $f(x)$ continuous at $x = 1$ $f(1)$ must be defined to be 1.

27. **B.** $f'(x) = 15x^4 - 30x^2$; $f''(x) = 60x^3 - 60x = 60x(x+1)(x-1)$; this equals 0 when $x = -1, 0$, or 1. Here are the signs within the intervals:

28. **C.** Note that $f'(x) = \dfrac{4 + x}{x^2 + 4}$, so f has a critical value at $x = -4$. As x passes through -4, the sign of f' changes from $-$ to $+$, so f has a local minimum at $x = -4$.

Part B

29. **B.** We are given that (1) $f'(a) > 0$; (2) $f''(a) < 0$; and (3) $G'(a) < 0$. Since $G'(x) = 2f(x) \cdot f'(x)$, therefore $G'(a) = 2f(a) \cdot f'(a)$. Conditions (1) and (3) imply that (4) $f(a) < 0$. Since $G''(x) = 2[f(x) \cdot f''(x) + (f'(x))^2]$, therefore $G''(a) = 2[f(a)f''(a) + (f'(a))^2]$. Then the sign of $G''(a)$ is $2[(-) \cdot (-) + (+)]$ or positive, where the minus signs in the parentheses follow from conditions (4) and (2).

30. **E.** Since $f'(x) = 1 - \dfrac{c}{x^2}$, it equals 0 for $x = \pm\sqrt{c}$. When $x = 3$, $c = 9$; this yields a minimum since $f''(3) > 0$.

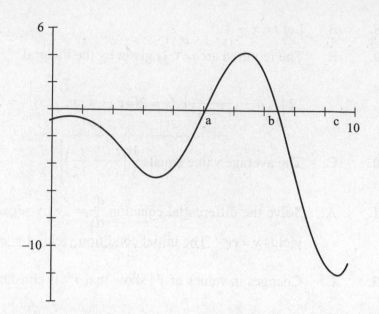

31. **E.** The calculator figure maps velocity Y_1 against time. Speed is the absolute value of velocity. The greatest deviation from $Y_1 = 0$ is at $t = c$. With a calculator, $c = 9.538$.

32. **D.** f' changes from increasing to decreasing, so f'' changes from positive to negative.

33. **D.** $H(3) = \int_0^3 f(t)\,dt = -2; \; H'(3) = f(3) = 2.$

34. **D.** The distance equals $\dfrac{1}{2}[30 + 2(22) + 2(12) + 0]$.

35. **D.** $\int_0^8 R(x)\,dx = 166.396.$

36. **A.** Selecting an answer for this question from your calculator graph is unwise. In some windows the graph may appear continuous; in others there may seem to be cusps, or a vertical asymptote. Put the calculator aside. Find

$$\lim_{x \to 1^+}\left(\arctan\left(\frac{1}{\ln x}\right)\right) = \frac{\pi}{2} \quad \text{and} \quad \lim_{x \to 1^-}\left(\arctan\left(\frac{1}{\ln x}\right)\right) = -\frac{\pi}{2}$$

These limits indicate the presence of a jump discontinuity in the function at $x = 1$.

37. **D.** $\dfrac{d}{du}\displaystyle\int_0^u \dfrac{1}{1+x^2}\,dx = \dfrac{1}{1+u^2}.$ When $u = t^2$,

$$\frac{d}{dt}\int_0^u \frac{1}{1+x^2}\,dx = \frac{1}{1+u^2}\frac{du}{dt} = \frac{1}{1+t^4}(2t).$$

38.   E. $\int (\sqrt{x} - 2)x^2 \, dx = \int (x^{5/2} - 2x^2) \, dx = \dfrac{2}{7}x^{7/2} - \dfrac{2}{3}x^3 + C.$

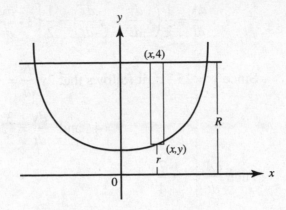

39.   E.   In the figure above, S is the region bounded by $y = \sec x$, the y axis, and $y = 4$. Send region S about the x-axis. Use washers; then $\Delta V = \pi(R^2 - r^2)\,\Delta x$. Symmetry allows you to double the volume generated by the first quadrant of S, so V is

$$2\pi \int_0^{\arccos\frac{1}{4}} (16 - \sec^2 x)\,dx$$

A calculator yields 108.177.

40.   A.   The curve falls when $f'(x) < 0$ and is concave up when $f''(x) > 0$

41.   B. $g'(y) = \dfrac{1}{f'(x)} = \dfrac{1}{5x^4}.$ To find $g'(0)$, find x such that $f(x) = 0$. By inspection,

$x = -1$, so $g'(0) = \dfrac{1}{5(-1)^4} = \dfrac{1}{5}.$

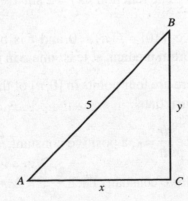

42. D. See the figure. It is given that $\dfrac{dx}{dt} = -2$; you want $\dfrac{dA}{dt}$, where $A = \dfrac{1}{2}xy$.

$$\frac{dA}{dt} = \frac{1}{2}\left(x\frac{dy}{dt} + y\frac{dx}{dt}\right) = \frac{1}{2}\left[3\cdot\frac{dy}{dt} + y\cdot(-2)\right].$$

Since $y^2 = 25 - x^2$, it follows that $2y\dfrac{dy}{dt} = -2x\dfrac{dx}{dt}$ and, when $x = 3$,

$$y = 4 \quad\text{and}\quad \frac{dy}{dt} = \frac{3}{2}.$$

Then $\dfrac{dA}{dt} = -\dfrac{7}{4}$.

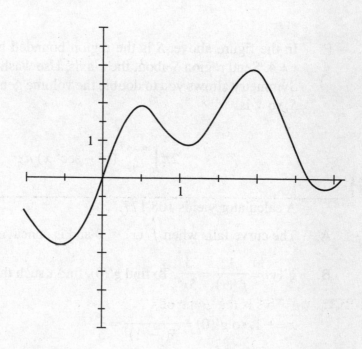

The function $f(x) = 2\sin x + \sin 4x$ is graphed above.

43. E. Since $f(0) = f(\pi) = 0$ and f is both continuous and differentiable, Rolle's theorem predicts at least one c in the interval such that $f'(c) = 0$.

There are four points in $[0,\pi]$ of the calculator graph above where the tangent is horizontal.

44. B. Since $\dfrac{dr}{dt} = k$, a positive constant, $\dfrac{dV}{dt} = 4\pi r^2\dfrac{dr}{dt} = 4\pi r^2 k = cr^2$, where c is a positive constant. Then $\dfrac{d^2V}{dt^2} = 2cr\dfrac{dr}{dt} = 2crk$, which is also positive.

45.   **C.**   If $Q(t)$ is the amount of contaminant in the tank at time t and Q_0 is the initial amount, then

$$\frac{dQ}{dt} = kQ \quad \text{and} \quad Q(t) = Q_0 e^{kt}.$$

Since $Q(1) = 0.8Q_0$, $0.8Q_0 = Q_0 e^{k \cdot 1}$, $0.8 = e^k$, and

$$Q(t) = Q_0(0.8)^t.$$

We seek t when $Q(t) = 0.02Q_0$. Thus,

$$0.02Q_0 = Q_0(0.8)^t$$

and

$$t \approx 17.53 \text{ min.}$$

Section II

Part A

AB/BC 1. (a) This is the graph of $f'(x)$.

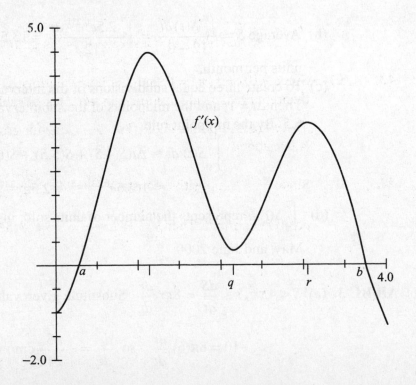

(b) f is increasing when $f'(x) > 0$. The graph shows this to be true in the interval $a < x < b$. Use the Solver to find a and b (where $f'(x) = 0$); then $a = 0.283 < x < 3.760 = b$.

(c) See signs analysis.

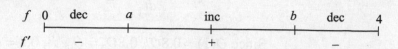

Thus f has local maxima at $x = 0$ and $x = 3.760$.

(d) See signs analysis.

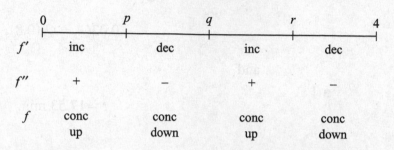

Since f changes concavity at p, q, and r, there are three points of inflection.

AB2. (a) S is measured in thousands of units sold per month, so the initial rate of 2500 units per month means that $S = 2.5$ when $t = 0$. Substitute these values in $S(t) = Ce^{kt}$ to see that $2.5 = Ce^{k \cdot 0}$ and thus $C = 2.5$. If the rate doubles every 3 months, then $S = 5$ when $t = 3$. Substitute these values and solve for k:

$$5 = 2.5e^{3k},$$
$$2 = e^{3k},$$
$$k = \frac{\ln 2}{3} \approx 0.231.$$

(b) Average $S = \dfrac{\int_0^{12} S(t)\,dt}{12 - 0} = \dfrac{\int_0^{12} 2.5e^{0.231t}\,dt}{12} = 13.525$ thousands, or 13,525 units per month.

(c) To create three equal subdivisions of the interval $[4,7]$, use $t = 5$ and $t = 6$. Then $\Delta t = 1$, and the midpoints of the subintervals are at $t = 4.5$, 5.5, and 6.5. By the midpoint rule

$$\int_4^7 S(t)\,dt \approx \Delta t (S(4.5) + S(5.5) + S(6.5))$$

$$\approx 1(2.5e^{0.231(4.5)} + 2.5e^{0.231(5.5)} + 2.5e^{0.231(6.5)}).$$

(d) $\int_4^7 S(t)\,dt$ represents the number of units sold, in thousands, during April, May, and June 2000.

AB/BC 3. (a) $S = 4\pi r^2$, so $\dfrac{dS}{dt} = 8\pi r \dfrac{dr}{dt}$. Substitute given values; then

$$-10 = 8\pi(6)\frac{dr}{dt}, \quad \text{so} \quad \frac{dr}{dt} = -\frac{5}{24\pi}\text{cm/min.}$$

Since $V = \dfrac{4}{3}\pi r^3$, therefore $\dfrac{dV}{dt} = 4\pi r^2 \dfrac{dr}{dt}$. Substituting known values

gives $\dfrac{dV}{dt} = 4\pi(6^2) \cdot \dfrac{-5}{24\pi} = -30 \text{ cm}^3/\text{min}$.

(b) Regions of consistent density are concentric spherical shells. The volume of each shell is approximated by its surface area ($4\pi x^2$) times its thickness (Δx). The weight of each shell is its density times its volume (g/cm^3·cm^3). If, when the snowball is 12 cm in diameter, ΔG is the weight of a spherical

shell x cm from the center, then $\Delta G = \dfrac{1}{1+\sqrt{x}} \cdot 4\pi x^2 \Delta x$, and the weight of

the snowball is

$$G = \int_0^6 \frac{1}{1+\sqrt{x}} \cdot 4\pi x^2 dx = 295.223 \text{ g}.$$

Part B

AB 4.   (a) At (2,5), $\dfrac{dy}{dx} = \dfrac{6(2^2)-4}{5} = 4$, so the tangent line is $y - 5 = 4(x - 2)$.

Solving for y yields $f(x) \approx 5 + 4(x - 2)$.

(b) $f(2.1) \approx 5 + 4(2.1 - 2) = 5.4$.

(c) The differential equation $\dfrac{dy}{dx} = \dfrac{6x^2 - 4}{5}$ is separable:

$$\int y\,dy = \int (6x^2 - 4)\,dx,$$

$$\frac{y^2}{2} = 2x^3 - 4x + C,$$

$$y = \pm\sqrt{4x^2 - 8x + c}, \text{ where } c = 2C.$$

Since f passes through (2,5), it must be true that $5 = \pm\sqrt{4(2^3) - 8(2) + c}$.

Thus $c = 9$, and the positive root is used.

The solution is $f(x) = \sqrt{4x^3 - 8x + 9}$.

(d) $f(2.1) = \sqrt{4(2.1^3) - 8(2.1) + 9} = 5.408$.

AB 5. (a) Draw a vertical element of area, as shown.

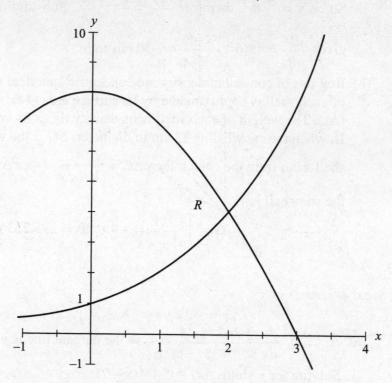

$$\Delta A = (y_{\text{top}} - y_{\text{bottom}})\,\Delta x = \left(8\cos\frac{\pi x}{6} - 2^x\right)\Delta x,$$

$$A = \int_0^2 \left(8\cos\frac{\pi x}{6} - 2^x\right)dx$$

$$= \frac{6}{\pi}\cdot 8\int_0^2 \cos\frac{\pi x}{6}\,dx - \int_0^2 2^x\,dx$$

$$= \frac{48}{\pi}\cdot\sin\frac{\pi x}{6}\bigg|_0^2 - \frac{2^x}{\ln 2}\bigg|_0^2$$

$$= \frac{48}{\pi}\left(\sin\frac{\pi}{3} - \sin 0\right) - \left(\frac{2^2}{\ln 2} - \frac{2^0}{\ln 2}\right)$$

$$= \frac{24\sqrt{3}}{\pi} - \frac{3}{\ln 2}.$$

(b) (i) Use washers; then

$$\Delta V = (r_2^2 - r_1^2)\,\Delta x = \pi(y_{\text{top}}^2 - y_{\text{bottom}}^2)\,\Delta x,$$

$$V = \pi\int_0^2 \left[\left(8\cos\frac{\pi x}{6}\right)^2 - (2^x)^2\right]dx.$$

(ii) See the figure above.

$$\Delta V = s^2 \Delta x = (y_{top} - y_{bottom})^2 \Delta x,$$

$$V = \int_0^2 \left(8\cos\frac{\pi x}{6} - 2^x \right)^2 dx.$$

AB/BC 6. (a) $f(x) = e^{2x}(x^2 - 2),$
  $f'(x) = e^{2x}(2x) + 2e^{2x}(x^2 - 2)$
  $= 2e^{2x}(x + 2)(x - 1)$
  $= 0$ at $x = -2, 1.$
  f is decreasing where $f'(x) < 0$, which occurs for $-2 < x < 1.$

  (b) f is decreasing on the interval $-2 < x < 1$, so there is a minimum at $(1, -e^2)$. Note that, as x approaches $\pm\infty$, $f(x) = e^{2x}(x^2 - 2)$ is always positive. Hence $(1, -e^2)$ is the global minimum.

  (c) As x approaches $+\infty$, $f(x) = e^{2x}(x^2 - 2)$ also approaches $+\infty$. There is no global maximum.

AB Practice Examination 3

Section I Multiple-Choice Questions _____

Part A

See instructions, page 455. Answers begin on page 508.

1. $\lim\limits_{x \to 2} \dfrac{x^2 - 2}{4 - x^2}$ is

   (A) -2 (B) -1 (C) $-\dfrac{1}{2}$ (D) 0   (E) nonexistent

2. $\lim\limits_{x \to \infty} \dfrac{\sqrt{x} - 4}{4 - 3\sqrt{x}}$ is

   (A) $-\dfrac{1}{3}$ (B) -1 (C) ∞ (D) 0   (E) $\dfrac{1}{3}$

3. If $y = \dfrac{e^{\ln u}}{u}$, then $\dfrac{dy}{du}$ equals

   (A) $\dfrac{e^{\ln u}}{u^2}$ (B) $e^{\ln u}$ (C) $\dfrac{2e^{\ln u}}{u^2}$ (D) 1   (E) 0

4. When the local linearization of $f(x) = \sqrt{9 + \sin(2x)}$ near 0 is used, an estimate of $f(0.06)$ is
   (A) 0.02   (B) 2.98   (C) 3.01   (D) 3.02   (E) 3.03

5. Air is escaping from a balloon at a rate of $R(t) = \dfrac{60}{1 + t^2}$ cubic feet per minute, where t is measured in minutes. How much air, in cubic feet, escapes during the first minute?

   (A) 15   (B) 15π (C) 30   (D) 30π (E) $30 \ln 2$

6. If $y = \sin^3 (1 - 2x)$, then $\dfrac{dy}{dx}$ is

   (A) $3 \sin^2 (1 - 2x)$ (B) $-2 \cos^3 (1 - 2x)$ (C) $-6 \sin^2 (1 - 2x)$
   (D) $-6 \sin^2 (1 - 2x) \cos (1 - 2x)$ (E) $-6 \cos^2 (1 - 2x)$

7. If $y = x^2 e^{1/x}$ $(x \neq 0)$, then $\dfrac{dy}{dx}$ is

(A) $xe^{1/x}(x+2)$ (B) $e^{1/x}(2x-1)$ (C) $\dfrac{-2e^{1/x}}{x}$ (D) $e^{-x}(2x-x^2)$
(E) none of these

8. A point moves along the curve $y = x^2 + 1$ so that the x-coordinate is increasing at the constant rate of $\dfrac{3}{2}$ units per second. The rate, in units per second, at which the distance from the origin is changing when the point has coordinates $(1, 2)$ is equal to

(A) $\dfrac{7\sqrt{5}}{10}$ (B) $\dfrac{3\sqrt{5}}{2}$ (C) $3\sqrt{5}$ (D) $\dfrac{15}{2}$ (E) $\sqrt{5}$

9. $\displaystyle\lim_{h \to 0} \dfrac{\sqrt{25+h}-5}{h}$

(A) $= 0$ (B) $= \dfrac{1}{10}$ (C) $= 1$ (D) $= 10$ (E) does not exist

10. The base of a solid is the first-quadrant region bounded by $y = \sqrt[4]{1-x^2}$. Each cross section perpendicular to the x-axis is a square with one edge in the xy-plane. The volume of the solid is

(A) $\dfrac{2}{3}$ (B) $\dfrac{\pi}{4}$ (C) 1 (D) $\dfrac{\pi}{2}$ (E) π

11. $\displaystyle\int \dfrac{x\,dx}{\sqrt{9-x^2}}$ equals

(A) $-\dfrac{1}{2}\ln\sqrt{9-x^2} + C$ (B) $\sin^{-1}\dfrac{x}{3} + C$ (C) $-\sqrt{9-x^2} + C$

(D) $-\dfrac{1}{4}\sqrt{9-x^2} + C$ (E) $2\sqrt{9-x^2} + C$

12. $\displaystyle\int \dfrac{(y-1)^2}{2y}\,dy$ equals

(A) $\dfrac{y^2}{4} - y + \dfrac{1}{2}\ln|y| + C$ (B) $y^2 - y + \ln|2y| + C$

(C) $y^2 - 4y + \dfrac{1}{2}\ln|2y| + C$ (D) $\dfrac{(y-1)^3}{3y^2} + C$ (E) $\dfrac{1}{2} - \dfrac{1}{2y^2} + C$

13. $\displaystyle\int_{\pi/6}^{\pi/2} \cot x\,dx$ equals

(A) $\ln\dfrac{1}{2}$ (B) $\ln 2$ (C) $-\ln(2-\sqrt{3})$ (D) $\ln(\sqrt{3}-1)$
(E) none of these

14. Given f' as graphed, which could be a graph of f?
(A) I only   (B) II only
(C) III only   (D) I and III
(E) none of these

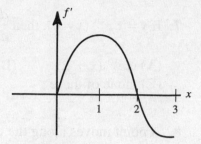

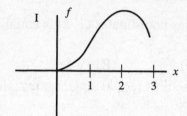

15. Women have only recently begun competing in marathons. The first woman officially timed was Violet Piercey of Great Britain in 1926. Her record of 3:40:22 stood until 1963, mostly because of a lack of women competitors. Soon after, times began dropping rapidly, but lately they have been declining at a much slower rate. Let $M(t)$ be the curve that best represents winning marathon times in year t. Which of the following is (are) negative?
I. $M(t)$ **II.** $M'(t)$ **III.** $M''(t)$
(A) I only   (B) II only   (C) III only   (D) II and III
(E) none of these

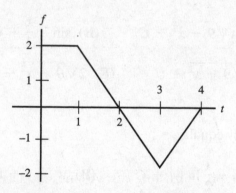

16. The graph of f is shown above. Let $G(x) = \int_0^x f(t)\,dt$ and $H(x) = \int_2^x f(t)\,dt$. Which of the following is true?

(A) $G(x) = H(x)$ (B) $G'(x) = H'(x + 2)$ (C) $G(x) = H(x + 2)$
(D) $G(x) = H(x) - 2$ (E) $G(x) = H(x) + 3$

17. The minimum value of $f(x) = x^2 + \dfrac{2}{x}$ on the interval $\dfrac{1}{2} \leqq x \leqq 2$ is

(A) $\dfrac{1}{2}$ (B) 1   (C) 3   (D) $4\dfrac{1}{2}$ (E) 5

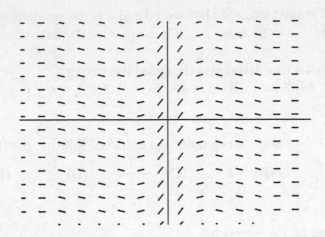

†18 Which function could be a particular solution of the differential equation whose slope field is shown above?

(A) $y = x^3$ (B) $y = \dfrac{2x}{x^2 + 1}$ (C) $y = \dfrac{x^2}{x^2 + 1}$ (D) $y = \sin x$ (E) $y = e^{-x^2}$

19. Which of the following functions could have the graph sketched below?

(A) $f(x) = xe^x$ (B) $f(x) = xe^{-x}$ (C) $f(x) = \dfrac{e^x}{x}$ (D) $f(x) = \dfrac{x}{x^2 + 1}$

(E) $f(x) = \dfrac{x^2}{x^3 + 1}$

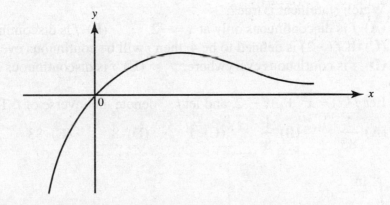

Use this graph, consisting of two line segments and a quarter-circle, for Questions 20–22. It shows the velocity of an object during a 6-second interval.

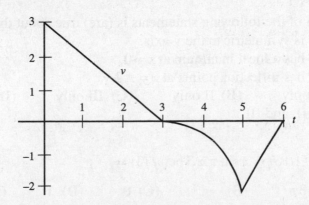

†The topic slope fields will *not* be tested on the AB Examination before 2004. See page x.

20. For how many values of t between 0 and 6 is the acceleration undefined?
(A) none   (B) one   (C) two   (D) three   (E) four

21. During what time interval is the speed increasing?
(A) $0 < t < 3$ (B) $3 < t < 5$ (C) $3 < t < 6$ (D) $5 < t < 6$
(E) never

22. What is the average acceleration (in units/sec^2) during the first 5 seconds?
(A) $-\dfrac{5}{2}$ (B) -1 (C) $-\dfrac{1}{5}$ (D) $\dfrac{1}{5}$ (E) $\dfrac{1}{2}$

23. The curve of $y = \dfrac{2x^2}{4 - x^2}$ has
(A) two horizontal asymptotes
(B) two horizontal asymptotes and one vertical asymptote
(C) two vertical but no horizontal asymptotes
(D) one horizontal and one vertical asymptote
(E) one horizontal and two vertical asymptotes

24. Suppose
$$f(x) = \begin{cases} x^2 & \text{if } x < -2, \\ 4 & \text{if } -2 < x \leq 1, \\ 6 - x & \text{if } x > 1. \end{cases}$$

Which statement is true?
(A) f is discontinuous only at $x = -2$.   (B) f is discontinuous only at $x = 1$.
(C) If $f(-2)$ is defined to be 4, then f will be continuous everywhere.
(D) f is continuous everywhere.   (E) f is discontinuous at $x = -2$ and at $x = 1$.

25. Let $f(x) = x^5 + 3x - 2$, and let f^{-1} denote the inverse of f. Then $(f^{-1})\,'(2)$ equals
(A) $\dfrac{1}{83}$ (B) $\dfrac{1}{8}$ (C) 1   (D) 8   (E) 83

26. $\displaystyle\int_1^e \dfrac{\ln^3 x}{x}\, dx =$
(A) $\dfrac{1}{4}$ (B) $\dfrac{1}{4}e$ (C) $\dfrac{1}{4}(e - 1)$ (D) $\dfrac{e^4}{4}$ (E) $\dfrac{e^4 - 1}{4}$

27. Which of the following statements is (are) true about the graph of $y = \ln(4 + x^2)$?
I. It is symmetric to the y-axis.
II. It has a local minimum at $x = 0$.
III. It has inflection points at $x = \pm 2$.
(A) I only   (B) II only   (C) III only   (D) I and II only
(E) I, II, and III

28. Let $\displaystyle\int_0^x f(t)\, dt = x \sin \pi x$. Then $f(3) =$
(A) -3π (B) -1 (C) 0   (D) 1   (E) 3π

Part B

See instructions, page 460. Answers begin on page 508.

29. The area bounded by the curve $x = 3y - y^2$ and the line $x = -y$ is represented by

(A) $\displaystyle\int_0^4 (2y - y^2)\, dy$ (B) $\displaystyle\int_0^4 (4y - y^2)\, dy$ (C) $\displaystyle\int_0^3 (3y - y^2)\, dy + \int_0^4 y\, dy$

(D) $\displaystyle\int_0^4 (y^2 - 4y)\, dy$ (E) $\displaystyle\int_0^3 (2y - y^2)\, dy$

30. The region bounded by $y = e^x$, $y = 1$, and $x = 2$ is rotated about the x-axis. The volume of the solid generated is given by the integral:

(A) $\displaystyle\pi\int_0^2 e^{2x}\, dx$ (B) $\displaystyle 2\pi\int_1^{e^2} (2 - \ln y)(y - 1)\, dy$ (C) $\displaystyle\pi\int_0^2 (e^{2x} - 1)\, dx$

(D) $\displaystyle 2\pi\int_0^{e^2} y\,(2 - \ln y)\, dy$ (E) $\displaystyle\pi\int_0^2 (e^x - 1)^2\, dx$

31. A particle moves on a straight line so that its velocity at time t is given by $v = 4s$, where s is its distance from the origin. If $s = 3$ when $t = 0$, then, when $t = \dfrac{1}{2}$, s equals

(A) $1 + e^2$ (B) $2e^3$ (C) e^2 (D) $2 + e^2$ (E) $3e^2$

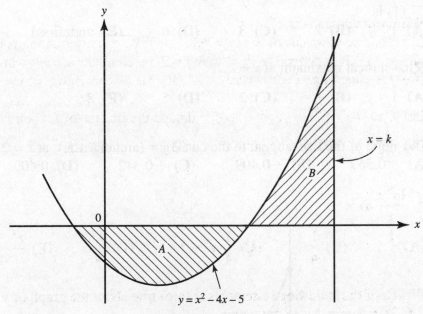

(This figure is not drawn to scale.)

32. The sketch shows the graphs of $f(x) = x^2 - 4x - 5$ and the line $x = k$. The regions labeled A and B have equal areas if $k =$
(A) 5   (B) 7.766   (C) 7.899   (D) 8   (E) 11

33. Bacteria in a culture increase at a rate proportional to the number present. An initial population of 200 triples in 10 hours. If this pattern of increase continues unabated, then the approximate number of bacteria after 1 full day is
(A) 1160   (B) 1440   (C) 2408   (D) 2793   (E) 8380

34. When the substitution $x = 2t - 1$ is used, the definite integral $\int_3^5 t\sqrt{2t - 1}\ dt$ may be expressed in the form $k\int_a^b (x + 1)\sqrt{x}\ dx$, where $\{k, a, b\} =$

(A) $\left\{\dfrac{1}{4}, 2, 3\right\}$ (B) $\left\{\dfrac{1}{4}, 3, 5\right\}$ (C) $\left\{\dfrac{1}{4}, 5, 9\right\}$ (D) $\left\{\dfrac{1}{2}, 2, 3\right\}$

(E) $\left\{\dfrac{1}{2}, 5, 9\right\}$

35. The curve defined by $x^3 + xy - y^2 = 10$ has a vertical tangent line when $x =$

(A) 0 or $-\dfrac{1}{3}$ (B) 1.037 (C) 2.074 (D) 2.096 (E) 2.154

Use the graph of f shown on $[0,7]$ for Questions 36 and 37. Let $G(x) = \int_2^{3x-1} f(t)\ dt$.

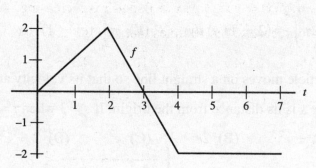

36. $G'(1)$ is
(A) 1 (B) 2 (C) 3 (D) 6 (E) undefined

37. G has a local maximum at $x =$

(A) 1 (B) $\dfrac{4}{3}$ (C) 2 (D) 5 (E) 8

38. The slope of the line tangent to the curve $y = (\arctan (\ln x))^2$ at $x = 2$ is
(A) -0.563 (B) -0.409 (C) -0.342 (D) 0.409 (E) 0.563

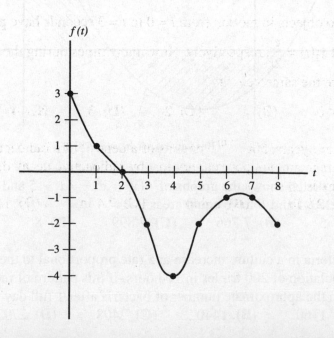

39. Using the left rectangular method and four subintervals, estimate $\int_0^8 |f(t)|\,dt$, where f is the function graphed on page 504.
(A) 4   (B) 5   (C) 8   (D) 15   (E) 16

40. Suppose $f(3) = 2$, $f'(3) = 5$, and $f''(3) = -2$. Then $\dfrac{d^2}{dx^2}\left(f^2(x)\right)$ at $x = 3$ is equal to
(A) -20 (B) 10   (C) 20   (D) 38   (E) 42

41. A particle moves along a line so that its acceleration, a, at time t is $a = -t^2$. If the particle is at the origin when $t = 0$ and 3 units to the right of the origin when $t = 1$, then its velocity at $t = 0$ is
(A) 0   (B) $\dfrac{1}{12}$ (C) $2\dfrac{11}{12}$ (D) $\dfrac{37}{12}$ (E) none of these

42. Suppose $f(x) = \dfrac{1}{3}x^3 + x$, $x > 0$, and x is increasing. The value of x for which the rate of increase of f is 10 times that of x is
(A) 1   (B) 2   (C) $\sqrt[3]{10}$ (D) 3   (E) $\sqrt{10}$

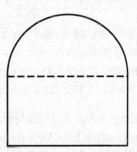

43. The figure above consists of a rectangle capped by a semicircle. Its area is 100 square yards. The minimum perimeter of the figure is
(A) 10.584 yd   (B) 28.284 yd   (C) 37.793 yd   (D) 38.721 yd
(E) 51.820 yd

44. Two objects in motion from $t = 0$ to $t = 3$ seconds have positions $x_1(t) = \cos(t^2 + 1)$ and $x_2(t) = \dfrac{e^t}{2t}$, respectively. How many times during the 3 seconds do the objects have the same velocity?

(A) 0   (B) 1   (C) 2   (D) 3   (E) 4

45. After t years, $50e^{-0.015t}$ pounds of a deposit of a radioactive substance remains. The average amount per year *not* lost by radioactive decay during the second hundred years is
(A) 2.9 lb   (B) 5.8 lb   (C) 7.4 ln   (D) 11.1 lb   (E) none of these

Section II

Part A

A graphing calculator is required for some of these problems.
See instructions on p. xi. Answers begin on p. 517.

1. Let function f be continuous and decreasing, with values as shown in the table:

x	2.5	3.0	3.5	4.0	4.5	5.0
$f(x)$	7.6	5.7	4.2	3.8	2.2	1.6

    (a) Use the trapezoid rule to estimate the area between f and the x-axis.
    (b) Find the average rate of change of f on this interval.
    (c) Estimate the instantaneous rate of change of f at $x = 2.5$.
    (d) If $g(x) = f^{-1}(x)$, estimate the slope of g at $x = 4$.

2. The town of East Newton has a water tower whose tank is a circular ellipsoid, formed by rotating an ellipse about its minor axis. Since the tank is 20 feet tall and 50 feet wide, the equation of the ellipse is $\dfrac{x^2}{625} + \dfrac{y^2}{100} = 1$.

    (a) If there are 7.48 gallons of water per cubic foot, what is the capacity of this tank to the nearest thousand gallons?
    (b) East Newton imposes water rationing whenever the tank is only one-quarter full. How deep is the water in the tank when rationing becomes necessary?

3. Newton's law of cooling states that the rate at which an object cools is proportional to the difference in temperature between the object and its surroundings.

    It is 9:00 P.M., time for your milk and cookies. The room temperature is 68° when you pour yourself a glass of 40° milk and start looking for the cookie jar. By 9:03 the milk has warmed to 43°, and the phone rings. It's your friend, with a fascinating calculus problem. Distracted by the conversation, you forget about the glass of milk. If you dislike milk warmer than 60°, how long, to the nearest minute, do you have to solve the calculus problem and still enjoy acceptably cold milk with your cookies?

Part B

No calculator is allowed for any of these problems. See instructions on p. xi.

This figure is for problem 4 on page 507.

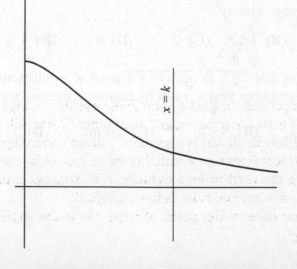

4. Consider the first-quadrant region bounded by the curve $y = \dfrac{18}{9+x^2}$, the coordinate axes, and the line $x = k$, as shown in the figure on page 506.
(a) For what value of k will the area of this region equal π?
(b) What is the average value of the function on the interval $0 \le x \le k$?
(c) What happens to the area of the region as the value of k increases?

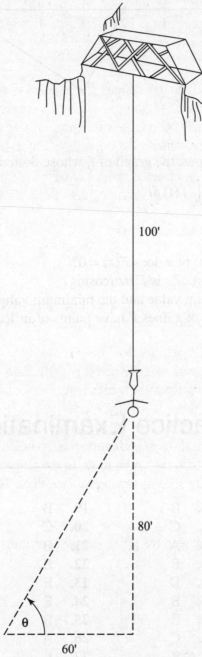

5. A bungee jumper has reached a point in her exciting plunge where the taut cord is 100 feet long with a 1/2-inch radius, and stretching. She is still 80 feet above the ground and is now falling at 40 feet per second. You are observing her jump from a spot on the ground 60 feet from the potential point of impact, as shown in the diagram above.
(a) Assuming the cord to be a cylinder with volume remaining constant as the cord stretches, at what rate is its radius changing?
(b) From your observation point, at what rate is the angle of elevation to the jumper changing?

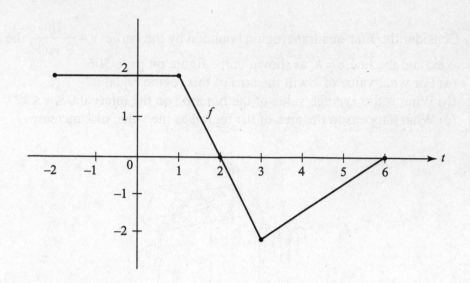

6. The figure above shows the graph of f, whose domain is the closed interval $[-2,6]$. Let $F(x) = \int_1^x f(t)\,dt$.

(a) Find $F(-2)$.
(b) Find $F(6)$.
(c) For what value(s) of x does $F(x) = 0$?
(d) For what value(s) of x is F increasing?
(e) Find the maximum value and the minimum value of F.
(f) At what value(s) of x does F have points of inflection? Justify your answer.

Answers to AB Practice Examination 3: Section I

1.	E	10.	B	19.	B	28.	A	37.	B
2.	A	11.	C	20.	C	29.	B	38.	D
3.	E	12.	A	21.	B	30.	C	39.	E
4.	D	13.	B	22.	B	31.	E	40.	E
5.	B	14.	D	23.	E	32.	D	41.	D
6.	D	15.	B	24.	E	33.	D	42.	D
7.	B	16.	E	25.	B	34.	C	43.	C
8.	B	17.	C	26.	A	35.	C	44.	E
9.	B	18.	B	27.	E	36.	D	45.	B

Part A

1. E. $\dfrac{x^2 - 2}{4 - x^2} \to + \infty$ as $x \to 2$.

2. A. Divide both numerator and denominator by $\sqrt{x}$.

3. E. Since $e^{\ln u} = u$, $y = 1$.

4. D. $f(0) = 3$, and $f'(x) = \dfrac{1}{2}(9 + \sin 2x)^{-1/2} \cdot (2 \cos 2x)$, so $f'(0) = \dfrac{1}{3}$; $y \approx \dfrac{1}{3}x + 3$.

5. B. $\displaystyle\int_0^1 \dfrac{60}{1 + t^2}\, dt = 60 \arctan 1 = 60 \cdot \dfrac{\pi}{4}$.

6. D. Here $y' = 3 \sin^2 (1 - 2x) \cos (1 - 2x) \cdot (-2)$.

7. B. $\dfrac{d}{dx}(x^2 e^{x^{-1}}) = x^2 e^{x^{-1}}\left(-\dfrac{1}{x^2}\right) + 2x e^{x^{-1}}$.

8. B. Let s be the distance from the origin: then

$$s = \sqrt{x^2 + y^2} \quad \text{and} \quad \dfrac{ds}{dt} = \dfrac{x\dfrac{dx}{dt} + y\dfrac{dy}{dt}}{\sqrt{x^2 + y^2}}.$$

Since

$$\dfrac{dy}{dt} = 2x\dfrac{dx}{dt} \quad \text{and} \quad \dfrac{dx}{dt} = \dfrac{3}{2},$$

$\dfrac{dy}{dt} = 3x$. Substituting yields $\dfrac{ds}{dt} = \dfrac{3\sqrt{5}}{2}$.

9. B. For $f(x) = \sqrt{x}$, this limit represents $f'(25)$.

10. B. $V = \displaystyle\int_0^1 y^2\, dy = \int_0^1 \sqrt{1 - x^2}\, dx$, which represents the area of a quadrant of the circle $x^2 + y^2 = 1$.

11. C. The integral is equivalent to $-\dfrac{1}{2}\displaystyle\int (9 - x^2)^{-1/2}(-2x\, dx)$. Use (3) on page 47 with $u = 9 - x^2$ and $n = -\dfrac{1}{2}$.

12. A. The integral is rewritten as

$$\int \dfrac{(y-1)^2}{2y}\, dy = \dfrac{1}{2}\int \dfrac{y^2 - 2y + 1}{y}\, dy$$

$$= \dfrac{1}{2}\int \left(y - 2 + \dfrac{1}{y}\right) dy$$

$$= \dfrac{1}{2}\left(\dfrac{y^2}{2} - 2y + \ln|y|\right) + C.$$

13.   B. $\displaystyle\int_{\pi/6}^{\pi/2} \cot x\, dx = \ln \sin x \Big|_{\pi/6}^{\pi/2} = 0 - \ln\frac{1}{2}.$

14.   D.   Note:

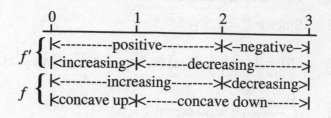

15.   B.   The winning times are positive, decreasing, and concave upward.

16.   E. $\displaystyle G(x) = H(x) + \int_0^2 f(t)\, dt.$

17.   C. $f'(x) = 0$ for $x = 1$ and $f''(1) > 0.$

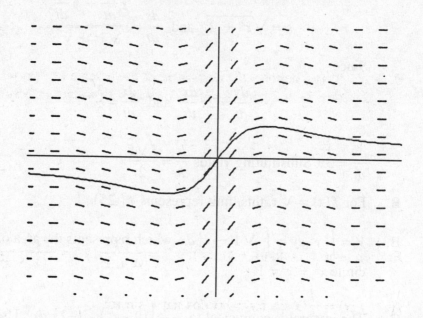

18.   B.   Solution curves appear to represent odd functions with a horizontal asymptote. In the figure above, the curve in (B) of the question has been superimposed on the slope field.

19.   B.   Note that

$$\lim_{x\to\infty} xe^x = \infty,\ \lim_{x\to\infty} \frac{e^x}{x} = \infty,\ \lim_{x\to-\infty} \frac{x}{x^2+1} = 0,\ \text{and}\ \frac{x^2}{x^3+1} \geqq 0\ \text{for}\ x > -1.$$

20.   C. v is not differentiable at $t = 3$ or $t = 5$.

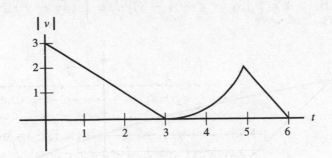

21.   B.   Speed is the magnitude of velocity; its graph is shown above.

22.   B. $\dfrac{v(5) - v(0)}{5 - 0} = \dfrac{-2 - 3}{5}$.

23.   E.   The curve has vertical asymptotes at $x = 2$ and $x = -2$ and a horizontal asymptote at $y = -2$.

24.   E.   The function is not defined at $x = -2$; $\lim\limits_{x \to 1^-} f(x) \neq \lim\limits_{x \to 1^+} f(x)$. Defining $f(-2) = 4$ will make f continuous at $x = -2$, but f will still be discontinuous at $x = 1$.

25.   B. $(f^{-1})\,'(y_0)$ denotes the derivative of the inverse of $f(x)$ when $y = y_0$. Setting $2 = x^5 + 3x - 2$ gives $4 = x^5 + 3x$, so $x = 1$ (when $y = 2$) by observation. Then

$$\left(f^{-1}\right)'(y) = \frac{1}{5x^4 + 3} \quad \text{and} \quad \left(f^{-1}\right)'(2) = \frac{1}{5 \cdot 1 + 3} = \frac{1}{8}.$$

26.   A. $\displaystyle\int_1^e \frac{\ln^3 x}{x}\, dx = \frac{1}{4} \ln^4 x \,\Big|_1^e = \frac{1}{4}(\ln^4 e - 0) = \frac{1}{4}$.

27.   E. $\ln(4 + x^2) = \ln(4 + (-x)^2)$; $y' = \dfrac{2x}{4 + x^2}$; $y'' = \dfrac{-2(x^2 - 4)}{(4 + x^2)^2}$.

28.   A. $f(x) = \dfrac{d}{dx}(x \sin \pi x) = \pi x \cos \pi x + \sin \pi x$.

Part B

29. **B.** $A = \int_0^4 [3y - y^2 - (-y)]\, dy = \int_0^4 (4y - y^2)\, dy.$

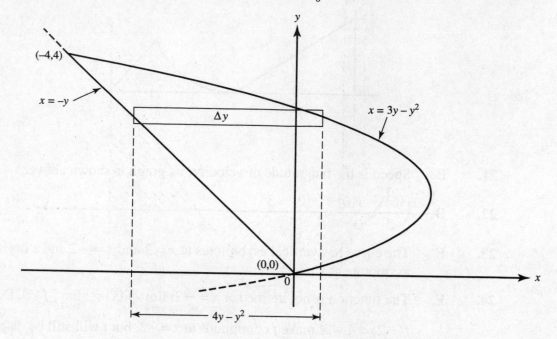

30. **C.** About the x-axis: Washer. $\Delta V = \pi(y^2 - 1^2)\,\Delta x,$

$$V = \pi \int_0^2 (e^{2x} - 1)\, dx.$$

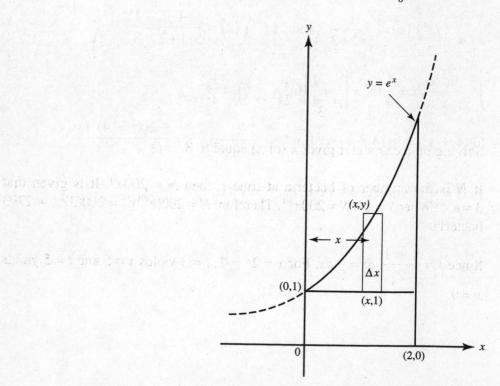

31. **E.** Since $v = \dfrac{ds}{dt} = 4s,\ \dfrac{ds}{s} = 4dt.$ Then $\ln s = 4t + C$, where, using $s(0) = 3,$

$$C = \ln 3.\ \text{Then, since } s = 3e^{4t},\ s\left(\frac{1}{2}\right) = 3e^2.$$

32.   D.

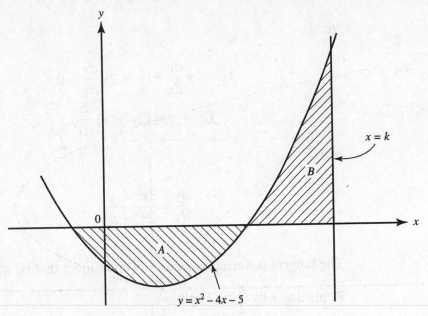

(This figure is not drawn to scale.)

The roots of $f(x) = x^2 - 4x - 5 = (x - 5)(x + 1)$ are $x = -1$ and 5. Since areas A and B are equal, therefore $\int_{-1}^{k} f(x)\, dx = 0$. Thus,

$$\left(\frac{x^3}{3} - 2x^2 - 5x \right)\Bigg|_{-1}^{k} = \left(\frac{k^3}{3} - 2k^2 - 5k \right) - \left(-\frac{1}{3} - 2 + 5 \right)$$

$$= \frac{k^3}{3} - 2k^2 - 5k - \frac{8}{3} = 0.$$

Solving on a calculator gives k (or x) equal to 8.

33.   D.   If N is the number of bacteria at time t, then $N = 200e^{kt}$. It is given that $3 = e^{10k}$. When $t = 24$, $N = 200e^{24k}$. Therefore $N = 200(e^{10k})^{2.4} = 200(3)^{2.4} \simeq 2793$ bacteria.

34.   C.   Since $t = \dfrac{x + 1}{2}$, $dt = \dfrac{1}{2}\, dx$. For $x = 2t - 1$, $t = 3$ yields $x = 5$ and $t = 5$ yields $x = 9$.

35.   **C.**   Using implicit differentiation on the equation

$$x^3 + xy - y^2 = 10 \qquad (1)$$

yields

$$3x^2 + x\frac{dy}{dx} + y - 2y\frac{dy}{dx} = 0,$$
$$3x^2 + y = (2y - x)\frac{dy}{dx},$$

and

$$\frac{dy}{dx} = \frac{3x^2 + y}{2y - x}.$$

The tangent is vertical when $\frac{dy}{dx}$ is undefined; that is, when $2y - x = 0$.
Replacing y by $\frac{x}{2}$ in (1) gives

$$x^3 + \frac{x^2}{2} - \frac{x^2}{4} = 10$$

or

$$4x^3 + x^2 = 40.$$

Let $y_1 = 4x^3 + x^2 - 40$. Inspection of the equation $y_1 = f(x) = 0$ reveals that there is a root near $x = 2$. Solving on a calculator yields $x = 2.074$.

36.   **D.** $G'(x) = f(3x - 1) \cdot 3.$

37.   **B.**   Since f changes from positive to negative at $t = 3$, G' does also where $3x - 1 = 3$.

38.   **D.**   Find $\frac{dy}{dx}$ from nDeriv $((\tan^{-1}(\ln x))^2, x, 2)$.

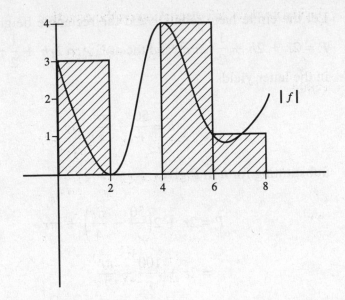

39.   E. $2(3) + 2(0) + 2(4) + 2(1)$.

See the figure above.

40.   E. $\dfrac{d}{dx}\left(f^2(x)\right) = 2f(x)f'(x),$

$$\dfrac{d^2}{dx^2}\left(f^2(x)\right) = 2[f(x)f''(x) + f'(x)f'(x)]$$

$$= 2[ff'' + (f')^2].$$

At $x = 3$, the answer is $2[2(-2) + 5^2] = 42$.

41.   D. $a = \dfrac{dv}{dt} = -t^2$ yields $v = -\dfrac{t^3}{3} + C_1$, and

$$s = -\dfrac{t^4}{12} + C_1 t + C_2.$$

Since $s(0) = 0$, $C_2 = 0$; and since $s(1) = 3$, $C_1 = \dfrac{37}{12}$. Thus $v(0) = \dfrac{37}{12}$.

42.   D. $\dfrac{df}{dt} = (x^2 + 1)\dfrac{dx}{dt}$. Find x when $\dfrac{df}{dt} = 10\dfrac{dx}{dt}$.

$$10\dfrac{dx}{dt} = (x^2 + 1)\dfrac{dx}{dt}$$

implies that $x = 3$.

43. **C.** Let the circle have radius r and the rectangle height h. Then the perimeter $P = 2r + 2h + \dfrac{1}{2}(2\pi r)$, and the area $A = 2rh + \dfrac{1}{2}\pi r^2 = 100$. Solving for h in the latter yields

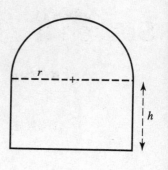

$$h = \frac{50}{r} - \frac{\pi r}{4}.$$

Substituting for h in P gives

$$P = 2r + 2\left(\frac{50}{r} - \frac{\pi r}{4}\right) + \pi r$$

$$= 2r + \frac{100}{r} + \frac{\pi r}{2}$$

$$P' = 2 - \frac{100}{r^2} + \frac{\pi}{2}.$$

Set $P' = 0$; then the Solver yields $r = 5.292$, and the calculator gives the minimum perimeter, for this r, as 37.793.

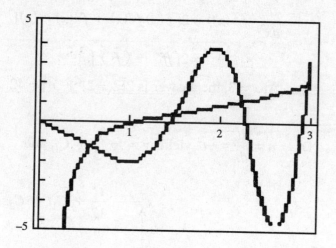

44. **E.** The velocity functions are

$$v_1 = -2t \sin(t^2 + 1)$$

and

$$v_2 = \frac{2t(e^t) - 2e^t}{(2t)^2} = \frac{e^t(t-1)}{2t^2}.$$

Let X be t, let $Y_1 = -2X\sin(X^2 + 1)$ and $Y_2 = (e^x)(X - 1)/2X^2$. Graph Y_1 and Y_2 in $[0, 3] \times [-5, 5]$. The graphs intersect four times during the first 3 sec, as shown in the figure above.

$$\frac{\displaystyle\int_{100}^{200} 50e^{-0.015t}dt}{100} \approx 5.78 \text{ lb.}$$

45.   **B.**

Section II

Part A

AB/BC1. (a) $T = \dfrac{h}{2}\left(y_0 + 2y_1 + 2y_2 + 2y_3 + 2y_4 + y_5\right)$

$= \dfrac{0.5}{2}\left(7.6 + 2(5.7) + 2(4.2) + 2(3.8) + 2(2.2) + 1.6\right) = 10.25.$

(b) $\dfrac{\Delta y}{\Delta x} = \dfrac{7.6 - 1.6}{2.5 - 5.0} = -2.4.$

(c) $f'(2.5) \approx \dfrac{5.7 - 7.6}{3.0 - 2.5} = -3.8.$

(d) To work with $g(x) = f^{-1}(x)$, interchange x and y:

x	7.6	5.7	4.2	3.8	2.2	1.6
$g(x)$	2.5	3.0	3.5	4.0	4.5	5.0

Now $g'(4) \approx \dfrac{4.0 - 3.5}{3.8 - 4.2} = -1.25.$

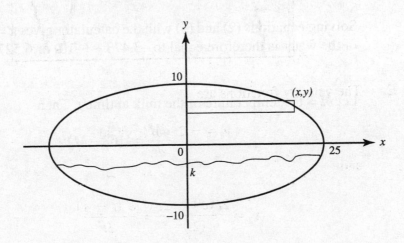

AB 2.   The figure above shows an elliptical cross section of the tank. Its equation is

$$\frac{x^2}{625} + \frac{y^2}{100} = 1.$$

(a)   The volume of the tank, using disks, is $V = 2\pi \int_0^{10} x^2\,dy$, where the ellipse's symmetry about the x-axis has been exploited. From the equation above it is evident that $x^2 = 6.25(100 - y^2)$, so

$$V = 12.5\pi \int_0^{10} \left(100 - y^2\right) dy.$$

Changing variables for the calculator (replacing y by X), evaluate

$$12.5\pi\,\mathtt{fnInt}\,(100\text{-}X^2,X,0,10) \to V$$

(store the answer as V to have it available for part b).

The capacity of the tank is $7.48V$, or 196,000 gal of water, rounded to the nearest 1000 gal.

(b)   Let k be the y-coordinate of the water level when the tank is one-fourth full. Then

$$6.25\pi \int_{-10}^{k} \left(100 - y^2\right) dy = \frac{V}{4} \qquad (1)$$

or

$$\int_{-10}^{k} \left(100 - y^2\right) dy = \frac{V}{25\pi}. \qquad (2)$$

Since

$$\left(100y - \frac{y^3}{3}\right)\Bigg|_{-10}^{k} = 100k - \frac{k^3}{3} - \left(-1000 + \frac{1000}{3}\right) = 100k - \frac{k^3}{3} + \frac{2000}{3} \qquad (3)$$

Solving equations (2) and (3) with the calculator gives $k = -3.473$. The depth of the water is therefore equal to $-3.473 - (-10)$ or 6.527 ft.

AB 3.   Let M = the temperature of the milk at time t. Then

$$\frac{dM}{dt} = k(68 - M).$$

The differential equation is separable:

$$\int \frac{dM}{68-M} = \int k \, dt,$$

$$-\ln|68-M| = kt + C \quad \text{(note that } 68-M > 0\text{)},$$

$$\ln(68-M) = -(kt+C),$$

$$68 - M = e^{-(kt+C)},$$

$$M = 68 - ce^{-kt},$$

where $c = e^{-C}$.

Find c, using the fact that $M = 40°$ when $t = 0$:

$$40 = 68 - ce^0 \quad \text{means} \quad c = 28.$$

Find k, using the fact that $M = 43°$ when $t = 3$:

$$43 = 68 - 28e^{-3k},$$

$$e^{-3k} = \frac{25}{28},$$

$$k = -\frac{1}{3}\ln\frac{25}{28}.$$

Hence $M = 68 - 28e^{\frac{1}{3}\ln\frac{25}{28}t}$.

Now find t when $M = 60°$:

$$60 = 68 - 28e^{\frac{1}{3}\ln\frac{25}{28}t},$$

$$e^{\frac{1}{3}\ln\frac{25}{28}t} = \frac{8}{28},$$

$$t = \frac{\ln\frac{8}{28}}{\frac{1}{3}\ln\frac{25}{28}} = 33.163.$$

Since the phone rang at $t = 3$, you have 30 min to solve the problem.

Part B

AB 4. (a)

See figure on page 506.

$$\int_0^k \frac{18}{9+x^2} \, dx = \pi,$$

$$3 \cdot \frac{18}{9} \int_0^k \frac{\frac{1}{3}dx}{1+\left(\frac{x}{3}\right)^2} = \pi,$$

$$6 \arctan \frac{x}{3}\Big|_0^k = \pi,$$

$$6 \arctan \frac{k}{3} - 6 \arctan \frac{0}{3} = \pi,$$

$$\frac{k}{3} = \tan\frac{\pi}{6},$$

$$k = \sqrt{3},$$

(b) The average value of a function on an interval is the area under the graph of the function divided by the interval width, here $\dfrac{\pi}{\sqrt{3}}$.

(c) From part (a) you know that the area of the region is given by

$$\int_0^k \frac{18}{9+x^2}\, dx = 6 \arctan \frac{k}{3}.$$ Since $\displaystyle\lim_{k \to \infty} 6 \arctan \frac{k}{3} = 6\left(\frac{\pi}{2}\right) = 3\pi$, as k

increases the area of the region approaches 3π.

AB/BC 5. (a) The volume of the cord is $V = \pi r^2 h$. Differentiate with respect to time, then substitute known values. (Be sure to use consistent units; here, all measurements have been converted to inches.)

$$\frac{dV}{dt} = \pi\left(r^2 \frac{dh}{dt} + 2rh\frac{dr}{dt}\right),$$

$$0 = \pi\left((\tfrac{1}{2})^2 \cdot 480 + \tfrac{1}{2}\cdot\tfrac{1}{2}\cdot 1200\frac{dr}{dt}\right),$$

$$\frac{dr}{dt} = -\frac{1}{10} \text{ in/sec}.$$

(b) Let θ represent the angle of elevation and h the height, as shown.

$$\tan\theta = \frac{h}{60}$$

$$\sec^2\theta\,\frac{d\theta}{dt} = \frac{1}{60}\frac{dh}{dt}$$

When $h = 80$, your distance to the jumper is 100 ft, as shown.

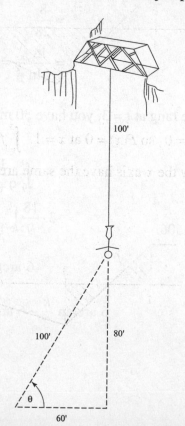

Then

$$\left(\frac{100}{60}\right)^2 \frac{d\theta}{dt} = \frac{1}{60}(-40),$$

$$\frac{d\theta}{dt} = \frac{6}{25} \text{ rad/sec}.$$

AB/BC 6.   (a) $F(-2) = \int_1^{-2} f(t)\,dt = -\int_{-2}^1 f(t)\,dt =$ the negative of the area of the shaded rectangle in the figure. Hence $F(-2) = -(3)(2) = -6$.

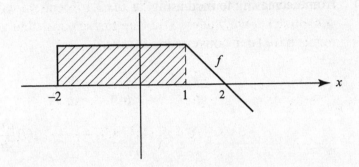

(b) $F(6) = \int_1^6 f(t)\,dt$ is represented by the shaded triangles in the figure.

$$\int_1^6 f(t)\,dt = \int_1^2 f(t)\,dt + \int_2^6 f(t)\,dt$$
$$= \tfrac{1}{2}(1)(2) - \tfrac{1}{2}(4)(2) = -3.$$

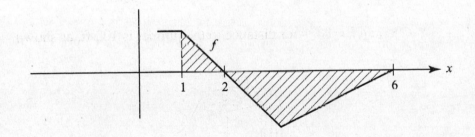

(c) $\int_1^1 f(t)\,dt = 0$, so $F(x) = 0$ at $x = 1$. $\int_1^3 f(t)\,dt = 0$ because the regions above and below the x-axis have the same area. Hence $F(x) = 0$ at $x = 3$.

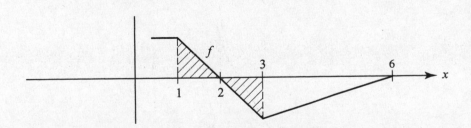

(d) F is increasing where $F' = f$ is positive: $-2 \le x < 2$.

(e)   The maximum value of F occurs at $x = 2$, where $F' = f$ changes from positive to negative. $F(2) = \int_1^2 f(t)\, dt = \frac{1}{2}(1)(2) = 1$.

   The minimum value of F must occur at one of the endpoints. Since $F(-2) = -6$ and $F(6) = -3$, the minimum is at $x = -2$.

(f) F has points of inflection where F'' changes sign, as occurs where $F' = f$ goes from decreasing to increasing, at $x = 3$.

AB Practice Examination 4

Section I Multiple-Choice Questions _____

Part A

See instructions, page 455. Answers begin on page 533.

1. $\lim\limits_{x \to 2}[x]$ (where $[x]$ is the greatest integer in x) is

    (A) 1   (B) 2   (C) 3   (D) ∞ (E) nonexistent

2. $\lim\limits_{h \to 0} \dfrac{\sin\left(\dfrac{\pi}{2}+h\right)-1}{h}$ is

    (A) 1   (B) -1 (C) 0   (D) ∞ (E) none of these

3. If $f(x) = x \ln x$, then $f'''(e)$ equals

    (A) $\dfrac{1}{e}$ (B) 0   (C) $-\dfrac{1}{e^2}$ (D) $\dfrac{1}{e^2}$ (E) $\dfrac{2}{e^3}$

4. The equation of the tangent to the curve $2x^2 - y^4 = 1$ at the point $(-1, 1)$ is

    (A) $y = -x$
    (B) $y = 2 - x$
    (C) $4y + 5x + 1 = 0$
    (D) $x - 2y + 3 = 0$
    (E) $x - 4y + 5 = 0$

5. On which interval(s) does the function $f(x) = x^4 - 4x^3 + 4x^2 + 6$ increase?

    (A) $x < 0$ and $1 < x < 2$ (B) $x > 2$ only   (C) $0 < x < 1$ and $x > 2$
    (D) $0 < x < 1$ only   (E) $1 < x < 2$ only

6. $\displaystyle\int \dfrac{\cos x}{4 + 2\sin x}\,dx$ equals

    (A) $\sqrt{4 + 2\sin x} + C$ (B) $-\dfrac{1}{2(4 + \sin x)} + C$

    (C) $\ln\sqrt{4 + 2\sin x} + C$ (D) $2\ln|4 + 2\sin x| + C$

    (E) $\dfrac{1}{4}\sin x - \dfrac{1}{2}\csc^2 x + C$

7. A relative maximum value of the function $y = \dfrac{\ln x}{x}$ is

   (A) 1   (B) e (C) $\dfrac{2}{e}$ (D) $\dfrac{1}{e}$ (E) none of these

8. If a particle moves on a line according to the law $s = t^5 + 2t^3$, then the number of times it reverses direction is

   (A) 4   (B) 3   (C) 2   (D) 1   (E) 0

†9. A particular solution of the differential equation whose slope field is shown above contains point P. This solution may also contain which other point?

   (A) A (B) B (C) C (D) D (E) E

10. Let $F(x) = \displaystyle\int_5^x \dfrac{dt}{1 - t^2}$. Which of the following statements is (are) true?

   I. The domain of F is $x \neq \pm 1$.   II. $F(2) > 0$.   III. F is concave upward.

   (A) none   (B) I only   (C) II only   (D) III only
   (E) II and III only

†The topic slope fields will *not* be tested on the AB examinations before 2004. See page x.

11. As the tides change, the water level in a bay varies sinusoidally. At high tide today at 8 A.M., the water level was 15 feet; at low tide, 6 hours later at 2 P.M., it was 3 feet. How fast, in feet per hour, was the water level dropping at noon today?

(A) 3   (B) $\dfrac{\pi\sqrt{3}}{2}$ (C) $3\sqrt{3}$ (D) $\pi\sqrt{3}$ (E) $6\sqrt{3}$

12. A smooth curve with equation $y = f(x)$ is such that its slope at each x equals x^2. If the curve goes through the point $(-1, 2)$, then its equation is

(A) $y = \dfrac{x^3}{3} + 7$ (B) $x^3 - 3y + 7 = 0$

(C) $y = x^3 + 3$ (D) $y - 3x^3 - 5 = 0$ (E) none of these

13. $\displaystyle\int \dfrac{e^u}{4 + e^{2u}}\, du$ is equal to

(A) $\ln(4 + e^{2u}) + C$ (B) $\dfrac{1}{2}\ln\left|4 + e^{2u}\right| + C$

(C) $\dfrac{1}{2}\tan^{-1}\dfrac{e^u}{2} + C$ (D) $\tan^{-1}\dfrac{e^u}{2} + C$

(E) $\dfrac{1}{2}\tan^{-1}\dfrac{e^{2u}}{2} + C$

14. Given $f(x) = \log_{10} x$ and $\log_{10}(102) \approx 2.0086$, which is closest to $f'(100)$?

(A) 0.0043   (B) 0.0086   (C) 0.01   (D) 1.0043   (E) 2

15. If $G(2) = 5$ and $G'(x) = \dfrac{10x}{9 - x^2}$, then an estimate of $G(2.2)$ using local linearization is approximately

(A) 5.4   (B) 5.5   (C) 5.8   (D) 8.8   (E) 13.8

16. The area bounded by the parabola $y = x^2$ and the lines $y = 1$ and $y = 9$ equals

(A) 8   (B) $\dfrac{84}{3}$ (C) $\dfrac{64}{3}\sqrt{2}$ (D) 32   (E) $\dfrac{104}{3}$

17. Suppose $f(x) = \dfrac{x^2 + x}{x}$ if $x \neq 0$ and $f(0) = 1$. Which of the following statements is (are) true of f?

  I. f is defined at $x = 0$.

  II. $\lim\limits_{x \to 0} f(x)$ exists.

  III. f is continuous at $x = 0$.

(A) I only   (B) II only   (C) I and II only

(D) None of the statements is true.   (E) All are true.

18. Which function could have the graph shown below?

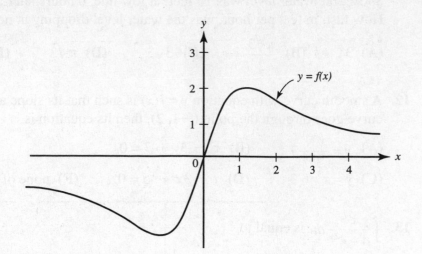

(A) $y = \dfrac{x}{x^2 + 1}$ (B) $y = \dfrac{4x}{x^2 + 1}$ (C) $y = \dfrac{2x}{x^2 - 1}$ (D) $y = \dfrac{x^2 + 3}{x^2 + 1}$

(E) $y = \dfrac{4x}{x + 1}$

19. Suppose the function f is both increasing and concave up, on $a \leqslant x \leqslant b$. Then, using the same number of subdivisions, and with L, R, M, and T denoting, respectively, left, right, midpoint, and trapezoid sums, it follows that
(A) $R \leqslant T \leqslant M \leqslant L$ (B) $L \leqslant T \leqslant M \leqslant R$
(C) $R \leqslant M \leqslant T \leqslant L$ (D) $L \leqslant M \leqslant T \leqslant R$
(E) none of these

20. $\displaystyle\lim_{x \to 3} \dfrac{x+3}{x^2 - 9}$ is

(A) $+\infty$ (B) 0 (C) $\dfrac{1}{6}$ (D) $-\infty$ (E) nonexistent

21. The only function that does not satisfy the Mean Value Theorem on the interval specified is

(A) $f(x) = x^2 - 2x$ on $[-3,1]$ (B) $f(x) = \dfrac{1}{x}$ on $[1,3]$

(C) $f(x) = \dfrac{x^3}{3} - \dfrac{x^2}{2} + x$ on $[-1, 2]$ (D) $f(x) = x + \dfrac{1}{x}$ on $[-1, 1]$

(E) $f(x) = x^{2/3} - 2x$ on $\left[\dfrac{1}{2}, \dfrac{3}{2}\right]$

22. Suppose $f'(x) = x(x - 2)^2(x + 3)$. Which of the following is (are) true?
   I. f has a local maximum at $x = -3$.
   II. f has a local minimum at $x = 0$.
   III. f has neither a local maximum nor a local minimum at $x = 2$.
   (A) I only   (B) II only   (C) III only   (D) I and II only
   (E) I, II, and III

23. If $y = \ln \dfrac{x}{\sqrt{x^2+1}}$, then $\dfrac{dy}{dx}$ is

   (A) $\dfrac{1}{x^2+1}$ (B) $\dfrac{1}{x(x^2+1)}$ (C) $\dfrac{2x^2+1}{x(x^2+1)}$ (D) $\dfrac{1}{x\sqrt{x^2+1}}$

   (E) $\dfrac{1-x^2}{x(x^2+1)}$

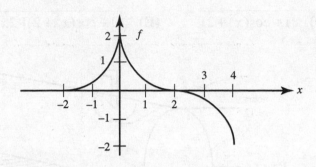

24. The graph of function f shown above consists of three quarter-circles.

   Which of the following is (are) equivalent to $\displaystyle\int_0^2 f(x)\,dx$?

   I. $\dfrac{1}{2}\displaystyle\int_{-2}^{2} f(x)\,dx$ **II.** $\displaystyle\int_4^2 f(x)\,dx$ **III.** $\dfrac{1}{2}\displaystyle\int_0^4 f(x)\,dx$

   (A) I only   (B) II only   (C) III only   (D) I and II only

   (E) all of these

25. The base of a solid is the first-quadrant region bounded by $y = \sqrt[4]{4-2x}$, and each cross section perpendicular to the x-axis is a semicircle with a diameter in the xy-plane. The volume of the solid is

   (A) $\dfrac{\pi}{2}\displaystyle\int_0^2 \sqrt{4-2x}\,dx$ (B) $\dfrac{\pi}{8}\displaystyle\int_0^2 \sqrt{4-2x}\,dx$ (C) $\dfrac{\pi}{8}\displaystyle\int_{-2}^{2} \sqrt{4-2x}\,dx$

   (D) $\dfrac{\pi}{4}\displaystyle\int_0^{\sqrt{2}} (4-y^4)^2\,dy$ (E) $\dfrac{\pi}{8}\displaystyle\int_0^{\sqrt[4]{4}} (4-y^4)^2\,dy$

26. The average value of $f(x) = 3 + |x|$ on the interval $[-2, 4]$ is

   (A) $2\dfrac{2}{3}$ (B) $3\dfrac{1}{3}$ (C) $4\dfrac{2}{3}$ (D) $5\dfrac{1}{3}$ (E) 6

27. $\displaystyle\lim_{x\to\infty} \dfrac{3+x-2x^2}{4x^2+9}$ is

   (A) $-\dfrac{1}{2}$ (B) $\dfrac{1}{2}$ (C) 1   (D) 3   (E) nonexistent

28. The area of the region in the xy-plane bounded by the curves $y = e^x$, $y = e^{-x}$, and $x = 1$ is equal to

   (A) $e + \dfrac{1}{e} - 2$ (B) $e - \dfrac{1}{e}$ (C) $e + \dfrac{1}{e}$ (D) $2e - 2$

   (E) none of the these

Part B

See instructions, page 460. Answers begin on page 538.

29. $f(x) = \int_0^{x^2+2} \sqrt{1+\cos t}\ dt$. Then $f'(x) =$

(A) $2x\sqrt{1+\cos(x^2+2)}$ (B) $2x\sqrt{1-\sin x}$ (C) $(x^2+2)\sqrt{1-\sin x}$

(D) $\sqrt{1+\cos(x^2+2)}$ (E) $\sqrt{[1+\cos(x^2+2)]\cdot 2x}$

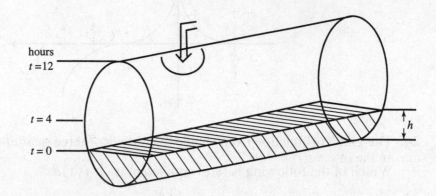

30. A cylindrical tank, shown in the figure above, is partially full of water at time $t = 0$, when more water begins flowing in at a constant rate. The tank becomes half full when $t = 4$, and is completely full when $t = 12$. Let h represent the height of the water at time t. During which interval is $\dfrac{dh}{dt}$ increasing?

(A) none   (B) $0 < t < 4$ (C) $0 < t < 8$ (D) $0 < t < 12$
(E) $4 < t < 12$

31. A particle moves on a line according to the law $s = f(t)$ so that its velocity $v = ks$, where k is a nonzero constant. Its acceleration is
(A) $k^2 v$ (B) $k^2 s$ (C) k (D) 0 (E) none of these

32. A cup of coffee placed on a table cools at a rate of $\dfrac{dH}{dt} = -0.05(H - 70)$ degrees per minute, where H represents the temperature of the coffee and t is time in minutes. If the coffee was at 120°F initially, what will its temperature be 10 minutes later?
(A) 73°F   (B) 95°F   (C) 100°F   (D) 118°F   (E) 143°F

33. An investment of \$4000 grows at the rate of $320e^{0.08t}$ dollars per year after t years. Its value after 10 years is approximately
(A) \$4902   (B) \$8902   (C) \$7122   (D) \$12,902
(E) none of these

34. If $f(x) = (1 + e^x)$ then the domain of $f^{-1}(x)$ is
(A) $(-\infty, \infty)$ (B) $(0, \infty)$ (C) $(1, \infty)$ (D) $\{x \mid x \geqslant 1\}$
(E) $\{x \mid x \geqslant 2\}$

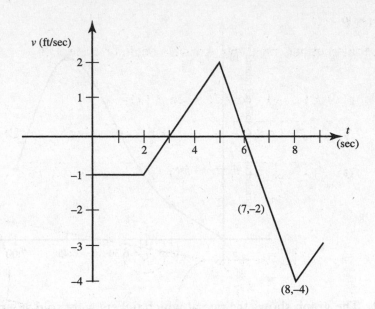

Use the graph shown for Questions 35 and 36. It shows the velocity of an object during the interval $0 \leqslant t \leqslant 9$.

35. The object attains its greatest speed at $t =$
   (A) 2   **(B)** 3   **(C)** 5   **(D)** 6   **(E)** 8

36. The object was at the origin at $t = 3$. It returned to the origin
   (A) at $t = 5$ **(B)** at $t = 6$ **(C)** during $6 < t < 7$
   (D) at $t = 7$ **(E)** during $7 < t < 8$

37. The region bounded by the y-axis, $y = e^x$, and $y = k$ is rotated around the x-axis to form a solid with volume π. Then $k =$
   (A) 1.732   **(B)** 1.895   **(C)** 1.911   **(D)** 2.048   **(E)** 2.149

38. If $\sqrt{x-2}$ is replaced by u, then $\displaystyle\int_3^6 \frac{\sqrt{x-2}}{x}\,dx$ is equivalent to

   (A) $\displaystyle\int_1^2 \frac{u\,du}{u^2+2}$ **(B)** $\displaystyle 2\int_1^2 \frac{u^2\,du}{u^2+2}$ **(C)** $\displaystyle\int_3^6 \frac{2u^2}{u^2+2}\,du$ **(D)** $\displaystyle\int_3^6 \frac{u\,du}{u^2+2}$

   (E) $\displaystyle\frac{1}{2}\int_1^2 \frac{u^2}{u^2+2}\,du$

39. A rectangular pigpen is to be built against a wall so that only three sides will require fencing. If p feet of fencing are to be used, the area of the largest possible pen is

   (A) $\dfrac{p^2}{2}$ **(B)** $\dfrac{p^2}{4}$ **(C)** $\dfrac{p^2}{8}$ **(D)** $\dfrac{p^2}{9}$ **(E)** $\dfrac{p^2}{16}$

40. How many points of inflection does the function f have on the interval $0 \leqslant x \leqslant 6$ if $f''(x) = 2 - 3\sqrt{x}\cos^3 x$?
   (A) 1   **(B)** 2   **(C)** 3   **(D)** 4   **(E)** 5

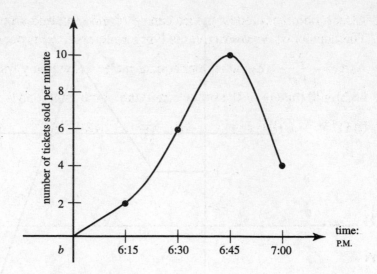

41. The graph shows the rate at which tickets were sold at a movie theater during the last hour before showtime. Using the right-rectangle method, estimate the size of the audience.

(**A**) 230   (**B**) 300   (**C**) 330   (**D**) 375   (**E**) 420

42. At what point of intersection of $f(x) = 4^{\sin x}$ and $g(x) = \ln(x^2)$ do their derivatives have the same sign?

(**A**) −5.2   (**B**) −4.0   (**C**) −1.2   (**D**) 2.6   (**E**) 7.8

43. Let C be the arc of $y = 8 - \dfrac{1}{2}x^2$ which lies above the x-axis. Find the longest line segment both of whose endpoints lie on C.

(**A**) 7.7   (**B**) 8   (**C**) 8.9   (**D**) 9.1   (**E**) 12.5

44. The graph of f' appears below. Which statement(s) about f must be true for $a < x < b$?

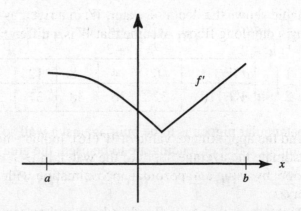

I. f is increasing.   **II.** f is continuous.   **III.** f is differentiable.

(**A**) I only   (**B**) II only   (**C**) I and II only   (**D**) I and III only

(**E**) all three

45. After a bomb explodes, pieces can be found scattered around the center of the blast. The density of bomb fragments lying x meters from ground zero is given by

$$N(x) = \frac{2x}{1 + x^{3/2}}$$ fragments per square meter. How many fragments will be found within 20 meters of the point where the bomb exploded?

(A) 13   **(B)** 278   **(C)** 556   **(D)** 712   **(E)** 4383

Section II

Part A

A graphing calculator is required for some of these problems. See instructions on p. xi. Answers begin on p. 542.

1. Let R represent the region bounded by $y = \sin x$ and $y = x^4$. Find:
(a) the area of R;
(b) the volume of the solid whose base is R if all cross sections perpendicular to the x-axis are isosceles triangles with height 3;
(c) the volume of the solid formed when R is rotated around the x-axis.

2. A curve is defined by $x^2 y - 3y^2 = 48$.

(a) Verify that $\dfrac{dy}{dx} = \dfrac{2xy}{6y - x^2}$.

(b) Write an equation of the linearization of this curve at $(5,3)$.
(c) Using your equation from part (a), estimate the y-coordinate of the point on the curve where $x = 4.93$.
(d) Find equations of all horizontal tangent lines.
(e) Find equations of all vertical tangent lines.

3. The table shows the depth of water, W, in a river, as measured at 4-hour intervals during a day-long flood. Assume that W is a differentiable function of time t.

t (hr)	0	4	8	12	16	20	24
$W(t)$ (ft)	32	36	38	37	35	33	32

(a) Find the approximate value of $W'(16)$. Indicate units of measure.
(b) Estimate the average depth of the water, in feet, over the time interval $0 \le t \le 24$ hours by using a trapezoidal approximation with subintervals of length $\Delta t = 4$ days.
(c) Scientists studying the flooding believe they can model the depth of the water with the function $F(t) = 35 - 3\cos\left(\dfrac{t+3}{4}\right)$, where $F(t)$ represents the depth of the water in feet after t hours. Find $F'(16)$ and explain the meaning of your answer, with appropriate units, in terms of the river depth.
(d) Use the function F to find the average depth of the water, in feet, over the time interval $0 \le t \le 24$ hours.

Part B

No calculator is allowed for any of these problems. See instructions on p. xi.

4. Two autos, P and Q, start from the same point and race along a straight road for 10 seconds. The velocity of P is given by $v_P(t) = 6\left(\sqrt{1+8t} - 1\right)$ feet per second. The velocity of Q is shown in the graph.

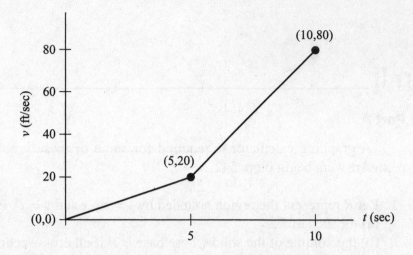

(a) At what time is P's actual acceleration equal to its average acceleration for the entire race?

(b) What is Q's acceleration then?

(c) At the end of the race, which auto was ahead? Explain.

5. Given the differential equation $\dfrac{dy}{dx} = x(4y^2 + 1)$.

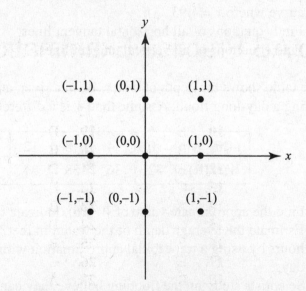

(a) Sketch the slope field for this differential equation at the points shown in the figure.

(b) Let f be the function that satisfies the differential equation and whose graph passes through $(0, \frac{1}{2})$. Express f as a function of x.

(c) Give one value of x for which $\dfrac{dy}{dx}$ does not exist.

6. The graph shown is for $F(x) = \int_0^x f(t)dt$.

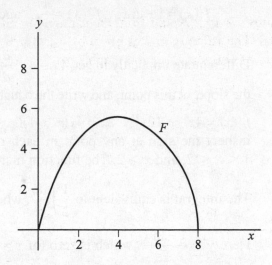

(a) What is $\int_0^2 f(t)dt$?

(b) What is $\int_2^7 f(t)\,dt$?

(c) At what value of x does $f(x) = 0$?
(d) Over what interval is $f'(x)$ negative?

(e) Let $G(x) = \int_2^x f(t)dt$. Sketch the graph of G on the same axes.

Answers to AB Practice Examination 4: Section I

1.	E	10.	E	19.	D	28.	A	37.	B
2.	C	11.	B	20.	E	29.	A	38.	B
3.	C	12.	B	21.	D	30.	E	39.	C
4.	A	13.	C	22.	E	31.	B	40.	A
5.	C	14.	A	23.	B	32.	C	41.	C
6.	C	15.	C	24.	D	33.	B	42.	B
7.	D	16.	E	25.	B	34.	C	43.	D
8.	E	17.	E	26.	C	35.	E	44.	E
9.	E	18.	B	27.	A	36.	E	45.	D

Part A

1.   E.   Here, $\lim_{x \to 2^-}[x] = 2$, while $\lim_{x \to 2^+}[x] = 2$.

2.   C.   The given limit equals $f'\left(\dfrac{\pi}{2}\right)$, where $f(x) = \sin x$.

3. **C.** Since $f(x) = x \ln x$,

$$f'(x) = 1 + \ln x, \quad f''(x) = \frac{1}{x}, \quad \text{and} \quad f'''(x) = -\frac{1}{x^2}.$$

4. **A.** Differentiate implicitly to get $4x - 4y^3 \dfrac{dy}{dx} = 0$. Substitute $(-1, 1)$ to find $\dfrac{dy}{dx}$, the slope, at this point, and write the equation of the tangent: $y - 1 = -1(x + 1)$.

5. **C.** $f'(x) = 4x^3 - 12x^2 + 8x = 4x(x - 1)(x - 2)$. To determine the signs of $f'(x)$, inspect the sign at any point in each of the intervals $x < 0$, $0 < x < 1$, $1 < x < 2$, and $x > 2$. The function increases whenever $f'(x) > 0$.

6. **C.** The integral is equivalent to $\dfrac{1}{2} \displaystyle\int \dfrac{du}{u}$, where $u = 4 + 2\sin x$.

7. **D.** Here $y' = \dfrac{1 - \ln x}{x^2}$, which is zero for $x = e$. Since the signs change from positive to negative as x increases through e, this critical value yields a relative maximum. Note that $f(e) = \dfrac{1}{e}$.

8. **E.** Since $v = \dfrac{ds}{dt} = 5t^4 + 6t^2$ never changes signs, there are no reversals in motion along the line.

9. **E.** The slope field suggests the curve shown above as a particular solution.

10.   E.   Since $f(x) = F'(x) = \dfrac{1}{1-x^2}$, f is discontinuous at $x = 1$; the domain of F is therefore $x > 1$. On $[2, 5]$ $f(x) < 0$, so $\displaystyle\int_5^2 f > 0$. $F''(x) = f'(x) = \dfrac{2x}{(1-x^2)^2}$, which is positive for $x > 1$.

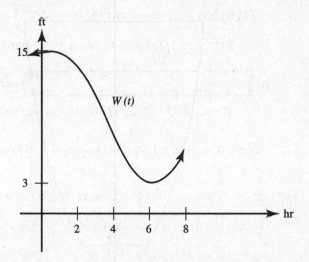

11.   B.   In the graph above, $W(t)$, the water level at time t, is a cosine function with amplitude 6 ft and period 12 hr:

$$W(t) = 6\cos\left(\frac{\pi}{6}t\right) + 9\,\text{ft},$$

$$W'(t) = -\pi\sin\left(\frac{\pi}{6}t\right)\text{ft/hr}.$$

Find $W'(4)$.

12.   B.   Solve the differential equation $\dfrac{dy}{dx} = x^2$, getting $y = \dfrac{x^3}{3} + C$. Use $x = -1$, $y = 2$ to determine C.

13.   C.   Note that the given integral is of the type

$$\int \frac{dv}{a^2 + v^2} = \frac{1}{a}\tan^{-1}\frac{v}{a} + C.$$

14.   A. $f'(100) \approx \dfrac{f(102) - f(100)}{102 - 100} = \dfrac{2.0086 - 2}{2}$.

15.   C. $G'(2) = 4$, so $G(x) \approx 4(x - 2) + 5$.

16.   E. $A = 2\int_1^9 x\,dy = 2\int_1^9 \sqrt{y}\,dy = \dfrac{104}{3}$. See the figure below.

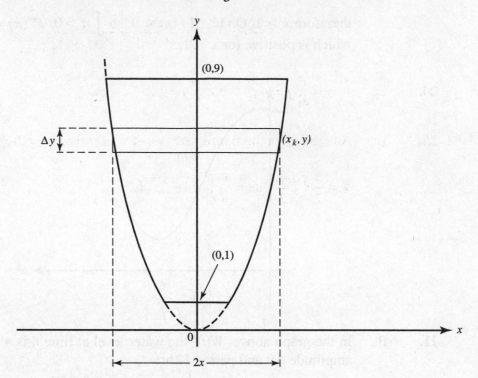

17.   E.   Note that $\lim\limits_{x\to 0} f(x) = f(0) = 1$.

18.   B.   Note that $(0, 0)$ is on the graph, as are $(1, 2)$ and $(-1, -2)$. So only (B) and (E) are possible. Since $\lim\limits_{x\to\infty} y = \lim\limits_{x\to-\infty} y = 0$, only (B) is correct.

19.   D.   See the figure.

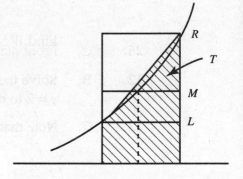

20.   E. $\dfrac{x+3}{x^2-9} = \dfrac{x+3}{(x+3)(x-3)} = \dfrac{1}{x-3}; \lim\limits_{x\to 3^+} \dfrac{1}{x-3} = +\infty; \lim\limits_{x\to 3^-} \dfrac{1}{x-3} = -\infty.$

21.   D.   In (D), $f(x)$ is not defined at $x = 0$. Verify that each of the other functions satisfies both conditions of the Mean Value Theorem.

22.   E.   The signs within the intervals bounded by the critical points are given below.

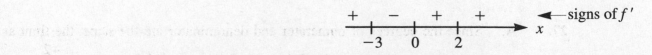

f has a local maximum at -3 and a local minimum at 0.

23.   **B.**   Since $\ln \dfrac{x}{\sqrt{x^2 + 1}} = \ln x - \dfrac{1}{2} \ln (x^2 + 1)$, then

$$\frac{dy}{dx} = \frac{1}{x} - \frac{1}{2} \cdot \frac{2x}{x^2 + 1} = \frac{1}{x(x^2 + 1)}.$$

24.   **D.** $\displaystyle\int_{-2}^{0} f = \int_{0}^{2} f = -\int_{2}^{4} f$, but $\displaystyle\int_{0}^{4} f = 0.$

25.   **B.**   As seen from the figure, $\Delta V = \dfrac{1}{2} \pi r^2 \Delta x$, where $y = 2r$,

$$V = \frac{\pi}{2} \int_{0}^{2} \left(\frac{y}{2}\right)^2 dx = \frac{\pi}{8} \int_{0}^{2} \sqrt{4 - 2x}\, dx.$$

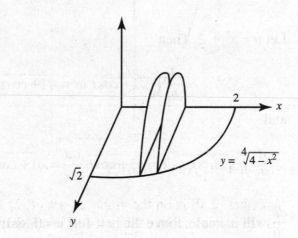

26.   **C.**   From the figure below, $\dfrac{\displaystyle\int_{-2}^{4} f}{6} = \dfrac{\dfrac{5+3}{2} \cdot 2 + \dfrac{3+7}{2} \cdot 4}{6} = \dfrac{28}{6}.$

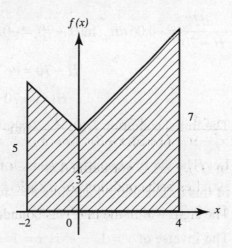

27.   **A.**   Since the degrees of numerator and denominator are the same, the limit as

$x \to \infty$ is the ratio of the coefficients of the terms of highest degree: $\dfrac{-2}{4}$.

28.   **A.**   We see from the figure that
$$\Delta A = (y_2 - y_1)\Delta x;$$

$$A = \int_0^1 (e^x - e^{-x})\, dx$$

$$= (e^x + e^{-x})\Big|_0^1 = e + \frac{1}{e} - 2.$$

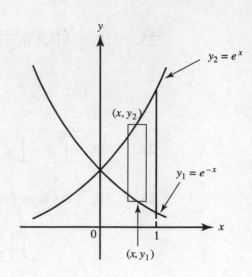

Part B

29.   **A.**   Let $u = x^2 + 2$. Then

$$\frac{d}{du}\int_0^u \sqrt{1+\cos t}\, dt = \sqrt{1+\cos u}$$

and

$$\frac{d}{dx}\int_0^u \sqrt{1+\cos t}\, dt = \sqrt{1+\cos u}\,\frac{du}{dx} = \sqrt{1+\cos(x^2+2)}\cdot(2x).$$

30.   **E.** $\dfrac{dh}{dt}$ will increase above the half-full level (that is, the height of the water will rise more rapidly) as the area of the cross section diminishes.

31.   **B.**   Since $v = ks = \dfrac{ds}{dt}$, then $a = \dfrac{d^2s}{dt^2} = k\,\dfrac{ds}{dt} = kv = k^2 s$.

32.   **C.** $\dfrac{dH}{H-70} = -0.05\,dt. \quad \ln|H - 70| = -0.05t + C$

$$H - 70 = ce^{-0.05t}$$

$$H(x) = 70 + ce^{-0.05t}$$

The initial condition $H(0) = 120$ shows $c = 50$. Evaluate $H(10)$.

33.   **B.**   Let P be the amount after t years. It is given that $\dfrac{dP}{dt} = 320e^{0.08t}$. The solution of this differential equation is $P = 4000e^{0.08t} + C$, where $P(0) = 4000$ yields $C = 0$. The answer is $4000e^{(0.08)\cdot 10}$.

34.   **C.**   The inverse of $y = 1 + e^x$ is $x = 1 + e^y$ or $y = \ln(x - 1)$; $(x - 1)$ must be positive.

35.   **E.**   Speed is the magnitude of velocity: $|v(8)| = 4$.

36.   **E.**   For $3 < t < 6$ the object travels to the right $\dfrac{1}{2}(3)(2) = 3$ units. At $t = 7$ it has returned 1 unit to the left; by $t = 8$, 4 units to the left.

37.   **B.**   Use washers; then

$$\Delta V = \pi(R^2 - r^2)\,\Delta x$$
$$= \pi(k^2 - (e^x)^2)\,\Delta x.$$

So

$$V = \pi \int_0^{\ln k} (k^2 - e^{2x})\,dx = \pi,$$

$$\left(k^2 x - \frac{e^{2x}}{2}\right)\Bigg|_0^{\ln k} = 1,$$

$$\left(k^2 \ln k - \frac{e^{2\ln k}}{2}\right) - \left(0 - \frac{1}{2}\right) = 1,$$

$$k^2 \ln k - \frac{k^2}{2} - \frac{1}{2} = 0.$$

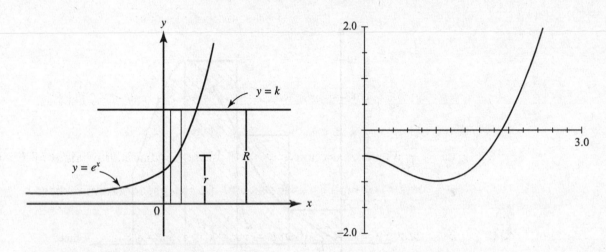

at the right above is a calculator graph for the function in k given by the left side of the east equation. There is a root for the equation near 2. Solving on a calculator yields $k = 1.895$.

38.   **B.**   If $u = \sqrt{x-2}$, then $u^2 = x - 2$, $x = u^2 + 2$, $dx = 2u\,du$. When $x = 3$, $u = 1$; when $x = 6$, $u = 2$.

39.   **C.**   Letting y be the length parallel to the wall and x the other dimension of the rectangle gives

$$p = 2x + y \text{ and } A = xy = x(p - 2x).$$

Then, for $x = \dfrac{p}{4}$, $\dfrac{dA}{dx} = 0$, which yields $y = \dfrac{p}{2}$. Note that $\dfrac{d^2 A}{dx^2} < 0$.

40. A. Graph f'' in $[0,6] \times [-5,10]$. The sign of f'' changes only at $x = a$ as seen in the figure.

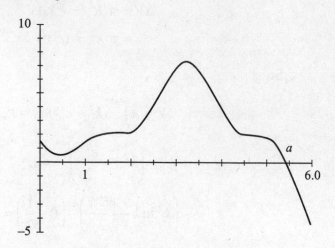

41. C. In the graph below, the first rectangle shows 2 tickets sold per minute for 15 min, or 30 tickets. Similarly, the total is $2(15) + 6(15) + 10(15) + 4(15)$.

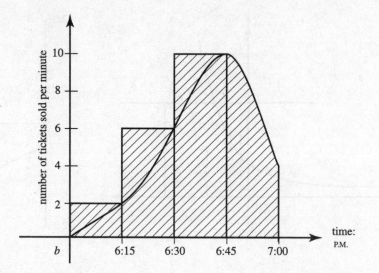

42. B. Graph both functions in $[-8,8] \times [-5,5]$. At point of intersection Q, both are decreasing. Tracing reveals $x \simeq -4$ at Q. If you zoom in on the curves at $x = T$, you will note that they do not actually intersect there.

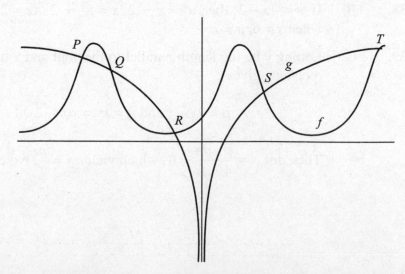

Note: Scales are unequal on x- and y-axes in the figure, so that OQ appears shorter than AB in the figure below, although they are equal. But the lengths given are correct.

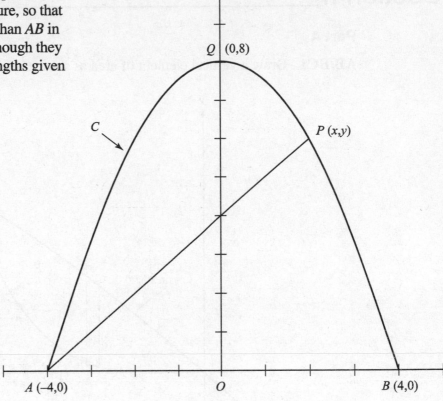

Q $(0,8)$

C

P (x,y)

A $(-4,0)$ O B $(4,0)$

43.   D. $AB = 8$, $AQ = \sqrt{4^2 + 8^2} = 8.944$. Could some segment AP be longer?

$$AP = \sqrt{(x+4)^2 + (y-0)^2}. \text{ Graph } Y_1 = \sqrt{\left((X+4)^2 + \left(8 - \frac{1}{2}X^2\right)^2\right)}$$

on $[0,1] \times [8,10]$, and note that its maximum occurs at $(0.586, 9.073)$.

44.   E. $f'(x) > 0$; the curve shows that f' is defined for all $a < x < b$, so f is differentiable and therefore continuous.

45.   D.   Consider the blast area as a set of concentric rings; one is shown in the figure. The area of this ring, which represents the region x meters from the center of the blast, may be approximated by the area of the rectangle shown. Since the number of particles in the ring is the area times the density, $\Delta P = 2\pi x \cdot \Delta x \cdot N(x)$. To find the total number of fragments within 20 m of the point of the explosion, integrate: $2\pi \int_0^{20} x \dfrac{2x}{1+x^{3/2}} dx \approx 711.575$.

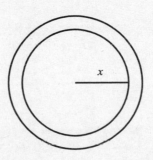

x

$2\pi x$

Δx

Section II

Part A

AB/BC1. Draw a vertical element of area as shown below.

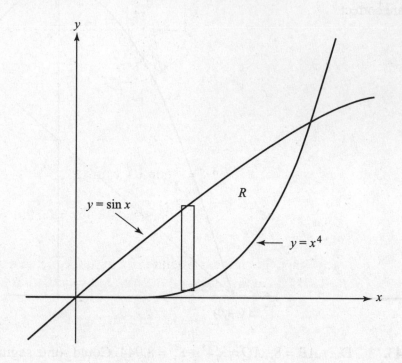

(a) Let a represent the positive point of intersection of $y = x^4$ and $y = \sin x$. Then $a^4 = \sin a$, and the Solver finds $a = 0.9496$.

$$\Delta A = (y_{\text{top}} - y_{\text{bottom}})\Delta x = (\sin x - x^4)\Delta x,$$

$$A = \int_0^a (\sin x - x^4)\, dx.$$

A calculator yields $A = 0.264$.

(b) Elements of volume are triangular prisms with height $h = 3$ and base $b = (\sin x - x^4)$, as shown on the next page.

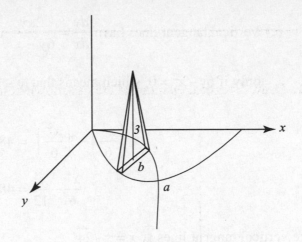

$$\Delta V = \frac{1}{2}(\sin x - x^4)(3)\Delta x,$$

$$V = \frac{3}{2}\int_0^a (\sin x - x^4)\,dx = 0.395.$$

(c) When R is rotated around the x-axis, the element generates washers. If r_1 and r_2 are the radii of the larger and smaller disks, respectively, then

$$\Delta V = \pi(r_1^2 - r_2^2)\Delta x = \pi\big((\sin x)^2 - (x^4)^2\big)\Delta x,$$

$$V = \pi\int_0^a (\sin^2 x - x^8)\,dx = 0.529.$$

AB2.   (a) Since $x^2 y - 3y^2 = 48$,

$$x^2\frac{dy}{dx} + 2xy - 6y\frac{dy}{dx} - 0,$$

$$(x^2 - 6y)\frac{dy}{dx} = -2xy,$$

$$\frac{dy}{dx} = \frac{2xy}{6y - x^2}.$$

(b) At $(5,3)$, $\dfrac{dy}{dx} = \dfrac{2\cdot 5\cdot 3}{6\cdot 3 - 5^2} = -\dfrac{30}{7}$, so the equation of the tangent line is

$$y - 3 = -\frac{30}{7}(x - 5).$$

(c) $y - 3 = -\dfrac{30}{7}(4.93 - 5) = 0.3$, so $y = 3.3$.

(d) Horizontal tangent lines have $\dfrac{dy}{dx} = \dfrac{2xy}{6y - x^2} = 0$. This could happen only if

$2xy = 0$, which means that $x = 0$ or $y = 0$.
If $x = 0$, $0y - 3y^2 = 48$, which has no real solutions.
If $y = 0$, $x^2 \cdot 0 - 3 \cdot 0^2 = 48$, which is impossible. Therefore, there are no horizontal tangents.

See page 544 for part (e).

(e) Vertical tangent lines have $\dfrac{dy}{dx} = \dfrac{2xy}{6y - x^2}$ undefined. This could happen

only if $6y - x^2 = 0$, which means that $y = \dfrac{x^2}{6}$. Substituting gives

$$x^2\left(\frac{x^2}{6}\right) - 3\left(\frac{x^2}{6}\right)^2 = 48,$$

$$\frac{x^4}{6} - \frac{x^4}{12} = 48.$$

There are vertical tangent lines at $x = \pm\sqrt{24}$.

AB/BC3. (a) $W'(16) \approx \dfrac{W(20) - W(16)}{20 - 16} = \dfrac{33 - 35}{4} = -\dfrac{1}{2}$ ft/hr.

(b) The average value of a function is the integral across the given interval

divided by the interval width. Here $\mathrm{Avg}(W) = \dfrac{\displaystyle\int_0^{24} W(t)\,dt}{24 - 0}$. Estimate the

value of the integral using trapezoid rule T with values from the table and $\Delta t = 4$:

$$T = \frac{\Delta t}{2}\big(W(0) + 2W(4) + 2W(8) + 2W(12) + 2W(16) + 2W(20) + W(24)\big)$$

$$= \frac{4}{2}(32 + 2\times36 + 2\times38 + 2\times37 + 2\times35 + 2\times33 + 32)$$

$$= 844.$$

Hence

$$\mathrm{Avg}(W) \approx \frac{844}{24} = 35.167\,\text{ft}.$$

(c) For $F(t) = 35 - 3\cos\left(\dfrac{t+3}{4}\right)$, use your calculator to evaluate

$F'(16) \approx -0.749$. After 16 hr, the river depth is dropping at the rate of 0.749 ft/hr.

(d) $\mathrm{Avg}(F) = \dfrac{\displaystyle\int_0^{24} F(t)\,dt}{24 - 0} \approx 35.116\,\text{ft}.$

Part B

AB/BC 4. (a) $v_P(t) = 6\left(\sqrt{1 + 8t} - 1\right)$, so $v(0) = 0$ and $v(10) = 48$.

The average acceleration is $\dfrac{\Delta v}{\Delta t} = \dfrac{48 - 0}{10 - 0} = \dfrac{24}{5}$ ft/sec^2.

Acceleration $a(t) = v'(t) = 6\cdot\dfrac{1}{2}(1 + 8t)^{-\frac{1}{2}}(8)$ ft/sec^2.

$$\frac{24}{\sqrt{1+8t}} = \frac{24}{5} \text{ when } t = 3\,\text{sec}.$$

(b) Since Q's acceleration, for all t in $0 \le t \le 5$, is the slope of its velocity graph, $a = \dfrac{20-0}{5-0} = 4$.

(c) Find the distance each auto has traveled. For P, the distance is

$$\int_0^{10} 6\left(\sqrt{1+8t}-1\right)dt,$$

$$6\left[\frac{1}{8}\int_0^{10}\sqrt{1+8t}\cdot 8\,dt - \int_0^{10}dt\right],$$

$$6\left(\frac{1}{8}\cdot\frac{2}{3}(1+8t)^{\frac{3}{2}} - t\right)\Bigg|_0^{10},$$

$$6\left(\frac{1}{12}\left(81^{\frac{3}{2}} - 1^{\frac{3}{2}}\right) - 10\right) = 304\,\text{ft}.$$

For auto Q, the distance is the total area of the triangle and trapezoid under the velocity graph shown below, namely,

$$\frac{1}{2}(5\cdot 20) + \frac{1}{2}(20+80)(5) = 300\,\text{ft}.$$

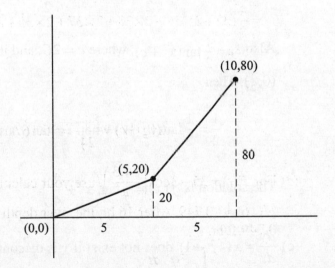

Auto P won the race.

AB5.   (a) Using the differential equation, evaluate the derivative at each point, then sketch a short segment having that slope. For example, at $(-1,-1)$, $\dfrac{dy}{dx} = -1(4(-1)^2 + 1) = -5$; draw a steeply decreasing segment at $(-1,-1)$.

Repeat this process at each of the other points. The result follows.

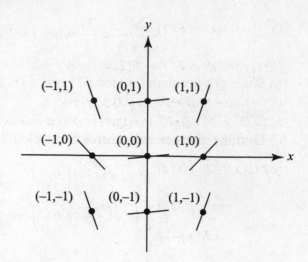

(b) The differential equation $\dfrac{dy}{dx} = x(4y^2 + 1)$ is separable:

$$\int \frac{dy}{4y^2 + 1} = \int x\, dx,$$

$$\frac{1}{2} \int \frac{2\, dy}{(2y)^2 + 1} = \int x\, dx,$$

$$\frac{1}{2}\arctan(2y) = \frac{x^2}{2} + C.$$

Also, $y = \dfrac{1}{2}\tan(x^2 + c)$, where $c = 2C$, and it is given that f passes through $(0, \tfrac{1}{2})$. Then

$$\frac{1}{2} = \frac{1}{2}\tan(0^2 + c) \text{ when } 1 = \tan(c), \text{ so } c = \frac{\pi}{4}.$$

The solution is $f(x) = \dfrac{1}{2}\tan\!\left(x^2 + \dfrac{\pi}{4}\right)$.

(c) $\dfrac{dy}{dx} = x(4y^2 + 1)$ does not exist if y is discontinuous. $y = \dfrac{1}{2}\tan\!\left(x^2 + \dfrac{\pi}{4}\right)$ is

discontinuous, for example, when $x^2 + \dfrac{\pi}{4} = \dfrac{\pi}{2}$, or $x = \dfrac{\sqrt{\pi}}{2}$.

AB 6.   (a) $\displaystyle\int_0^2 f(t)\,dt = F(2) = 4.$

(b) $\displaystyle\int_2^7 f(t)\,dt = F(7) - F(2) = 2 - 4 = -2.$

(c) $f(x) = F'(x)$; $F'(x) = 0$ at $x = 4$.

(d) $f'(x) = F''(x)$. F'' is negative when F is concave downward, which is true for the entire interval $0 < x < 8$.

(e) $\displaystyle G(x) = \int_2^x f(t)\,dt$

$\displaystyle\qquad = \int_0^x f(t)\,dt - \int_0^2 f(t)\,dt$

$\displaystyle\qquad = F(x) - 4.$

Then the graph of G is the graph of F translated downward 4 units.

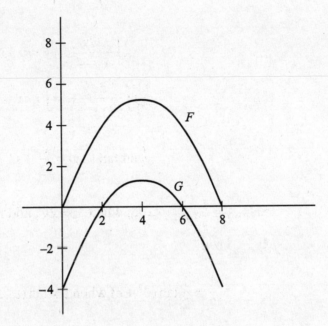

BC PRACTICE
EXAMINATIONS

BC Practice Examination 1

Section I Multiple-Choice Questions _____

Part A

The use of calculators is *not* permitted for this part of the examination.

There are 28 questions in Part A, for which 55 minutes are allowed. To compensate for possible guessing, the grade on this part is determined by subtracting one-fourth of the number of wrong answers from the number answered correctly.

Choose the best answer for each question.

Answers begin on page 561.

1. $\lim\limits_{x \to \infty} \dfrac{3x^2 - 4}{2 - 7x - x^2}$ is

    (A) 3 (B) 1 (C) −3 (D) ∞ (E) 0

2. $\lim\limits_{h \to 0} \dfrac{\cos\left(\dfrac{\pi}{2} + h\right)}{h}$ is

    (A) 1 (B) nonexistent (C) 0 (D) −1 (E) none of these

3. If, for all x, $f'(x) = (x - 2)^4(x - 1)^3$, it follows that the function f has
    (A) a relative minimum at $x = 1$
    (B) a relative maximum at $x = 1$
    (C) both a relative minimum at $x = 1$ and a relative maximum at $x = 1$
    (D) neither a relative maximum nor a relative minimum
    (E) relative minima at $x = 1$ and at $x = 2$

4. Let $F(x) = \displaystyle\int_0^x \dfrac{10}{1 + e^t}\, dt$. Which of the following statements is (are) true?
    I. $F'(0) = 5$.   II. $F(2) < F(6)$.   III. F is concave upward.
    (A) I only (B) II only (C) III only (D) I and II (E) I and III

5. If $f(x) = 10^x$ and $10^{1.04} \approx 10.96$, which is closest to $f'(1)$?
    (A) 0.24 (B) 0.92 (C) 0.96 (D) 10.5 (E) 24

6. If f is differentiable, we can use the line tangent to f at $x = a$ to approximate values of f near $x = a$. Suppose this method always underestimates the correct values. If so, then at $x = a$, f must be

(A) positive   (B) increasing   (C) decreasing
(D) concave upward   (E) concave downward

7. The region in the first quadrant bounded by the x-axis, the y-axis, and the curve of $y = e^{-x}$ is rotated about the x-axis. The volume of the solid obtained is equal to

(A) π (B) 2π (C) $\dfrac{1}{2}$ (D) $\dfrac{\pi}{2}$ (E) none of these

8. $\displaystyle\int_0^1 \dfrac{x\,dx}{x^2 + 1}$ is equal to

(A) $\dfrac{\pi}{4}$ (B) $\ln\sqrt{2}$ (C) $\dfrac{1}{2}(\ln 2 - 1)$ (D) $\dfrac{3}{2}$ (E) $\ln 2$

9. $\displaystyle\lim_{x \to 0^+} x^x$

(A) $= 0$ (B) $= 1$ (C) $= e$ (D) $= \infty$ (E) does not exist

Use the table, which shows the values of differentiable functions f and g, for Questions 10 and 11.

x	f	f'	g	g'
1	2	$\dfrac{1}{2}$	-3	5
2	3	1	0	4
3	4	2	2	3
4	6	4	3	$\dfrac{1}{2}$

10. If $P(x) = g^2(x)$, then $P'(3)$ equals
(A) 4   (B) 6   (C) 9   (D) 12   (E) 18

11. If $H(x) = f^{-1}(x)$, then $H'(3)$ equals
(A) $-\dfrac{1}{16}$ (B) $-\dfrac{1}{8}$ (C) $-\dfrac{1}{2}$ (D) $\dfrac{1}{2}$ (E) 1

12. $\displaystyle\int_0^1 xe^x\,dx$ equals

(A) 1   (B) -1 (C) $2 - e$ (D) $\dfrac{e^2}{2} - e$ (E) $e - 1$

13. The curve of $y = \dfrac{1-x}{x-3}$ is concave up when

    (A) $x > 3$ (B) $1 < x < 3$ (C) $x > 1$ (D) $x < 1$ (E) $x < 3$

14. The area of the largest isosceles triangle that can be drawn with one vertex at the origin and with the others on a line parallel to and above the x-axis and on the curve $y = 27 - x^2$ is

    (A) $12\sqrt{3}$ (B) 27 (C) $24\sqrt{3}$ (D) 54 (E) 108

15. The length of the curve $y = 2x^{3/2}$ between $x = 0$ and $x = 1$ is equal to

    (A) $\dfrac{2}{27}(10^{3/2})$ (B) $\dfrac{2}{27}(10^{3/2} - 1)$ (C) $\dfrac{2}{3}(10^{3/2})$ (D) $\dfrac{4}{5}$

    (E) none of these

16. If $\dfrac{dx}{dt} = kx$, and if $x = 2$ when $t = 0$ and $x = 6$ when $t = 1$, then k equals

    (A) $\ln 4$ (B) 8 (C) e^3 (D) 3 (E) none of these

17. If $y = x^2 \ln x \; (x > 0)$, then y'' is equal to

    (A) $3 + \ln x$ (B) $3 + 2\ln x$ (C) $3\ln x$ (D) $3 + 3\ln x$

    (E) $2 + x + \ln x$

18. A particle moves along the curve given parametrically by $x = \tan t$ and $y = \sec t$. At the instant when $t = \dfrac{\pi}{6}$, the particle's speed equals

    (A) $\sqrt{2}$ (B) $2\sqrt{7}$ (C) $\dfrac{2\sqrt{5}}{3}$ (D) $\dfrac{2\sqrt{13}}{3}$ (E) none of these

19. Suppose $\dfrac{dy}{dx} = \dfrac{10x}{x+y}$ and $y = 2$ when $x = 0$. Use Euler's method with two steps to estimate y at $x = 1$.

    (A) 1 (B) 2 (C) 3 (D) 4 (E) $5\dfrac{1}{3}$

    The graph shown is for Questions 20 and 21. It consists of a quarter-circle and two line segments, and represents the velocity of an object during the 6-second interval.

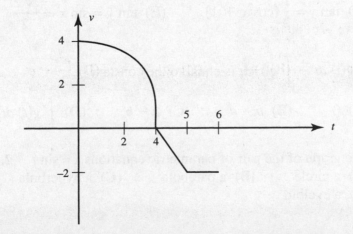

20. The object's average speed during the 6-second interval is

(A) $\dfrac{4\pi + 3}{6}$ (B) $\dfrac{4\pi - 3}{6}$ (C) -1 (D) $-\dfrac{1}{3}$ (E) 1

21. The object's acceleration at $t = 2$ is

(A) -1 (B) $-\dfrac{1}{2}$ (C) $-\dfrac{1}{3}$ (D) $-\dfrac{1}{\sqrt{3}}$ (E) $-\sqrt{3}$

22. Which of the following equations can be a solution of the differential equation whose slope field is shown above?

(A) $2xy = 1$ (B) $2x + y = 1$ (C) $2x^2 + y^2 = 1$ (D) $2x^2 - y^2 = 1$
(E) $y = 2x^2 + 1$

23. If y is a differentiable function of x, then the slope of the curve of $xy^2 - 2y + 4y^3 = 6$ at the point where $y = 1$ is

(A) $-\dfrac{1}{18}$ (B) $-\dfrac{1}{26}$ (C) $\dfrac{5}{18}$ (D) $-\dfrac{11}{18}$ (E) 2

24. If $\dfrac{dy}{dx} = \cos x \cos^2 y$ and $y = \dfrac{\pi}{4}$ when $x = 0$, then

(A) $\tan y = \sin x + 1$ (B) $\tan y = -\sin x + 1$ (C) $\sec^2 y = \sin x + 2$

(D) $\tan y = \dfrac{1}{2}(\cos^2 x + 1)$ (E) $\tan y = \sin x - \dfrac{\sqrt{2}}{2}$

25. $\displaystyle\int_a^x g(t)\, dt - \int_b^x g(t)\, dt$ is equal to the constant

(A) 0 (B) $b - a$ (C) $a - b$ (D) $\displaystyle\int_a^b g(t)\, dt$ (E) $g(b) - g(a)$

26. The graph of the pair of parametric equations $x = \sin t - 2$, $y = \cos^2 t$ is part of
(A) a circle   (B) a parabola   (C) a hyperbola   (D) a line
(E) a cycloid

27. The base of a solid is the region bounded by the parabola $y^2 = 4x$ and the line $x = 2$. Each plane section perpendicular to the x-axis is a square. The volume of the solid is
(A) 6   (B) 8   (C) 10   (D) 16   (E) 32

28. Which of the following could be the graph of $y = \dfrac{x^2}{e^x}$?

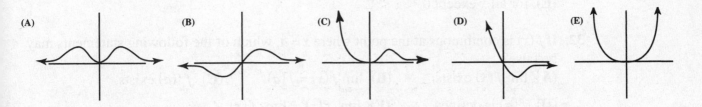

(A)   (B)   (C)   (D)   (E)

Part B

Some questions in this part of the examination require the use of a graphing calculator.

There are 17 questions in Part B, for which 50 minutes will be allowed. The penalty for guessing on this part is the same as that for Part A.

If the exact numerical value of the correct answer is not listed as a choice, select the choice that is closest to the exact numerical answer.

Answers begin on page 562.

29. When partial fractions are used, the decomposition of $\dfrac{x - 1}{x^2 + 3x + 2}$ is equal to

(A) $\dfrac{2}{x + 1} - \dfrac{3}{x + 2}$ (B) $-\dfrac{2}{x + 1} + \dfrac{3}{x + 2}$ (C) $\dfrac{3}{x + 1} - \dfrac{2}{x + 2}$

(D) $\dfrac{2}{x + 1} + \dfrac{3}{x + 2}$ (E) $-\dfrac{2}{x + 1} - \dfrac{3}{x + 2}$

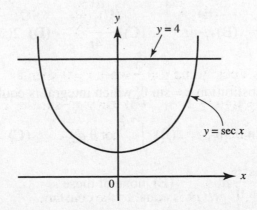

30. The region S in the figure is bounded by $y = \sec x$ and $y = 4$. What is the volume of the solid formed when S is rotated about the x-axis?
(A) 0.304   (B) 39.867   (C) 53.126   (D) 54.088   (E) 108.177

31. The series

$$(x - 1) - \frac{(x - 1)^2}{2!} + \frac{(x - 1)^3}{3!} - \frac{(x - 1)^4}{4!} + \cdots$$

converges
(A) for all real x **(B)** if $0 \leqq x < 2$ **(C)** if $0 < x \leqq 2$ **(D)** only if $x = 1$
(E) for all x except $0 < x < 2$

32. If $f(x)$ is continuous at the point where $x = a$, which of the following statements may be false?
(A) $\lim\limits_{x \to a} f(x)$ exists. **(B)** $\lim\limits_{x \to a} f(x) = f(a)$. **(C)** $f'(a)$ exists.
(D) $f(a)$ is defined. **(E)** $\lim\limits_{x \to a^-} f(x) = \lim\limits_{x \to a^+} f(x)$.

33. A Maclaurin polynomial is to be used to approximate $y = \sin x$ on the interval $-\pi \leqq x \leqq \pi$. What is the least number of terms needed to guarantee no error greater than 0.1?
(A) 3 **(B)** 4 **(C)** 5 **(D)** 6 **(E)** none of these

34. Find the area bounded by the y-axis and the curve defined parametrically by $x(t) = 4 - t^2$, $y(t) = 2^t$.
(A) 6.328 **(B)** 8.916 **(C)** 10.667 **(D)** 12.190 **(E)** 74.529

35. Which series diverges?
(A) $\sum\limits_{n=1}^{\infty} \frac{(-1)^n}{5^n}$ **(B)** $\sum\limits_{n=1}^{\infty} \frac{(-1)^n}{n^5}$ **(C)** $\sum\limits_{n=1}^{\infty} \frac{(-1)^n}{\sqrt[5]{n}}$ **(D)** $\sum\limits_{n=1}^{\infty} \frac{(-1)^n}{5n + 1}$
(E) $\sum\limits_{n=1}^{\infty} \frac{(-1)^n \cdot n}{5n + 1}$

36. If $x = 2t - 1$ and $y = 3 - 4t^2$, then $\frac{dy}{dx}$ is

(A) $4t$ **(B)** $-4t$ **(C)** $-\frac{1}{4t}$ **(D)** $2(x + 1)$ **(E)** $-4(x + 1)$

37. For the substitution $x = \sin \theta$, which integral is equivalent to $\int_0^1 \frac{\sqrt{1 - x^2}}{x} \, dx$?

(A) $\int_{\pi/4}^0 \cot \theta \, d\theta$ **(B)** $\int_0^{\pi/2} \cot \theta \, d\theta$ **(C)** $\int_0^{\pi/2} \frac{\cos^2 \theta}{\sin \theta} \, d\theta$

(D) $\int_0^1 \frac{\cos^2 \theta}{\sin \theta} \, d\theta$ **(E)** none of these

38. If the area under $y = \sin x$ is equal to the area under $y = x^2$ between $x = 0$ and $x = k$, then $k =$
(A) -1.105 **(B)** 0.877 **(C)** 1.105 **(D)** 1.300 **(E)** 1.571

39. The rate at which a rumor spreads across a campus of college students is given by $\frac{dP}{dt} = 0.16\,(1200 - P)$, where $P(T)$ represents the number of students who have heard the rumor after t days. If 200 students heard the rumor today, how many will have heard it by midnight the day after tomorrow?

(A) 320   (B) 474   (C) 494   (D) 520   (E) 726

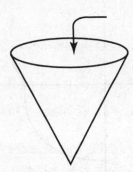

40. Water is poured at a constant rate into the conical reservoir shown above. If the depth of the water is graphed as a function of time, the graph is
(A) decreasing   (B) constant
(C) linear   (D) concave upward
(E) concave downward

41. A 26-foot ladder leans against a building so that its foot moves away from the building at the rate of 3 feet per second. When the foot of the ladder is 10 feet from the building, the top is moving down at the rate of r feet per second, where r is

(A) $\frac{46}{3}$ (B) $\frac{3}{4}$ (C) $\frac{5}{4}$ (D) $\frac{5}{2}$ (E) $\frac{4}{5}$

42. The coefficient of x^3 in the Taylor series of $\ln\,(1 - x)$ about $x = 0$ (the so-called Maclaurin series) is

(A) $-\frac{2}{3}$ (B) $-\frac{1}{2}$ (C) $-\frac{1}{3}$ (D) 0   (E) $\frac{1}{3}$

43. The graph shows an object's acceleration (in ft/sec²). It consists of a quarter-circle and two line segments. If the object was at rest at $t = 5$ seconds, what was its initial velocity?

(**A**) -2 ft/sec   (**B**) $3 - \pi$ ft/sec   (**C**) 0 ft/sec

(**D**) $\pi - 3$ ft/sec   (**E**) $\pi + 3$ ft/sec

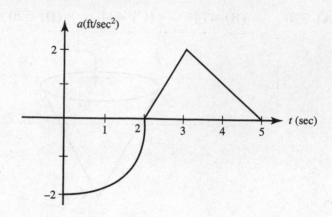

44. Water is leaking from a tank at the rate of $R(t) = 5 \arctan\left(\dfrac{t}{5}\right)$ gallons per hour, where t is the number of hours since the leak began. How many gallons will leak out during the first day?

(**A**) 7   (**B**) 82   (**C**) 124   (**D**) 141   (**E**) 164

45. Find the first-quandrant area inside the rose $r = 3 \sin 2\theta$ but outside the circle $r = 2$.

(**A**) 0.393   (**B**) 0.554   (**C**) 0.790   (**D**) 1.328   (**E**) 2.657

Section II

Part A

A graphing calculator is required for some of these problems. See instructions on p. xi. Answers begin on p. 565.

1. (a) For what *positive* values of x does $f(x) = \sum_{n=1}^{\infty} (-1)^{n+1} \dfrac{x^n}{\ln(n+1)}$ converge?

(b) How many terms are needed to estimate $f(0.5)$ to within 0.01?

(c) Would an estimate for $f(-0.5)$ using the same number of terms be more accurate, less accurate, or the same? Explain.

2. An object in motion along the x-axis has velocity $v(t) = (t + e^t)\sin t^2$ for $1 \le t \le 3$.

(a) Sketch the graph of velocity as a function of time in the window $[1, 3] \times [-15, 20]$.

(b) When is the object moving to the left?

(c) Give one value of t from the interval in part (b) at which the speed of the object is increasing.

(d) At $t = 1$ this object's position was $x = 10$. Where is it when $t = 3$?

3. The sides of a watering trough are made by folding a sheet of metal 24 inches wide and 5 feet long at an angle of 60°, as shown in the following figure. Ends are added, and then the trough is filled with water.

NOTE: Scales are different on the three figures.

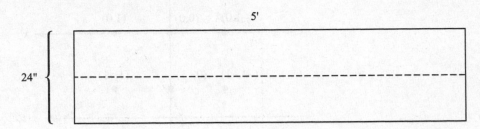

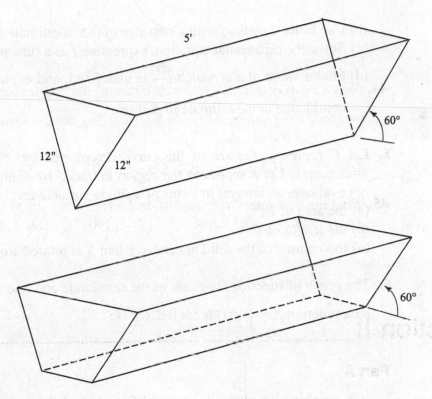

(a) If water pours into the trough at the rate of 600 cubic inches per minute, how fast is the water level rising when the water is 4 inches deep?

(b) Is this the biggest trough that can be made from these materials? Suppose instead, the sheet of metal is folded twice, keeping the sides of equal height and inclined at an angle of 60°, as shown. Where should the folds be in order to maximize the volume of the trough? Justify your answer.

Part B

No calculator is allowed for any of these problems. See instructions on p. xi.

4. Let f be the function satisfying the differential equation $\dfrac{dy}{dx} = x(4y^2 + 1)$ and passing through $\left(0, \dfrac{1}{2}\right)$.

(a) Sketch the slope field for this differential equation at the points shown.

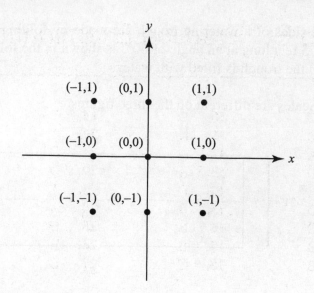

(b) Use Euler's method with a step size of 0.5 to estimate $f(1)$.

(c) Solve the differential equation, expressing f as a function of x.

(d) Find a value of x at which $\dfrac{dy}{dx}$ is undefined, and explain how this influences your confidence in the estimate in part (b).

5. Let C represent the arc of the curve determined by $P(t) = \left(9 - t^2, 2^t\right)$ between its y-intercepts. Let R represent the region bounded by C and the y-axis. Set up, but do not evaluate, an integral in terms of a single variable for:

(a) the area of R;

(b) the length of C;

(c) the volume of the solid generated when R is rotated around the y-axis.

6. The graph of function f consists of the semicircle and line segment shown. Define area function $A = \displaystyle\int_0^x f(t)\,dt$ for $0 \le x \le 18$.

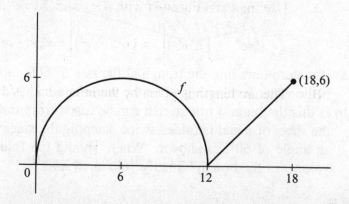

(a) Find $A(6)$ and $A(18)$.

(b) What is the average value of f on the given interval?

(c) Write the equation of the line tangent to A at $x = 6$.

(d) Use this line to estimate the area between f and the x-axis on $[0, 7]$.

(e) Give the coordinates of any points of inflection on the graph of A. Justify your answer.

Answers to BC Practice Examination 1:
Section I

1. C	10. D	19. C	28. C	37. C
2. D	11. E	20. A	29. B	38. D
3. A	12. A	21. D	30. E	39. B
4. D	13. E	22. D	31. A	40. E
5. E	14. D	23. A	32. C	41. C
6. D	15. B	24. A	33. B	42. C
7. D	16. E	25. D	34. B	43. D
8. B	17. B	26. B	35. E	44. C
9. B	18. C	27. E	36. B	45. D

The explanations for questions not given below will be found in the answer section for AB Practice Examination 1 on pages 465 to 472. Identical questions in Section I of Practice Examinations AB1 and BC1 have the same number. For example, explanations of the answers for Questions 1–6, not given below, will be found in Section I of Examination AB1, Answers 1–6, page 463.

7. D. The volume is given by $\lim_{k\to\infty} \pi \int_0^k e^{-2x} \, dx = \frac{\pi}{2}$.

9. B. Let $y = x^x$ and take logarithms. $\ln y = x \ln x = \frac{\ln x}{1/x}$. As $x \to 0^+$, this function has the indeterminate form ∞/∞. Apply L'Hôpital's rule:

$$\lim_{x\to 0^+} \ln y = \lim_{x\to 0^+} \frac{1/x}{-1/x^2} = \lim_{x\to 0^+} (-x) = 0.$$

So $y \to e^0$ or 1.

12. A. Use the Parts Formula with $u = x$ and $dv = e^x \, dx$. Then $du = dx$ and $v = e^x$, and

$$(xe^x - \int e^x \, dx)\Big|_0^1 = (xe^x - e^x)\Big|_0^1 = (e - e) - (0 - 1).$$

15. B. The arc length is given by the integral $\int_0^1 \sqrt{1 + 9x}$, which is equal to

$$\frac{1}{9}\cdot\frac{2}{3}(1 + 9x)^{3/2}\Big|_0^1 = \frac{2}{27}(10^{3/2} - 1).$$

16. E. Separating variables yields $\frac{dx}{x} = k \, dt$. Integrating gives $\ln x = kt + C$. Since $x = 2$ when $t = 0$, $\ln 2 = C$. Then $\ln\frac{x}{2} = kt$. Using $x = 6$ when $t = 1$, it follows that $\ln 3 = k$.

17. **B.** $y' = x + 2x \ln x$ and $y'' = 3 + 2 \ln x.$

18. **C.** $|\mathbf{v}| = \sqrt{\left(\dfrac{dx}{dt}\right)^2 + \left(\dfrac{dy}{dt}\right)^2} = \sqrt{(\sec^2 t)^2 + (\sec t \tan t)^2}.$ At $t = \dfrac{\pi}{6},$

$$|\mathbf{v}| = \sqrt{\left(\dfrac{2}{\sqrt{3}}\right)^4 + \left(\dfrac{2}{\sqrt{3}} \cdot \dfrac{1}{\sqrt{3}}\right)^2}.$$

19. **C.** At $(0, 2),$ $\dfrac{dy}{dx} = 0.$ With step size $\Delta x = \dfrac{1}{2},$ the first step gives $\left(\dfrac{1}{2}, 2\right),$

where $\dfrac{dy}{dx} = \dfrac{5}{2\frac{1}{2}} = 2;$ so the next step produces $\left(1, 2 + \dfrac{1}{2}(2)\right).$

24. **A.** Separate the variables to obtain $\dfrac{dy}{\cos^2 y} = \cos x \, dx$ and solve $\displaystyle\int \sec^2 y \, dy = \displaystyle\int \cos x \, dx.$ Use the given condition to obtain the particular solution.

26. **B.** Since $x + 2 = \sin t$ and $y = \cos^2 t,$ then

$$(x + 2)^2 + y = 1,$$

where $-3 \le x \le -1$ and $0 \le y \le 1.$

Part B

29. **B.** Set

$$\frac{x - 1}{x^2 + 3x + 2} = \frac{x - 1}{(x + 1)(x + 2)} = \frac{A}{x + 1} + \frac{B}{x + 2}.$$

$$x - 1 = A(x + 2) + B(x + 1);$$

$$x = -2 \text{ implies } -3 = -B, \text{ or } B = 3;$$

$$x = -1 \text{ implies } -2 = A, \text{ or } A = -2.$$

30. **E.** S is the region bounded by $y = \sec x,$ the y-axis, and $y = 4.$

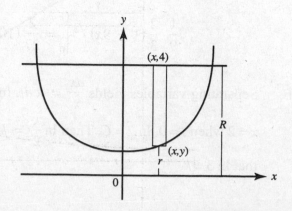

We send region S about the x-axis. Using washers, $\Delta V = \pi(R^2 - r^2)\,\Delta x$. Symmetry allows us to double the volume generated by the first-quadrant portion of S. So for V we have

$$2\pi \int_0^{arc\,\cos\frac{1}{4}} \left(16 - \sec^2 x\right)dx$$

A calculator yields 108.177.

31.   A.   Use the Ratio Test, page 350:

$$\lim_{n\to\infty} \left| \frac{(x-1)^{n+1}}{(n+1)!} \cdot \frac{n!}{(x-1)^n} \right| = \lim_{n\to\infty} \frac{1}{n+1} \left| x - 1 \right|,$$

which equals zero if $x \neq 1$. The series also converges if $x = 1$.

32.   C.   The absolute value function $f(x) = |x|$ is continuous at $x = 0$, but $f'(0)$ does not exist.

33.   B.   The Maclaurin series is

$$\sin x = x - \frac{x^3}{3!} + \frac{x^5}{5!} - \frac{x^7}{7!} + \frac{x^9}{9!} - \;\colon\colon\colon$$

When the sum of an alternating series is approximated by using a finite number of terms, the error is less than the first term omitted. On the interval $-\pi \leq x \leq \pi$, the maximum error (numerically) occurs when $x = \pi$. Since

$$\frac{\pi^7}{7!} < 0.6 \quad \text{and} \quad \frac{\pi^9}{9!} < 0.09,$$

four terms will suffice to assure no error greater than 0.1.

34.   B.

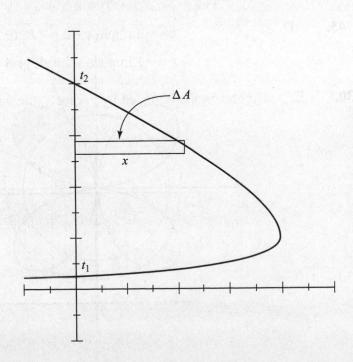

(cont'd)   Letting X_{1T} be $4 - T^2$ and Y_{1T} be 2^T, graph the parametric equations in the following window: T in $[-3, 3]$, (X_{1T}, Y_{1T}) in $[-1, 5] \times [-1, 5]$. Now $\Delta A = x\Delta y$; the limits of integration are the two points where the curve cuts the y-axis, that is, where $x = 0$. In terms of t, these are $t_1 = -2$ and $t_2 = +2$. So

$$A = \int_{t = t_1}^{t = t_2} x \, dy = \int_{-2}^{2} (4 - t^2)2^t \ln 2 \, dt.$$

Using the calculator, the answer is 8.916.

35.   E. $\lim\limits_{n \to \infty} \dfrac{n}{5n + 1} = \dfrac{1}{5} \neq 0.$

36.   B. $\dfrac{dx}{dt} = 2$ and $\dfrac{dy}{dt} = -8t$. Use the fact that $\dfrac{dy}{dx} = \dfrac{dy/dt}{dx/dt}.$

37.   C. $\sqrt{1 - \sin^2 \theta} = \cos\theta$, $dx = \cos\theta \, d\theta$, $\sin^{-1} 0 = 0$, $\sin^{-1} 1 = \dfrac{\pi}{2}.$

39.   B.   Solve by separation of variables; then

$$\frac{dP}{1200 - P} = 0.16 \, dt,$$

$$-\ln(1200 - P) = 0.16t + C,$$

$$1200 - P = ce^{-0.16t}.$$

Use $P(0) = 200$; then $c = 1000$, so $P(x) = 1200 - 1000e^{-0.16t}$. Now $P(2) = 473.85.$

40.   E.   As the water gets deeper, the rate of change of depth decreases: $d^2h/dt^2 < 0.$

42.   C.   The power series for $\ln(1 - x)$, if $x < 1$, is $-x - \dfrac{x^2}{2} - \dfrac{x^3}{3} - \cdots.$

45.   D.

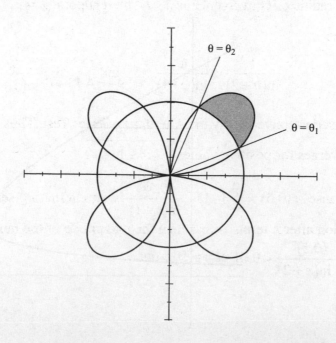

(cont'd).   Using the polar mode, let $r_1 = 3 \sin 2\theta$, $r_2 = 2$, and graph r_1 and r_2 in $[-6, 6] \times [-4, 4]$ for θ in $[0, 2\pi]$. The limits of integration are θ_1 and θ_2, where

$$\theta_1 \text{ is the solution of } (r_1 - r_2) \text{ near } 0.5,$$

and

$$\theta_2 \text{ is the solution of } (r_1 - r_2) \text{ near } 1.$$

To find the shaded area A, subtract the area in the circle from that in the rose. Then

$$A = \frac{1}{2}\int_{\theta_1}^{\theta_2} r_1^2 \, d\theta - \frac{1}{2}\int_{\theta_1}^{\theta_2} r_2^2 \, d\theta = \frac{1}{2}\int_{\theta_1}^{\theta_2} \left(r_1^2 - r_2^2\right) d\theta$$

The answer is 1.328.

Section II

Part A

1. (a) Use the Ratio Test:

$$\lim_{n\to\infty}\left|\frac{x^{n+1}}{\ln(n+2)} \cdot \frac{\ln(n+1)}{x^n}\right| < 1,$$

$$|x|\lim_{n\to\infty}\left|\frac{\ln(n+1)}{\ln(n+2)}\right| < 1,$$

$$|x| < 1.$$

The radius of convergence is 1. At the endpoint $x = 1$, $f(1) = \sum_{n=1}^{\infty}(-1)^{n+1}\frac{1}{\ln(n+1)}$.

Since

$$\frac{1}{\ln(n+2)} < \frac{1}{\ln(n+1)} \qquad \text{and} \qquad \lim_{n\to\infty}\frac{1}{\ln(n+1)} = 0,$$

this series converges by the Alternating Series Test. Thus $f(x) = \sum_{n=1}^{\infty}(-1)^{n+1}\frac{x^n}{\ln(n+1)}$ converges for positive values $0 < x \le 1$.

(b) Because $f(0.5) = \sum_{n=1}^{\infty}(-1)^{n+1}\frac{(0.5)^n}{\ln(n+1)}$ is an alternating series, the error in approximation after n terms is less than the magnitude of the next term. The Solver shows that $\frac{(0.5)^{n+1}}{\ln(n+2)} < 0.01$ at $n = 5$ terms.

(c) $f(-0.5) = \sum_{n=1}^{\infty} (-1)^{n+1} \frac{(-0.5)^n}{\ln(n+1)} = \sum_{n=1}^{\infty} \frac{-(0.5)^n}{\ln(n+1)}$ is a negative series. Therefore the error will be larger than the magnitude of the first omitted term, and thus less accurate than the estimate for $f(0.5)$.

2. See solution for AB-2, page 473.

3. See solution for AB-3, pages 474–475.

Part B

4. (a) Using the differential equation, evaluate the derivative at each point, then sketch a short segment having that slope. For example, at $(-1, -1)$, $\frac{dy}{dx} = -1\left(4(-1)^2 + 1\right) = -5$; draw a segment at $(-1, -1)$ that decreases steeply. Repeat this process at each of the other points. The result is shown below.

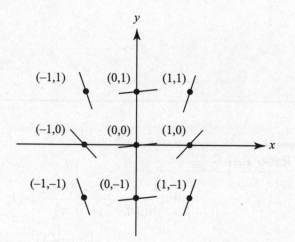

(b) At $(0, 0.5)$, $\frac{dy}{dx} = 0 \cdot \left(4(0.5)^2 + 1\right) = 0$. For $\Delta x = 0.5$ and $\frac{\Delta y}{\Delta x} = 0$, $\Delta y = 0$, so move to $(0 + 0.5, 0.5 + 0) = (0.5, 0.5)$.

At $(0.5, 0.5)$, $\frac{dy}{dx} = 0.5\left(4(0.5)^2 + 1\right) = 1$. Thus, for $\Delta x = 0.5$ and $\frac{\Delta y}{\Delta x} = 1$, $\Delta y = 0.5$. Move to $(0.5 + 0.5, 0.5 + 0.5) = (1, 1)$, then $f(0) \approx 1$.

(c) The differential equation $\frac{dy}{dx} = x\left(4y^2 + 1\right)$ is separable:

$$\int \frac{dy}{4y^2 + 1} = \int x\, dx,$$

$$\frac{1}{2} \int \frac{2dy}{(2y^2) + 1} = \int x\, dx,$$

$$\frac{1}{2} \arctan(2y) = \frac{x^2}{2} + C$$

$y = \frac{1}{2} \tan\left(x^2 + c\right)$ (where $c = 2C$), and it is given that f passes through $\left(0, \frac{1}{2}\right)$: $\frac{1}{2} = \frac{1}{2} \tan\left(0^2 + c\right)$ when $1 = \tan c$, so $c = \frac{\pi}{4}$.

The solution is $f(x) = \dfrac{1}{2}\tan\left(x^2 + \dfrac{\pi}{4}\right)$.

(d) $\dfrac{dy}{dx} = x(4y^2 + 1)$ does not exist if f is discontinuous. $f(x) = \dfrac{1}{2}\tan\left(x^2 + \dfrac{\pi}{4}\right)$ is

discontinuous when $x^2 + \dfrac{\pi}{4} = \dfrac{\pi}{2}$, or $x = \dfrac{\sqrt{\pi}}{2}$. Since this occurs between $x = 0$ and

$x = 1$, Euler's method cannot be used to estimate $f(1)$. See page 309.

5. (a) To find the y-intercepts of the graph of $P(t) = (9 - t^2, 2^t)$, let $x = 9 - t^2 = 0$, and

solve: $t = -3, 3$. Then $P - 3 = \left(0, \dfrac{1}{8}\right)$ and $P(-3) = (0, 8)$.

Draw a horizontal element of area as shown in the graph.   Then:

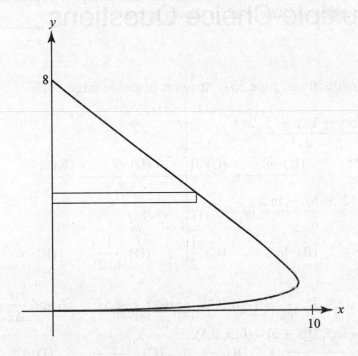

$$\Delta A = x\,\Delta y,$$

$$A = \int x\,dy = \int_{-3}^{3}(9 - t^2)(2^t \ln 2\,dt).$$

(b) $L = \int \sqrt{\left(\dfrac{dx}{dy}\right)^2 + \left(\dfrac{dy}{dt}\right)^2}\,dt = \int_{-3}^{3}\sqrt{(-2t)^2 + (2^t \ln 2)^2}\,dt.$

(c) Use disks. Then $\Delta V = \pi x^2\,\Delta y,$

$$V = \pi \int_{-3}^{3}(9 - t^2)^2(2^t \ln 2\,dt).$$

6. See solution for AB-6, page 476.

BC Practice Examination 2

Section I Multiple-Choice Questions

Part A

See instructions, page 551. Answers begin on page 577.

1. $\lim\limits_{x \to \infty} \dfrac{20x^2 - 13x + 5}{5 - 4x^3}$ is

   (A) -5 (B) ∞ (C) 0 (D) 5 (E) 1

2. $\lim\limits_{h \to 0} \dfrac{\ln(2+h) - \ln 2}{h}$ is

   (A) 0 (B) $\ln 2$ (C) $\dfrac{1}{2}$ (D) $\dfrac{1}{\ln 2}$ (E) ∞

3. If $x = \sqrt{1 - t^2}$ and $y = \sin^{-1} t$, then $\dfrac{dy}{dx}$ equals

   (A) $-\dfrac{\sqrt{1 - t^2}}{t}$ (B) $-t$ (C) $\dfrac{t}{1 - t^2}$ (D) 2 (E) $-\dfrac{1}{t}$

Use the table shown for Questions 4 and 5. The differentiable functions f and g have the values shown.

x	f	f'	g	g'
1	2	$\dfrac{1}{2}$	-3	5
2	3	1	0	4
3	4	2	2	3
4	6	4	3	$\dfrac{1}{2}$

4. The average rate of change of function *f* on [1, 4] is
   (A) 7/6   (B) 4/3   (C) 15/8   (D) 9/4   (E) 8/3

5. If $h(x) = g(f(x))$ then $h'(3) =$
   (A) 1/2   (B) 1   (C) 4   (D) 6   (E) 9

6. $\int_{1}^{2} (3x - 2)^3 \, dx$ is equal to

   (A) $\dfrac{16}{3}$ (B) $\dfrac{63}{4}$ (C) $\dfrac{13}{3}$ (D) $\dfrac{85}{4}$ (E) none of these

7. The maximum value of the function $f(x) = xe^{-x}$ is

   (A) $\dfrac{1}{e}$ (B) *e*   (C) 1   (D) −1   (E) none of these

8. A rectangle of perimeter 18 inches is rotated about one of its sides to generate a right circular cylinder. The rectangle that generates the cylinder of largest volume has an area, in square inches, of

   (A) 14   (B) 20   (C) $\dfrac{81}{4}$ (D) 18   (E) $\dfrac{77}{4}$

9. Which equation has the slope field shown below?
   (A) $\dfrac{dy}{dx} = \dfrac{5}{y}$ (B) $\dfrac{dy}{dx} = \dfrac{5}{x}$ (C) $\dfrac{dy}{dx} = \dfrac{x}{y}$ (D) $\dfrac{dy}{dx} = 5y$ (E) $\dfrac{dy}{dx} = x + y$

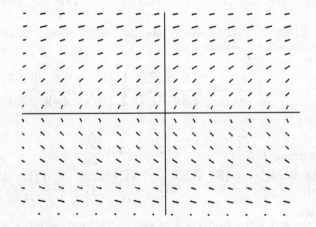

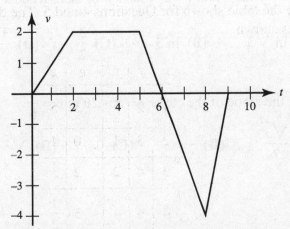

The graph above shows the velocity of an object moving along a line, for $0 \leqslant t \leqslant 9$. It is for questions 10 and 11 on page 570.

10. At what time does the object attain its maximum acceleration?
   (A) $2 < t < 5$ (B) $5 < t < 8$ (C) $t = 6$ (D) $t = 8$ (E) $8 < t < 9$

11. The object is farthest from the starting point at $t =$
   (A) 2   (B) 5   (C) 6   (D) 8   (E) 9

12. If $x = 2 \sin \theta$, then $\displaystyle\int_0^2 \frac{x^2 \, dx}{\sqrt{4 - x^2}}$ is equivalent to:

   (A) $\displaystyle 4 \int_0^1 \sin^2 \theta \, d\theta$ (B) $\displaystyle \int_0^{\pi/2} 4 \sin^2 \theta \, d\theta$ (C) $\displaystyle \int_0^{\pi/2} 2 \sin \theta \tan \theta \, d\theta$

   (D) $\displaystyle \int_0^2 \frac{2 \sin^2 \theta}{\cos^2 \theta} \, d\theta$ (E) $\displaystyle 4 \int_{\pi/2}^0 \sin^2 \theta \, d\theta$

13. $\displaystyle\int_{-1}^1 (1 - |x|) \, dx$ equals

   (A) 0   (B) $\dfrac{1}{2}$ (C) 1   (D) 2   (E) none of these

14. $\displaystyle\lim_{x \to \infty} x^{1/x}$

   (A) $= 0$ (B) $= 1$ (C) $= e$ (D) $= \infty$ (E) does not exist

15. A differentiable function has the values shown in this table:

x	2.0	2.2	2.4	2.6	2.8	3.0
$f(x)$	1.39	1.73	2.10	2.48	2.88	3.30

Estimate $f'(2.1)$.
   (A) 0.34   (B) 0.59   (C) 1.56   (D) 1.70   (E) 1.91

16. If $\dfrac{dy}{dx} = y \tan x$ and $y = 3$ when $x = 0$, then, when $x = \dfrac{\pi}{3}$, $y =$

   (A) $\ln \sqrt{3}$ (B) $\ln 3$ (C) $\dfrac{3}{2}$ (D) $\dfrac{3\sqrt{3}}{2}$ (E) 6

17. Find the slope of the curve $r = \cos 2\theta$ at $\theta = \dfrac{\pi}{6}$.

   (A) $\dfrac{\sqrt{3}}{7}$ (B) $\dfrac{1}{\sqrt{3}}$ (C) 0   (D) $\sqrt{3}$ (E) $-\sqrt{3}$

18. $\displaystyle\int_0^6 f(x-1)\,dx =$

(A) $\displaystyle\int_{-1}^{7} f(x)\,dx$ (B) $\displaystyle\int_{-1}^{5} f(x)\,dx$ (C) $\displaystyle\int_{-1}^{5} f(x+1)\,dx$

(D) $\displaystyle\int_{1}^{5} f(x)\,dx$ (E) $\displaystyle\int_{1}^{7} f(x)\,dx$

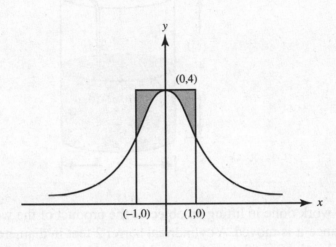

19. The equation of the curve shown above is $y = \dfrac{4}{1 + x^2}$. What does the area of the shaded region equal?

(A) $4 - \dfrac{\pi}{4}$ (B) $8 - 2\pi$ (C) $8 - \pi$ (D) $8 - \dfrac{\pi}{2}$ (E) $2\pi - 4$

20. The length of one arch of the cycloid $x = \theta - \sin\theta$, $y = 1 - \cos\theta$ is given by the integral

(A) $\displaystyle\int_0^{\pi} \sin\frac{\theta}{2}\,d\theta$ (B) $2\displaystyle\int_0^{\pi} \sin\frac{\theta}{2}\,d\theta$ (C) $2\displaystyle\int_0^{2\pi} \sin\frac{\theta}{2}\,d\theta$

(D) $\sqrt{2}\displaystyle\int_0^{\pi} \sqrt{1 - \cos\theta}\,d\theta$ (E) none of these

21. A particle moves along a line with velocity, in feet per second, $v = t^2 - t$. The total distance, in feet, traveled from $t = 0$ to $t = 2$ equals

(A) $\dfrac{1}{3}$ (B) $\dfrac{2}{3}$ (C) 2   (D) 1   (E) $\dfrac{4}{3}$

22. The general solution of the differential equation $\dfrac{dy}{dx} = \dfrac{1 - 2x}{y}$ is a family of

(A) straight lines   (B) circles   (C) hyperbolas
(D) parabolas   (E) ellipses

23. The curve $x^3 + x\tan y = 27$ passes through $(3, 0)$. Use local linearization to estimate the value of y at $x = 3.1$. The value is

(A) -2.7 (B) -0.9 (C) 0   (D) 0.1   (E) 3.0

24. $\int x \cos x \, dx =$

    (A) $x \sin x + \cos x + C$ **(B)** $x \sin x - \cos x + C$

    (C) $\dfrac{x^2}{2} \sin x + C$ **(D)** $\dfrac{1}{2} \sin x^2 + C$ **(E)** none of these

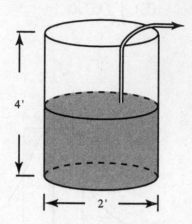

25. The work done in lifting an object is the product of the weight of the object and the distance it is moved. A cylindrical barrel 2 feet in diameter and 4 feet high is half-full of oil weighing 50 pounds per cubic feet. How much work is done, in foot-pounds, in pumping the oil to the top of the tank?

    (A) 100π **(B)** 200π **(C)** 300π **(D)** 400π **(E)** 1200π

26. The coefficient of the $(x - 8)^2$ in the Taylor polynomial for $y = x^{2/3}$ around $x = 8$ is

    (A) $-\dfrac{1}{144}$ **(B)** $-\dfrac{1}{72}$ **(C)** $-\dfrac{1}{9}$ **(D)** $\dfrac{1}{144}$ **(E)** $\dfrac{1}{6}$

27. If $f'(x) = h(x)$ and $g(x) = x^3$, then $\dfrac{d}{dx} f(g(x)) =$

    (A) $h(x^3)$ **(B)** $3x^2 h(x)$ **(C)** $h'(x)$ **(D)** $3x^2 h(x^3)$ **(E)** $x^3 h(x^3)$

28. $\displaystyle\int_0^\infty e^{-x/2} \, dx =$

    (A) $-\infty$ **(B)** -2 **(C)** 1 **(D)** 2 **(E)** ∞

Part B

See instructions, page 555. Answers begin on page 580.

29. The path of a satellite is given by the parametric equations

$$x = 4 \cos t + \cos 12t,$$

$$y = 4 \sin t + \sin 12t.$$

The upward velocity at $t = 1$ equals

    (A) 2.829   **(B)** 3.005   **(C)** 3.073   **(D)** 3.999   **(E)** 12.287

30. A population of rabbits grows according to the differential equation
$\dfrac{dR}{dt} = 0.001R(200 - R)$, where $R(t)$ is the number of rabbits after t months. If there
were initially 25 rabbits, approximately how many months will it take the population
to double?
(A) 2.7   (B) 3.5   (C) 4.2   (D) 6.9   (E) 8.4

31. An object moving along a line has velocity $v(t) = t \cos t - \ln (t + 2)$, where
$0 \le t \le 10$. The object achieves its maximum speed when $t =$
(A) 3.743   (B) 5.107   (C) 6.419   (D) 7.550   (E) 9.538

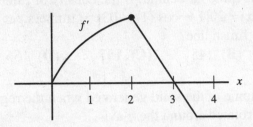

32. The graph of f', which consists of a quarter-circle and two line segments, is shown
above. At $x = 2$ which of the following statements is true?
(A) f is not continuous.
(B) f is continuous but not differentiable.
(C) f has a relative maximum.
(D) f has a point of inflection.
(E) none of these

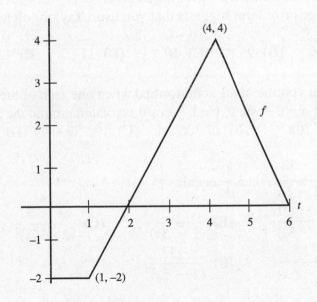

33. Let $H(x) = \displaystyle\int_0^x f(t)\, dt$, where f is the function whose graph appears above.

The local linearization of $H(x)$ near $x = 3$ is $H(x) \simeq$
(A) $-2x + 8$ (B) $2x - 4$ (C) $-2x + 4$ (D) $2x - 8$ (E) $2x - 2$

34. The table shows the speed of an object, in feet per second, during a 3-second period.

time (sec)	0	1	2	3
speed (ft/sec)	30	22	12	0

Estimate the distance the object travels, using the trapezoid method.
(A) 34 ft　　(B) 45 ft　　(C) 48 ft　　(D) 49 ft　　(E) 64 ft

35. In a marathon, when the winner crosses the finish line many runners are still on the course, some quite far behind. If the density of runners x miles from the finish line is given by $R(x) = 20[1 - \cos(1 + .03x^2)]$ runners per mile, how many are within 8 miles of the finish line?
(A) 30　　(B) 145　　(C) 157　　(D) 166　　(E) 195

36. Find the volume of the solid generated when the region bounded by the y-axis, $y = e^x$, and $y = 2$ is rotated around the y-axis.
(A) 0.296　　(B) 0.592　　(C) 2.427　　(D) 3.998　　(E) 27.577

37. If $f(t) = \displaystyle\int_0^{t^2} \frac{1}{1 + x^2}\, dx$, then $f'(t)$ equals

(A) $\dfrac{1}{1 + t^2}$　　(B) $\dfrac{2t}{1 + t^2}$　　(C) $\dfrac{1}{1 + t^4}$　　(D) $\dfrac{2t}{1 + t^4}$　　(E) $\tan^{-1} t^2$

38. You wish to estimate e^x, over the interval $|x| < 2$, with an error less than 0.001. The Lagrange error term suggests that you use a Taylor polynomial at 0 with degree at least
(A) 6　　(B) 9　　(C) 10　　(D) 11　　(E) 12

39. Find the volume of the solid formed when one arch of the cycloid defined parametrically by $x = \theta - \sin\theta$, $y = 1 - \cos\theta$ is rotated around the x-axis.
(A) 15.708　　(B) 17.306　　(C) 19.739　　(D) 29.609　　(E) 49.348

40. If $y = \dfrac{x - 3}{2 - 5x}$, then $\dfrac{dy}{dx}$ equals

(A) $\dfrac{17 - 10x}{(2 - 5x)^2}$　　(B) $\dfrac{13}{(2 - 5x)^2}$　　(C) $\dfrac{x - 3}{(2 - 5x)^2}$

(D) $\dfrac{17}{(2 - 5x)^2}$　　(E) $\dfrac{-13}{(2 - 5x)^2}$

41. For which function is $\displaystyle\sum_{n=0}^{\infty} \frac{(-1)^n x^{2n}}{(2n)!}$ the Taylor series about 0?
(A) e^x　　(B) e^{-x}　　(C) $\sin x$　　(D) $\cos x$　　(E) $\ln(1 + x)$

42. The hypotenuse AB of a right triangle ABC is 5 feet, and one leg, AC, is decreasing at the rate of 2 feet per second. The rate, in square feet per second, at which the area is changing when $AC = 3$ is

(A) $\dfrac{25}{4}$ (B) $\dfrac{7}{4}$ (C) $-\dfrac{3}{2}$ (D) $-\dfrac{7}{4}$ (E) $-\dfrac{7}{2}$

43. If $f'(x)$ exists on the closed interval $[a, b]$, then it follows that
(A) $f(x)$ is constant on $[a, b]$
(B) there exists a number c, $a < c < b$, such that $f'(c) = 0$
(C) the function has a maximum value on the open interval (a, b)
(D) the function has a minimum value on the open interval (a, b)
(E) the Mean Value Theorem applies

44. Which one of the following series converges?

(A) $\displaystyle\sum_{n=1}^{\infty} \dfrac{1}{\sqrt{n}}$ (B) $\displaystyle\sum_{n=1}^{\infty} \dfrac{1}{n}$ (C) $\displaystyle\sum_{n=1}^{\infty} \dfrac{1}{2n + 1}$

(D) $\displaystyle\sum_{n=1}^{\infty} \dfrac{n}{n^2 + 1}$ (E) $\displaystyle\sum_{n=1}^{\infty} \dfrac{1}{n^2 + 1}$

45. The rate at which a purification process can remove contaminants from a tank of water is proportional to the amount of contaminant remaining. If 20% of the contaminant can be removed during the first minute of the process and 98% must be removed to make the water safe, approximately how long will the decontamination process take?

(A) 2 min   (B) 5 min   (C) 18 min   (D) 20 min   (E) 40 min

Section II

Part A

A graphing calculator is required for some of these problems.
See instructions on p. xi. Answers begin on p. 582.

1. A function f is defined on the interval $[0,4]$, and its derivative is
$f'(x) = e^{\sin(x)} - 2\cos 3x$.
   (a) Sketch f' in the window $[0,4] \times [-2,5]$.
       (Note that the following questions refer to f.)
   (b) On what interval is f increasing?
   (c) At what value(s) of x does f have local maxima? Justify your answer.
   (d) How many points of inflection does f have? Justify your answer.

2. The velocity of an object in motion in the plane for $0 \le t \le 1$ is given by the vector
$$\mathbf{v}(t) = \left(\frac{1}{\sqrt{4-t^2}}, \frac{t}{\sqrt{4-t^2}} \right).$$
   (a) When is this object at rest?
   (b) If this object was at the origin when $t = 0$, what are its speed and position when $t = 1$?
   (c) Find an equation of the curve the object follows, expressing y as a function of x.

3. (a) A spherical snowball melts so that its surface area shrinks at the rate of 10 square centimeters per minute. What is the rate of change of volume when the snowball is 12 centimeters in diameter?

   (b) The snowball is packed most densely nearest the center. Suppose that, when it is 12 centimeters in diameter, its density x centimeters from the center is given by $d(x) = \dfrac{1}{1+\sqrt{x}}$ grams per cubic centimeter. How much does the snowball weigh then?

Part B

No calculator is allowed for any of these problems. See instructions on p. xi.

4. (a) Write the Maclaurin series (including the general term) for $f(x) = \ln(e + x)$.
   (b) What is the radius of convergence?
   (c) Use the first three terms of that series to write an expression that estimates the value of $\displaystyle\int_0^1 \ln(e + x^2)\,dx$.

5. After pollution-abatement efforts, conservation researchers introduce 100 trout into a small lake. The researchers predict that after m months the rate of growth, F, of the trout population will be modeled by the differential equation $\dfrac{dF}{dm} = 0.0002 F(600 - F)$.
   (a) How large is the trout population when it is growing the fastest?
   (b) Solve the differential equation, expressing F as a function of m.
   (c) How long after the lake was stocked will the population be growing the fastest?

6. Given the function $f(x) = e^{2x}(x^2 - 2)$
   (a) For what values of x is f decreasing?
   (b) Does this decreasing arc reach a local or a global minimum? Justify your answer.
   (c) Set up, but do not evaluate, a definite integral in terms of a single variable for the length of this decreasing arc of the curve.

Answers to BC Practice Examination 2: Section I

1.	C	10.	E	19.	B	28.	D	37.	D
2.	C	11.	C	20.	C	29.	E	38.	C
3.	E	12.	B	21.	D	30.	C	39.	E
4.	B	13.	C	22.	E	31.	E	40.	E
5.	B	14.	B	23.	B	32.	D	41.	D
6.	D	15.	D	24.	A	33.	D	42.	D
7.	A	16.	E	25.	C	34.	D	43.	E
8.	D	17.	A	26.	A	35.	D	44.	E
9.	A	18.	B	27.	D	36.	B	45.	C

The explanation for questions not given below will be found in the answer section for AB Practice Examination 2 on pages 487 to 493. Identical questions in Section I of Practice Examinations AB2 and BC2 have the same number. For example, an explanation of the answer to Question 1, not given below, will be found in Section I of Examination AB2, Answer 1, page 487.

Part A

3. **E.** Here,

$$\frac{dy}{dx} = \frac{\dfrac{dy}{dt}}{\dfrac{dx}{dt}} = \frac{\dfrac{1}{\sqrt{1 - t^2}}}{\dfrac{1}{2} \dfrac{(-2t)}{\sqrt{1 - t^2}}} = -\frac{1}{t}.$$

6. **D.** Evaluate $\dfrac{1}{12} (3x - 2)^4 \Big|_{1}^{2}$.

7. **A.** Here, $f'(x)$ is $e^{-x} (1 - x)$; f has maximum value when $x = 1$.

12. **B.** Note that, when $x = 2 \sin \theta$, $x^2 = 4 \sin^2 \theta$, $dx = 2 \cos \theta \, d\theta$, and $\sqrt{4 - x^2} = 2 \cos \theta$. Also,

$$\text{when } x = 0, \theta = 0;$$

$$\text{when } x = 2, \theta = \frac{\pi}{2}.$$

13. **C.** The given integral is equivalent to

$$\int_{-1}^{0} (1 + x) \, dx + \int_{0}^{1} (1 - x) \, dx.$$

The figure shows the graph of

$$f(x) = 1 - |x| \text{ on } [-1,1].$$

The area of triangle PQR is equal to $\displaystyle\int_{-1}^{1} (1 - |x|) \, dx.$

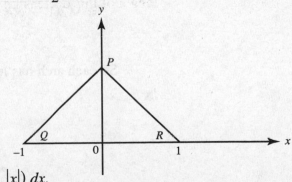

14. B. Let $y = x^{1/x}$; then take logarithms. $\ln y = \dfrac{\ln x}{x}$. As $x \to \infty$, the fraction is of the form ∞/∞. $\displaystyle\lim_{x\to\infty} \ln y = \lim_{x\to\infty} \dfrac{1/x}{1} = 0$. So $y \to e^0$ or 1.

16. E. Separating variables yields $\dfrac{dy}{y} = \tan x$, so $\ln y = -\ln \cos x + C$. With $y = 3$ when $x = 0$, $C = \ln 3$. The general solution is therefore $(\cos x)\, y = 3$. When $x = \dfrac{\pi}{3}$,

$$\cos x = \frac{1}{2} \quad \text{and} \quad y = 6.$$

17. A. Represent the coordinates parametrically as $(r \cos \theta, r \sin \theta)$. Then

$$\frac{dy}{dx} = \frac{\dfrac{dy}{d\theta}}{\dfrac{dx}{d\theta}} = \frac{r \cos \theta + \dfrac{dr}{d\theta} \cdot \sin \theta}{-r \sin \theta + \dfrac{dr}{d\theta} \cdot \cos \theta}.$$

Note that $\dfrac{dr}{d\theta} = -2 \sin 2\theta$, and evaluate $\dfrac{dy}{dx}$ at $\theta = \dfrac{\pi}{6}$. (Alternatively, write $x = \cos 2\theta \cos \theta$ and $y = \cos 2\theta \sin \theta$ to find $\dfrac{dy}{dx}$ from $\dfrac{dy/d\theta}{dx/d\theta}$.)

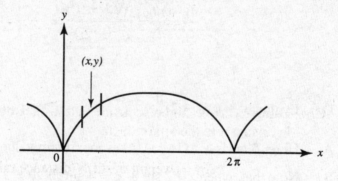

20. C. $\dfrac{ds}{d\theta} = \sqrt{\left(\dfrac{dx}{d\theta}\right)^2 + \left(\dfrac{dy}{d\theta}\right)^2} = \sqrt{(1 - \cos \theta)^2 + (\sin \theta)^2}$

$$= \sqrt{2(1 - \cos \theta)} = \sqrt{2 \cdot 2 \sin^2 \frac{\theta}{2}} = 2 \sin \frac{\theta}{2}.$$

So each arch has length $2 \displaystyle\int_0^{2\pi} \sin \frac{\theta}{2}\, d\theta.$

21.   D.   Note that v is negative from $t = 0$ to $t = 1$, but positive from $t = 1$ to $t = 2$. Thus the distance traveled is given by

$$-\int_0^1 (t^2 - t)\, dt + \int_1^2 (t^2 - t)\, dt.$$

22.   E.   Separating variables yields $y\, dy = (1 - 2x)\, dx$. Integrating gives

$$\frac{1}{2} y^2 = x - x^2 + C \text{ or } y^2 = 2x - 2x^2 + k \text{ or } 2x^2 + y^2 - 2x = k.$$

24.   A.   Use parts; then $u = x$, $dv = \cos x\, dx$; $du = dx$, $v = \sin x$. Thus,

$$\int x \cos x\, dx = x \sin x - \int \sin x\, dx.$$

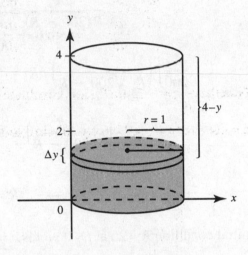

25.   C.   Using the above figure, consider a thin slice of the oil, and the work Δw done in raising it to the top of the tank:

$$\Delta w = (\text{weight of oil}) \times (\text{distance raised})$$

$$= (50 \cdot \pi \cdot 1^2 \Delta y)(4 - y)$$

The total work is thus $50\pi \displaystyle\int_0^2 (4 - y)\, dy.$

26.   A.   The Taylor polynomial of degree 2 at $x = 8$ for $f(x) = x^{2/3}$ is

$$x^{2/3} = 4 + \frac{1}{3}(x - 8) - \frac{1}{72 \cdot 2!}(x - 8)^2.$$

27.   D.   Here,

$$\frac{d}{dx} f(g(x)) = f'(g(x))g'(x) = h(g(x))g'(x) = h(x^3) \cdot 3x^2.$$

28. D. Evaluate

$$\lim_{b \to \infty} \int_0^b e^{-x/2}\, dx = - \lim_{b \to \infty} 2e^{-x/2}\bigg|_0^b = -2(0-1).$$

Part B

29. E. The vertical component of velocity is

$$\frac{dy}{dt} = 4\cos t + 12\cos 12t.$$

Evaluate at $t = 1$.

30. C. Solve using separation of variables and partial fractions.

$$\frac{dR}{R(200-R)} = 0.001\, dt,$$

$$\frac{1}{200}\int \left(\frac{1}{R} + \frac{1}{200-R}\right) dR = 0.001\, dt,$$

$$\frac{1}{200}\ln\left(\frac{R}{200-R}\right) = 0.001t + C,$$

$$\frac{200-R}{R} = ce^{-0.2t} \quad \to \quad R = \frac{200}{1+ce^{-0.2t}}.$$

Initial condition $R = 25$ at $t = 0$ yields $c = 7$. Now solve

$$50 = \frac{200}{1+7e^{-0.2t}} \quad \text{to find } t \simeq 4.236.$$

36. B.

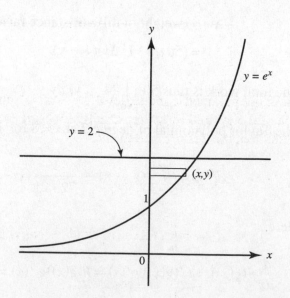

Use disks; then $\Delta V = \pi R^2 H = \pi (\ln y)^2 \, \Delta y$. Note that the limits of the definite integral are 1 and 2. Evaluate

$$\int_1^2 \pi (\ln x)^2 / dx.$$

The required volume is 0.592.

38.   **C.**   The Maclaurin expansion is

$$e^x = 1 + x + \frac{x^2}{2!} + \frac{x^3}{3!} + \cdots + \frac{x^n}{n!} + \cdots.$$

The Lagrange remainder R, after n terms, for some c in the interval $|x| \leqq 2$, is

$$R = \frac{f^{(n+1)}(c) \cdot c^{n+1}}{(n+1)!} = \frac{e^c c^{n+1}}{(n+1)!}.$$

Since R is greatest when $c = 2$, n needs to satisfy the inequality

$$\frac{e^2 2^{n+1}}{(n+1)!} < 0.001.$$

Set Y_1 equal to $(\text{e}^2)(2^{(\text{X}+1)}) / (\text{X}+1)!$ Evaluating Y_1 successively at various integral values of X gives $\text{Y}_1(8) > 0.01$, $\text{Y}_1(9) > 0.002$, $\text{Y}_1(10) < 3.8 \times 10^{-4}$ < 0.0004. Thus we achieve the desired accuracy with a Taylor polynomial at 0 of degree at least 10.

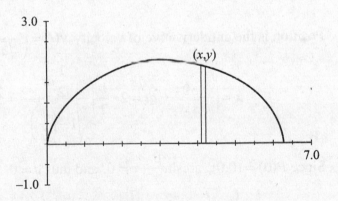

39.   **E.**   Using the parametric mode, let $\text{X}_{1\text{T}} = \text{T} - \sin \text{T}$ and $\text{Y}_{1\text{T}} = 1 - \cos \text{T}$. Graph one arch of the cycloid by setting T in $[0, 2\pi]$ and (X, Y) in $[0, 7] \times [-1, 3]$. Use disks; then the desired volume is

$$V = \pi \int_{t=0}^{t=2\pi} y^2 \, dx$$

$$= \pi \int_0^{2\pi} (1 - \cos t)^2 (1 - \cos t) \, dt$$

$$= \pi \int_0^{2\pi} (1 - \cos t)^3 \, dt,$$

which yields 49.348.

41.   D.   See series (3) on page 371.

43.   E.   The Mean Value Theorem holds if (1) $f(x)$ is continuous on $[a, b]$ and (2) $f'(x)$ exists on (a, b). The condition given in the question assures both of these. Find counterexamples for (A) through (D).

44.   E.   Each is essentially a p-series, $\sum \dfrac{1}{n^p}$. Such a series converges only if $p > 1$.

Section II

Part A

1. See solution for AB-1, pages 493–494.

2. (a) Because $\dfrac{dx}{dt} = \dfrac{1}{\sqrt{4-t^2}}$, which never equals zero, the object is never at rest.

(b) $\mathbf{v}(1) = \left(\dfrac{1}{\sqrt{4-1^2}}, \dfrac{1}{\sqrt{4-1^2}} \right) = \left(\dfrac{1}{\sqrt{3}}, \dfrac{1}{\sqrt{3}} \right)$, so the object's speed is

$$|\mathbf{v}(1)| = \sqrt{\left(\dfrac{1}{\sqrt{3}} \right)^2 + \left(\dfrac{1}{\sqrt{3}} \right)^2} = \sqrt{\dfrac{2}{3}}.$$

Position is the antiderivative of velocity $\mathbf{v}(t) = \left(\dfrac{1}{\sqrt{4-t^2}}, \dfrac{t}{\sqrt{4-t^2}} \right)$.

$$x = \int \dfrac{1}{\sqrt{4-t^2}}\, dt = 2 \cdot \dfrac{1}{2} \int \dfrac{\frac{1}{2}\,dt}{\sqrt{1 - \left(\frac{t}{2} \right)^2}} = \arcsin \dfrac{t}{2} + c.$$

Since $P(0) = (0,0)$, $\arcsin \dfrac{0}{2} + c = 0$, and thus $c = 0$.

$$y = \int \dfrac{t}{\sqrt{4-t^2}}\, dt = -\dfrac{1}{2} \int \left(4 - t^2 \right)^{-1/2} (-2t\,dt) = -\sqrt{4-t^2} + c.$$

Since $P(0) = (0,0)$, $-\sqrt{4-0^2} + c = 0$, and thus $c = 2$.

Then

$$P(t) = \left(\arcsin \dfrac{t}{2}, 2 - \sqrt{4-t^2} \right),$$

$$P(1) = \left(\arcsin \dfrac{1}{2}, 2 - \sqrt{4-t^2} \right) = \left(\dfrac{\pi}{6}, 2 - \sqrt{3} \right).$$

(c) Solving $x = \arcsin\dfrac{t}{2}$ for t yields $t = 2\sin x$. Therefore

$$y = 2 - \sqrt{4 - (2\sin x)^2} = 2 - 2\sqrt{1 - \sin^2 x} = 2 - 2|\cos x|.$$

Since $0 \le t \le 1$ means $0 \le x \le \dfrac{\pi}{6}$, then $\cos x > 0$, so $y = 2 - 2\cos x$.

3. See solution for AB-3, pages 494–495.

Part B

4. (a) To write the Maclaurin series for $f(x) = \ln(e + x)$, use Taylor's theorem at $x = 0$.

n	$f^{(n)}(x)$	$f^{(n)}(0)$	$a_n = \dfrac{f^{(n)}(0)}{n!}$
0	$\ln(e+x)$	1	1
1	$\dfrac{1}{e+x}$	$\dfrac{1}{e}$	$\dfrac{1}{e}$
2	$-(e+x)^{-2}$	$-\dfrac{1}{e^2}$	$-\dfrac{1}{2e^2}$
3	$2(e+x)^{-3}$	$\dfrac{2}{e^3}$	$\dfrac{1}{3e^3}$

$$f(x) = 1 + \frac{x}{e} - \frac{x^2}{2e^2} + \frac{x^3}{3e^3} \cdots + \frac{(-1)^{n+1} x^n}{ne^n} + \cdots.$$

(b) By the Ratio Test, the series converges when

$$\lim_{n \to \infty} \left| \frac{x^{n+1}}{(n+1)e^{n+1}} \cdot \frac{ne^n}{x^n} \right| < 1,$$

$$|x| \lim_{n \to \infty} \frac{1}{e} < 1,$$

$$|x| < e.$$

Thus, the radius of convergence is e.

(c) $\displaystyle \int_0^1 \ln(e + x^2)\, dx \approx \int_0^1 1 + \frac{x^2}{e} - \frac{(x^2)^2}{2e^2}\, dx,$

$\displaystyle \int_0^1 \ln(e + x^2)\, dx \approx \left(x + \frac{x^3}{3e} - \frac{x^5}{5 \cdot 2e^2} \right)\Big|_0^1 = 1 + \frac{1}{3e} - \frac{1}{10e^2}.$

5. (a) To find the maximum rate of growth, first find the derivative of

$$\frac{dF}{dm} = 0.0002F(600 - F) = 0.0002(600F - F^2).$$

$$\frac{d^2F}{dm^2} = 0.0002(600 - 2F), \text{ which equals } 0 \text{ when } F = 300.$$

A signs analysis shows that $\dfrac{d^2F}{dm^2}$ changes from positive to negative there, con-

firming that $\dfrac{dF}{dm}$ is at its maximum when there are 300 trout.

(b) The differential equation $\dfrac{dF}{dm} = 0.0002F(600 - F)$ is separable.

$$\int \frac{dF}{F(600 - F)} = 0.0002 \int dm.$$

To integrate the left side of this equation, use the method of partial fractions.

$$\frac{1}{F(600 - F)} = \frac{A}{F} + \frac{B}{600 - F},$$

$$1 = A(600 - F) + B(F).$$

Let $F = 0$; then $A = \dfrac{1}{600}$.

Let $F = 600$; then $B = \dfrac{1}{600}$.

$$\frac{1}{600}\int \left(\frac{1}{F} + \frac{1}{600 - F}\right) dF = 0.0002 \int dm,$$

$$\int \frac{1}{F}\,dF + (-1)\int \frac{-1}{600 - F}\,dF = 0.12 \int dm,$$

$$\ln F - \ln(600 - F) = 0.12m + C,$$

$$\ln\left(\frac{F}{600 - F}\right) = 0.12m + C,$$

$$\frac{F}{600 - F} = e^{0.12m + C}$$

$$= ce^{0.12m}, \text{ where } c = e^C,$$

$$F = \frac{600}{1 + ce^{-0.12m}}.$$

Since $F = 100$ when $m = 0$, $100 = \dfrac{600}{1 + C}$, so $c = 5$.

The solution is $F = \dfrac{600}{1 + 5e^{-0.12m}}$.

(c) In (a) the population was found to be growing the fastest when $F = 300$. Then:

$$300 = \frac{600}{1 + 5e^{-0.12m}},$$

$$e^{-0.12m} = \tfrac{1}{5},$$

$$m = \frac{\ln \tfrac{1}{5}}{-0.12} \text{ months.}$$

6. (a) See solution for AB-6(a), page 497.

(b) See solution for AB-6(b), page 497.

(c) Find the arc length using $L = \int \sqrt{1 + \left(\dfrac{dy}{dx}\right)^2}\, dx$:

$$L = \int_{-2}^{1} \sqrt{1 + \left[2e^{2x}(x+2)(x-1)\right]^2}\, dx.$$

BC Practice Examination 3

Section I Multiple-Choice Questions _____

Part A

See instructions, page 551. Answers begin on page 595.

1. A function $f(x)$ equals $\dfrac{x^2 - x}{x - 1}$ for all x except $x = 1$. For the function to be continuous at $x = 1$, the value of $f(1)$ must be
   (A) 0   (B) 1   (C) 2   (D) ∞ (E) none of these

2. $\displaystyle\lim_{x \to 0} \dfrac{\sin^2 \dfrac{x}{2}}{x^2}$ is

   (A) 4   (B) 0   (C) $\dfrac{1}{4}$ (D) 2   (E) nonexistent

3. The first four terms of the Taylor series about $x = 0$ of $\sqrt{1 + x}$ are
   (A) $1 - \dfrac{x}{2} + \dfrac{x^2}{4 \cdot 2} - \dfrac{3x^3}{8 \cdot 6}$ (B) $x + \dfrac{x^2}{2} + \dfrac{x^3}{8} + \dfrac{x^4}{48}$
   (C) $1 + \dfrac{x}{2} - \dfrac{x^2}{8} + \dfrac{x^3}{16}$ (D) $1 + \dfrac{x}{4} - \dfrac{x^2}{24} + \dfrac{x^3}{32}$ (E) $-1 + \dfrac{x}{2} - \dfrac{x^2}{8} + \dfrac{x^3}{16}$

4. When the local linearization of $f(x) = \sqrt{9 + \sin(2x)}$ near 0 is used, an estimate of $f(0.06)$ is
   (A) 0.02   (B) 2.98   (C) 3.01   (D) 3.02   (E) 3.03

5. Air is escaping from a balloon at a rate of $R(t) = \dfrac{60}{1 + t^2}$ cubic feet per minute, where t is measured in minutes. How much air, in cubic feet, escapes during the first minute?
   (A) 15   (B) 15π (C) 30   (D) 30π (E) $30\ln 2$

6. The motion of a particle in a plane is given by the pair of equations $x = e^t \cos t$, $y = e^t \sin t$. The magnitude of its acceleration at any time t equals
   (A) $\sqrt{x^2 + y^2}$ (B) $2e^t \sqrt{\cos 2t}$ (C) $2e^t$ (D) e^t (E) $2e^{2t}$

7. By differentiating term by term the series

$$(x - 1) + \frac{(x-1)^2}{4} + \frac{(x-1)^3}{9} + \frac{(x-1)^4}{16} + \cdots,$$

the interval of convergence obtained is
(A) $0 \leqq x \leqq 2$ **(B)** $0 \leqq x < 2$ **(C)** $0 < x \leqq 2$ **(D)** $0 < x < 2$
(E) only $x = 1$

8. A point moves along the curve $y = x^2 + 1$ so that the x-coordinate is increasing at the constant rate of $\frac{3}{2}$ units per second. The rate, in units per second, at which the distance from the origin is changing when the point has coordinates $(1, 2)$ is equal to

(A) $\frac{7\sqrt{5}}{10}$ **(B)** $\frac{3\sqrt{5}}{2}$ **(C)** $3\sqrt{5}$ **(D)** $\frac{15}{2}$ **(E)** $\sqrt{5}$

9. $\lim\limits_{h \to 0} \dfrac{\sqrt{25 + h} - 5}{h}$

(A) $= 0$ **(B)** $= \frac{1}{10}$ **(C)** $= 1$ **(D)** $= 10$ **(E)** does not exist

10. $\displaystyle\int_{\pi/4}^{\pi/3} \sec^2 x \tan^2 x \, dx$ equals

(A) 5 **(B)** $\sqrt{3} - 1$ **(C)** $\frac{8}{3} - \frac{2\sqrt{2}}{3}$ **(D)** $\sqrt{3}$ **(E)** $\sqrt{3} - \frac{1}{3}$

11. $\displaystyle\int_1^e \ln x \, dx$ equals

(A) $\frac{1}{2}$ **(B)** $e - 1$ **(C)** $e + 1$ **(D)** 1 **(E)** -1

12. $\displaystyle\int \frac{(y-1)^2}{2y} dy$ equals

(A) $\frac{y^2}{4} - y + \frac{1}{2} \ln |y| + C$ **(B)** $y^2 - y + \ln |2y| + C$

(C) $y^2 - 4y + \frac{1}{2} \ln |2y| + C$ **(D)** $\frac{(y-1)^3}{3y^2} + C$ **(E)** $\frac{1}{2} - \frac{1}{2y^2} + C$

13. $\displaystyle\int \frac{x-6}{x^2 - 3x} dx =$
(A) $\ln |x^2 (x - 3)| + C$ **(B)** $-\ln |x^2 (x - 3)| + C$

(C) $\ln \left| \dfrac{x^2}{x-3} \right| + C$ **(D)** $\ln \left| \dfrac{x-3}{x^2} \right| + C$ **(E)** none of these

14. Given f' as graphed, which could be a graph of f?

(A) I only   (B) II only   (C) III only   (D) I and III

(E) none of these

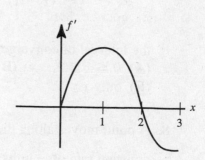

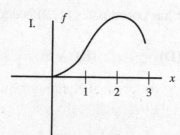 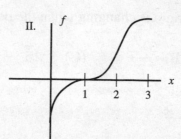 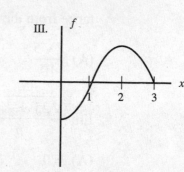

15. Women have just recently begun competing in marathons. The first woman officially timed was Violet Piercey of Great Britain in 1926. Her record of 3:40:22 stood until 1963, mostly because of a lack of women competitors. Soon after, times began dropping rapidly, but lately they have been declining at a much slower rate. Let $M(t)$ be the curve which best represents winning marathon times in year t. Which of the following is negative?

I. $M(t)$ II. $M'(t)$ III. $M''(t)$

(A) I only   (B) II only   (C) III only

(D) II and III   (E) none of these

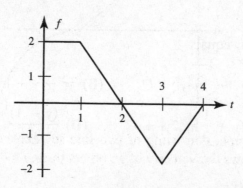

16. The graph of f is shown above. Let $G(x) = \int_0^x f(t)\, dt$ and $H(x) = \int_2^x f(t)\, dt$. Which of the following is true?

(A) $G(x) = H(x)$ (B) $G'(x) = H'(x + 2)$

(C) $G(x) = H(x + 2)$ (D) $G(x) = H(x) - 2$ (E) $G(x) = H(x) + 3$

17. Which one of the following improper integrals converges?

(A) $\int_{-1}^{1} \dfrac{dx}{(x+1)^2}$ (B) $\int_{1}^{\infty} \dfrac{dx}{\sqrt{x}}$ (C) $\int_{0}^{\infty} \dfrac{dx}{(x^2+1)}$ (D) $\int_{1}^{3} \dfrac{dx}{(2-x)^3}$

(E) none of these

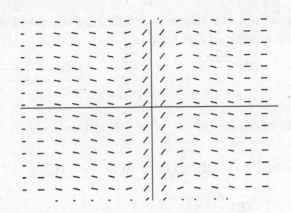

18. Which function could be a particular solution of the differential equation whose slope field is shown above?

(A) $y = x^3$ (B) $y = \dfrac{2x}{x^2+1}$ (C) $y = \dfrac{x^2}{x^2+1}$ (D) $y = \sin x$

(E) $y = e^{-x^2}$

19. A particular solution of the differential equation $\dfrac{dy}{dx} = x + y$ passes through the point (2,1). Using Euler's method with $\Delta x = 0.1$, estimate its y-value at $x = 2.2$.

(A) 0.34   (B) 1.30   (C) 1.34   (D) 1.60   (E) 1.64

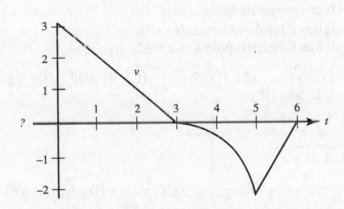

    Use this graph, consisting of two line segments and a quarter circle, for Questions 20 and 21. It shows the velocity of an object during a 6-second interval.

20. For how many values of t is the acceleration undefined?

(A) none   (B) one   (C) two   (D) three   (E) four

21. During what time interval is the speed increasing?

(A) $0 < t < 3$ (B) $3 < t < 5$ (C) $3 < t < 6$ (D) $5 < t < 6$

(E) never

22. If $\dfrac{dy}{dx} = \dfrac{y}{x}$ $(x > 0, y > 0)$ and $y = 3$ when $x = 1$, then

    (A) $x^2 + y^2 = 10$ (B) $y = x + \ln 3$ (C) $y^2 - x^2 = 8$
    (D) $y = 3x$ (E) $y^2 - 3x^2 = 6$

23. A solid is cut out of a sphere of radius 2 by two parallel planes each 1 unit from the center. The volume of this solid is

    (A) 8π (B) $\dfrac{32\pi}{3}$ (C) $\dfrac{25\pi}{3}$ (D) $\dfrac{22\pi}{3}$ (E) $\dfrac{20\pi}{3}$

24. The length of the arc of $y = \dfrac{1}{2}x^2 - \dfrac{1}{4}\ln x$ from $x = 1$ to $x = 4$ is

    (A) $\dfrac{15}{2} + \ln 2$ (B) $\dfrac{1}{2}(15 + \ln 2)$ (C) $\dfrac{1}{2}(17 + \ln 2)$

    (D) $3\dfrac{15}{64}$ (E) $\dfrac{45}{16}$

25. Let $f(x) = x^5 + 3x - 2$, and let f^{-1} denote the inverse of f. Then $(f^{-1})'(2)$ equals

    (A) $\dfrac{1}{83}$ (B) $\dfrac{1}{8}$ (C) 1 (D) 8 (E) 83

26. The curve with parametric equations $x = \sqrt{t - 2}$ and $y = \sqrt{6 - t}$ is
    (A) part of a circle   (B) a parabola   (C) a straight line
    (D) part of a hyperbola   (E) none of these

27. Which of the following statements is (are) true about the graph of $y = \ln(4 + x^2)$?
    I. It is symmetric to the y-axis.
    II. It has a local minimum at $x = 0$.
    III. It has inflection points at $x = \pm 2$.

    (A) I only   (B) II only   (C) III only   (D) I and II only
    (E) I, II, and III

28. $\displaystyle\int_1^2 \dfrac{dx}{\sqrt{4 - x^2}}$ is

    (A) $-\dfrac{\pi}{3}$ (B) $\dfrac{\pi}{6}$ (C) $\dfrac{\pi}{4}$ (D) $\dfrac{\pi}{3}$ (E) nonexistent

Part B

See instructions, page 555. Answers begin on p. 598.

29. The area bounded by the curve $x = 3y - y^2$ and the line $x = -y$ is represented by

    (A) $\displaystyle\int_0^4 (2y - y^2)\, dy$ (B) $\displaystyle\int_0^4 (4y - y^2)\, dy$ (C) $\displaystyle\int_0^3 (3y - y^2)\, dy + \int_0^4 y\, dy$

    (D) $\displaystyle\int_0^3 (y^2 - 4y)\, dy$ (E) $\displaystyle\int_0^3 (2y - y^2)\, dy$

30. Find the area bounded by the spiral $r = \ln \theta$ on the interval $\pi \leq \theta \leq 2\pi$.
    (A) 2.405   (B) 2.931   (C) 3.743   (D) 4.810   (E) 7.487

31. Find the slope of the curve defined parametrically by $x = e^t$, $y = t - \dfrac{t^3}{9}$ at its smallest x-intercept.
    (A) -40.171 (B) -2 (C) 0.050   (D) 0.999   (E) 1

32. Which infinite series converge(s)?
    I. $\displaystyle\sum_{n=1}^{\infty}\frac{3^n}{n!}$ II. $\displaystyle\sum_{n=1}^{\infty}\frac{3^n}{n^3}$ III. $\displaystyle\sum_{n=1}^{\infty}\frac{3n^2}{n^3 + 1}$
    (A) I only   (B) II only   (C) III only   (D) I and III
    (E) none of these

33. Bacteria in a culture increase at a rate proportional to the number present. An initial population of 200 triples in 10 hours. If this pattern of increase continues unabated, then the approximate number of bacteria after 1 full day is
    (A) 1160   (B) 1440   (C) 2408   (D) 2793   (E) 8380

34. When the substitution $x = 2t - 1$ is used, the definite integral $\displaystyle\int_3^5 t\sqrt{2t - 1}\,dt$ may be expressed in the form $k\displaystyle\int_a^b (x + 1)\sqrt{x}\,dx$, where $\{k, a, b\} =$

    (A) $\left\{\dfrac{1}{4}, 2, 3\right\}$ (B) $\left\{\dfrac{1}{4}, 3, 5\right\}$ (C) $\left\{\dfrac{1}{4}, 5, 9\right\}$ (D) $\left\{\dfrac{1}{2}, 2, 3\right\}$
    (E) $\left\{\dfrac{1}{2}, 5, 9\right\}$

35. The curve defined by $x^3 + xy - y^2 = 10$ has a vertical tangent line when $x =$

    (A) 0 or $-\dfrac{1}{3}$ (B) 1.037   (C) 2.074   (D) 2.096   (E) 2.154

Use the graph of f shown on $[0,7]$ for Questions 36 and 37. Let $G(x) = \displaystyle\int_2^{3x-1} f(t)\,dt$.

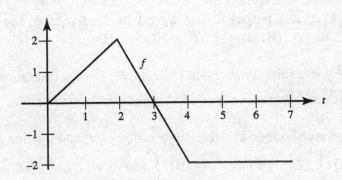

36. $G'(1)$ is
    (A) 1   (B) 2   (C) 3   (D) 6   (E) undefined

37. G has a local maximum at $x =$

(A) 1   (B) $\dfrac{4}{3}$ (C) 2   (D) 5   (E) 8

38. If the half-life of a radioactive substance is 8 years, how long will it take, in years, for two thirds of the substance to decay?

(A) 4.68   (B) 7.69   (C) 12   (D) 12.21   (E) 12.68

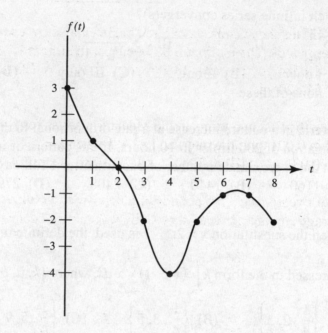

39. Using the left rectangular method and four subintervals, estimate $\displaystyle\int_0^8 |f(t)|\, dt$, where f is the function graphed above.

(A) 4   (B) 5   (C) 8   (D) 15   (E) 16

40. The area in the first quadrant bounded by the curve with parametric equations $x = 2a \tan\theta$ and $y = 2a \cos^2\theta$, and the lines $x = 0$ and $x = 2a$, is equal to

(A) πa^2 (B) $2\pi a^2$ (C) $\dfrac{\pi a}{4}$ (D) $\dfrac{\pi a}{2}$ (E) none of these

41. The base of a solid is the region bounded by $x^2 = 4y$ and the line $y = 2$, and each plane section perpendicular to the y-axis is a square. The volume of the solid is

(A) 8   (B) 16   (C) 20   (D) 32   (E) 64

42. An object initially at rest at (3,3) moves with acceleration $a(t) = (2, e^{-t})$. Where is the object at $t = 2$?

(A) $(4, e^{-2})$ (B) $(4, e^{-2} + 2)$ (C) $(7, e^{-2})$ (D) $(7, e^{-2} + 2)$

(E) $(7, e^{-2} + 4)$

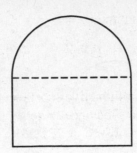

43. The figure above consists of a rectangle capped by a semicircle. Its area is 100 square yards. The minimum perimeter of the figure is
(A) 10.584 yd   (B) 28.284 yd   (C) 37.793 yd   (D) 38.721 yd
(E) 51.820 yd

44. Using the first two terms in the Maclaurin series for $y = \cos x$ yields accuracy to within 0.001 over the interval $|x| < k$ when $k =$
(A) 0.032   (B) 0.394   (C) 0.786   (D) 0.788   (E) 1.570

45. After t years, $50e^{-0.015t}$ pounds of a deposit of a radioactive substance remains. The average amount per year *not* lost by radioactive decay during the second hundred years is
(A) 2.9 lb   (B) 5.8 lb   (C) 7.4 lb   (D) 11.1 lb   (E) none of these

Section II

Part A.

A graphing calculator is required for some of these problems. See instructions on p. xi. Answers begin on p. 600.

1. Let function f be continuous and decreasing, with values as shown in the table:

x	2.5	3.0	3.5	4.0	4.5	5.0
$f(x)$	7.6	5.7	4.2	3.8	2.2	1.6

(a) Use the trapezoid rule to estimate the area between f and the x-axis.
(b) Find the average rate of change of f on this interval.
(c) Estimate the instantaneous rate of change of f at $x = 2.5$.
(d) If $(g)x = f^{-1}(x)$, estimate the slope of g at $x = 4$.

2. An object starts at point $(1,3)$, and moves along the parabola $y = x^2 + 2$ for $0 \le t \le 2$, with the horizontal component of its velocity given by $\dfrac{dx}{dt} = \dfrac{4}{t^2 + 4}$.

(a) Find the object's position at $t = 2$.
(b) Find the object's speed at $t = 2$.
(c) Find the distance the object traveled during this interval.

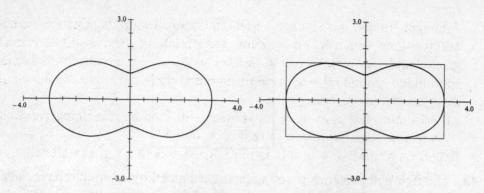

3. Let R be the region bounded by $r = 2 + \cos 2\theta$, as shown above.
    (a) Find the dimensions of the smallest rectangle that contains R and has sides parallel to the x- and y-axes.
    (b) Find the area of R.

Part B

No calculator is allowed for any of these problems. See instructions on p. xi.

4. Given a function f such that $f(3) = 1$ and $f^{(n)}(3) = \dfrac{(-1)^n n!}{(2n+1)2^n}$.

    (a) Write the first four nonzero terms and the general term of the Taylor series for f around $x = 3$.
    (b) Find the radius of convergence of the Taylor series.
    (c) Show that the third-degree Taylor polynomial approximates $f(4)$ to within 0.01.

5. The figure at the right is for the question that follows on page 595.

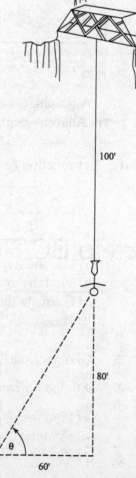

A bungee jumper has reached a point in her exciting plunge where the taut cord is 100 feet long with a 1/2-inch radius, and stretching. She is still 80 feet above the ground and is now falling at 40 feet per second. You are observing her jump from a spot on the ground 60 feet from the potential point of impact, as shown in the diagram on the preceding page.

(a) Assuming the cord to be a cylinder with volume remaining constant as the cord stretches, at what rate is its radius changing?

(b) From your observation point, at what rate is the angle of elevation to the jumper changing?

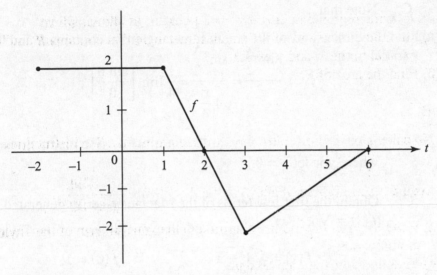

6. The figure above shows the graph of f, whose domain is the closed interval

[-2,6]. Let $F(x) = \int_{1}^{x} f(t)\,dt$.

(a) Find $F(-2)$.

(b) Find $F(6)$.

(c) For what value(s) of x does $F(x) = 0$?

(d) For what value(s) of x is F increasing?

(e) Find the maximum value and the minimum value of F.

(f) At what value(s) of x does F have points of inflection? Justify your answer.

Answers to BC Practice Examination 3: Section I

1. B	10. E	19. E	28. D	37. B
2. C	11. D	20. C	29. B	38. E
3. C	12. A	21. B	30. C	39. E
4. D	13. C	22. D	31. A	40. A
5. B	14. D	23. D	32. A	41. D
6. C	15. B	24. B	33. D	42. E
7. B	16. E	25. B	34. C	43. C
8. B	17. C	26. A	35. C	44. B
9. B	18. B	27. E	36. D	45. B

The explanations for questions not given below will be found in the answer sections for AB Practice Examination 3, on pages 509 to 517. Identical questions in Section I of Practice Examinations AB3 and BC3 have the same number. For example, explanations of the answers to Questions 4 and 5 not given below will be found in Section I of Examination AB3, Answers 4 and 5, page 509.

Part A

1. **B.** Since $\lim_{x \to 1} f(x) = 1$, to render $f(x)$ continuous at $x = 1$, define $f(1)$ to be 1.

2. **C.** Note that

$$\frac{\sin^2 \frac{x}{2}}{x^2} = \frac{\sin^2 \frac{x}{2}}{4 \frac{x^2}{4}} = \frac{1}{4} \lim_{\theta \to 0} \left[\frac{\sin \theta}{\theta} \right]^2,$$

where you let $\frac{x}{2} = \theta$.

3. **C.** Obtain the first few terms of the Maclaurin series generated by $f(x) = \sqrt{1 + x}$:

$$f(x) = \sqrt{1 + x}; \qquad\qquad f(0) = 1;$$

$$f'(x) = \frac{1}{2}(1 + x)^{-1/2}; \qquad\qquad f'(0) = \frac{1}{2};$$

$$f''(x) = -\frac{1}{4}(1 + x)^{-3/2}; \qquad f''(0) = -\frac{1}{4};$$

$$f'''(x) = \frac{3}{8}(1 + x)^{-5/2}. \qquad\qquad f'''(0) = \frac{3}{8}.$$

So $\sqrt{1 + x} = 1 + \frac{x}{2} - \frac{1}{4} \cdot \frac{x^2}{2} + \frac{3}{8} \cdot \frac{x^3}{6} - \cdots.$

6. **C.** Here,

$$\frac{dx}{dt} = e^t(\cos t - \sin t), \quad \frac{dy}{dt} = e^t(\sin t + \cos t),$$

and

$$\frac{d^2x}{dt^2} = -2(\sin t)e^t, \quad \frac{d^2y}{dt^2} = 2(\cos t)e^t;$$

and the magnitude of the acceleration, $|\mathbf{a}|$, is given by

$$|\mathbf{a}| = \sqrt{\left(\frac{d^2x}{dt^2}\right)^2 + \left(\frac{d^2y}{dt^2}\right)^2} = 2e^t.$$

7.   B.   The new series is

$$1 + \frac{x-1}{2} + \frac{(x-1)^2}{3} + \frac{(x-1)^3}{4} + \cdots.$$

Use the Ratio Test, page 350; then the series converges when $0 < x < 2$. Be sure to check the endpoints!

10.   E.   The integral is equal to $\dfrac{\tan^3 x}{3}\Big|_{\pi/4}^{\pi/3} = \dfrac{1}{3}(3\sqrt{3} - 1)$.

11.   D. $\displaystyle\int_1^e \ln x \, dx$ can be integrated by parts to yield $(x \ln x - x)\Big|_1^e$, which equals

$$e \ln e - e - (1 \ln 1 - 1) = e - e - (0 - 1) = 1.$$

13.   C.   Use the method of partial fractions, letting

$$\frac{x-6}{x(x-3)} = \frac{A}{x} + \frac{B}{x-3}.$$

Then $A = 2$ and $B = -1$.

17.   C. $\displaystyle\int_0^\infty \frac{dx}{x^2 + 1} = \lim_{b \to \infty} \tan^{-1} x \Big|_0^b = \frac{\pi}{2}$. The integrals in (A), (B), and (D) all diverge to infinity.

19.   E.   At $(2,1)$, $\dfrac{dy}{dx} = 3$. Use $\Delta x = 0.1$; then Euler's method moves to $(2.1, 1 + 3(0.1))$.

At $(2.1, 1.3)$, $\dfrac{dy}{dx} = 3.4$, so the next point is $(2.2, 1.3 + 3.4(0.1))$.

22.   D.   Separate variables to get $\dfrac{dy}{y} = \dfrac{dx}{x}$, and integrate to get $\ln y = \ln x + C$. Since $y = 3$ when $x = 1$, $C = \ln 3$.

23.   D.   The generating circle has equation $x^2 + y^2 = 4$. The volume, V, is given by

$$V = \pi \int_{-1}^1 x^2 \, dy = 2\pi \int_0^1 (4 - y^2) \, dy.$$

24.   B.   The arc length is

$$\int_1^4 \sqrt{1 + \left(\frac{dy}{dx}\right)^2} \, dx = \int_1^4 \sqrt{1 + \left(x - \frac{1}{4x}\right)^2} \, dx$$

$$= \int_1^4 \sqrt{\left(x + \frac{1}{4x}\right)^2} \, dx = \int_1^4 \left(x + \frac{1}{4x}\right) dx$$

$$= \frac{15}{2} + \frac{1}{4} \ln 4 = \frac{15}{2} + \frac{1}{2} \ln 2.$$

26.   A.   Here $x^2 + y^2 = 4$, whose locus is a circle; but since the given equations imply x and y both nonnegative, the curve defined is in the first quadrant.

28.   D. $\displaystyle\int_1^2 \frac{1}{\sqrt{4-x^2}}\,dx = \lim_{h\to 2^-}\int_1^h \frac{1}{\sqrt{4-x^2}}\,dx = \lim_{h\to 2^-}\sin^{-1}\frac{x}{2}\bigg|_1^h$

$\displaystyle = \lim_{h\to 2^-}\left(\sin^{-1}\frac{h}{2} - \sin^{-1}\frac{1}{2}\right) = \frac{\pi}{2} - \frac{\pi}{6} = \frac{\pi}{3}.$

Part B

30.   C.   Since the equation of the spiral is $r = \ln\theta$, use the polar mode. The formula for area in polar coordinates is

$$\frac{1}{2}\int_{\theta_1}^{\theta_2} r^2\,d\theta$$

Therefore, calculate

$$0.5\int_{\pi}^{2\pi}\ln^2\theta\,d\theta.$$

The result is 3.743.

31.   A.   The curve is given parametrically by

$$x = e^t, \quad y = t - \frac{t^3}{9}.$$

Since the x-intercepts occur where $y = 0$, solve the equation $t - \dfrac{t^3}{9} = 0$, getting $t = -3, 0$, and 3. The smallest x-intercept is for $t = -3$. At that point, the slope

$$\frac{dy}{dx} = \frac{\dfrac{dy}{dt}}{\dfrac{dx}{dt}} = \frac{1 - \dfrac{t^2}{3}}{e^t}.$$

At $t = -3$, $\dfrac{dy}{dx} = \dfrac{-2}{e^{-3}} = -40.171.$

32. **A.** $\lim\limits_{n\to\infty}\dfrac{3^n}{n^3}=\infty$; $\dfrac{3n^2}{n^3+1}>\dfrac{1}{n}$.

38. **E.** If Q_0 is the initial amount of the substance and Q is the amount at time t, then

$$Q = Q_0 e^{-kt}.$$

When $t = 8$, $Q = \dfrac{1}{2}Q_0$, so

$$\frac{1}{2}Q_0 = Q_0 e^{-8k}$$

and $\dfrac{1}{2} = e^{-8k}$. Thus $k = 0.08664$. Using a calculator, find t when $Q = \dfrac{1}{3}Q_0$.

$\dfrac{1}{3} = e^{-0.08664t}$, so $t \approx 12.68$. Don't round off k too quickly.

40. **A.** See figure below.

$$A = \int_0^{2a} y\, dx = \int_{\theta=0}^{\theta=\frac{\pi}{4}} 2a\cos^2\theta \cdot 2a\sec^2\theta\, d\theta = 4a^2\, \theta\,\Big|_0^{\pi/4} = \pi a^2$$

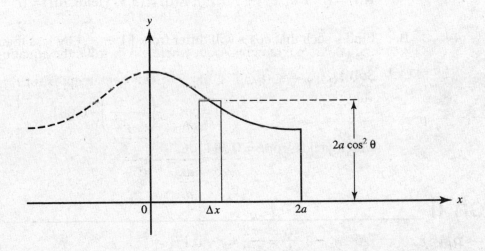

41. D. See figure below.

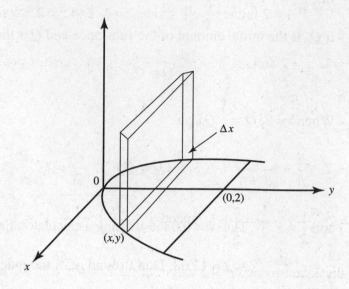

$$\Delta V = (2x)^2 \, \Delta y = 4x^2 \, \Delta y$$

$$= 16y \, \Delta y,$$

$$V = \int_0^2 16y \, dy.$$

42. E. $\mathbf{v}(t) = (2t + c_1, -e^{-t} + c_2)$; $\mathbf{v}(0) = (0,0)$ yields $c_1 = 0$ and $c_2 = 1$.
$\mathbf{R}(t) = (t^2 + c_3, e^{-t} + t + c_4)$; $\mathbf{R}(0) = (3,3)$ yields $\mathbf{R}(t) = (t^2 + 3, e^{-t} + t + 2)$.

44. B. Find k such that $\cos x$ will differ from $\left(1 - \dfrac{x^2}{2}\right)$ by less than 0.001 at $x = k$.
Solve

$$0 = \cos x - \left(1 - \frac{x^2}{2}\right) - 0.001,$$

which yields x or $k = 0.394$.

Section II

Part A

1. See solution for AB-1, page 517.

2. (a) Position is the antiderivative of $\dfrac{dx}{dt} = \dfrac{4}{t^2 + 4}$:

$$x = \int \frac{4}{t^2 + 4} \, dt = \frac{4}{4} \int \frac{1}{\left(\dfrac{t}{2}\right)^2 + 1} \, dt = 2 \int \frac{\dfrac{1}{2} dt}{\left(\dfrac{t}{2}\right)^2 + 1}$$

$$= 2 \arctan \frac{t}{2} + c.$$

To find c, substitute the initial condition that $x = 1$ when $t = 0$:

$$1 = 2 \arctan \frac{0}{2} + c \text{ shows } c = 1, \text{ and } x = 2 \arctan \frac{t}{2} + 1.$$

At $t = 2$, $x = 2 \arctan \frac{2}{2} + 1 = 2 \cdot \frac{\pi}{4} + 1 = \frac{\pi}{2} + 1$, and the position of the object

is $\left(\frac{\pi}{2} + 1, \left(\frac{\pi}{2} + 1 \right)^2 + 2 \right)$.

(b) At $t = 2$, $\dfrac{dx}{dt} = \dfrac{4}{2^2 + 4} = \dfrac{1}{2}$. Since $y = x^2 + 2$, $\dfrac{dy}{dt} = 2x \dfrac{dx}{dt}$, and at $t = 2$,

$$\frac{dy}{dt} = 2\left(\frac{\pi}{2} + 1 \right) \cdot \frac{1}{2} = \frac{\pi}{2} + 1, \text{ speed} = \sqrt{ \left(\frac{dx}{dt} \right)^2 + \left(\frac{dy}{dt} \right)^2 } = \sqrt{ \left(\frac{1}{2} \right)^2 + \left(\frac{\pi}{2} + 1 \right)^2 }.$$

(c) The distance traveled is the length of the arc of $y = x^2 + 2$ in the interval
$1 < x < \dfrac{\pi}{2} + 1$:

$$L = \int \sqrt{ 1 + \left(\frac{dy}{dx} \right)^2 } \, dx = \int_1^{\pi/2+1} \sqrt{ 1 + (2x)^2 } \, dx = 5.839.$$

3. (a) To find the smallest rectangle with sides parallel to the x- and y-axes, you need a rectangle formed by vertical and horizontal tangents as shown in the figure. The vertical tangents are at the x-intercepts, $x = \pm 3$. The horizontal tangents are at the points where y (not r) is a maximum. You need, therefore, to maximize

$$y = r \sin \theta = (2 + \cos 2\theta) \sin \theta,$$

$$\frac{dy}{d\theta} = (2 + \cos 2\theta) \cos \theta + \sin \theta (-2 \sin 2\theta).$$

Use the calculator to find that $\dfrac{dy}{d\theta} = 0$ when $\theta = 0.7854$. Therefore, $y = 1.414$, so the desired rectangle has dimensions 6×2.828.

(b) Since the polar formula for the area is $\dfrac{1}{2} \displaystyle\int_{\theta_1}^{\theta_2} r^2 \, d\theta$, the area of R (enclosed by r) is
$4 \cdot \dfrac{1}{2} \displaystyle\int_0^{\pi/2} r^2 \, d\theta$, which is 14.137.

Part B

4. (a)

n	$f^{(n)}(3) = \dfrac{(-1)^n n!}{(2n+1)2^n}$	$a_n = \dfrac{f^{(n)}(3)}{n!}$
0	1	1
1	$\dfrac{-1}{3 \cdot 2}$	$\dfrac{-1}{3 \cdot 2}$
2	$\dfrac{2!}{5 \cdot 2^2}$	$\dfrac{1}{5 \cdot 2^2}$
3	$\dfrac{-3!}{7 \cdot 2^3}$	$\dfrac{-1}{7 \cdot 2^3}$

$$f(x) = 1 - \frac{1}{6}(x-3) + \frac{1}{20}(x-3)^2 - \frac{1}{56}(x-3)^3 + \cdots + \frac{(-1)^n (x-3)^n}{(2n+1) \cdot 2^n} + \cdots.$$

(b) By the Ratio Test, the series converges when

$$\lim_{x \to \infty} \left| \frac{(x-3)^{n+1}}{(2n+3) \cdot 2^{n+1}} \cdot \frac{(2n+1) \cdot 2^n}{(x-3)^n} \right| < 1,$$

$$|x-3| \lim_{x \to \infty} \left| \frac{(2n+1)}{(2n+3) \cdot 2} \right| < 1,$$

$$|x-3| < 2.$$

Thus, the radius of convergence is 2.

(c) $f(4) \approx 1 - \dfrac{1}{6}(4-3) + \dfrac{1}{20}(4-3)^2 - \dfrac{1}{56}(4-3)^3 \cdots$ is an alternating series, so the

error is less than the magnitude of the first omitted term:

$$\text{error} < \frac{(4-3)^4}{(2 \cdot 4 + 1) \cdot 2^4} = \frac{1}{144} < 0.01$$

5. See solution for AB-5, pages 520–521.

6. See solution for AB-6, page 521.

BC Practice Examination 4

Section I Multiple-Choice Questions _____

Part A

See instructions, page 551. Answers are given on page 614.

1. $\lim_{x \to 2}[x]$ (where $[x]$ is the greatest integer in x) is

    (A) 1   (B) 2   (C) 3   (D) ∞ (E) nonexistent

2. $\lim_{h \to 0} \dfrac{\sin\left(\dfrac{\pi}{2} + h\right) - 1}{h}$ is

    (A) 1   (B) -1 (C) 0   (D) ∞ (E) none of these

3. The set of all x for which the power series $\displaystyle\sum_{n=0}^{\infty} \dfrac{x^n}{(n+1) \cdot 3^n}$ converges is

    (A) $\{-3, 3\}$ (B) $|x| < 3$ (C) $|x| > 3$ (D) $-3 \leqq x < 3$
    (E) $-3 < x \leqq 3$

4. The equation of the tangent to the curve $2x^2 - y^4 = 1$ at the point $(-1, 1)$ is
    (A) $y = -x$ (B) $y = 2 - x$ (C) $4y + 5x + 1 = 0$
    (D) $x - 2y + 3 = 0$ (E) $x - 4y + 5 = 0$

5. The nth term of the Taylor series expansion about $x = 0$ of the function $f(x) = \dfrac{1}{1+2x}$ is

    (A) $(2x)^n$ (B) $2x^{n-1}$ (C) $\left(\dfrac{x}{2}\right)^{n-1}$ (D) $(-1)^{n-1}(2x)^{n-1}$

    (E) $(-1)^n (2x)^{n-1}$

6. When the method of partial fractions is used to decompose $\dfrac{2x^2 - x + 4}{x^3 - 3x^2 + 2x}$, one of the fractions obtained is

(A) $-\dfrac{5}{x-1}$ (B) $-\dfrac{2}{x-1}$ (C) $\dfrac{1}{x-1}$ (D) $\dfrac{2}{x-1}$ (E) $\dfrac{5}{x-1}$

7. A relative maximum value of the function $y = \dfrac{\ln x}{x}$ is

(A) 1   (B) e (C) $\dfrac{2}{e}$ (D) $\dfrac{1}{e}$ (E) none of these

8. When a series is used to approximate $\displaystyle\int_0^{0.3} e^{-x^2}\, dx$, the value of the integral, to two decimal places, is

(A) -0.09 (B) 0.29   (C) 0.35   (D) 0.81   (E) 1.35

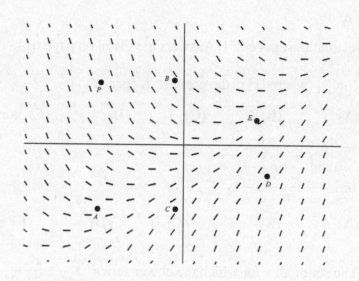

9. A particular solution of the differential equation whose slope field is shown above contains point P. This solution may also contain which other point?

(A) A   (B) B   (C) C   (D) D   (E) E

10. Let $F(x) = \displaystyle\int_5^x \dfrac{dt}{1 - t^2}$. Which of the following statements is (are) true?

I. The domain of F is $x \neq \pm 1$.   II. $F(2) > 0$.   III. F is concave upward.

(A) none   (B) I only   (C) II only   (D) III only

(E) II and III only

11. As the tides change, the water level in a bay varies sinusoidally. At high tide today at 8 A.M., the water-level was 15 feet; at low tide, 6 hours later at 2 P.M. it was 3 feet. How fast, in feet per hour, was the water level dropping at noon today?

(A) 3   (B) $\dfrac{\pi\sqrt{3}}{2}$ (C) $3\sqrt{3}$ (D) $\pi\sqrt{3}$ (E) $6\sqrt{3}$

12. Let $\int_0^x f(t)dt = x\sin\pi x$. Then $f(3) =$

(A) -3π (B) -1 (C) 0 (D) 1 (E) 3π

13. $\int \dfrac{e^u}{4+e^{2u}}du$ is equal to

(A) $\ln(4 + e^{2u}) + C$ (B) $\dfrac{1}{2}\ln\left|4+e^{2u}\right|+C$ (C) $\dfrac{1}{2}\tan^{-1}\dfrac{e^u}{2}+C$

(D) $\tan^{-1}\dfrac{e^u}{2}+C$ (E) $\dfrac{1}{2}\tan^{-1}\dfrac{e^{2u}}{2}+C$

14. Given $f(x) = \log_{10}x$ and $\log_{10}(102) \approx 2.0086$, which is closest to $f'(100)$?

(A) 0.0043 (B) 0.0086 (C) 0.01 (D) 1.0043 (E) 2

15. If $G(2) = 5$ and $G'(x) = \dfrac{10x}{9-x^2}$, then an estimate of $G(2.2)$ using local lincarization is approximately

(A) 5.4 (B) 5.5 (C) 5.8 (D) 8.8 (E) 13.8

16. The area bounded by the parabola $y = x^2$ and the lines $y = 1$ and $y = 9$ equals

(A) 8 (B) $\dfrac{84}{3}$ (C) $\dfrac{64}{3}\sqrt{2}$ (D) 32 (E) $\dfrac{104}{3}$

17. The first-quadrant region bounded by $y = \dfrac{1}{\sqrt{x}}, y = 0, x = q\ (0 < q < 1)$, and $x = 1$ is rotated about the x-axis. The volume obtained as $q \to 0^+$ equals

(A) $\dfrac{2\pi}{3}$ (B) $\dfrac{4\pi}{3}$ (C) 2π (D) 4π (E) none of these

18. A curve is given parametrically by the equations

$$x = 3 - 2\sin t \text{ and } y = 2\cos t - 1.$$

The length of the arc from $t = 0$ to $t = \pi$ is

(A) $\dfrac{\pi}{2}$ (B) π (C) $2 + \pi$ (D) 2π (E) 4π

19. Suppose the function f is both increasing and concave up on $a \leqslant x \leqslant b$. Then, using the same number of subdivisions, and with L, R, M, and T denoting, respectively, left, right, midpoint, and trapezoid sums, it follows that

(A) $R \leqslant T \leqslant M \leqslant L$ (B) $L \leqslant T \leqslant M \leqslant R$ (C) $R \leqslant M \leqslant T \leqslant L$

(D) $L \leqslant M \leqslant T \leqslant R$ (E) none of these

20. Which of the following statements about the graph of $y = \dfrac{x^2}{x-2}$ is (are) true?

    I. The graph has no horizontal asymptote.
    II. The line $x = 2$ is a vertical asymptote.
    III. The line $y = x + 2$ is an oblique asymptote.
    (A) I only   (B) II only   (C) I and II only
    (D) I and III only   (E) all three

21. The only function that does not satisfy the Mean Value Theorem on the interval specified is
    (A) $f(x) = x^2 - 2x$ on $[-3, 1]$

    (B) $f(x) = \dfrac{1}{x}$ on $[1, 3]$

    (C) $f(x) = \dfrac{x^3}{3} - \dfrac{x^2}{2} + x$ on $[-1, 2]$

    (D) $f(x) = x + \dfrac{1}{x}$ on $[-1, 1]$

    (E) $f(x) = x^{2/3}$ on $\left[\dfrac{1}{2}, \dfrac{3}{2}\right]$

22. $\displaystyle\int_0^1 x^2 e^x \, dx =$
    (A) $-3e - 1$ (B) $-e$ (C) $e - 2$ (D) $3e$ (E) $4e - 1$

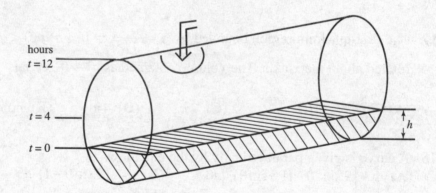

23. A cylindrical tank, shown in the figure above, is partially full of water at time $t = 0$, when more water begins flowing in at a constant rate. The tank becomes half full when $t = 4$, and is completely full when $t = 12$. Let h represent the height of the water at time t. During which interval is $\dfrac{dh}{dt}$ increasing?
    (A) none   (B) $0 < t < 4$ (C) $0 < t < 8$ (D) $0 < t < 12$
    (E) $4 < t < 12$

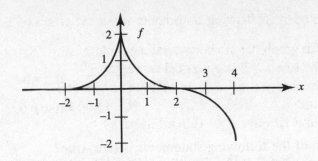

24. The graph of function f shown above consists of three quarter-circles. Which of the following is (are) equivalent to $\int_0^2 f(x)dx$?

I. $\dfrac{1}{2}\int_{-2}^2 f(x)dx$ II. $\int_4^2 f(x)dx$ III. $\dfrac{1}{2}\int_0^4 f(x)dx$

(A) I only   (B) II only   (C) III only   (D) I and II only
(E) all three

25. The base of a solid is the first-quadrant region bounded by $y = \sqrt[4]{4-2x}$, and each cross section perpendicular to the x-axis is a semicircle with a diameter in the xy-plane. The volume of the solid is

(A) $\dfrac{\pi}{2}\int_0^2 \sqrt{4-2x}\,dx$ (B) $\dfrac{\pi}{8}\int_0^2 \sqrt{4-2x}\,dx$ (C) $\dfrac{\pi}{8}\int_{-2}^2 \sqrt{4-2x}\,dx$

(D) $\dfrac{\pi}{4}\int_0^{\sqrt{2}} \left(4-y^4\right)^2 dy$ (E) $\dfrac{\pi}{8}\int_0^{\sqrt[4]{4}} \left(4-y^4\right)^2 dy$

26. The average value of $f(x) = 3 + |x|$ on the interval $[-2, 4]$ is

(A) $2\dfrac{2}{3}$ (B) $3\dfrac{1}{3}$ (C) $4\dfrac{2}{3}$ (D) $5\dfrac{1}{3}$ (E) 6

27. The area inside the circle $r = 3\sin\theta$ and outside the cardioid $r = 1 + \sin\theta$ is given by

(A) $\displaystyle\int_{\pi/6}^{\pi/2}\left[9\sin^2\theta - (1+\sin\theta)^2\right]d\theta$ (B) $\displaystyle\int_{\pi/6}^{\pi/2}(2\sin\theta - 1)^2\,d\theta$

(C) $\dfrac{1}{2}\displaystyle\int_{\pi/6}^{5\pi/6}\left(8\sin^2\theta - 1\right)d\theta$ (D) $\dfrac{9\pi}{4} - \dfrac{1}{2}\displaystyle\int_{\pi/6}^{5\pi/6}(1+\sin\theta)^2\,d\theta$

(E) none of these

28. Let

$$f(x) = \begin{cases} \dfrac{x^2 - 36}{x-6} & \text{if } x \neq 6, \\ 12 & \text{if } x = 6. \end{cases}$$

Which of the following statements is (are) true?
  I. f is defined at $x = 6$.
  II. $\lim\limits_{x \to 6} f(x)$ exists.
  III. f is continuous at $x = 6$.
  (A) I only   (B) II only   (C) I and II only   (D) I, II, and III
  (E) none of the statements

Part B

See instructions, page 555. Answers begin on p. 613.

29. Two objects in motion from $t = 0$ to $t = 3$ seconds have positions, $x_1(t) = \cos(t^2 + 1)$
  and $x_2(t) = \dfrac{e^t}{2t}$, respectively. How many times during the 3 seconds do the objects
  have the same velocity?
  (A) 0   (B) 1   (C) 2   (D) 3   (E) 4

30. The table below shows values of $f''(x)$ for various values of x:

x	-1	0	1	2	3
$f''(x)$	-4	-1	2	5	8

The function f could be
  (A) a linear function   (B) a quadratic function   (C) a cubic function
  (D) a fourth-degree function   (E) an exponential function

31. Where, in the first quadrant, does the rose $r = \sin 3\theta$ have a vertical tangent?
  (A) nowhere   (B) $\theta = 0.39$ (C) $\theta = 0.47$
  (D) $\theta = 0.52$ (E) $\theta = 0.60$

32. A cup of coffee placed on a table cools at a rate of $\dfrac{dH}{dt} = -0.05(H - 70)$ degrees per
  minute, where H represents the temperature of the coffee and t is time in minutes. If
  the coffee was at $120°\,$F initially, what will its temperature be 10 minutes later?
  (A) $73°\,$F   (B) $95°\,$F   (C) $100°\,$F   (D) $118°\,$F   (E) $143°\,$F

33. An investment of \$4000 grows at the rate of $320e^{0.08t}$ dollars per year after t years. Its
  value after 10 years is approximately
  (A) \$4902   (B) \$8902   (C) \$7122   (D) \$12,902
  (E) none of these

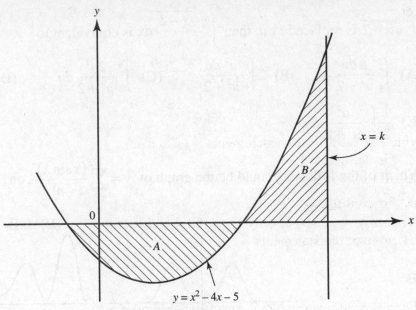

$y = x^2 - 4x - 5$

(This figure is not drawn to scale.)

34. The sketch shows the graphs of $f(x) = x^2 - 4x - 5$ and the line $x = k$. The regions labeled A and B have equal areas if $k =$

(A) 5   (B) 7.766   (C) 7.899   (D) 8   (E) 11

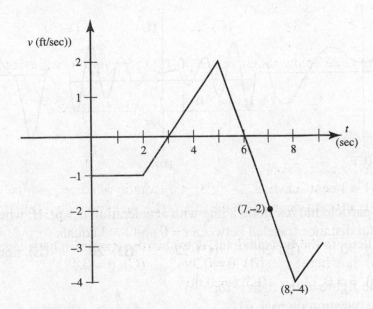

Use the graph above for Questions 35 and 36. It shows the velocity of an object during the interval $0 \le t \le 9$.

35. The object attains its greatest speed at $t =$

(A) 2   (B) 3   (C) 5   (D) 6   (E) 8

36. The object was at the origin at $t = 3$. It returned to the origin

(A) at $t = 5$ (B) at $t = 6$ (C) during $6 < t < 7$

(D) at $t = 7$ (E) during $7 < t < 8$

37. An object in motion in the plane has acceleration vector $\mathbf{a}(t) = (\sin t, e^{-t})$ for $0 \le t \le 5$. It is at rest when $t = 0$. What is the maximum speed it attains?

(A) 1.022   (B) 1.414   (C) 2.217   (D) 2.958   (E) 3.162

38. If $\sqrt{x-2}$ is replaced by u, then $\int_3^6 \frac{\sqrt{x-2}}{x}\, dx$ is equivalent to

(A) $\int_1^2 \frac{u\,du}{u^2+2}$ (B) $2\int_1^2 \frac{u^2\,du}{u^2+2}$ (C) $\int_3^6 \frac{2u^2}{u^2+2}\,du$ (D) $\int_3^6 \frac{u\,du}{u^2+2}$

(E) $\frac{1}{2}\int_1^2 \frac{u^2}{u^2+2}\,du$

39. Which of the following could be the graph of $y = \sum_{n=0}^{9} \frac{(x\sin x)^n}{n!}$ on $-2\pi \leqq x \leqq 2\pi$?

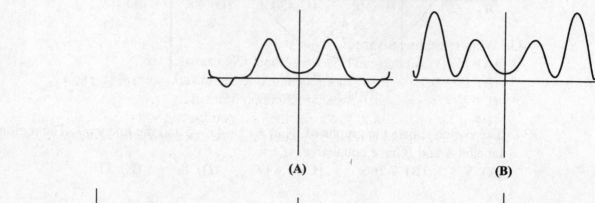

(A)   (B)

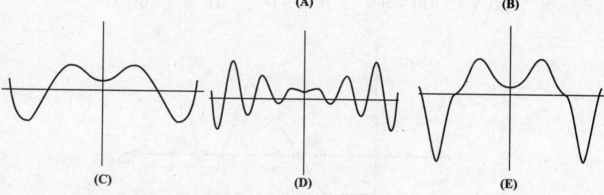

(C)   (D)   (E)

40. A particle moves along a line with acceleration $a = 6t$. If, when $t = 0$, $v = 1$, then the total distance traveled between $t = 0$ and $t = 3$ equals

(A) 30   (B) 28   (C) 27   (D) 26   (E) none of these

41. The figure at the right is for the question on page 611.

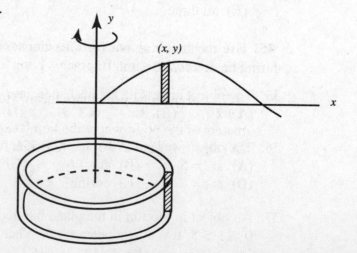

41. The region bounded by the first arc of $y = \sin x$ and the x-axis is rotated around the y-axis. The element shown on the preceding page forms a cylindrical shell whose volume is the product of its surface area and its thickness. The total volume of solid formed (the sum of the volumes of all such shells) is given by

   (A) $\pi \int_0^\pi \sin^2 x \, dx$ (B) $\pi \int_0^\pi x \sin x \, dx$ (C) $2\pi \int_0^\pi x \sin x \, dx$

   (D) $2\pi \int_0^\pi \sin x \, dx$ (E) none of these

42. Suppose $f(3) = 2$, $f'(3) = 5$, and $f''(3) = -2$. Then $\dfrac{d^2}{dx^2}\left(f^2(x)\right)$ at $x = 3$ is equal to

   (A) -20 (B) 10 (C) 20 (D) 38 (E) 42

43. Which statement is true?
   (A) If $f(x)$ is continuous at $x = c$, then $f'(c)$ exists.
   (B) If $f'(c) = 0$, then f has a local maximum or minimum at $(c, f(c))$.
   (C) If $f''(c) = 0$, then f has an inflection point at $(c, f(c))$.
   (D) If f is differentiable at $x = c$, then f is continuous at $x = c$.
   (E) If f is continuous on (a, b), then f maintains a maximum value on (a, b).

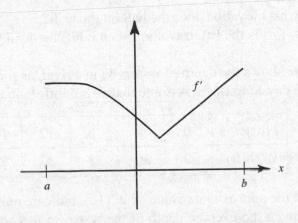

44. The graph of f' is shown above. Which statements about f must be true for $a < x < b$?
   I. f is increasing.   II. f is continuous.   III. f is differentiable.
   (A) I only   (B) II only   (C) I and II only   (D) I and III only
   (E) all three

45. After a bomb explodes, pieces can be found scattered around the center of the blast. The density of bomb fragments lying x meters from ground zero is given by $N(x) =$

   $\dfrac{2x}{1 + x^{3/2}}$ fragments per square meter. How many fragments will be found within 20 meters of the point where the bomb exploded?
   (A) 13   (B) 278   (C) 556   (D) 712   (E) 4383

Section II

Part A

A graphing calculator is required for some of these problems. See instructions on p. xi. Answers begin on p. 619.

1. Let R represent the region bounded by $y = \sin x$ and $y = x^4$. Find:
   (a) the area of R;
   (b) the volume of the solid whose base is R, if all cross sections perpendicular to the x-axis are isosceles triangles with height 3;
   (c) the volume of the solid formed when R is rotated around the x-axis.

2. The Boston Red Sox play in Fenway Park, notorious for its Green Monster, a wall 37 feet tall and 315 feet from home plate at the left-field foul line. Suppose a batter hits a ball 2 feet above home plate, driving the ball down the left-field line at an initial angle of 30° above the horizontal, with initial velocity of 120 feet per second. (Since Fenway is near sea level, assume that the acceleration due to gravity is $-32.172 \, \text{ft/sec}^2$.)
   (a) Write the parametric equations for the location of the ball t seconds after it has been hit.
   (b) At what elevation does the ball hit the wall?
   (c) How fast is the ball traveling when it hits the wall?

3. The table shows the depth of water, W, in a river, as measured at 4-hour intervals during a day-long flood. Assume that W is a differentiable function of time t.

t (hr)	0	4	8	12	16	20	24
$W(t)$ (ft)	32	36	38	37	35	33	32

   (a) Find the approximate value of $W'(16)$. Indicate units of measure.
   (b) Estimate the average depth of the water, in feet, over the time interval $0 \le t \le 24$ hours by using a trapezoidal approximation with subintervals of length $\Delta t = 4$ days.
   (c) Scientists studying the flooding believe they can model the depth of the water with the function $F(t) = 35 - 3\cos\left(\dfrac{t+3}{4}\right)$, where $F(t)$ represents the depth of the water in feet after t hours. Find $F'(16)$ and explain the meaning of your answer, with appropriate units, in terms of the river depth.
   (d) Use the function F to find the average depth of the water, in feet, over the time interval $0 \le t \le 24$ hours.

Part B

No calculator is allowed for any of these problems. See instructions on p. xi.

4. Two autos, P and Q, start from the same point and race along a straight road for 10 seconds. The velocity of P is given by $v_p(t) = 6\left(\sqrt{1 + 8t} - 1\right)$ feet per second. The velocity of Q is shown in the graph.

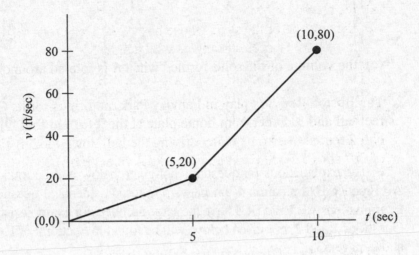

(a) At what time is P's actual acceleration equal to its average acceleration for the entire race?
(b) What is Q's acceleration then?
(c) At the end of the race, which auto was ahead? Explain.

5. Given that a function f is continuous and differentiable throughout its domain, and that $f(5) = 2, f'(5) = -2, f''(5) = -1$, and $f'''(5) = 6$.
(a) Write a Taylor polynomial of degree 3 that approximates f around $x = 5$.
(b) Use your answer to estimate $f(5.1)$.
(c) Let $g(x) = f(2x + 5)$. Write a cubic Maclaurin polynomial approximation for g.

6. Let f be the function that contains the point $(-1,8)$ and satisfies the differential equation $\dfrac{dy}{dx} = \dfrac{10}{x^2 + 1}$.
(a) Write the linearization of f at $x = -1$.
(b) Using your answer to part (a), estimate $f(0)$.
(c) Using Euler's method with a step size of 0.5, estimate $f(0)$.
(d) Estimate $f(0)$ using an integral.

Answers to BC Practice Examination 4: Section I

1.	E	10.	E	19.	D	28.	D	37.	C
2.	C	11.	B	20.	E	29.	E	38.	B
3.	D	12.	A	21.	D	30.	C	39.	A
4.	A	13.	C	22.	C	31.	C	40.	A
5.	D	14.	A	23.	E	32.	C	41.	C
6.	A	15.	C	24.	D	33.	B	42.	E
7.	D	16.	E	25.	B	34.	D	43.	D
8.	B	17.	E	26.	C	35.	E	44.	E
9.	E	18.	D	27.	A	36.	E	45.	D

The explanations for questions not given below will be found in the answer sections for AP Practice Examination 4, on pages 533 to 541. Identical questions in Section I of Practice Examinations AB4 and BC4 have the same number. For example, explanations of the answers to Questions 1 and 2, not given below, will be found in Section I of Examination AB4, Answers 1 and 2, page 533.

Part A

3.   **D.**   Use the Ratio Test:

$$\lim_{n\to\infty}\left|\frac{x^{n+1}}{(n+2)\cdot 3^{n+1}}\cdot\frac{(n+1)\cdot 3^n}{x^n}\right|=\lim_{n\to\infty}\frac{n+1}{n+2}\cdot\frac{1}{3}|x|=\frac{|x|}{3},$$

which is less than 1 if $-3 < x < 3$. When $x = -3$, the convergent alternating harmonic series is obtained.

5.   **D.**   The series is $1-2x+4x^2-8x^3+16x^4-\cdots$.

6.   **A.**   Assume that

$$\frac{2x^2-x+4}{x(x-1)(x-2)}=\frac{A}{x}+\frac{B}{x-1}+\frac{C}{x-2}.$$

Then

$$2x^2-x+4=A(x-1)(x-2)+Bx(x-2)+Cx(x-1).$$

Since you are looking for B, let $x = 1$:

$$2(1)-1+4=0+B(-1)+0; B=-5.$$

8.   **B.**   Since $e^x \simeq 1+x, e^{-x^2} \simeq 1-x^2$. So

$$\int_0^{0.3}(1-x^2)dx=x-\frac{x^3}{3}\Big|_0^{0.3}=0.3-\frac{0.027}{3}$$

12.   A. $f(x) = \dfrac{d}{dx}(x \sin \pi x) = \pi x \cos \pi x + \sin \pi x.$

17.   E.

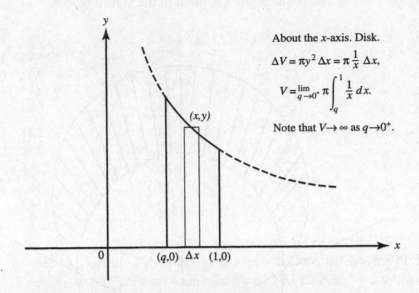

About the x-axis. Disk.

$\Delta V = \pi y^2 \, \Delta x = \pi \dfrac{1}{x} \, \Delta x,$

$V = \lim_{q \to 0^+} \pi \displaystyle\int_q^1 \dfrac{1}{x} \, dx.$

Note that $V \to \infty$ as $q \to 0^+$.

18.   D.   See the figure below, which shows that the length of a semicircle of radius 2 is needed here. The answer can, of course, be found by using the formula for arc length:

$$s = \int_0^\pi \sqrt{\left(\dfrac{dx}{dt}\right)^2 + \left(\dfrac{dy}{dt}\right)^2} \, dt.$$

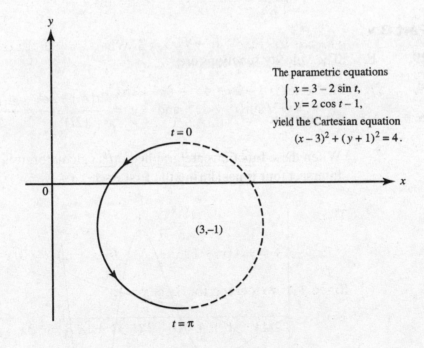

The parametric equations

$\begin{cases} x = 3 - 2 \sin t, \\ y = 2 \cos t - 1, \end{cases}$

yield the Cartesian equation

$(x - 3)^2 + (y + 1)^2 = 4.$

20.   E. $\lim_{x \to \infty} f(x) = \infty;\ \lim_{x \to 2^-} f(x) = -\infty;\ y = x + 2 + \dfrac{4}{x - 2}.$

22.   C.   Using parts twice yields the antiderivative $x^2 e^x - 2xe^x + 2e^x.$

23.   E. $\dfrac{dh}{dt}$ will increase above the half-full level (that is, the height of the water will rise more rapidly) as the area of the cross section diminishes.

27.   A.   The required area is lined in the figure below.

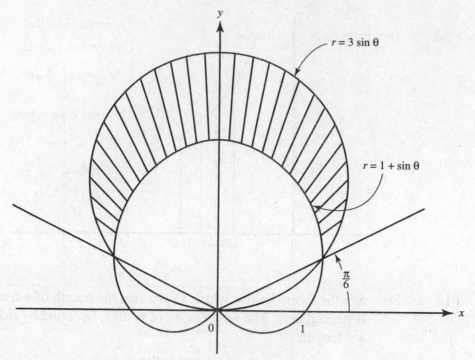

28.   D.   Note that $f(x) = x + 6$ if $x \ne 6$, that $f(6) = 12$, and that $\lim\limits_{x \to 6} f(x) = 12$. So f is continuous at $x = 6$, which implies the truth of I and II.

Part B

29.   E.   The velocity functions are

$$v_1 = -2t \sin(t^2 + 1) \quad \text{and} \quad v_2 = \frac{2t(e^t) - 2e^t}{(2t)^2} = \frac{e^t(t-1)}{2t^2}$$

When these functions are graphed on a calculator, it is clear that they intersect four times during the first 3 sec.

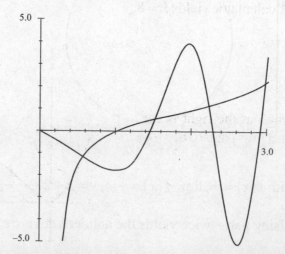

30.   C.   Changes in values of f'' show that f''' is constant.

31.   C.   Expressed parametrically, $x = \sin 3\theta \cos \theta$, $y = \sin 3\theta \sin \theta$. Where

$$\frac{dx}{d\theta} = -\sin 3\theta \sin \theta + 3\cos 3\theta \cos \theta = 0, \ \frac{dy}{dx} \text{ is undefined. Use the Solver}$$

to find θ.

34.   D.   See the figure below.

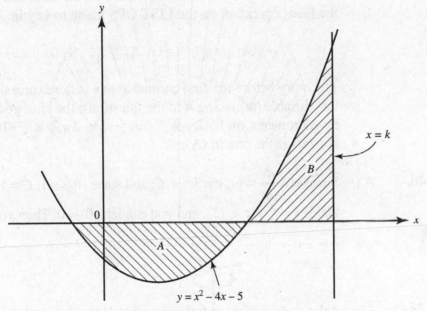

(This figure is not drawn to scale.)

The roots of $f(x) = x^2 - 4x - 5 = (x - 5)(x + 1)$ are $x = -1$ and 5. Since areas A and B are equal, therefore $\int_{-1}^{k} f(x)(dx) = 0$. Thus,

$$\left(\frac{x^3}{3} - 2x^2 - 5x \right)\Bigg|_{-1}^{k} = \left(\frac{k^3}{3} - 2k^2 - 5k \right) - \left(-\frac{1}{3} - 2 + 5 \right)$$

$$= \frac{k^3}{3} - 2k^2 - 5k - \frac{8}{3} = 0.$$

A calculator yields $k = 8$.

The graph at the right is for
answer 37 on page 618.

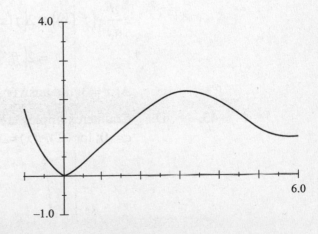

37.   C.   (See graph on page 617). It is given that $a(t) = (\sin t, e^{-t})$. An antiderivative is $v(t) = (-\cos t + c_1, -e^{-t} + c_2)$. Since $v(0) = (0,0)$, the constants are $c_1 = c_2 = 1$. The object's speed is

$$|v(t)| = \sqrt{(-\cos t + 1)^2 + (-e^{-t} + 1)^2}$$

Use a calculator to find that the object's maximum speed is 2.217.

39.   A.   On the TI-82 use the [sum] operation on the LIST MATH menu followed by the [seq] operation on the LIST OPS menu to key in

```
Y₁=sum seq((Xsin X)ᴺ/N!,N,0,9,1).
```

The entry before the first comma above is the expression to be summed; N is the variable (replacing n in the question); the sum goes from N = 0 to N = 9 by increments of 1. Graph Y_1 in $[-2\pi, 2\pi] \times [-10, 10]$. The graph that appears is the one in (A).

40.   A.   Since $a = \dfrac{dv}{dt} = 6t$, $v = 3t^2 + C$; and since $v(0) = 1$, $C = 1$. Then $v = \dfrac{ds}{dt} = 3t^2 + 1$ yields $s = t^3 + t + C'$, and you can let $s(0) = 0$. Then you want $s(3)$.

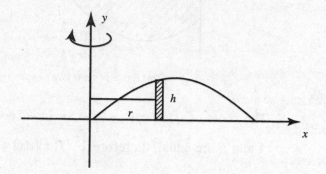

41.   C.   We note from the figure above that $\Delta V = $ (lateral surface area)(thickness) $= (2\pi rh)(\Delta x) = 2\pi xy \bullet \Delta x$. Therefore,

$$V = 2p \int_0^p x \sin x \, dx.$$

42.   E. $\dfrac{d}{dx}\left(f^2(x)\right) = 2f(x)f'(x)$,

$$\dfrac{d^2}{dx^2}\left(f^2(x)\right) = 2\left[f(x)f''(x) + f'(x)f'(x)\right]$$

$$= 2\left[ff'' + (f')^2\right]$$

At $x = 3$; the answer is $2[2(-2) + 5^2] = 42$.

43.   D.   Counterexamples are, respectively: for (A), $f(x) = |x|$, $c = 0$; for (B), $f(x) = x^3$, $c = 0$; for (C), $f(x) = x^4$, $c = 0$; for (E), $f(x) = x^2$ on $(-1, 1)$.

Section II

Part A

1. See solution for AB-1, page 542.

2.

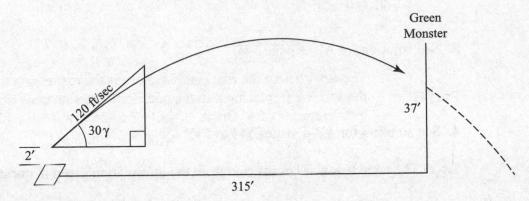

(a) The following table shows x- and y-components of acceleration, velocity, and position:

	HORIZONTAL	**VERTICAL**
acceleration components:	$a_x = 0$	$a_y = -32.172$
Velocity $= \int a\,dt$:	$v_x = c_1$	$v_y = -32.172t + c_2$
initial velocities:	$v_x(0) = 60\sqrt{3} = c_1$	$v_y(0) = 60 = c_2$
velocity components:	$v_x(t) = 60\sqrt{3}$	$v_y(t) = -32.172t + 60$
position, $= \int v\,dt$:	$x(t) = 60\sqrt{3}t + c_3$	$y(t) = -16.086t^2 + 60t + c_4$
initial position $(0,2)$:	$x(0) = 0 = c_3$	$y(0) = 2 = c_4$
position components:	$x(t) = 60\sqrt{3}t$	$y(t) = -16.086t^2 + 60t + 2$

The last line in the table is the answer to part (a).

(b) To determine how far above the ground the ball is when it hits the wall, find out when $x = 315$, and evaluate y at that time.

$$60\sqrt{3}t = 315 \text{ yields } t = \frac{315}{60\sqrt{3}} \approx 3.03109 \text{ sec}.$$

$$y\left(\frac{315}{60\sqrt{3}}\right) \approx 36.075 \text{ ft}.$$

(c) The ball's speed at the moment of impact in part (b) is $|v(t)|$ evaluated at $t = \dfrac{315}{60\sqrt{3}}$.

$$|v(t)| = \sqrt{\left(v_x(t)\right)^2 + \left(v_y(t)\right)^2}$$

$$= \sqrt{\left(60\sqrt{3}\right)^2 + (-32.172t + 60)^2} \text{ when } x = 315,$$

$$\left|v\left(\frac{315}{60\sqrt{3}}\right)\right| \approx 110.487 \text{ ft/sec.}$$

3. See solution for AB-3, page 544.

Part B

4. See solution for AB-4, pages 544 to 545.

5. (a) The table below is constructed from the information given in question 5 on page 613.

n	$f^{(n)}(5)$	$a_n = \dfrac{f^{(n)}(5)}{n!}$
0	2	2
1	-2	-2
2	-1	$-\dfrac{1}{2}$
3	6	1

$$f(x) \approx 2 - 2(x-5) - \frac{1}{2}(x-5)^2 + (x-5)^3.$$

(b) $f(5.1) \approx 2 - 2(5.1-5) - \dfrac{1}{2}(5.1-5)^2 + (5.1-5)^3,$

$$\approx 2 - 0.2 - 0.005 + 0.001 = 1.796.$$

(c) Use Taylor's theorem around $x = 0$.

n	$g^{(n)}(x)$	$g^{(n)}(0)$	$a_n = \dfrac{g^{(n)}(0)}{n!}$
0	$f(2x+5)$	$f(5) = 2$	2
1	$2f'(2x+5)$	$2f'(5) = 2(-2) = -4$	-4
2	$4f''(2x+5)$	$4f''(5) = 4(-1) = -4$	-2
3	$8f'''(2x+5)$	$8f'''(5) = 8(6) = 48$	8

$$g(x) \approx 2 - 4x - 2x^2 + 8x^3.$$

6. (a) At $(-1,8)$, $\dfrac{dy}{dx} = \dfrac{10}{x^2+1} = \dfrac{10}{2} = 5$, so the tangent line is

$y - 8 = 5(x-(-1))$. Therefore $f(x) \approx 8 + 5(x+1)$.

(b) $f(3) \approx 8 + 5(0+1) = 13$.

(c) At $(-1,8)$, $\dfrac{dy}{dx} = 5 \approx \dfrac{\Delta y}{\Delta x}$. For $\Delta x = 0.5$, $\Delta y = 0.5(5) = 2.5$, so move to

$(-1+0.5,\, 8+2.5) = (-0.5, 10.5)$.

At $(-0.5, 10.5)$, $\dfrac{dy}{dx} = \dfrac{10}{\left(-\dfrac{1}{2}\right)^2+1} = \dfrac{10}{\dfrac{5}{4}} = 8 \approx \dfrac{\Delta y}{\Delta x}$. For $\Delta x = 0.5$, $\Delta y = 0.5(8) = 4$,

so move to $(-0.5+0.5,\, 10.5+4)$.

Thus $f(0) \approx 14.5$.

(d) $\displaystyle\int_{-1}^{0} \dfrac{10}{x^2+1}\, dx = f(0) - f(-1)$, so $f(0) = 8 + \displaystyle\int_{-1}^{0} \dfrac{10}{x^2+1}\, dx$

$= 8 + \operatorname{arctan}(x)\big|_{-1}^{0}$

$= 8 + 10\big(\operatorname{arctan}(0) - \operatorname{arctan}(-1)\big)$

$= 8 + \dfrac{5\pi}{2}$.

Appendix: Formulas and Theorems for Reference

Algebra

1. QUADRATIC FORMULA.   The roots of the quadratic equation

    $$ax^2 + bx + c = 0 \ (a \neq 0)$$

    are given by

    $$x = \frac{-b \pm \sqrt{b^2 - 4ac}}{2a}.$$

2. BINOMIAL THEOREM.   If n is a positive integer, then

    $$(a + b)^n = a^n + na^{n-1}b + \frac{n(n-1)}{1 \cdot 2}\ a^{n-2}b^2 + \frac{n(n-1)(n-2)}{1 \cdot 2 \cdot 3}\ a^{n-3}b^3$$

    $$+ \cdots + nab^{n-1} + b^n.$$

3. REMAINDER THEOREM.   If the polynomial $Q(x)$ is divided by $(x - a)$ until a constant remainder R is obtained, then $R = Q(a)$. In particular, if a is a root of $Q(x) = 0$, then $Q(a) = 0$.

Geometry

The sum of the angles of a triangle is equal to a straight angle (180°).

PYTHAGOREAN THEOREM

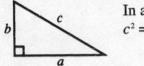

In a right triangle,
$c^2 = a^2 + b^2$.

In the following formulas,

A	is	area		B	is	area of base
S		surface area		r		radius
V		volume		C		circumference
b		base		l		arc length
h		height or altitude		θ		central angle (in radians)
s		slant height				

4. Triangle: $A = \dfrac{1}{2}bh.$

5. Trapezoid: $A = \dfrac{1}{2}(b_1 + b_2)h.$

6. Parallelogram: $A = bh.$

7. Circle: $C = 2\pi r;\ A = \pi r^2.$

8. Circular sector: $A = \dfrac{1}{2}r^2\theta.$

9. Circular arc: $l = r\theta.$

10. Cylinder:

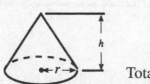

$$V = \pi r^2 h = Bh.$$
$$S\ (\text{lateral}) = 2\pi rh.$$
$$\text{Total surface area} = 2\pi r^2 + 2\pi rh.$$

11. Cone:

$$V = \dfrac{1}{3}\pi r^2 h = \dfrac{1}{3}Bh.$$
$$S\ (\text{lateral}) = \pi r\sqrt{r^2 + h^2}.$$
$$\text{Total surface area} = \pi r^2 + \pi r\sqrt{r^2 + h^2}.$$

12. Sphere: $V = \dfrac{4}{3}\pi r^3.$
$S = 4\pi r^2.$

Trigonometry

BASIC IDENTITIES
13. $\sin^2\theta + \cos^2\theta = 1.$
14. $1 + \tan^2\theta = \sec^2\theta.$
15. $1 + \cot^2\theta = \csc^2\theta.$

SUM AND DIFFERENCE FORMULAS
16. $\sin(\alpha \pm \beta) = \sin\alpha\cos\beta \pm \cos\alpha\sin\beta.$
17. $\cos(\alpha \pm \beta) = \cos\alpha\cos\beta \mp \sin\alpha\sin\beta.$
18. $\tan(\alpha \pm \beta) = \dfrac{\tan\alpha \pm \tan\beta}{1 \mp \tan\alpha\tan\beta}.$

DOUBLE-ANGLE FORMULAS

19. $\sin 2\alpha = 2 \sin \alpha \cos \alpha$.

20. $\cos 2\alpha = \cos^2 \alpha - \sin^2 \alpha = 2 \cos^2 \alpha - 1 = 1 - 2 \sin^2 \alpha$.

21. $\tan 2\alpha = \dfrac{2 \tan \alpha}{1 - \tan^2 \alpha}$.

HALF-ANGLE FORMULAS

22. $\sin \dfrac{\alpha}{2} = \pm\sqrt{\dfrac{1 - \cos \alpha}{2}}$; $\sin^2 \alpha = \dfrac{1}{2} - \dfrac{1}{2} \cos 2\alpha$.

23. $\cos \dfrac{\alpha}{2} = \pm\sqrt{\dfrac{1 + \cos \alpha}{2}}$; $\cos^2 \alpha = \dfrac{1}{2} + \dfrac{1}{2} \cos 2\alpha$.

REDUCTION FORMULAS

24. $\sin(-\alpha) = -\sin \alpha$: $\cos(-\alpha) = \cos \alpha$.

25. $\sin\left(\dfrac{\pi}{2} - \alpha\right) = \cos \alpha$; $\cos\left(\dfrac{\pi}{2} - \alpha\right) = \sin \alpha$.

26. $\sin\left(\dfrac{\pi}{2} + \alpha\right) = \cos \alpha$; $\cos\left(\dfrac{\pi}{2} + \alpha\right) = -\sin \alpha$.

27. $\sin(\pi - \alpha) = \sin \alpha$; $\cos(\pi - \alpha) = -\cos \alpha$.

28. $\sin(\pi + \alpha) = -\sin \alpha$; $\cos(\pi + \alpha) = -\cos \alpha$.

If a, b, c are the sides of triangle ABC, and A, B, C are respectively the opposite interior angles, then:

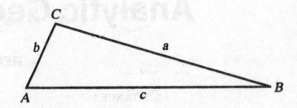

29. LAW OF COSINES. $c^2 = a^2 + b^2 - 2ab \cos C$.

30. LAW OF SINES. $\dfrac{a}{\sin A} = \dfrac{b}{\sin B} = \dfrac{c}{\sin C}$.

31. The area $A = \dfrac{1}{2} ab \sin C$.

GRAPHS OF TRIGONOMETRIC FUNCTIONS

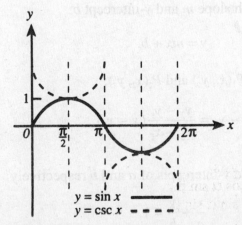

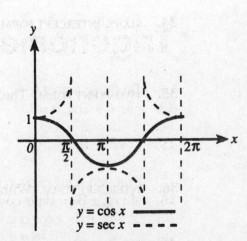

The four functions sketched above, sin, cos, csc, and sec, all have period 2π.

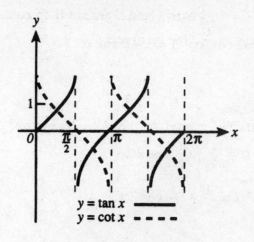

The two functions tan and cot have period π.

$y = \tan x$ ────

$y = \cot x$ ‑ ‑ ‑ ‑

INVERSE TRIGONOMETRIC FUNCTIONS

$y = \sin^{-1}$ $x = \arcsin x$ implies $x = \sin y,$ where $-\dfrac{\pi}{2} \leqq y \leqq \dfrac{\pi}{2}.$

$y = \cos^{-1}$ $x = \arccos x$ implies $x = \cos y,$ where $0 \leqq y \leqq \pi.$

$y = \tan^{-1}$ $x = \arctan x$ implies $x = \tan y,$ where $-\dfrac{\pi}{2} < y < \dfrac{\pi}{2}.$

Analytic Geometry

RECTANGULAR COORDINATES

DISTANCE

32. The distance d between two points, $P_1(x_1, y_1)$ and $P_2(x_2, y_2)$, is given by

$$d = \sqrt{(x_2 - x_1)^2 + (y_2 - y_1)^2}.$$

EQUATIONS OF THE STRAIGHT LINE

33. POINT-SLOPE FORM. Through $P_1(x_1, y_1)$ and with slope m:

$$y - y_1 = m(x - x_1).$$

34. SLOPE-INTERCEPT FORM. With slope m and y-intercept b:

$$y = mx + b.$$

35. TWO-POINT FORM. Through $P_1(x_1, y_1)$ and $P_2(x_2, y_2)$:

$$y - y_1 = \frac{y_2 - y_1}{x_2 - x_1}(x - x_1).$$

36. INTERCEPT FORM. With x- and y-intercepts of a and b respectively:

$$\frac{x}{a} + \frac{y}{b} = 1.$$

37. GENERAL FORM. $Ax + By + C = 0$, where A and B are not both zero. If $B \neq 0$, the slope is $-\dfrac{A}{B}$; the y-intercept, $-\dfrac{C}{B}$; the x-intercept, $-\dfrac{C}{A}$.

DISTANCE FROM POINT TO LINE

38. Distance d between a point $P(x_1, y_1)$ and the line $Ax + By + C = 0$ is

$$d = \left| \frac{Ax_1 + By_1 + C}{\sqrt{A^2 + B^2}} \right|.$$

EQUATIONS OF THE CONICS

CIRCLE

39. With center at $(0, 0)$ and radius r: $x^2 + y^2 = r^2$.
40. With center at (h, k) and radius r: $(x - h)^2 + (y - k)^2 = r^2$.

PARABOLA

41. With vertex at $(0, 0)$ and focus at $(p, 0)$: $y^2 = 4px$.
42. With vertex at $(0, 0)$ and focus at $(0, p)$: $x^2 = 4py$.

With vertex at (h, k) and axis

43. parallel to x-axis, focus at $(h + p, k)$: $(y - k)^2 = 4p(x - h)$.
44. parallel to y-axis, focus at $(h, k + p)$: $(x - h)^2 = 4p(y - k)$.

ELLIPSE

With major axis of length $2a$, minor axis of length $2b$, and distance between foci of $2c$:

45. Center at $(0, 0)$, foci at $(\pm c, 0)$, and vertices at $(\pm a, 0)$:

$$\frac{x^2}{a^2} + \frac{y^2}{b^2} = 1.$$

46. Center at $(0, 0)$, foci at $(0, \pm c)$, and vertices at $(0, \pm a)$:

$$\frac{y^2}{a^2} + \frac{x^2}{b^2} = 1.$$

47. Center at (h, k), major axis horizontal, and vertices at $(h \pm a, k)$:

$$\frac{(x - h)^2}{a^2} + \frac{(y - k)^2}{b^2} = 1.$$

48. Center at (h, k), major axis vertical, and vertices at $(h, k \pm a)$:

$$\frac{(y - k)^2}{a^2} + \frac{(x - h)^2}{b^2} = 1.$$

For the ellipse, $a^2 = b^2 + c^2$, and the eccentricity $e = \dfrac{c}{a}$, which is *less* than 1.

HYPERBOLA

With real (transverse) axis of length $2a$, imaginary (conjugate) axis of length $2b$, and distance between foci of $2c$:

49. Center at $(0, 0)$, foci at $(\pm c, 0)$, and vertices at $(\pm a, 0)$:

$$\frac{x^2}{a^2} - \frac{y^2}{b^2} = 1.$$

50. Center at $(0, 0)$, foci at $(0, \pm c)$, and vertices at $(0, \pm a)$:

$$\frac{y^2}{a^2} - \frac{x^2}{b^2} = 1.$$

51. Center at (h, k), real axis horizontal, vertices at $(h \pm a, k)$:

$$\frac{(x - h)^2}{a^2} - \frac{(y - k)^2}{b^2} = 1.$$

52. Center at (h, k), real axis vertical, vertices at $(h, k \pm a)$:

$$\frac{(y - k)^2}{a^2} - \frac{(x - h)^2}{b^2} = 1.$$

For the hyperbola, $c^2 = a^2 + b^2$, and the eccentricity $e = \dfrac{c}{a}$, which is *greater* than 1.

POLAR COORDINATES

RELATIONS WITH RECTANGULAR COORDINATES

53. $x = r \cos \theta$;
$y = r \sin \theta$;
$r^2 = x^2 + y^2$;
$\tan \theta = \dfrac{y}{x}$.

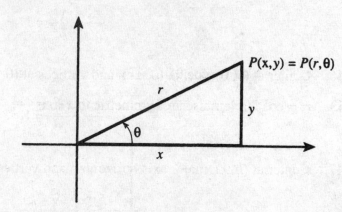

SOME POLAR EQUATIONS

54. $r = a$ circle, center at pole, radius a.
55. $r = 2a \cos \theta$ circle, center at $(a, 0)$, radius a.
56. $r = 2a \sin \theta$ circle, center at $(0, a)$, radius a.
57. $\left.\begin{array}{l} r = a \sec \theta \\ \text{or } r \cos \theta = a \end{array}\right\}$ line, $x = a$.
58. $\left.\begin{array}{l} r = b \csc \theta \\ \text{or } r \sin \theta = b \end{array}\right\}$ line, $y = b$.

roses (four leaves)

59. $r = \cos 2\theta$.

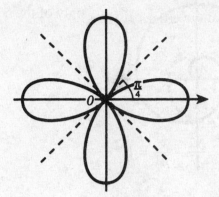

60. $r = \sin 2\theta$.

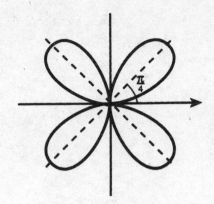

cardioids (specific examples below)

61. $r = a\,(1 \pm \cos \theta)$.

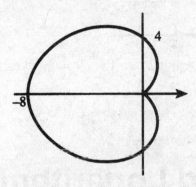

$r = 4\,(1 - \cos \theta)$

62. $r = a\,(1 \pm \sin \theta)$.

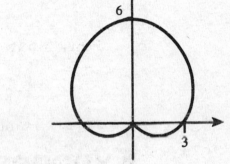

$r = 3\,(1 + \sin \theta)$

63. $r^2 = \cos 2\theta$, lemniscate, symmetric to x-axis.

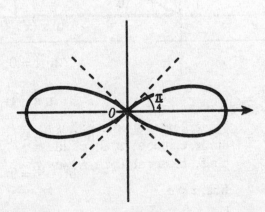

$(r^2 = \sin 2\theta$ is a lemniscate symmetric to $y = x$.)

64. $r = \theta$, (double) spiral of Archimedes

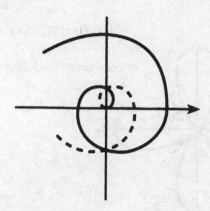

For $\theta > 0$, the curve consists only of the solid spiral.

65. $r\theta = a$ ($\theta > 0$), hyperbolic (or reciprocal) spiral

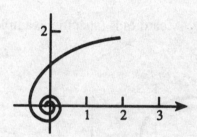

This curve is for $r\theta = 2$.
Note that $y = 2$ is an asymptote.

Exponential and Logarithmic Functions

PROPERTIES

e^x	$\ln x \ (x > 0)$
$e^0 = 1$;	$\ln 1 = 0$;
$e^1 = e$;	$\ln e = 1$;
$e^{x_1} \cdot e^{x_2} = e^{x_1 + x_2}$;	$\ln (x_1 \cdot x_2) = \ln x_1 + \ln x_2$;
$\dfrac{e^{x_1}}{e^{x_2}} = e^{x_1 - x_2}$;	$\ln \dfrac{x_1}{x_2} = \ln x_1 - \ln x_2$;
$e^{-x} = \dfrac{1}{e^x}$.	$\ln x^r = r \ln x$ (r real).

INVERSE PROPERTIES

$f(x) = e^x$ and $f^{-1}(x) = \ln x$ are inverses of each other:

$$f^{-1}(f(x)) = f(f^{-1}(x)) = x;$$

$$\ln e^x = e^{\ln x} = x \ (x > 0).$$

GRAPHS

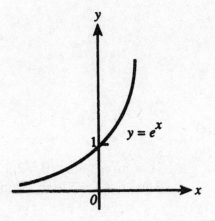

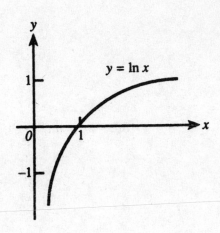

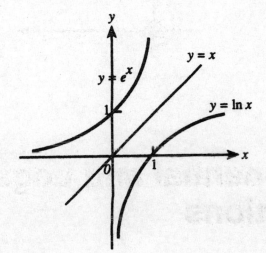

Index